ENGINEERING
OUR DIGITAL FUTURE

THE INFINITY PROJECT℠

ENGINEERING
OUR DIGITAL FUTURE

Geoffrey C. Orsak
Southern Methodist University, Dallas, TX

Sally L. Wood
Santa Clara University, Santa Clara, CA

Scott C. Douglas
Southern Methodist University, Dallas, TX

David C. Munson, Jr.
University of Illinois at Urbana-Champaign, Urbana, IL

John R. Treichler
Applied Signal Technology, Sunnyvale, CA

Ravindra Athale
Defense Advanced Research Projects Agency, Arlington, VA
on leave from *George Mason University*

Mark A. Yoder
Rose-Hulman Institute of Technology, Terre Haute, IN

Upper Saddle River, NJ 07458

THE INFINITY PROJECT℠

Library of Congress Cataloging-in-Publication Data on file

Vice President and Editorial Director, ECS: Marcia J. Horton
Publisher: Tom Robbins
Associate Editor: Alice Dworkin
Vice President and Director of Production and Manufacturing, ESM: David W. Riccardi
Executive Managing Editor: Vince O'Brien
Managing Editor: David A. George
Production Editor: Scott Disanno
Director of Creative Services: Paul Belfanti
Creative Director: Carole Anson
Interior Design: Pearson Education Development Group; John Christiana; Geoffrey Cassar
Cover Illustration: Kevin Jones and Associates; Geoffrey Cassar
Art Director and Cover Manager: John Christiana
Managing Editor, AV Management and Production: Patricia Burns
Art Editor: Xiaohong Zhu
Manufacturing Manager: Trudy Pisciotti
Manufacturing Buyer: Lynda Castillo
Marketing Manager: Holly Stark

Pearson Prentice Hall
Pearson Education, Inc.
Upper Saddle River, NJ 07458

Printed in the United States of America

10 9 8

ISBN: 0-13-035482-1

ISBN: 0-13-184828-3

ISBN: 0-13-143643-0

Pearson Education Ltd., London
Pearson Education Australia Pty. Ltd., Sydney
Pearson Education Singapore, Pte. Ltd.
Pearson Education North Asia Ltd., Hong Kong
Pearson Education Canada, Inc., Toronto
Pearson Educación de Mexico, S.A. de C.V.
Pearson Education—Japan, Tokyo
Pearson Education Malaysia, Pte. Ltd.
Pearson Education, Inc., Upper Saddle River, New Jersey

Brief Table of Contents

chapter 1

The World of Modern Engineering 2

Creating Digital Music 32

chapter 2

chapter 3

Making Digital Images 102

Math You Can See 180

chapter 4

chapter 5

Digitizing the World 248

Coding Information for Storage and Secrecy 296

Communicating with Ones and Zeros 358

Networks from the Telegraph to the Internet 408

chapter 8

The Big Picture of Engineering 450

Preface

ABOUT THE INFINITY PROJECT

As we move into the 21st century, engineering and technology will have an ever-increasing impact on our daily activities. Yet as our lives have become more and more dependent on technology, public awareness and knowledge about technology-related issues has declined. All of this is compounded by the fact that young students today continue to see little relevance in traditional math and science curricula—sadly suggesting that this unfortunate trend may continue into the foreseeable future, resulting in a reduced ability of our population to deal with society's challenges.

The Infinity Project was created to address just this problem by developing an innovative approach to applying fundamental science and mathematics concepts to solving contemporary engineering problems. This nationwide program, designed by leading college engineering professors in cooperation with education experts, is sponsored and run by the Institute for Engineering Education at SMU, with generous support from Texas Instruments, the National Science Foundation, and the Department of Education.

The Infinity Project engineering and technology curriculum encourages students to be curious about math and science by connecting their relevance to prized personal technologies such as MP3, CD, and DVD players; cellular phones; pagers; and handheld video devices. The perennial question "Why do I need to learn this?" is answered in ways that are both relevant and fun. The Infinity Project curriculum sharpens math- and science-based problem-solving skills, and encourages students to be innovative, to go beyond what is, and to dream of what can be.

The Infinity Project supplies schools and teachers with a complete turnkey solution that includes this first-of-its-kind engineering textbook. *Engineering Our Digital Future* covers a selection of topics and hands-on activities to inspire and excite students. The Infinity Project curriculum encourages young people to learn about engineering, inspires them to understand the relevance of technology and the importance of mathematics and science, and shows how these concepts can lead to rewarding, challenging, and creative career opportunities. And although we emphasize the current leading-edge digital technologies that are important and exciting to today's youth, the approach to problem solving emphasized throughout the book applies to all fields of engineering and many other professions as well.

The Infinity Project provides a complete answer for effectively and easily incorporating engineering and technology into standard curricula today: stimulating, well-thought-out content; comprehensive teacher training; cutting-edge classroom technology; lab materials and lab activities; and an outstanding supplements package. On-line Web support guarantees that you are never alone.

CURRICULUM

The Infinity Project curriculum is typically covered in a yearlong class. Students learn how to apply math and science concepts to design new technologies involving digital music and images, special effects for films, personal communication devices such as cell phones, and the Internet—all while clearly understanding how information in the digital era is collected, stored, processed, and moved around the globe.

The curriculum is significantly enhanced by many hands-on experiments that are carefully integrated with the course materials. The classroom and lab equipment produced by the Infinity Project, in partnership with Texas Instruments and Hyperception, Inc., is based on new cutting-edge digital signal processing technology and has been made available by our industrial partners as the Infinity Technology Kit. This very low cost kit converts standard PCs found in classrooms and laboratories into a modern engineering design platform and allows instructors to clearly demonstrate engineering design in the digital era. The modern design tools that are part of the Infinity Technology Kit allow instructors and students to design and build remarkably capable and complex systems from simple function blocks.

Engineering Our Digital Future is designed for students who have completed mathematics through a second course in algebra (Algebra II) and who have had at least one laboratory science course. This innovative engineering and technology course allows students to see firsthand the applications of math and science to engineering and technology early enough in their studies to encourage them to pursue more advanced math and science courses and to begin to consider future careers in technology and engineering. The book focuses squarely on the math and science fundamentals of engineering during the information revolution and teaches students how engineers create, design, test, and improve the technology around them. Applications are drawn from a wide array of modern devices and systems seen today.

Scope and Focus of Content: Engineering is an exceptionally broad field—so broad, in fact, that no single course or book could adequately capture all the application areas that engineers are working in today. However, what makes *Engineering Our Digital Future* directly relevant to every area of engineering is the application of math and science concepts to the creative aspects of engineering design.

The members of the Infinity Project had to make a number of difficult decisions in choosing which topical areas of engineering to emphasize in this book. Our choice was made all the easier by talking to students and teachers, who stated clearly that they love high technology, particularly those areas that touch students' everyday lives. So, the Infinity Project assembled one of the best teams in the country to create this innovative engineering textbook. The content within the book focuses on the engineering applications of basic math and science concepts used by engineers to dream up, design, and build many of the new high-tech innovations that are changing our world. While we wish we could have shared the full breadth of engineering applications with students, we believe that the material contained in this book shows clearly how engineers, armed with knowledge of math, science, and technology, have the ability to impact nearly every aspect of our lives.

Engineering Our Digital Future is composed of nine separate chapters, each emphasizing different application areas of engineering and each using different areas of math and science commonly seen in high school and early college.

Chapter 1 introduces students to the engineering design process and the basics of modern technology, including integrated circuits, computer chips, and mathematical concepts such as Moore's Law, binary numbers, and simple exponential functions that describe constant growth rates.

Chapter 2 exposes students to some of the most important engineering ideas associated with the creation of digital music. Students learn how basic ideas drawn from the right triangle, such as sines and cosines, are fundamental to making computer music.

Chapter 3 develops the basic concepts behind digital imaging technologies, including capturing and storing digital pictures. Various mathematical concepts are developed, explored, and shown to be interesting and relevant to manipulating these images.

Chapter 4 extends the ideas in Chapter 3 to using digital images and video in several interesting human applications. Additional mathematics are developed wherein images are treated as matrices, and operations to improve image quality or extract information from images are defined in terms of simple matrix operations.

Chapter 5 focuses on the general ideas associated with the digitization of a wide range of information, from text in books and magazines, to speech and music, to images and video. Students learn the details of how all these types of information are captured and stored in digital form. They also learn about the various practical trade-offs when real-world, or "analog", information is converted to numbers and stored with finite precision.

Chapter 6 focuses on some of the more interesting opportunities for coding information when the information is stored on a computer, using only bits. Problems in computer security and encryption as well as redundancy of numbers and data compression are both highly relevant and interesting to students. The basic concepts of detecting and correcting errors in digital data are discussed. This chapter gives students some very interesting applications of simple polynomials and random numbers.

Chapter 7 exposes students to the basic ideas behind wireless and radio communications. Students see firsthand how sines and cosines enable all wireless communications. They learn such important fundamental concepts as bandwidth and data rate.

Chapter 8 gives a very good overview of computer networks and the Internet from both a modern and an historical perspective. Students will be fascinated to learn how similar the Internet is to many other networks, including the U.S. Postal Service and the telegraph system. In this chapter, students have the opportunity to undertake some very interesting mathematical calculations involving simple economic trade-offs in system and network design.

Chapter 9 looks at the big picture of engineering. By examining the engineering concepts and social implications of ten important engineering feats throughout history, this chapter shows how these accomplishments changed they way people live, work, and play. Various fields within engineering are introduced, and certain myths and misconceptions of engineering are discussed and dispelled.

Course Options

A typical course using *Engineering Our Digital Future* begins with Chapter 1 and ends with Chapter 9, with the instructor judiciously selecting a subset of chapters from the book in accordance with the anticipated pace of the class. The well-paced classroom can expect to complete the entire book taken in order. Classrooms with a more conservative pace will want to select an appropriate subset of chapters that will ensure a high-impact course.

In selecting chapters that will be of interest to students, the instructor should consider the mathematical level of the students taking the class. If the students have had a successful experience with mathematics up to a second course in algebra and have had reasonable exposure to the most basic ideas from geometry and trigonometry, then all of the chapters can be selected without any fear of the students not having the appropriate background. In this case, the selection criteria should then be based on the level of student interest in the various chapters. For classes with a high interest in music and video, the educator should focus on Chapters 1–5 and 9. For classes with an interest in cell phones and the Internet, the educator should focus on Chapters 1 and 5–9.

For classes with larger percentages of students with little exposure to trigonometric ideas, some topics in Chapter 2 and in Chapter 7 might come across as a bit challenging. However, the material is relevant and the engineering designs are exciting, so the educator might want to consider covering only the first few sections of these chapters. Similarly, if a large percentage of the class has had little exposure to mathematical abstraction, some parts of Chapters 5 and 6 may be omitted without losing continuity.

For classes that want to sample the entire book, but don't feel that they have the time to thoroughly cover the full scope of the material, teachers should feel comfortable covering the first few sections of each chapter without fear that they are leaving important material out that will be necessary for future chapters. Each chapter is fairly well contained, and important supporting material, if it is needed, is easy to find.

PEDAGOGICAL FEATURES

- Hundreds of four-color illustrations that demystify engineering and technology concepts
- Real-world examples that are actually fun
- Notes and facts in the margins that emphasize important points and interesting facts
  - Definitions
  - Interesting Facts
  - Key Concepts
  - Keep in Mind

- Infinity Project experiments fully integrated within the text that are designed specifically for use with the Infinity Technology Kit
- Infinity Technology Kit—a complete engineering design platform for today's most relevant and interesting new technologies
- Interesting applications presented in boxes within the text
- Exercises divided by pedagogical approach
  - Mastering the Concepts
  - Try This
  - In the Laboratory
  - Back of the Envelope
  - Master Design Problems
- Chapter review and summary
  - Big Ideas
  - Math & Science Concepts Learned
  - Important Equations
  - Building Your Knowledge Library
- Comprehensive glossary

INFINITY TECHNOLOGY KIT

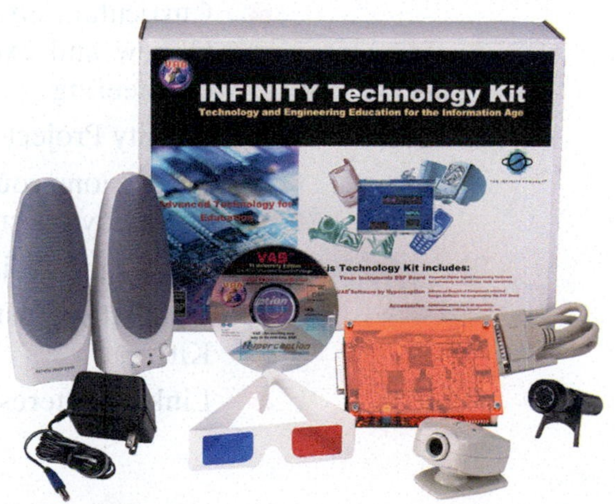

Engineering is about doing things. Therefore, throughout this book, students will have the opportunity to master engineering and technology concepts by building and testing new designs—while using digital technologies, including video, audio, and graphics, that are engaging for students and teachers. To aid in the classroom or laboratory, the team behind the Infinity Project has created the cutting-edge Infinity Technology Kit, a multimedia hardware and software system for converting standard PCs into engineering design environments. The Infinity Technology Kit brings to life the engineering concepts taught in The Infinity Project engineering curriculum. The predesigned lab experiments and engineering designs allow students to experience firsthand the full range of the engineering design process of envisioning, designing, building, and testing modern technology.

The technology used within the Infinity Technology Kit is based upon Texas Instruments' Digital Signal Processor (DSP) chips and a new and innovative graphical programming environment called Visual Application Builder, designed and developed by one of the Infinity Project partners, Hyperception, Inc. With the Infinity Technology Kit, students with limited experience can act like real engineers—creating innovations that are relevant and exciting. No previous experience with any programming languages is required.

Components:

Hardware: High-performance digital signal processing board based on TI DSP technology.

Software: Easy-to-use, yet powerful, graphical computer programming software created and produced by Hyperception, Inc.

Accessories include Web camera, PC powered speakers, PC microphone, and all the necessary cables for easy installation

SUPPLEMENTS

The Infinity Project supplements offer comprehensive additional resources and classroom support:

Instructor's manual containing

Microsoft PowerPoint lecture notes

Lab exercise handouts

Extensive test-item file with complete solutions

Laboratory manual

WEBSITE

The Infinity Project website, at http://www.infinity-project.org, provides ongoing classroom support and resources:

- Curriculum updates, including additional on-line chapters related to new and exciting topics such as robotics and the physics of engineering
- Infinity Project engineering experiments and updates
- Discussion groups for teachers and educators, monitored and supported by the staff of the Infinity Project
- Training materials
- Installation instructions and support for the Infinity Technology Kit
- Links to interesting and related websites

Acknowledgments

The authors would like to acknowledge the generous support of the many individuals and organizations that made the Infinity Project and this book possible.

First, we would like to thank SMU President R. Gerald Turner; Provost Ross Murfin, School of Engineering; Dean Steve Szygenda; and past Electrical Engineering Department Chair Jerry Gibson for their financial support and long-term commitment to the Infinity Project.

We would also like to thank Texas Instruments, Inc., for its generous underwriting and corporate vision in helping ensure that engineering is part of a student's standard educational experience. In particular, we are deeply indebted to Torrence Robinson for his continuous involvement, support, and insights. The involvement of Philip Ritter, Greg Delagi, Jeffrey McCreary, Leon Adams, Renee Hartshorn, Kim Quirk, Steve Leven, Gene Frantz, and Thomas J. Engibous was also instrumental in our success. We would also like to acknowledge that additional financial support for the Infinity Project has been provided by the National Science Foundation and the Department of Education.

The authors are also deeply indebted to the very talented and tireless efforts of the entire publishing team at Prentice Hall. Tom Robbins, our publisher, has been a visionary throughout the conceptualization and delivery of this book. Also, many of the innovative ideas and concepts that have been so critical to the philosophy and shape of this book are due to the remarkable efforts of David A. George, managing editor; Scott Disanno, production editor; Alice Dworkin, associate editor; and Holly Stark, marketing manager. We would also like to sincerely thank John Christiana, Xiaohong Zhu, and Kevin Jones Associates for their inventive and creative contributions in producing the great look and feel of the pages and engineering artwork.

Hands-on engineering design is a fundamental component of this book and the Infinity Project curriculum. Many thanks go to Jim Zachman and the entire team at Hyperception, Inc., for developing a superb software environment and the cutting-edge hardware associated with the Infinity Technology Kit.

We are also very much indebted to the wonderful group of educators who piloted this new engineering curriculum during the 2000–2001 academic year and whose insights and feedback were valuable in improving the scope and sequence of the content. We would particularly like to thank Andrew Brown, Sue Kile, Richard Taylor, Doug Rummel, Susan Cinque, Bill Miller, Don Ruggles, Sylvia San Pedro, John Spikerman, and Kurt Oehler. In addition, we would like to express our deep appreciation to the teachers who participated in the detailed roundtable discussions for their significant contributions in sharing with us their insights, experiences, and enthusiasm: Karen Donathan, Mark Connor, Aurelia Weil, Sylvia San Pedro, and Debbie Thompson.

Also, many of the engineering design projects were completed by an outstanding collection of graduate students at SMU who worked tirelessly for many hours on short deadlines. These students include Sumant Paranjpe, Amitabh Dixit, J. D. Norris, Mark Westerman, and Joji Phillip.

And last, but most certainly not least, all of us associated with the Infinity Project are deeply indebted to the coordinator of the Infinity Project, Felicia Hopson, for her tireless efforts in keeping us organized, on time, and on schedule.

Geoffrey C. Orsak
Southern Methodist University

Sally L. Wood
Santa Clara University

Scott C. Douglas
Southern Methodist University

David C. Munson, Jr.
University of Illinois

John R. Treichler
Applied Signal Technology

Ravindra Athale
Defense Advanced Research Projects Agency (DARPA)
on leave from *George Mason University*

Mark A. Yoder
Rose-Hulman Institute of Technology

About the Authors

Geoffrey C. Orsak is Associate Dean for Research and Development and a Professor with the Department of Electrical Engineering, Southern Methodist University. He also serves as the Executive Director of the federally supported Institute for Engineering Education at SMU and Director of the Infinity Project. Dr. Orsak is widely regarded as one of the nation's leaders in K-12 engineering education.

In addition to these activities, he has served as an advisor on matters associated with the science and technology of national defense to the U.S. Department of Defense; the National Academy of Engineering; the Institute for Defense Analysis, among others. Dr. Orsak received the B.S.E.E., M.E.E., and Ph.D. degrees in electrical and computer engineering from Rice University, Houston, TX.

Sally L. Wood is a Professor with the Department of Electrical Engineering, Santa Clara University, the oldest university in California. For the past six years, she has also been the head of the department. She was born in Swansea MA, raised in Georgia, and graduated from high school in Washington state. After she received the B.S. degree from Columbia University, she worked for five years in the northeast and in Europe designing systems to automatically read printed text. She then returned to graduate school at Stanford University and earned a Ph.D. degree for research on medical imaging. She enjoys teaching classes from the freshman level to the graduate level in digital logic and signal processing and continues her research on video and image processing. She is a past vice president of the IEEE Signal Processing Society.

Scott C. Douglas is an Associate Professor with the Department of Electrical Engineering, Southern Methodist University. He is also the Associate Director for the Institute for Engineering Education at SMU. Before attending college, he was drawn to engineering through his love of music and the arts, and he has performed in numerous orchestras, bands, and musical groups as a saxophonist and singer throughout his life. He received the B.S., M.S., and Ph.D. degrees from Stanford University. Afterward, he became a professor, educator, and engineering researcher. He regularly consults with companies all over the world on topics related to his research interests, which focus on the processing of acoustic signals for sound and vibration control, speech enhancement, and spatial understanding.

David C. Munson, Jr. is both the Chairman and a Professor with the Department of Electrical and Computer Engineering, University of Michigan. Prior to this, he was a Professor with the Department of Electrical and Computer Engineering, University of Illinois, Urbana-Champaign. His research focuses on computer algorithms for tomography and synthetic aperture radar. In addition to his research, Dr. Munson particularly enjoys being in the classroom, where he has taught thousands of students their first course in digital signal processing. He received the B.S. degree from the University of Delaware and the M.S.,

M.A., and Ph.D. degrees from Princeton University, all in electrical engineering. Dr. Munson is a past-president of the IEEE Signal Processing Society and the founding editor-in-chief of the *IEEE Transactions on Image Processing*.

John R. Treichler is the Chief Technical Officer of Applied Signal Technology, Inc., an engineering company that builds specialized electronic equipment for the U.S. government and its friends overseas. His research focuses on the application of advanced technology to communications systems. Dr. Treichler has been a professor at both Cornell University and Stanford University. He received the B.A. and Masters of Electrical Engineering degrees from Rice University and the Ph.D. degree from Stanford University. He served aboard ships as an officer in the U.S. Navy, and continues to work at the company he co-founded.

Ravindra Athale is currently a Program Manager of Photonics at the Defense Advanced Research Projects Agency (DARPA), on leave from George Mason University. Prior to this, he has worked at various government research labs and in private industry. He has been fascinated by everything optical—lasers, holography, liquid crystals, optical computers—for his entire career. He is the co-inventor of HoloSpex glasses, which is the first mass scale consumer product based on the esoteric technology of far-field computer-generated holography. Dr. Athale received the B.Sc. and M.Sc. degrees in physics in India and the Ph.D. degree from the University of California, San Diego, in electrical engineering. During his career as a professor, he has developed and taught a course that teaches principles of information technology to non-science and engineering students.

Mark A. Yoder is a Professor with the Department of Electrical and Computer Engineering, Rose-Hulman Institute of Technology, Terre Haute, IN. Dr. Yoder received the B.S. degree in 1980 and the Ph.D. degree in 1984, both in electrical engineering and both from Purdue University. While working as a research scientist, he discovered that teaching engineering was most enjoyable, prompting him to join Rose-Hulman Institute of Technology. He is a national leader in teaching digital signal processing to young college students, using symbolic algebra systems in electrical engineering education and in developing engineering curriculums for high school students.

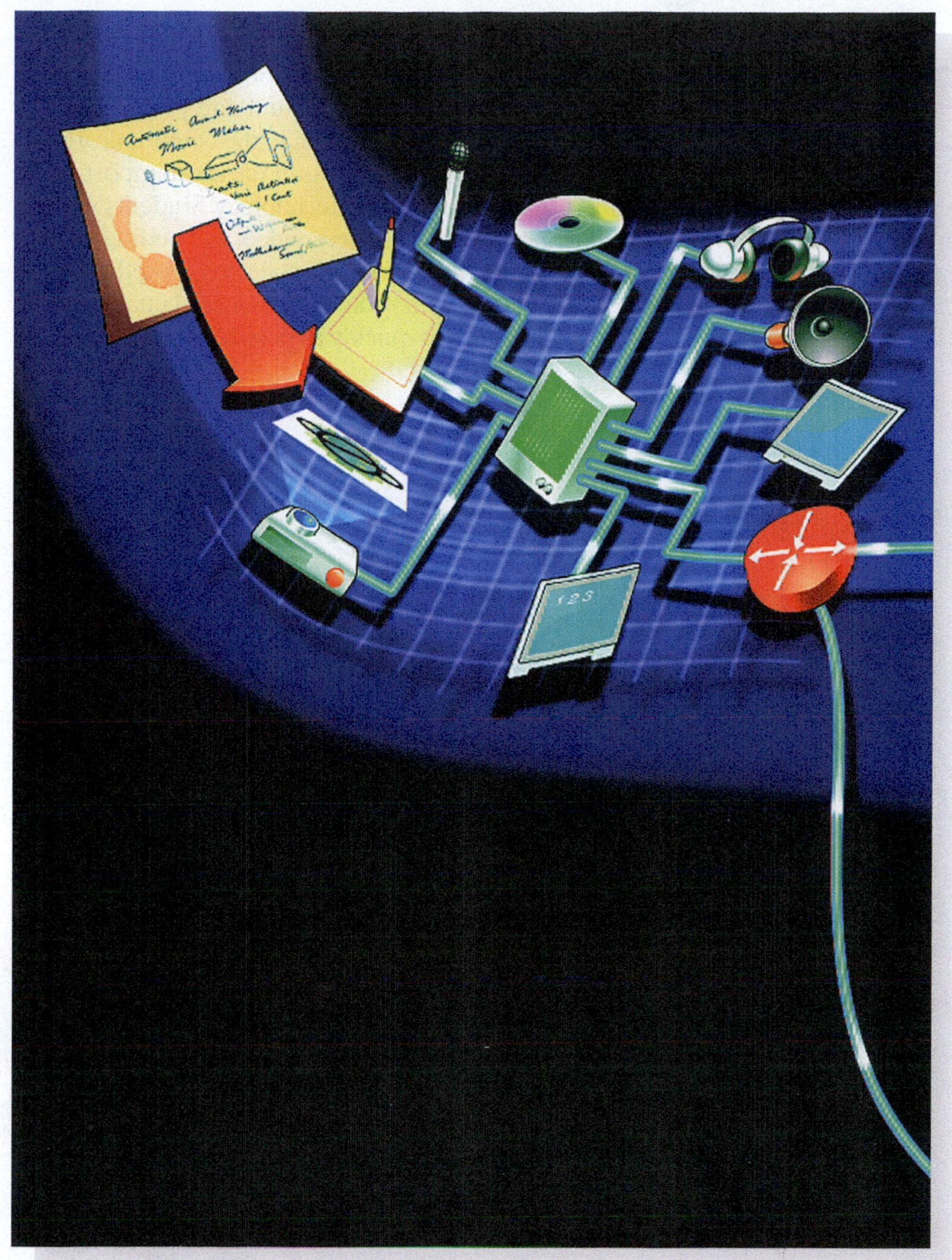
Automatic Award-Winning
Movie Maker
123

A Guide to Using this Text

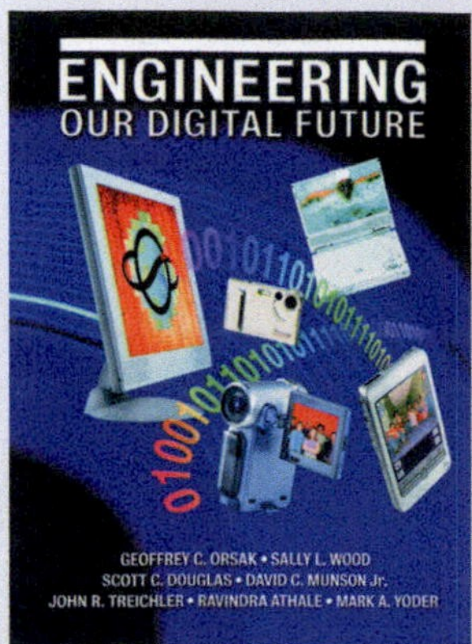

Engineering Our Digital Future, developed as the textbook companion for the Infinity Project, is an integral part of the curriculum and presents innovative approaches to teaching modern engineering in the communications and information age. The book is designed to engage students in the most important, relevant, and interesting ideas of modern engineering and technology. Students will learn and design new Internet technologies involving images, video, and audio, as well as learn to analyze and understand how digital information is collected, stored, processed, and moved. The organization and presentation of the material enable students to understand and apply the fundamentals as quickly as possible.

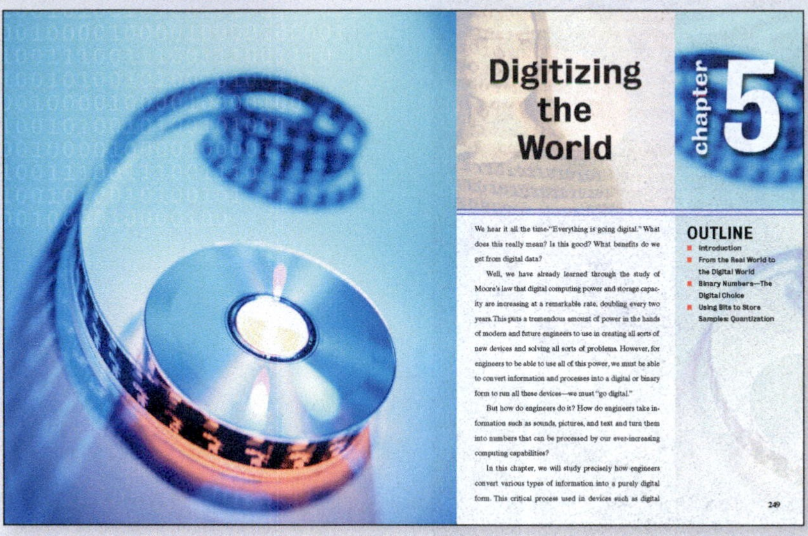

Digitizing the World

chapter 5

We hear it all the time-"Everything is going digital." What does this really mean? Is this good? What benefits do we get from digital data?

Well, we have already learned through the study of Moore's law that digital computing power and storage capacity are increasing at a remarkable rate, doubling every two years. This puts a tremendous amount of power in the hands of modern and future engineers to use in creating all sorts of new devices and solving all sorts of problems. However, for engineers to be able to use all of this power, we must be able to convert information and processes into a digital or binary form to run all these devices—we must "go digital."

But how do engineers do it? How do engineers take information such as sounds, pictures, and text and turn them into numbers that can be processed by our ever-increasing computing capabilities?

In this chapter, we will study precisely how engineers convert various types of information into a purely digital form. This critical process used in devices such as digital

OUTLINE

- Introduction
- From the Real World to the Digital World
- Binary Numbers—The Digital Choice
- Using Bits to Store Samples: Quantization

249

▲ Informative Chapter Openers

Every chapter opens with an eye-catching graphic to encourage the student's curiosity about the material presented in the chapter.

Hundreds of Illustrations and Photos

The four-color art program has been carefully designed and drafted to maximize the impact and clarity of each chapter's subject matter. The art complements the subject matter and offers additional information and guidance to the student and instructor. The photographs illustrate historically significant events and present fascinating real-world applications. ▶

Design Objective: A Digital Backpack

You do a lot of things at school, such as take notes, listen to lectures, read books, hand in homework, eat lunch, and socialize with your friends. To keep your school life going, you probably have used a bag or backpack to hold important information and items. A traditional backpack is just a container for your stuff. But what if the backpack could do much more? Consider the following:

1. Suppose it could "take notes" for you by recording the lecture and taking pictures of the chalkboard or whiteboard. Given that you are in school every day, the amount of audio and images that you would need to store could get huge. How would you store all that information efficiently?
2. It could hand in your homework for you when you walk into the classroom by "beaming" your assignment to the teacher's backpack. How would you make sure that others wouldn't be able to copy your work before or when you hand it in?
3. If you lose your backpack, all your private information might not be so private anymore. How could you keep your personal information safe in case it falls into the wrong hands?

In this design, we assume that we already have a way to digitize the information that we need in a simple manner, so that we can use all the ideas that we learned in Chapter 5 to help us. Digitizing information, if done right, gives us a nearly perfect copy of the original. It also has the advantage of being easy to transport, because new generations of digital devices continually are getting smaller and more efficient. For this design, we must ask ourselves the following questions:

- **What problem are we trying to solve**? We want to design a digital scheme by which large amounts of information, in binary form, can be stored efficiently and securely so that other individuals can access the information only when we want them to.
- **How do we formulate the underlying engineering design problem**? Some of the important features of the digital backpack include the following:
  1. It should preserve our information in an accurate and efficient manner, using the fewest possible number of bits.
  2. It should have some way of releasing our personal information only when we need to use that information.
  3. It should enable the user to define a level of security by which the information can be accessed or used.

◀ Design Objectives

Each chapter begins with a discussion of the design objectives for the particular subject covered by that section. The Design Objectives section list, a number of questions and issues the student should keep in mind while studying the chapter material, including questions such as the following:

- What problem are we trying to solve?
- How do we formulate the underlying engineering design problem?
- What are the rewards if our design satisfies our needs?
- How will we test our design?

Real-World Examples ▶

Each chapter presents numerous examples of real-world applications of the subject presented. These well-constructed and clearly written examples provide students with step-by-step solutions containing both numerical and verbal explanations.

> **EXAMPLE 6.14 How Much MP3 Audio Will Fit on a Single CD-ROM?**
>
> The MPEG Audio Level III standard is known to achieve good quality at bit rates of about 128 kB per second for most popular music. How many hours of MP3-encoded audio could be stored on one CD at this bit rate?
>
> **Solution**
>
> We need to set up the proper ratios for the calculation. Any compact disc can store up to 700 MB of information. By converting bytes to bits, we can see that the number of bits stored on a CD is
>
> $$N\frac{\text{bits}}{\text{CD}} = 700 \times 10^6 \frac{\text{bytes}}{\text{CD}} \times \frac{8\text{ bits}}{1\text{ byte}} = 5.6 \times 10^9 \frac{\text{bits}}{\text{CD}}$$
>
> The rate at which the MP3 music is to be read off the encoded CD is
>
> $$r_{\text{MP3}} = 128 \times 10^3\text{ bits/s} = 1.28 \times 10^5\text{ bits/s}$$
>
> The number of seconds of audio that can be stored can be found by dividing these two numbers:
>
> $$T_{\text{MP3}}\frac{\text{s}}{\text{CD}} = N\frac{\text{bits}}{\text{CD}} \times \frac{1}{r_{\text{MP3}}\text{bits/s}}$$
>
> Plugging in the values, we get
>
> $$T_{\text{MP3}}\frac{\text{s}}{\text{CD}} = 5.6 \times 10^9 \frac{\text{bits}}{\text{CD}} \times \frac{1}{1.28 \times 10^5\text{ bits/s}} = 4.375 \times 10^4 \frac{\text{s}}{\text{CD}}$$
>
> Finally, converting the time units into hours, we obtain
>
> $$T_{\text{MP3}}\frac{\text{s}}{\text{CD}} = 43{,}750 \frac{\text{s}}{\text{CD}} \times \frac{1\text{ hour}}{3600\text{ s}} = 12.15 \frac{\text{hour}}{\text{CD}}$$
>
> In other words, a CD can hold over 12 hours of audio in MP3 format at near-CD quality. If you repeat the calculation for a single-layer single-sided DVD, the amount of music that such a DVD can hold is about 82 hours!

▼ Infinity Project Experiments

Designed for use with the Infinity Technology Kit, computer-based designs and experiments are boxed throughout the text. Indicated by an icon and color, these experiments challenge the students to apply the concepts presented in the text to real-world design or application problems.

> **Infinity Project Experiment: How a Facsimile Works**
>
> 
>
> The function of a facsimile machine is simple: Take a black-and-white image, scan it pixel by pixel, and transmit a one or a zero for each pixel, depending on its color. A demonstration of the fax machine's operation shows how the transmission works. Verify that the transmitter works by watching the scanning window and observing that a tone burst is created whenever a dark pixel is exposed. Use the sliders to increase the scanning rate, and observe the increase in transmission rate. Substitute your own binary image as a file, and listen to it as it is transmitted. Can you deduce any characteristics of the image by how the output sounds?

Marginal Notes

Each chapter contains four different types of marginal notes designed for clarification, reinforcement, and the provision of historical and interesting information about the subject matter. These notes include the following:

> **Transmitter:** A device or circuit that converts a communication signal into a form that can be conveyed to a distant physical location. For example, a signal might be converted into sound vibrations in the air or an electrical signal on a wire.

◀ *Definitions*: Important terms are set in boldface within the text. The term's complete definition is presented in the definition marginal box as close to the term's text reference as possible. This box not only provides clarification of the term, but also acts as a study aid for students when reviewing chapter material.

> **INTERESTING FACT:**
> Sound travels at 341 meters per second at sea level, at a temperature of 15°C. As the altitude increases, the atmospheric pressure drops, and so does the velocity of sound. The velocity is also affected by humidity, temperature, and pressure. For example, at the top of Mt. Everest at a temperature of −25°C, the speed of sound would be 310 meters per second.

◀ *Interesting Fact*: These marginal notes are provided to further stimulate the student's interest in the subject matter or to provide informative or historical information.

> **KEY CONCEPT**
> When receiving a signal in the presence of noise or interference, more confident detection and more accurate determination of frequency will be possible if the burst carrying each letter is allowed to remain on for a longer time. While this approach will usually improve the accuracy of reception, it also slows down the rate of transmission. This trade-off between speed of transmission and accuracy of reception is a fundamental one in telecommunications systems.

> **KEEP IN MIND**
> A tone is specified completely by three parameters. The frequency of the sinusoid or cosinusoid is represented by f, the maximum amplitude by a, and the phase angle by ϕ.

◀ *Keep in Mind*: These notes are presented as helpful hints or reminders of important points.

Key Concept: These notes are provided to further clarify and reinforce important concepts presented within the text.

EXERCISES 6.7

Mastering the Concepts

1. Which is a more secure encryption method, rotational encoding or permutation encoding? Why?
2. How many entries with a value of 1 does a permutation table have when encoding and decoding letters?
3. A permutation table has all but one row already specified. Can you figure out the last row? Try it and see!

Try This

4. ENCODE THIS SENTENCE USING A ROTATION OF TWELVE.
5. Suppose you wanted to encode the numbers 0 through 9 by using rotational encoding. Draw the rotational-encoder wheels for a shift of four values, and then write an equation for it.
6. ENCODE THIS SENTENCE USING THE FOLLOWING KEY:

$$K = [(1,24), (2,8), (3,21), (4,25), (5,15), (6,26), (7,2), (8,22), (9,16), (10,19), (11,10), (12,7), (13,4), (14,3), (15,11), (16,14), (17,6), (18,12), (19,23), (20,9), (21,5), (22,20), (23,18), (24,13), (25,17), (26,1)]$$

7. Give the five-by-five permutation table that describes the following permutation: GREAT becomes TEAGR.
8. Consider a slightly modified rotational encoder that reverses the order of the letters in the inner wheel of Figure 6.15. That is, the outer ring of letters goes around clockwise, but the inner ring of letters goes around counterclockwise. Draw this rotational encoder for shifts of 5, −5, and 12 characters. Then, use this rotational encoder to encode a simple message, and compare the encoded version with that of the original rotational encoder (which doesn't reverse the letter order). Which one of the two encoding methods provides more protection of the message?
9. It is possible to develop permutation encoders that are easier to describe compared with the completely random encoding table in Figure 6.17. Consider the key given by

$$K = [(1,4), (2,8), (3,12), (4,16), (5,20), (6,24), (7,2), (8,6), (9,10), (10,14), (11,18), (12,22), (13,26), (14,3), (15,7), (16,11), (17,15), (18,19), (19,23), (20,1), (21,5), (22,9), (23,13), (24,17), (25,21), (26,25)]$$

Draw the encoding table for this permutation encoder. Just write the 1's in the table to save time. Do you notice any structure to the encoding method?

10. Consider the effects of double encoding.
    (a) If a rotational encoder is used with key K_1, and then the encoded result is passed through a second rotational encoding

◀ End-of-Section Exercises

Exercises are included at the end of each section rather than ganged together at the end of the chapter. These exercises are designed to reinforce the concepts presented within the section before the student moves on to the next section and subject. The exercises are presented in four subsets:

Mastering the Concepts: These exercises are presented as a review of the important concepts discussed in the section.

Try This: These exercises ask students to apply the concepts discussed in the section to theoretical applications.

In the Laboratory: These exercises require students to apply their newfound knowledge of the subject matter to various experiments.

Back of the Envelope: These exercises help students explore additional design challenges.

Master Design Problems 

Presented as end-of-chapter material, the Master Design Problems refer back to the Design Objectives and require the student to use the concepts presented in the entire chapter to solve real-world design problems.

Master Design Problem

Design an image and video system to record local athletic events and theater performances. You will need to consider the following basic questions for each specific application:

- What subjects do you want to photograph with still images and video?
- What is the maximum required field of view in the horizontal direction? What is the maximum required field of view in the vertical direction?
- What is the size of the smallest object or feature that you want to see clearly in your photographs?
- How many cameras are needed, and where should they be placed?
- What kind of lens should be used? Note that three general lens specifications were used in this chapter, but that other lenses with different ratios of field of view to distance can be used.
- What video frame rate is needed?

The answers to these questions are not independent of each other and will depend on the specifics of the application. For example, if multiple cameras are used, a lower resolution image might be acceptable, because the closest camera could be used, and not all cameras would have to capture the smallest important details. If multiple cameras are used, then the field-of-view requirement might be reduced as well. With suitable positioning of cameras, images from two different cameras could be joined together to provide a wider field of view than either camera could provide independently.

1. For the athletic events, consider football, soccer, and basketball games and track and gymnastics meets. You will want to be able to get good images of individual athletes, but you also will want to be able to see the whole field of play.
   - For each athletic event, what is the field of view of the whole game or meet? For a wide-angle lens, what is the *minimum* distance the camera must be from the field?
   - What is the size of the smallest feature that you should be able to see when viewing the whole field? For example, is it enough to be able to count the athletes? Or should you be able to see the markings on the uniforms or the facial features of the athletes? Consider a digital sensor that is 1000 pixels in width. What is the *maximum* distance the camera can be from the athletes? If the sensor is 2000 pixels wide, how will this distance change?
   - Based on the limits of the camera position for appropriate detail and a complete view of the meet or game, how many cameras would you use and how would you place them?

Big Ideas ▶

The Big Ideas section is presented at the end of each chapter as a helpful review and study guide. This section is broken down further to include the following subsections:

Math and Science Concepts Learned: This subsection is a summary of the important math and science concepts discussed within the chapter. It can be used as a study tool to aid the student in reviewing the chapter material.

Important Equations: This subsection presents all of the important equations presented in the chapter. It can be used as both a review and a study guide.

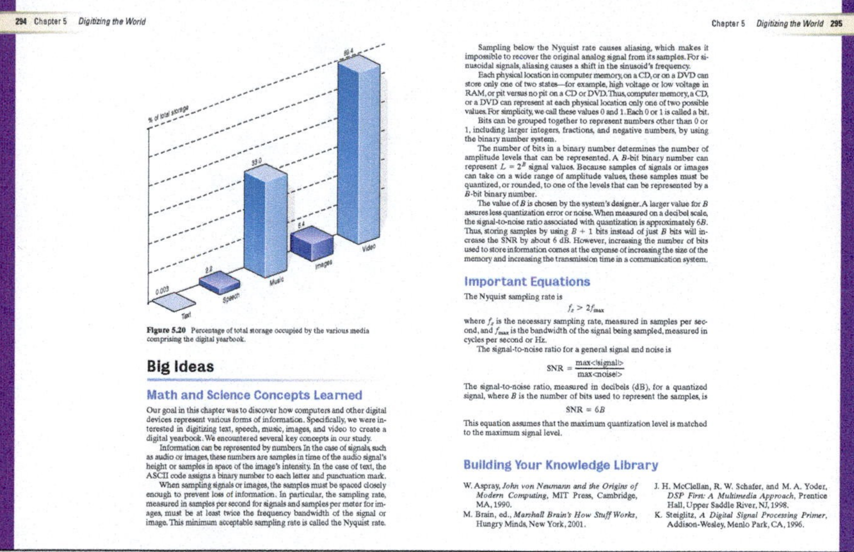

294 Chapter 5 *Digitizing the World*

Figure 5.20 Percentage of total storage occupied by the various media comprising the digital yearbook.

Big Ideas

Math and Science Concepts Learned

Our goal in this chapter was to discover how computers and other digital devices represent various forms of information. Specifically, we were interested in digitizing text, speech, music, images, and video to create a digital yearbook. We encountered several key concepts in our study.

Information can be represented by numbers. In the case of signals, such as audio or images, these numbers are samples in time of the audio signal's height or samples in space of the image's intensity. In the case of text, the ASCII code assigns a binary number to each letter and punctuation mark.

When sampling signals or images, the samples must be spaced closely enough to prevent loss of information. In particular, the sampling rate, measured in samples per second for signals and samples per meter for images, must be at least twice the frequency bandwidth of the signal or image. This minimum acceptable sampling rate is called the Nyquist rate.

Chapter 5 *Digitizing the World* 295

Sampling below the Nyquist rate causes aliasing, which makes it impossible to recover the original analog signal from its samples. For sinusoidal signals, aliasing causes a shift in the sinusoid's frequency.

Each physical location in computer memory, on a CD, or on a DVD can store only one of two states—for example, high voltage or low voltage in RAM, or pit versus no pit on a CD or DVD. Thus, computer memory, a CD, or a DVD can represent at each physical location only one of two possible values. For simplicity, we call these values 0 and 1. Each 0 or 1 is called a bit.

Bits can be grouped together to represent numbers other than 0 or 1, including larger integers, fractions, and negative numbers, by using the binary number system.

The number of bits in a binary number determines the number of amplitude levels that can be represented. A B-bit binary number can represent $L = 2^B$ signal values. Because samples of signals or images can take on a wide range of amplitude values, these samples must be quantized, or rounded, to one of the levels that can be represented by a B-bit binary number.

The value of B is chosen by the system's designer. A larger value for B assures less quantization error or noise. When measured on a decibel scale, the signal-to-noise ratio associated with quantization is approximately $6B$. Thus, storing samples by using $B + 1$ bits instead of just B bits will increase the SNR by about 6 dB. However, increasing the number of bits used to store information comes at the expense of increasing the size of the memory and increasing the transmission time in a communication system.

Important Equations

The Nyquist sampling rate is

$$f_s > 2f_{max}$$

where f_s is the necessary sampling rate, measured in samples per second, and f_{max} is the bandwidth of the signal being sampled, measured in cycles per second or Hz.

The signal-to-noise ratio for a general signal and noise is

$$\mathrm{SNR} = \frac{\max\langle|signal|\rangle}{\max\langle|noise|\rangle}$$

The signal-to-noise ratio, measured in decibels (dB), for a quantized signal, where B is the number of bits used to represent the samples, is

$$\mathrm{SNR} = 6B$$

This equation assumes that the maximum quantization level is matched to the maximum signal level.

Building Your Knowledge Library

W. Aspray, *John von Neumann and the Origins of Modern Computing*, MIT Press, Cambridge, MA, 1990.

M. Brain, ed., *Marshall Brain's How Stuff Works*, Hungry Minds, New York, 2001.

J. H. McClellan, R. W. Schafer, and M. A. Yoder, *DSP First: A Multimedia Approach*, Prentice Hall, Upper Saddle River, NJ, 1998.

K. Steiglitz, *A Digital Signal Processing Primer*, Addison-Wesley, Menlo Park, CA, 1996.

Building Your Knowledge Library: This subsection includes references for further reading. A short review of each title is provided to help clarify the contents of each reference.

◀ Glossary

Presented at the end of the book, the glossary defines all the important terms presented in the entire text. Each definition includes a reference to the chapter and section where the term was first presented.

Glossary

Access: A user's entry point into a data network. (8.3)

Access Server: A set of equipment located at an Internet Service Provider through which users gain access to the network and its services. An access server is commonly composed of a large number of voiceband or DSL modems, a controlling computer, and a router that aggregates the data from the modems and sends them into the network. (8.3)

Additive Synthesis: A synthesis technique that creates a periodic signal by adding sinusoids together. (2.4)

Algorithm: A step-by-step process to achieve a goal. In the case of this book, our goal is to describe and illustrate how to design new products or devices. (1.1)

Aliasing: An undesired effect, caused by undersampling, where one signal can masquerade as another. (5.2)

American Standard Code for Information Interchange (ASCII): American Standard Code for Information Interchange, which is currently used on most computer systems. This binary representation of the letters and other characters on a keyboard can represent 128 different symbols, using 7 bits of information. (5.3, 7.5)

Amplitude: The size of a signal. (2.2)

Analog: Used to describe an electrical circuit or system in which the signals are direct analogs of physical processes such as distance, sound pressure, and velocity. It implies a continuously evolving or continuously changing signal, since that is the way that the physical processes are found in nature. (1.2)

Analog-to-Digital Conversion: Sampling of an analog signal followed by quantization of the samples to binary numbers. Also called digitization. (5.4)

Animation: Creation of moving images by creating individual frames that produce the illusion of smooth motion when shown in quick succession. (3.1)

Antialiasing Filter: An analog low-pass filter preceding the sampler to assure that the analog signal is band limited prior to sampling. (5.2)

Approximation: A calculation or procedure that produces a value close to the actual or perfectly computed one. (2.2)

Audio Sampling: The process of recording values (samples) of a signal at distinct points in time or space. (5.2)

Average: The average of a set of values is the sum of all the values, divided by the number of values. It gives us a general idea about the values in the set, but it does not tell us any specific value. (4.3)

Average Codeword Length: The average number of coded symbols it takes to store an original symbol sequence, using a particular code. (6.3)

Backbone Network: An Internet Service Provider whose sole function is to connect other Internet Service Providers to each other. A backbone network tends to use high-capacity transmission systems and very fast routers in order to handle the traffic load imposed by the numerous client ISPs. (8.3)

Band-Limited Signal: A signal whose highest frequency falls below some finite value. (5.2)

Bandpass Filter: A filter that has a strong response to frequencies in a specific range or band and has minimal response to all other frequencies; an "ideal" bandpass filter would have zero response to all frequencies outside the particular band. (7.2)

Bandwidth: The difference between the highest and lowest frequencies contained in a signal. Often, the lowest frequency is zero. In this case, the bandwidth is just the highest frequency in the signal. (5.2)

Bass Clef: Also called F clef. Indicates that the note on the fourth line is F below middle C. (2.2)

Baudot Code: A five-bit code (Tables 7.5, 7.6, and 7.7) used with a keyboard for international communications until the mid-1970s; represents 32 different symbols by using five bits of information per symbol or character. (7.5)

Binary: A term referring to a quantity that takes on one of two different values. Often used when referring to base-two, or binary, numbers. (1.2)

473

ENGINEERING
OUR DIGITAL FUTURE

The World of Modern Engineering

chapter 1

OUTLINE

Where did all this great "stuff" come from? Who designed and created the TV you watch, the car you ride in, the computer you use to surf the Internet, and your cell phone? Nearly everything you touch had to be thought of, designed, and built (Figure 1.1). Who did all of this? And how did they do it?

The answer is simple: *Engineers, armed with their ingenuity*.

It might come as a surprise to learn that engineers, as a group, are some of the most creative and inventive people working today. Society calls upon engineers not only to envision what the world will look like tomorrow, but, more importantly, to make "tomorrow" happen.

Can you imagine what your life would be like if engineers hadn't designed and built TVs,

Figure 1.1 People use all sorts of technology to enhance their lives.

radios, recording studios, stereos, and automobiles? What would your life be like if you couldn't call or e-mail your friends at night, or if there were no radios, CDs, or DVDs to entertain you? What if there were no X rays and CAT scans to help doctors diagnose injuries and illnesses? What if the only way for you to get to school were to walk or ride a horse?

We often take today's great creations for granted, again thanks to engineers. What new high-tech health treatment, communications device, digital entertainment experience, transportation vehicle, or manufacturing method will we all take for granted in the coming years? Computers that talk to us or even "think" for us? Cars that drive themselves? Engineers are already working on these devices today!

INTERESTING FACT:

The word "engineer" stems from the French word *ingénieur*, which means "ingenuity." Contrary to popular belief, its root is not the English word "engine."

1.1 What Exactly Do Scientists and Engineers Do?

Making Dreams a Reality

Let's go back in time to the 1950s, before humans ever ventured into space. Back then, there were science fiction movies that suggested what life could be like in space, what our spaceships might look like, what the surfaces of planets might look like, and even what aliens might look like. Yet, until the 1960s, these images were all largely figments of Hollywood's imagination.

However, on May 25, 1961, then-president John F. Kennedy proclaimed that the United States would put a person on the Moon "in this decade." Who did he think was actually going to make this dream a reality? Politicians? Bankers? Lawyers? No, he knew it was going to be engineers and scientists.

Did engineers and scientists know how they were going to achieve this remarkable goal at the time of President Kennedy's speech? No, but they had confidence that, by working together, breaking the problem of space travel into manageable pieces, solving these smaller individual problems, putting all the components together, and then testing the final solution, they would have a very good shot at placing a human on the Moon before 1970.

Well, the engineers and scientists were right. Through their hard work and with the help of many, on July 20, 1969, the Apollo 11 mission placed Neil Armstrong and Buzz Aldrin on the surface of the Moon (Figure 1.2), culminating all their efforts into one of the greatest achievements in all of human history and registering a triumph for engineering and science.

Figure 1.2 Photo of Buzz Aldrin on the surface of the Moon, taken by Neil Armstrong, July 20, 1969.

What Makes Engineers Different from Scientists?

What makes engineers unique? And how are engineers different from scientists and mathematicians? You have had years of experience taking math and science courses. These classes have helped you understand and describe the world around you. As you have probably already learned from these courses, the primary purpose of science and math is to help humans *understand and describe* their world: How do cells divide? What makes objects fall to the ground? What are the basic building blocks of life? What is the distance between the Earth and the Moon?

To answer these questions, scientists throughout history have followed the **scientific method**. This five-step process, or **algorithm**, is the basic roadmap for discovery and understanding. Scientists who have sought to answer the fundamental questions about our world have used the scientific method as their guidepost. The five steps are as follows:

Scientific Method: The five-step process by which scientists explain natural phenomena.

Algorithm: A step-by-step process to achieve a goal.

1. Observe some aspect of the universe.
2. Invent a tentative description (hypothesis) consistent with what you have observed.
3. Use the hypothesis to make predictions.
4. Test those predictions by experiments or further observations, and modify the hypothesis in light of your results.
5. Repeat Steps 3 and 4 until there are no discrepancies between theory and experiment and/or observation.

While scientists seek to explain how the world works, engineers attempt to create new objects and devices that are important to humans and society, such as cutting-edge medical devices, innovative video games, safer cars, and high-tech communication devices. While scientists rely on the scientific method for discovery, engineers rely upon the **engineering design algorithm** to create nearly every object around you. The engineering design algorithm includes the following nine steps:

Engineering Design Algorithm: A nine-step process followed by engineers to create new objects or systems.

1. Identify the problem or design objective.
2. Define the goals and identify the **constraints**.
3. Research and gather information.
4. Create potential design solutions.
5. Analyze the viability of solutions.
6. Choose the most appropriate solution.
7. Build or implement the design.
8. Test and evaluate the design.
9. Repeat ALL steps as necessary.

Constraints: Limits that are placed on the design problem. For example, a constraint might be that the final design should not cost more than $X or weigh more than Y pounds.

As indicated by Step 1 of this list, the engineer must first answer the fundamental question: *What do I want to create today?* Very few professions place such a high premium on the creative spirit of the individual.

EXAMPLE 1.1 Designing the Cell Phone of Today

As a way of understanding the engineering design algorithm, let's apply it to a piece of existing technology—the cell phone—as if we were the engineers about to begin its design process.

Step 1: *Identify the design objective.* We want to build something that will allow humans to communicate with one another between any two locations on the globe at any time.

Step 2: *Define goals and constraints.* Some of the design goals and constraints for this device include the following:

- *Movement:* The device should not be connected physically to anything else that would limit our movement when using it. For example, it shouldn't need to be plugged into a wall outlet or a network jack.
- *Size:* The device should be small and portable so that we can carry it in our hand, a pocket, a bag, or backpack.
- *Form:* It should be large enough to be easy to hold in our hand, since devices that are too small are hard to grip. It should also provide a way for us to talk into it and for us to hear the caller at the other end of the call.
- *Energy use:* It shouldn't require too much energy in its operation, or else we'll need to change or recharge its energy source too often.
- *Cost:* It should be inexpensive enough so that people will buy it.

Step 3: *Research and gather information.* Has anyone ever done something like this before? Wireless radio telephones were being researched by the American Telephone and Telegraph Company in the 1930s, but these systems were more like modern family radio systems and "walkie-talkies" than cell phones. The Citizen Band (CB) radio craze of the 1970s brought point-to-point, two-way radio communications to large numbers of automobile travelers. But neither system reaches around the globe, and neither can be used easily to contact a wide variety of individuals. It wasn't until the 1990s that a system such as the one described in Step 1 was put into widespread use.

Steps 4 to 8: *Create, analyze, choose, build, and test.* We are all fortunate that engineers at international companies such as AT&T's Bell Telephone Laboratories, Nokia, Ericsson, Qualcomm, and Motorola completed these steps and designed, built, and tested a wide variety of cell phones that meet the design goals and constraints.

Solution

How well do you think current cell phones meet the objectives specified in Step 1 and satisfy the constraints given in Step 2? Are you pleased with current cell phone technology? Would you change anything about the goals or constraints?

EXERCISES 1.1

Mastering the Concepts

1. Identify five items designed by engineers. What did these items do that was new and innovative at the time of their creation? What items did these new creations replace? How is the world a better place because of these designs?
2. Identify five items that you suspect were not designed by engineers. How do they differ from those designed by engineers?
3. Apply the engineering design algorithm to the following processes:
   a. Making the family dinner
   b. Creating new laws
   c. Treating illness in a patient
   Be specific about each step in the design algorithm.
4. Determine the likely constraints applied by the engineer in designing these items:
   a. Cash register
   b. Bicycle
   c. Office lamp
   d. Sneakers
   e. Calculator

Back of the Envelope

5. Select a device designed by an engineer. Discuss each step of the engineering design algorithm, and describe the likely path taken by engineers in creating the device.
6. Evaluate the effectiveness of the following engineering designs:
   a. Conventional telephone
   b. Medical CAT scan
   c. Desktop computer
   d. Cell phone
   e. MP3 player
   f. PDA
   Make sure to describe the strengths and weaknesses of the designs. Try to guess what capabilities these technologies will have in the future as engineers continue to improve the current designs.

1.2 Birth of the Digital Age

Before Digital There Was Analog

To understand where engineers will be taking our world in the future, it is important that we briefly look back. Up until the middle of the 20th century, the technology designed by engineers was primarily **analog**; more specifically, the devices and systems that engineers created relied primarily upon physical forces and matter for their basic operation rather than some abstract quantity, such as numbers.

Analog: Phenomena that are characterized by fluctuating, evolving, or continually changing physical quantities, such as force or mass.

For example, analog audio records introduced in the first part of the 20th century use the physical bumps and indentations in the grooves on vinyl discs (albums) to store audio data. The stylus at the end of the tone arm of a turntable rides in these grooves and vibrates nearly identically to the original sound waves of the audio. This mechanical motion is then converted to an identical electrical version of the audio that is subsequently amplified and played through speakers. The entire process of re-creating audio from bumps and indentations is purely physical. Never is the sound waveform converted to numbers to be stored or manipulated; in other words, the system is analog.

Analog systems like turntables and albums are quite functional. However, like all designs, they suffer from a number of shortcomings:

- Analog systems can be large.
- Analog systems can consume lots of energy.
- Analog systems are not easily modified to solve new or different tasks.
- Analog systems are prone to breakdowns due to their physical operation.

Building Blocks for Analog Designs

Vacuum Tube: An early technology that was used in nearly every piece of electronics. Today, it is rare to find vacuum tubes other than in very high-end audio systems, guitar amps, and radar devices.

Early electronic technology was built using an important analog device called a **vacuum tube**. Figure 1.3 shows a vacuum tube. Such tubes were used to control the electrical current and voltage in systems such as radios, radar, and very early computers. Unfortunately, like lightbulbs, these vacuum tubes got very hot, and burned out regularly. Your older family members might remember how often TVs used to break down due to vacuum tubes burning out.

The Mathematics and Engineering That Gave Birth to the Digital Age

During the middle of the 20th century, mathematicians and engineers discovered a process for converting most physical quantities found in the world (such as sound waves, light intensity, forces, voltage, current, or charge) to *numbers or digits.* This discovery should not be surprising, since scientists had been using mathematics to describe the physical world for centuries. This remarkable, yet simple, discovery was the mathematical foundation that gave birth to the **digital age**.

Digital Age: The era born with the creation of the transistor. The digital age is generally thought to have reached full maturity at the time that computers gained widespread use, during the mid-1980s.

Digital: Describing technology or phenomena that are characterized by numbers.

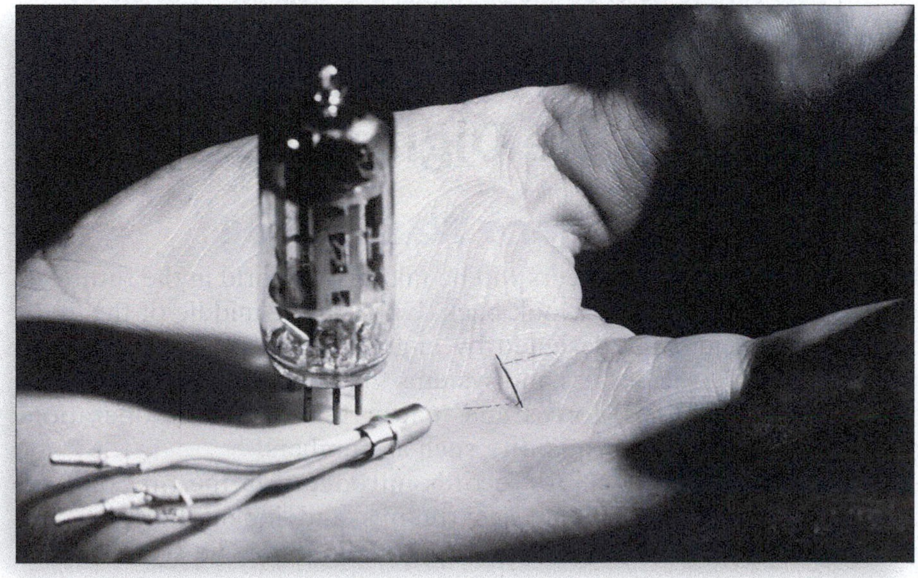

Figure 1.3 A vacuum tube next to an early transistor in the palm of a hand.

There are many advantages to "digitizing" analog quantities. For example:

1. Numbers are much less sensitive to physical problems caused by the physical nature of the device used to store or manipulate them.
2. Numbers are easier to store than an equivalent physical "amount" of something.
3. Numbers can be moved through space, using electronic, optical, or acoustic means.

To illustrate these advantages, let's revisit the re-creation of audio systems as discussed earlier in this section. Unlike analog systems, today the sounds of most audio are converted to numbers at the recording studio and stored on a compact disc (CD) or DVD. A CD player simply reads the numbers from the CD and then converts these numbers back to the original audio. We will learn about the details of this process in Chapter 5, *Digitizing The World*. If you have ever compared the quality of audio between an average turntable and an average CD player, you should have little doubt that digital technology is significantly better than the earlier analog technology. Also, can you imagine trying to jog or walk with a turntable strapped to your waist or inside your backpack? Table 1.1 lists some analog devices and the corresponding digital devices. Which do you prefer to use?

Still a Long Way To Go

Unfortunately, there was a major problem in building new digital devices when they were first conceived. Engineers just didn't have the right parts to build new digital systems. Not to be deterred, engineers working during the first half of the 20th century tried the smart and reasonable thing: They attempted to use readily available vacuum tubes as basic digital building blocks. Following this approach, in 1945, engineers successfully produced the first digital computer, called the ENIAC. It was built out of more than 17,000 vacuum tubes, weighed 30 tons, and filled a 30-by-50–foot room, as seen in Figure 1.4. Just think of the heat produced by 17,000 lightbulbs all burning in the same room!

Figure 1.4 The ENIAC computer.

INTERESTING FACT:

You should not be surprised to learn that the first computer programmers were actually women. Early pioneers, like Grace Hopper, Kay McNulty, Betty Snyder, Marilyn Wescoff, Betty Jean Jennings, Ruth Lichterman, and Frances Bilas, programmed the ENIAC to calculate trajectories of missiles during World War II. Previously, a single trajectory's calculations took 20 to 40 person-hours.

Table 1.1 Analog versus Digital Devices

Analog Devices	Digital Devices
LPs	CDs
Film cameras	Digital cameras
Dial watches	Digital watches
Standard TV	HDTV
VHS camcorders	Digital camcorders

INTERESTING FACT:
The ENIAC, short for Electronic Numerical Integrator and Computer, was capable of multiplying 5000 numbers every second. Compare this rate with that of today's computers, which can execute billions of mathematical operations every second and can fit into your backpack.

While primitive by today's standards, the ENIAC was a major advance in engineering and technology. Never before in human history could we do math so fast and so accurately. While the ENIAC opened up new digital horizons for society, these first computers were so large and so expensive that only governments and the largest of companies could ever hope to own or even use one.

The Transistor Replaced the Vacuum Tube

Transistor: A switch that regulates the voltage or current flow through electrical circuits. It is the basic building block for all digital electronic technology.

What the digital age needed was a truly digital component that could replace the vacuum tube. It would have to run fast, use much less power than the vacuum tube, and, most importantly, be small and inexpensive. Fortunately, in 1947 engineers at AT&T's Bell Laboratories developed that component, called the **transistor** (Figure 1.5). Its creation changed the world forever. Bill Shockley, Walter Brattain, and John Bardeen won the Nobel Prize in 1956 for their joint discovery and development of the transistor, which, within a few decades, had completely replaced the vacuum tube in nearly every piece of technology.

Now, engineers could unleash their imaginations to create smaller, portable devices that could run on the relatively small amounts of energy contained in batteries and were rugged in normal use. For this reason, many people believe that the transistor is the most important invention of the 20th century. Just look around you today to see the nearly infinite array of small gadgets and pieces of technology built from transistors.

The Integrated Circuit (IC)

Integrated Circuit: A single computer chip that is built from many different components. Typically, nearly all of the individual components on an IC are transistors.

As engineers designed devices for more complex tasks, such as in robotics or medicine, the resulting systems required ever more transistors. This push for more transistors made the devices large and hard to wire together. The next critical step forward into the digital age was the ability to put many transistors onto a single small part that could be used for these increasingly complex tasks. Jack Kilby accomplished this remarkable feat in 1958 at Texas Instruments when he invented the **integrated circuit**, or IC,

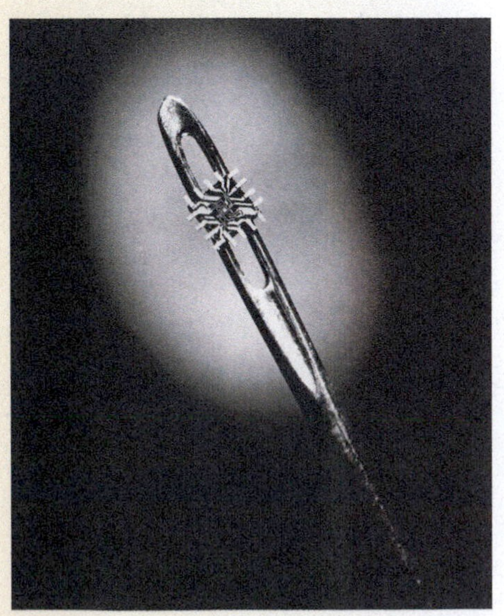

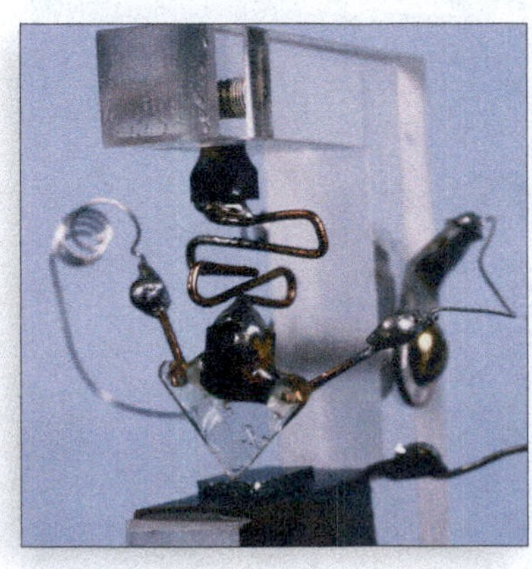

Figure 1.5 (a) The first transistor. (b) A vacuum tube, a modern transistor, and an integrated circuit.

Figure 1.6 (a) The first integrated circuit produced by Jack Kilby. (b) A modern integrated circuit.

shown in Figure 1.6(a). For this discovery, Kilby was awarded the 2000 Nobel Prize in physics. This groundbreaking invention was coined the "integrated circuit" because it cleverly integrated many parts, typically transistors, into a single small package like that shown in Figure 1.6(b).

With the invention of the IC, engineers were able to undertake more complicated designs, because they now had modern digital parts that could do significantly more complicated math on the newly digitized version of the real analog world. Interestingly, the integrated circuit has become so pervasive in devices from computers to anti-lock brakes that it is difficult to find individual transistors in modern technology today—they are now all part of integrated circuits.

KEY CONCEPT

What do transistors actually do? Transistors behave just like electronic doors in digital circuits. They are either open or closed; they either allow current to pass through or prevent it from doing so. So from a mathematical perspective, we can assign the number 1 to the state when the transistor is open (current doesn't flow) and the number 0 to the state when the transistor is closed (current does flow).

Why are Bits So Important?

Modern engineers and computer scientists always seem to be making reference to **bits**. What are they, and why are they so important? As we discussed previously, technology and engineering are steadily moving away from an analog world into a digital world. The advantage to doing this is that engineers can create devices that are smaller, faster, more reliable, and more powerful than their predecessors. The basis behind this shift from analog to digital is the engineer's ability to convert the physical or real world into numbers—the same process used to convert audio into numbers and then store it on a CD or DVD. No matter how complex, these numbers are all stored on computers, calculators, or any other digital technology by using bits instead of digits.

Bits: "Bit" is short for ***bi**nary digi**t***. A bit takes only the value of 0 or 1.

Base-10 Arithmetic Nearly all numbers that are used in society today are expressed in the base-10 system of numbers—the traditional number system adopted by Western society. This number system uses powers of 10 and the digits 0, 1, 2, . . . , 9 to express all numeric quantities. For example, the number 361 really means the following:

3×10^2, or three 100s

6×10^1, or six 10s

1×10^0, or one 1s

INTERESTING FACT:

A typical human hair is 50 to 100 micrometers (microns) in diameter. While this may seem small, it is quite large when compared with the size of modern transistors. Today, transistors on computer chips are much smaller than a millionth of meter (micron) on a side.

If we add up these powers of 10 appropriately multiplied by 3, 6, and 1, we get $(3 \times 100) + (6 \times 10) + (1 \times 1) = 361$. So the number 361 in base-10 is really shorthand for three 100s, plus six 10s, plus one 1.

This choice of base-10 for our number system was completely arbitrary and likely driven by the fact that humans have 10 fingers, or digits. To understand why early mathematicians chose base 10, try this simple experiment: Pick a three-digit number like 361 and attempt to describe this number to a friend using only your hands, and without talking or writing anything down. You will discover that it is hard to efficiently convey the number without understanding base-10 numbers.

Now, if humans had evolved with only one finger on each hand, they would have likely chosen base 2 for their everyday number system. In this case, we would have used only the numbers 0 and 1 when expressing quantities, rather than the familiar 0, 1, 2, 3, . . . , 9.

Transistors, the building blocks of the digital age, have only two states in digital circuits: on and off. So, it made perfect sense to the earliest digital engineers to choose the base-2 number system for digital technology, rather than the base-10 system. These "binary digits" were coined "bits," and the concept of a bit was born. Eight consecutive bits equal one **byte**.

Byte: Another very common computer term, a byte is eight consecutive bits, such as 10010011.

Now, this doesn't mean that you and I need to switch from the familiar base-10 system to the base-2 system. Computers and other digital systems today automatically convert digits to bits and back again. Think about your calculator: You type in digits because you are familiar with them. These digits are converted to bits, or base-2 numbers, and then the appropriate **binary**, or base-2, mathematical function is applied to these bits in accordance with your wishes, such as addition, subtraction, or taking the square root. The base-2 result stored within the calculator is then converted back to base-10 digits and displayed for your convenience.

Binary: A term referring to a quantity that takes on one of two different values.

It is not difficult to change a number from one base to another. In fact, many modern calculators can do it automatically for you with just a click of a few buttons. However, to be a modern digital engineer, it is critical that you understand how to do this simple mathematical operation by hand. Chapter 5 of this book has a thorough treatment of simple binary mathematics, but we can give you a preview here so that you can understand the basics of **binary numbers**.

Binary Numbers: Mathematical representation of numbers using the base-2 system of numbers rather than the familiar base-10 system.

You will remember that the quantity "361" in base 10 means that there are three 10^2's, or hundreds, six 10^1's, or tens, and one 10^0's, or ones. Well, in the base-2 number system, we follow the same process, but express the binary number in terms of 2^N, such as $8(2^3)$, $4(2^2)$, $2(2^1)$, and $1(2^0)$, rather than 10^N, such as $1000(10^3)$, $100(10^2)$, $10(10^1)$, and $1(10^0)$.

EXAMPLE 1.2 Converting a Base-2 Number to a Base-10 Number

What is the binary number 1110 in base 10?

Solution

To determine the base-10 value of 1110, we need to scale the appropriate powers of 2 by the values 1, 1, 1, and 0:

$$1 \times 2^3, \quad \text{or} \quad \text{one } 8;$$
$$1 \times 2^2, \quad \text{or} \quad \text{one } 4;$$
$$1 \times 2^1, \quad \text{or} \quad \text{one } 2;$$
$$0 \times 2^0, \quad \text{or} \quad \text{zero } 1.$$

Summing these results together, we get

$$(1 \times 2^3) + (1 \times 2^2) + (1 \times 2^1) + (0 \times 2^0) = (1 \times 8) + (1 \times 4) + (1 \times 2) + (0 \times 1) = 14$$

KEY CONCEPT

Bits are the numbers of the digital age.

This simple process can be reversed to convert base-10 numbers into binary numbers. To demonstrate this, let's consider the following examples:

EXAMPLE 1.3 Representing a Decimal Number in Binary Form

Represent the decimal number 361 in binary form, or base 2.

Solution

You can try it yourself, starting with $N = 8$. See if you can demonstrate that

$$361 \text{ (base 10)} = 101101001 \text{ (base 2)}$$

And, with just a little bit of arithmetic, it is easy to show that

$$(1 \times 256) + (0 \times 128) + (1 \times 64) + (1 \times 32) + (0 \times 16) + (1 \times 8) + (0 \times 4) + (0 \times 2) + (1 \times 1) = 361$$

The process of converting base-10 numbers to binary numbers can seem a little tedious, but it is easy if you follow the rules.

EXAMPLE 1.4 Converting a Base-10 Number to a Base-2 Number

What is the base-2 version of the number 5?

Solution

We need to write the number 5 as a sum of powers of 2, such as 8, 4, 2, and 1, to determine the number of each that add up to 5. It is very easy to do once you understand the approach.

Step 1: If we start with $N = 3$, then we first need to determine the whole number of 2^3, or eights, in 5. Since there is not a whole 8 in 5, the answer is 0, giving us the largest (or most significant bit) as a 0.

Step 2: Proceeding to $N = 2$, we determine the whole number of 2^2, or fours, in 5. The answer is 1, with a remainder of 1.

Step 3: Carrying the remainder forward, we now let $N = 1$ and determine the whole number of 2^1, or twos, in the number 1, which, of course, is 0 with a remainder of 1.

Step 4: The final step is to let $N = 0$ and therefore determine the whole number of 2^0, or ones, in the final remainder of 1, which is 1.

Step 5: So, putting it all together, starting with the most significant bit on the left, the number 5 is 0101 in base 2. As an important style note, if the leftmost bit value (the most significant bit) is a zero, we typically drop it in order to conserve space when we write the number down on paper. We do the same thing with base-10 numbers; for example, we don't write 54 as 054. Thus, the bit representation most commonly used for the number 5 is 101—one four, plus zero twos, plus one one.

EXERCISES 1.2

Mastering the Concepts

1. Why is modern technology based on binary mathematics?
2. State three differences between vacuum tubes and transistors.
3. Why was Jack Kilby's creation so important to the advancement of technology?

Try This

4. Write the following binary numbers in base-10 form:
   a. 111
   b. 01001
   c. 10010001
   d. 100100100
5. Write the following base-10 numbers in binary form, or base 2:
   a. 6
   b. 27
   c. 42
   d. 18
   e. 167

Back of the Envelope

6. Determine whether the following designs are digital or analog:
   a. Car speedometer
   b. Car radio
   c. TV
   d. VCR
7. Identify five modern systems that are completely analog. Do you think these systems will stay analog or be converted to digital systems?

1.3 Moore's Law

Predicting the Future

A recent newspaper headline reads as follows: "Engineers predict that in 10 years computers will be able to talk to us." How do they know this? What is the basis for their predictions?

Well, we are fortunate that history has shown that technological evolution often follows a fairly predictable path. Engineers routinely take advantage of this condition when determining if and when to design and build some new piece of fantastic technology.

Moore's Law and the Future Growth of Technology

Is there a systematic way to predict the future growth of technology? Well, in 1965, Gordon Moore, co-founder of Fairchild and the Intel Corporation, made a startling observation. Moore looked back in time and noticed that every two years, the number of transistors on his company's integrated circuits had doubled. This meant that his company's ICs were roughly twice as powerful or twice as fast as they had been two years earlier. And, as it turned out, this remarkable observation was true for nearly every computer chip manufacturer's products, irrespective of whether, the IC was a microprocessor or digital signal processor (DSP). This insightful observation has since been known as **Moore's law** and is used as one of the strongest principles for predicting the future of technology.

What was particularly bold about this prediction was that Moore said that the doubling of speed, computing power, or number of transistors on digital ICs would continue indefinitely and, as such, computers would continue to get faster and faster. To fully understand the remarkable implication of Moore's law, we need first to understand the power of doubling, which is at the heart of his observation.

Moore's Law: The number of transistors on an IC will double every two years. Equivalently stated, the computing power of ICs doubles every two years.

The Power of Doubling

Have you ever heard someone say "double or nothing" when gambling? If you have watched someone bet like this, then you no doubt noticed that the gambler either wins or loses a lot of money very fast. Doubling gets you to large numbers quite quickly. This observation can be quantified by developing some simple mathematical relationships that describe the power of doubling.

INTERESTING FACT:

Gordon Moore stated Moore's law when there were fewer than 100 transistors on an integrated circuit. Today, there are many millions of transistors on an integrated computer chip!

Assume that you begin with $\$X$, where X is any number you choose. If you double it, you will have $\$2X$. If you double it again, you will have $\$(2 \times 2)X$, or $\$2^2X$. Doubling once more gives you $\$(2 \times 2 \times 2)X$, or $\$2^3X$. From this set of results, it follows that if you double $\$X$, exactly N times, you will have $\$2^NX$.

To make this example more concrete, let's assume that we start with \$1, or $X = \$1$. If N were 5, that is, we double \$1 five times, we would have $\$2^5 = \$32$. If N were 10, that is, we double \$1 ten times, we would have $\$2^{10} = \$1024$. If N were 50, we would have $\$2^{50} = \$1.125 \times 10^{15}$. That is more than \$1 quadrillion. Wow!

INTERESTING FACT:

Periodically, people predict the death of Moore's law. They state that Moore's law eventually will end because of some future technological or scientific barrier. However, to date, engineers and scientists have found a way around these problems, and Moore's law continues to be an accurate means of predicting the future development of technology.

Doubling just 50 times turns \$1 into a staggering amount of money. So, the "power of doubling" takes small numbers (e.g., corresponding to dollars or numbers of transistors on an IC) and turns them into absolutely gigantic numbers very fast.

We can plot the value of \$1 doubled N times for increasing values of N to see firsthand how fast money or the number of transistors grows through the power of doubling. In Figure 1.7, we have plotted the value of \$1 doubled N times against increasing values of N, using two different types of plots: One is the standard linear scale plot that simply plots the value 2^N versus N, and the other is a semilog plot that plots 2^N on a logarithmic scale versus N. Both plots show the exact same data, but depict them in slightly different ways.

You will notice in Figure 1.7 that the semilog version of this plot shows that the increase in monetary value appears to be linear (following a straight line). Be aware that, while the graph is in fact a straight line, the data read from this graph do not grow linearly; on this plot, the y-axis is actually on a logarithmic scale.

To gain some practice reading the plots, see if you can confirm that at $N = 2$, the y value on both plots is 4, and at $N = 7$ the y value is $2^7 = 128$. The semilog version of the graph demonstrates an important fact: Quantities that grow by doubling over time always appear as straight lines on a semilog plot (plots with the y-axis on a logarithmic scale) with the slope of the line being controlled by the rate of doubling; the more frequent the doubling, the larger is the slope. So, if you see a straight line on a semilog plot, you immediately know that the quantity being plotted is actually doubling at some regular rate that you can measure.

What does the power of doubling tell us about Moore's law and its impact on digital engineering or computer technology? No matter how

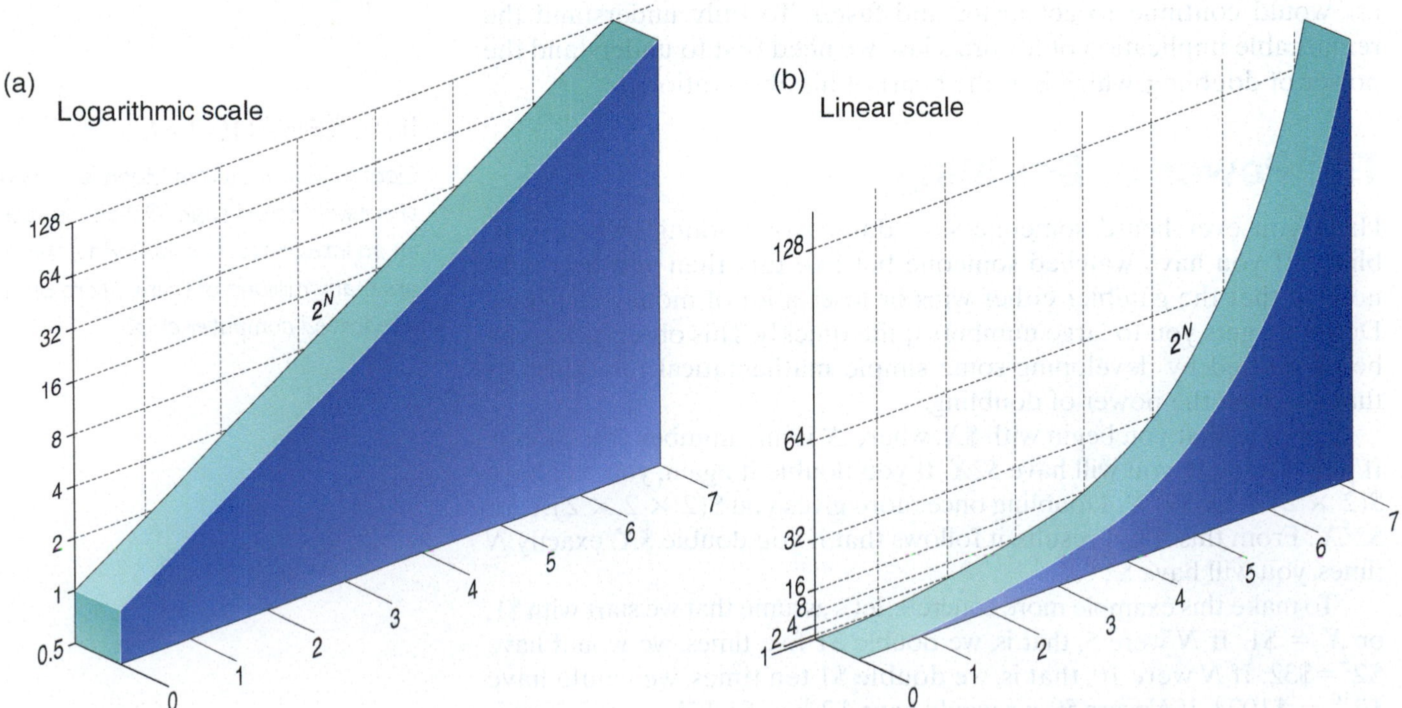

Figure 1.7 Plots of 2^N versus N for increasing values of N on (a) a logarithmic scale, and (b) a linear scale.

limited computer chips were when they were first introduced in the early 1960s, by doubling their computing power every two years in accordance with Moore's Law, we arrive at staggering amounts of digital power that will enable engineers to design systems that can do almost anything. If it can be dreamt of, Moore's' law tells us that we can build it at some predictable time in the near future.

The Mathematics of Moore's Law

Now that we have a good understanding of the power of doubling, we can translate Moore's law into simple mathematics to solve some of our own technological prediction problems. Let's say that in the year Y_1, there happen to be exactly N_1 transistors on a computer company's chip. Then, following Moore's law, two years later, or in year $Y_1 + 2$, there should be $2N_1$ transistors on the next version of the chip. We can generalize this relationship to determine the number of transistors on an IC in a later year Y_2, given that we know the number of transistors in year Y_1.

First, let's calculate the number of "doubles" that will occur between years Y_1 and Y_2:

$$\text{Number of doubles} = \frac{Y_2 - Y_1}{2} \tag{1.1}$$

KEY CONCEPT

The number of "doubles" used in Moore's law does not have to be an integer, like 1, 2, 3, or 4. The equation works even when the number of doubles is 3.3, 4.5, or any other positive number, for that matter.

For example, if Y_1 is the year 2004 and Y_2 is the year 2010, then there would be $(2010 - 2004)/2 = 3$ "doubles" in the number of transistors on computer chips between the years 2004 and 2010. And, if there happened to be X transistors on a computer chip in the year 2004, then there would be

$$2 \times 2 \times 2 \times X = 2^3X = 8X$$

transistors in the year 2010.

From this exercise, we are now able to identify the general relationship that predicts the number N_2 of transistors, or equivalent computing power, in any given year Y_2 if we happen to know the number N_1 of transistors, or computing power, in any other year Y_1:

$$N_2 = 2^{\frac{Y_2 - Y_1}{2}} \times N_1 \tag{1.2}$$

Gordon Moore made these same calculations roughly 40 years ago. To verify his result, Moore plotted the actual number of transistors on his computer chips for various years and compared these values with those obtained from the previous formula. What he found was absolutely remarkable and still holds true today: The predictions made by Moore's law were nearly identical to the actual values.

To see how accurate Moore's law has turned out to be over the last three decades, look at Figure 1.8, in which we compare Moore's predictions of the number of transistors over time with the actual numbers on a semilog plot. Remember that straight lines on a semilog plot imply that the number of transistors on a computer chip is actually doubling at some regular interval—in this case, every two years.

Moore's startling observation means that from the first days of digital technology, computers and other digital devices have been increasing in speed and power at a staggering and predictable rate, giving us computer chips with such enormous power that nearly any design will be possible in the future.

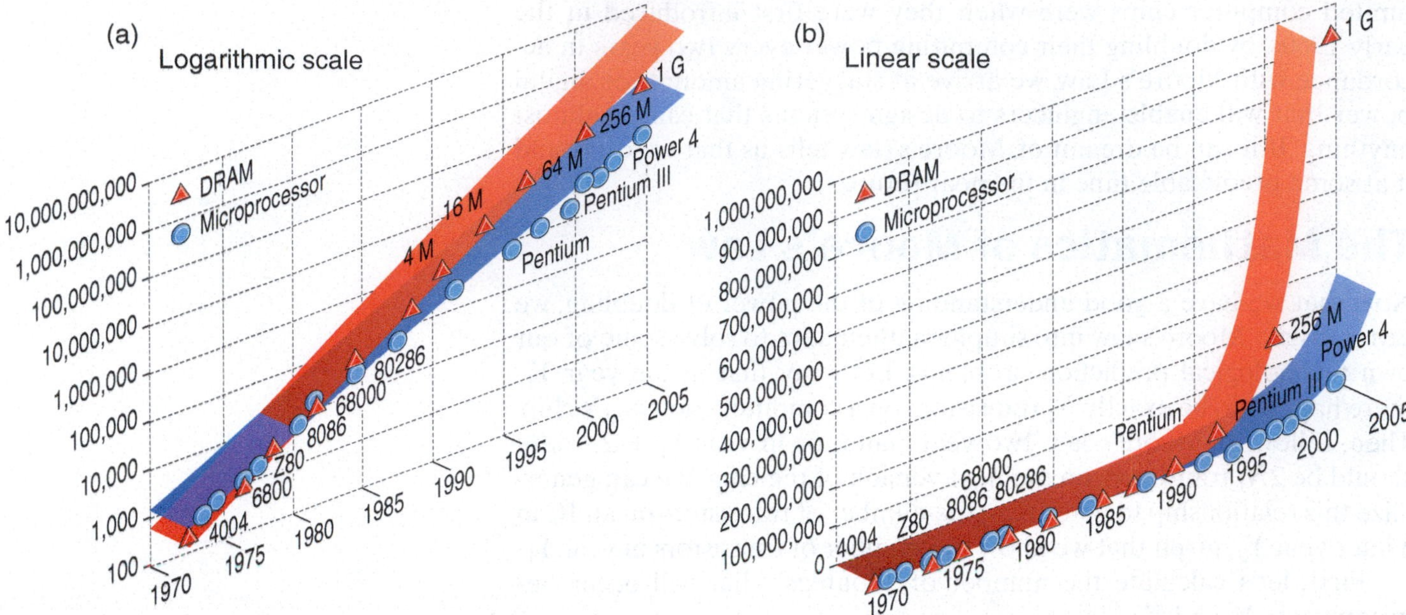

Figure 1.8 Moore's law for various digital components. The lines represent the predictions; and the marks represent the actual numbers of transistors. DRAM are memory chips, microprocessors are the chips used inside computers. The same growth rate is seen in graphics chips, and DSP or digital signal processors used in cell phones and MP3 players. All are forms of ICs.

INTERESTING FACT:

The time interval over which the number of transistors doubles can be different for different kinds of integrated circuits. Moore's Law tells us it is two years for standard computer chips, but the doubling interval can be as short as 18 months for simpler integrated circuits like memory circuits.

EXAMPLE 1.5 How Many Transistors Will Be on a Chip in the Future?

The Pentium 4 microprocessor (IC) made in 2001 had 47 million transistors in it. How many transistors should be in the Pentium chip made

a. 10 years later?

b. 30 years later?

Solution

The result is easy to find by applying Moore's law:

$$Y_1 = 2001 \text{ (the date of manufacture of the Pentium 4)}$$
$$N_1 = 47{,}000{,}000$$

For part (a), $Y_2 = 2011$, and for part (b) $Y_2 = 2031$. From the formula, we have for part (a)

$$N_2 = 2^{\left(\frac{2011-2001}{2}\right)} \times 47{,}000{,}000 = 2^5 \times 47{,}000{,}000 \approx 1{,}503{,}000{,}000$$

and for part (b)

$$N_2 = 2^{\left(\frac{2031-2001}{2}\right)} \times 47{,}000{,}000 = 2^{15} \times 47{,}000{,}000 \approx 1{,}503{,}000{,}000$$

Moore's law predicts that 30 years from now, we will have $2^{15} = 32{,}768$ times as much computing power on a single microprocessor as we have today, putting the power of more than 30,000 computers into a single device.

The Units of Modern Technology

As you have just learned from the previous section, the numbers that modern engineers deal with can be very small (such as the size of a transistor) or very large (such as the number of transistors on a single IC). Carrying around all those digits can make the mathematics tedious and lead to errors. So, to make engineering mathematics easier, engineers rely upon names for different-sized numbers.

You are no doubt familiar with the terms "mega" and "giga," but have you heard of "peta" and "femto"? Table 1.2 lists the size of various numbers as a multiplication factor, along with the name or prefix, the mathematical symbol, and the definition (ranging from the exceptionally small to the remarkably large).

For example, a typical DVD can store 37,600,000,000 bits. Using the terminology presented in Table 1.2, engineers would say that a DVD holds 37.6 gigabits, or 4.7 gigabytes, of data, where you should remember, a byte equals eight bits. Isn't it easier to say "giga" than to say 1,000,000,000?

Converting Numbers Just as it was important for us to learn how to convert base-10 numbers to binary numbers, it is important that engineers be able to convert numbers from one format or scale to another.

EXAMPLE 1.6 Converting Nanoseconds to Minutes

How many nanoseconds are in a minute?

Solution

From Table 1.2, we observe that one nanosecond lasts $1/10^9$ of a second, or is 10^{-9} seconds long. This means that in 1 second, we have $1/10^{-9}$, or 10^9; nanoseconds. Each minute has 60 seconds; therefore, one minute has $60 \text{ sec} \times 10^9 \text{ nsec/sec} = 60 \times 10^9 \text{ nsec} = 6.0 \times 10^{10} \text{ nsec}$.

Table 1.2 Number Sizes, Prefixes, and Symbols

Multiplication Factor	Prefix	Symbol	American Term
10^{18}	exa	E	One quintillion
10^{15}	peta	P	One quadrillion
10^{12}	tera	T	One trillion
10^{9}	giga	G	One billion
10^{6}	mega	M	One million
10^{3}	kilo	K	One thousand
10^{2}	hecto	h	One hundred
10^{1}	deka	da	Ten
10^{-1}	deci	d	One tenth
10^{-2}	centi	c	One hundredth
10^{-3}	milli	m	One thousandth
10^{-6}	micro	μ	One millionth
10^{-9}	nano	n	One billionth
10^{-12}	pico	p	One trillionth
10^{-15}	femto	f	One quadrillionth
10^{-18}	atto	a	One quintillionth

EXAMPLE 1.7 How Many People Would It Take to Reach the Moon?

The Moon is 384,402 km from the Earth. How many humans standing on each other's shoulders would it take to reach the Moon?

Solution

Let's assume that the distance from an average person's shoulders to the ground is five feet. So a thousand people standing on each other's shoulders would reach 5000 feet high. To determine the number of people required in order to reach the moon, we need only follow the conversion rule given below:

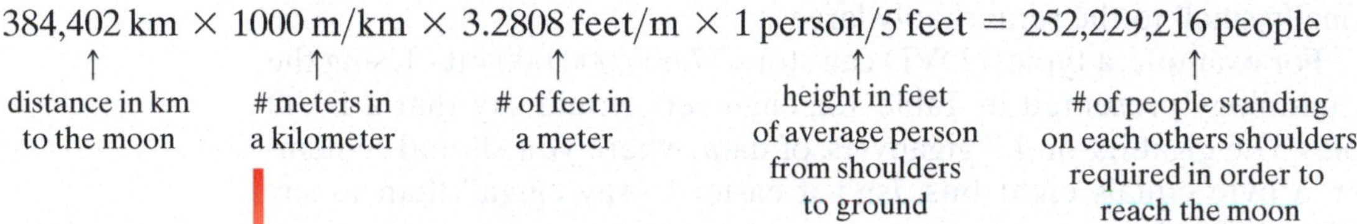

384,402 km × 1000 m/km × 3.2808 feet/m × 1 person/5 feet = 252,229,216 people
distance in km to the moon
meters in a kilometer
of feet in a meter
height in feet of average person from shoulders to ground
# of people standing on each others shoulders required in order to reach the moon

Interestingly, that is about 4% of the world's population and a little less than the entire U.S. population in the year 2000.

Putting These Numbers into Human Terms We can see from Table 1.2 that peta means 10^{15}. However, it is very hard to put numbers of this scale into any human context. How big is 10^{15}? How small is 10^{-15}?

Figure 1.9 will help you better understand such big and small numbers by relating items from your daily lives with some of these numbers.

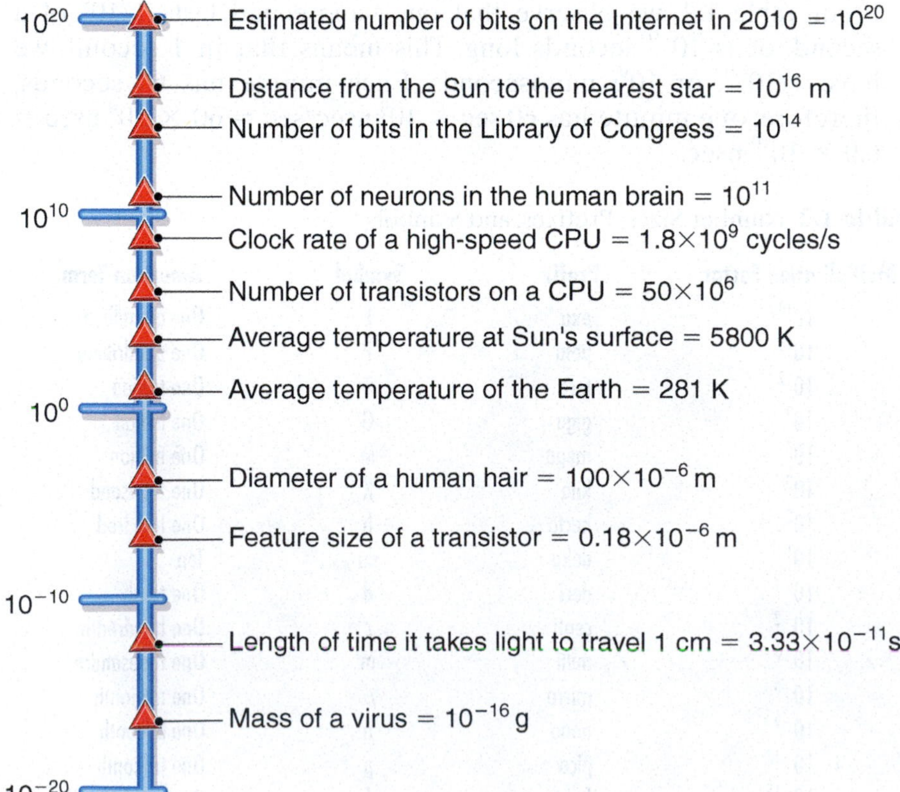

Figure 1.9 How big is big? How small is small?

EXERCISES 1.3

Mastering the Concepts

1. According to Moore's law, how many months does it take engineers to double the number of transistors on a computer chip.
2. Does it make sense to measure the distance between two cities in millimeters? Is it wise to measure the size of a transistor in meters? Why or why not?
3. How many picoseconds are in an hour?
4. How many pounds are in a megaton?

Try This

5. What is more valuable, one penny doubled 30 times or $1 million?
6. Using the Pentium 4 as your reference point, determine the number of transistors on the version of the Pentium to be released in 2015.
7. Make the following conversions:
   a. Convert 0.34 seconds to milliseconds.
   b. Convert 18 miles to inches.
8. Convert 0.00005 milligrams to nanograms.
9. How many times does $1000 fit into a terra of dollars?
10. How many nanometers are in a hectometer?
11. How heavy is a typical virus in terms of femtograms?
12. Determine the equation that tells you the correct monetary value in terms of dollars of a penny doubled N times.
13. How much money would you have at the end of a year if you had $1 on January 1 and
    a. you doubled it once a month;
    b. you doubled it once a week;
    c. you doubled it every day.
14. If Moore's law said that the number of transistors would double every three years, how many transistors would be on the Pentium computer chip to be released in 2015?
15. Derive equations that predict the number of transistors on computer chips if
    a. Moore's law said that the number of transistors tripled every two years.
    b. Moore's law said that the number of transistors doubled every three years.

Back of the Envelope

16. Assume that you want to build a computer system that can carry on a natural conversation with you. Engineers predict that it will take a very fast computer to achieve this goal. One prediction says that the computer will have to be able to execute 10^{12} instructions per second. Predict the year that this will happen, assuming that the number of instructions per second

also doubles every two years and that computers today run at 500 million instructions per second.

17. Moore's law predicts that the number of transistors on an integrated circuit will double every two years. Assume that the size of the IC chip stays the same over time and that transistors are square. Show that if a transistor is X_1 micrometers (microns) per side in year Y_1, then it will be $X_2 = 2^{-1/4(Y_2-Y_1)}X_1$ microns per side in year Y_2.
18. Assume that the side of a transistor is 1.6 microns in 2002 and that the diameter of a human hair is 100 microns. Using Moore's law, determine the number of transistors that would fit across a human hair in the year 2010.
19. What is the diameter of the following atoms?
    a. Hydrogen
    b. Silicon
    c. Iron
    d. Uranium
20. How big are transistors today, in square micrometers? Assuming that transistors are square, how many transistors would fit on a U.S. postal stamp?
21. Estimate the number of people required in order to circumnavigate the globe at the equator if they were all to hold hands. *Hint*: Assume the average human arm span is four feet.
22. Assuming that Moore's law continues to be accurate, predict the year that transistors will be the size of the atoms in Exercise 1.3.19 given that they are 1.5 square micrometers in area in 2003.
23. Assuming Moore's law continues, do you think engineers will be able to build systems that create
    a. Academy Award–winning movies?
    b. Prize-winning literature?

    Explain your answers.
24. How has engineering technology changed popular music?
25. How has engineering technology changed popular films?
26. How has engineering technology changed the way in which people purchase products such as clothing?
27. Imagine three new technologies that you would like to see in the cars of the future. Can you predict when these technologies might be implemented?
28. An atom of sodium is a cube approximately 0.4 nm on each side (nm = nanometer). How many sodium atoms are in a cubic millimeter.
29. How far does light travel in a single nanosecond? *Hint*: The speed of light is 3×10^8 m/s.

1.4 Block Diagrams—Organizing Engineering Designs

Suppose that engineers want to design a relatively complicated system, such as a high-performance video game player or the flight control system on the space shuttle. Do you think they would just gather together some parts and then sit down at a workbench, emerging some time later from the lab with a working system? No! Most systems designed and built by engineers today are so complex that one person or even one team of people couldn't reasonably design the whole thing alone.

Real-world, cutting-edge designs typically are created by breaking the complete system into collections of simpler elements that are then organized through a **block diagram**. Individuals or individual teams are typically responsible for one of the elements in the block diagram. Usually, there is another team of so-called systems engineers who have the responsibility of making sure that all of the elements of the block diagram operate together to create the final total system.

Block Diagrams: Block diagrams graphically describe how a particular system works or how a particular activity is to proceed. Block diagrams are used in nearly all creative endeavors, including advertising, manufacturing, computer science, medicine, and engineering.

As a good example, video game engineers created the block diagram shown in Figure 1.10 as the first step in designing a recent video game console.

As you can see in Figure 1.10, this new video game console has a variety of individual components or elements with separate names. Each of these components has a different responsibility in the overall design. The components are connected to one another in the block diagram with arrows that describe the "flow" of information or activity in the system.

As a general rule, each element, or block in the block diagram, has an **input**, an **output**, and a job to do. The inputs control the actions of the various blocks in the system, and the outputs are the resulting actions produced by the individual blocks. It is interesting to note that one block's output is often another block's input. Knowing this, it is very easy to read block diagrams if you just follow the flow of activity as directed by the arrows in the block diagram.

Input: Instructions or data used by a system to carry out a task.

Output: The final product of a system or device.

Let's try it: If you view the block diagram in Figure 1.10 as a whole, you will notice that the overall video game has three inputs and four outputs. The three inputs allow the user to actually play and interact with the game in various different ways. Let's just focus on the most obvious input—the game controller. The controller is what the player actually holds in his or her hands to play the game.

Notice that the controller is one of three possible inputs into the block called the I/O processor. ("I/O" stands for "input and output.") The I/O processor converts the raw information from the controller into information useful to the actual game program.

The three video game outputs come from two separate blocks: the video synthesizer and the sound synthesizer. These two blocks produce the video images and the sounds of the game, respectively.

The other blocks within the video game console system serve various functions that allows the game to operate:

1. RAM is a form of memory to store the software and the data.

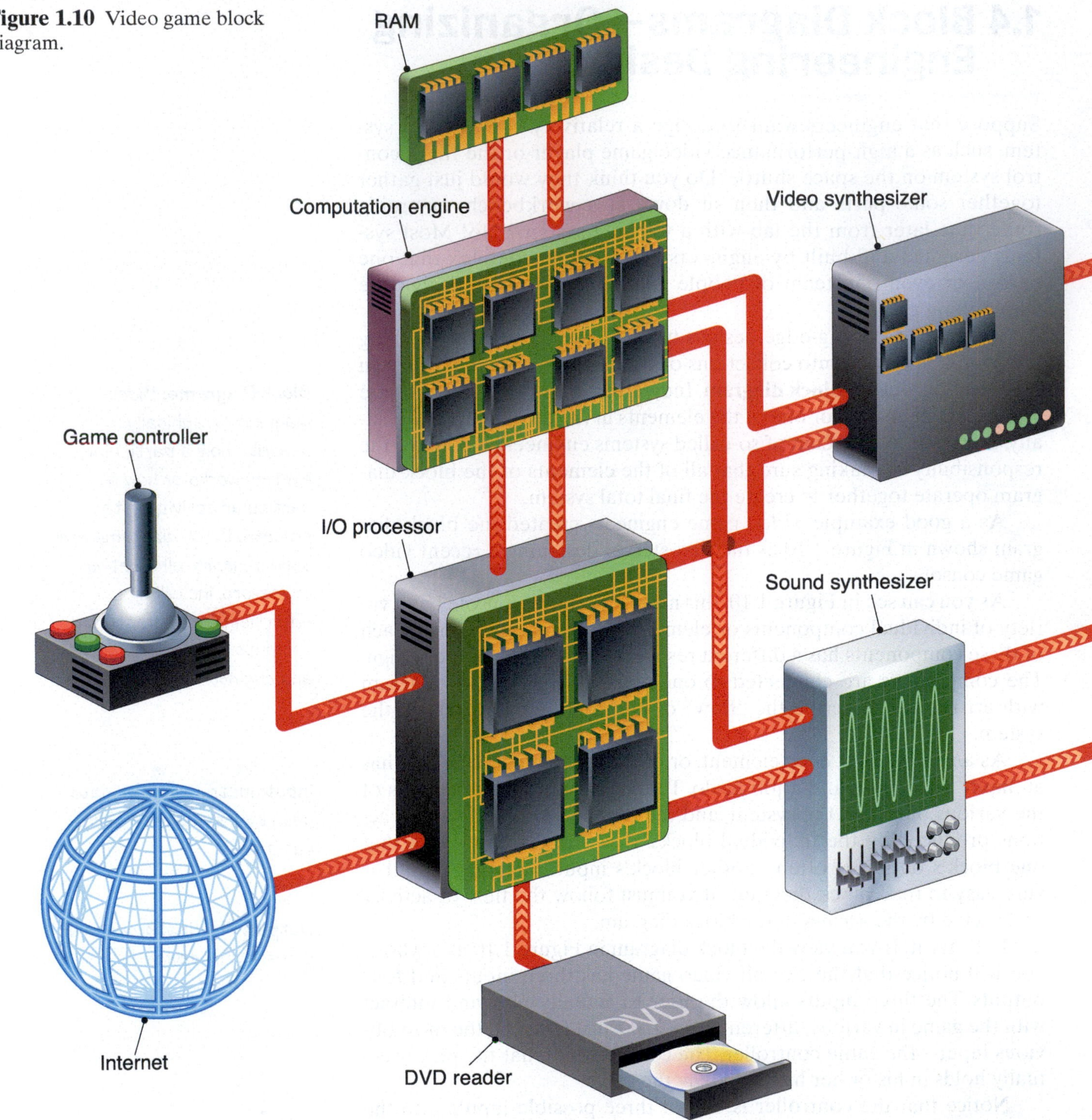

Figure 1.10 Video game block diagram.

2. The computation engine is the digital computer that determines where to place objects featured in the game, such as people, cars, and buildings, and determines how they are to move.
3. The video synthesizer takes the information from the computation engine and creates the corresponding digital images for the video display.

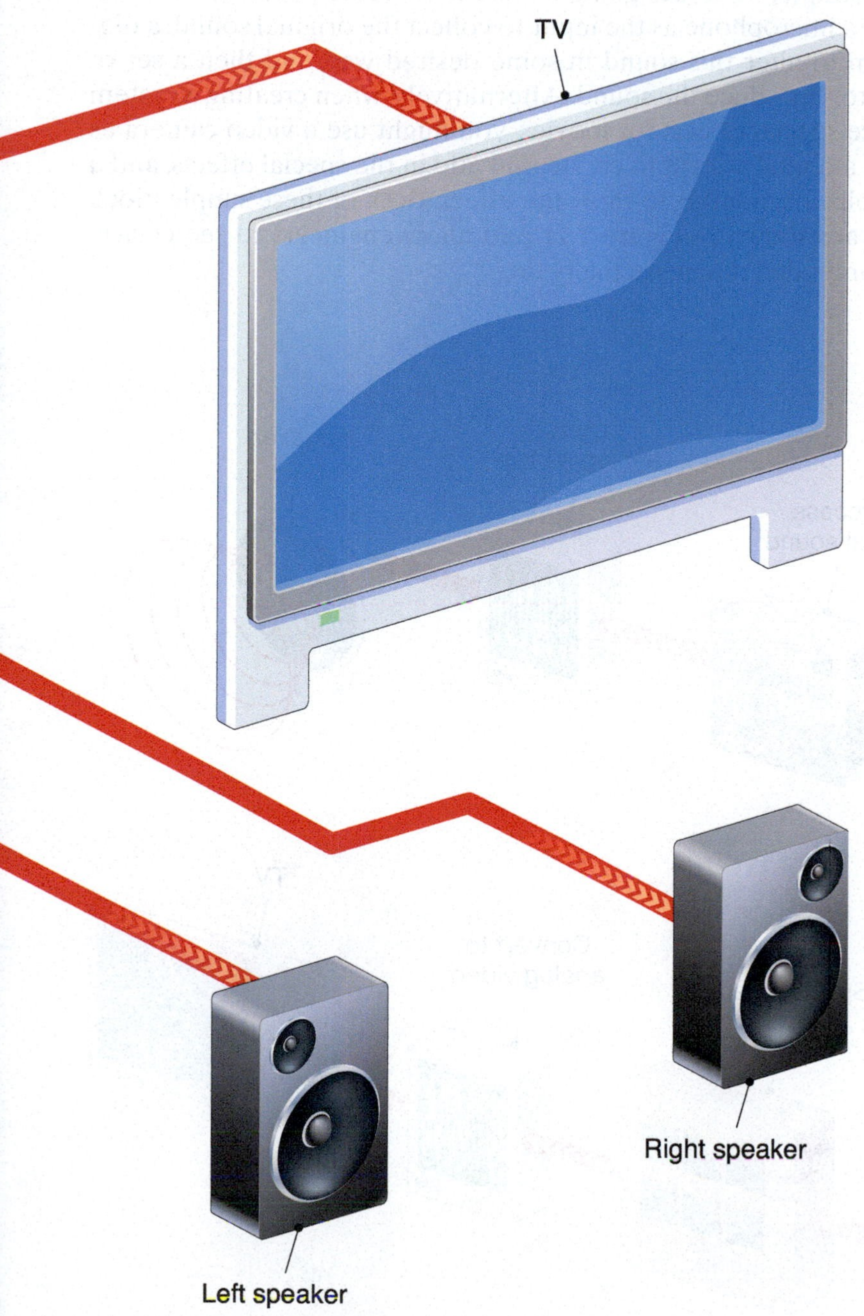

4. The sound synthesizer creates the sounds of the game.
5. The I/O processor controls the inputs and the outputs.
6. The DVD reader extracts game information from the DVD.

Throughout this book, you will have the opportunity to create your own high-tech designs and devices by following the engineering design process. During the actual design phase, you will create your designs by first constructing block diagrams of the overall system. These block diagrams will serve as a visual description of the total design and will allow you to clearly identify the inputs and outputs of the systems.

For example, when designing a system to create sound effects, you might use a microphone as the input to collect the original sound, a digital system to alter this sound in some desired way, and then a set of speakers to reproduce the sound. Alternatively, when creating a system to produce special effects for movies, you might use a video camera as the input, a digital system to create and add in the special effects, and a video display or monitor to show the effect. Both of these simple block diagrams are shown in Figure 1.11 and allow engineers to see clearly what the intended designs actually do.

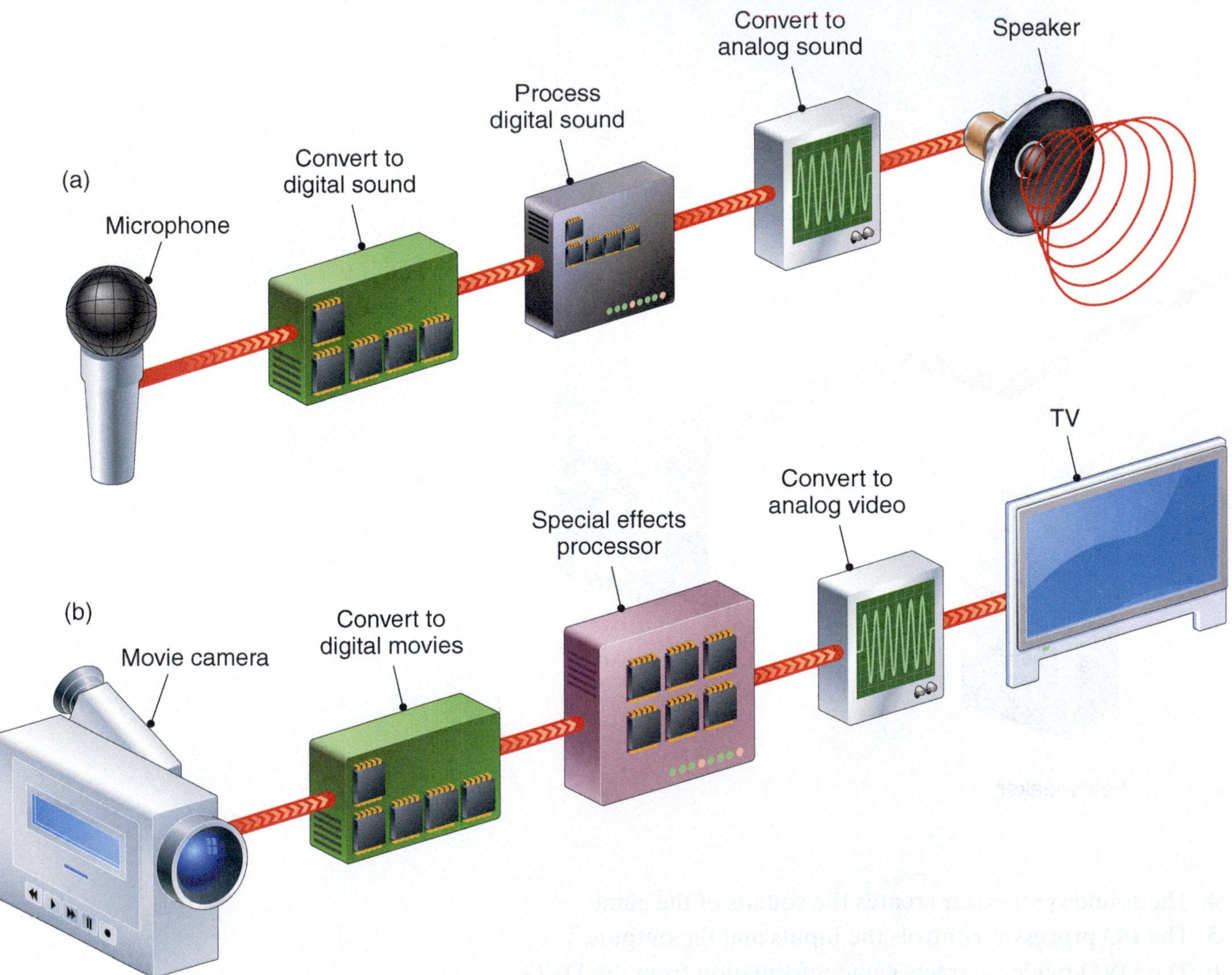

Figure 1.11 Block diagrams describing two systems: (a) one for creating sound effects and (b) one for creating special effects in movies.

Infinity Project Experiment: High-Tech Demos

Using block diagrams and modern technology, engineers can create prototypes of many different engineering designs quickly and easily. Some of these designs might include systems for creating sound or visual effects, for automatically counting the money in your pocket, or for tracking objects that are moving in space.

Explore these different designs, and ask yourself the following questions:

- What is the problem that they are each trying to solve?
- What are the constraints on the design?
- How effective are these prototypes in meeting the design objectives?
- How would you improve the design?
- How will Moore's law impact these designs in the future?

From Design Concept to Prototype

One of the many remarkable advances in digital technology that will become increasingly important in coming years is the ability for engineering manufacturers to produce a **prototype**, or first working system design, directly from the block diagram. What does this mean? Engineers can rapidly produce a working design to test and evaluate directly from the block diagram. So, in the near future, the block diagram will become the actual design itself.

Prototype: An original model of a design. Engineers use prototypes of systems to prove that the systems work.

EXERCISES 1.4

Try This

1. Construct a block-diagram description for the following activities:
   a. Getting dressed in the morning
   b. Cooking dinner at night
   c. Preparing for an exam

Back of the Envelope

2. Let's analyze a simple system: a portable CD player. Draw a block diagram that describes the functionality of this system.
   a. Describe the input and outputs of all the blocks.
   b. Imagine that it is your job to improve this system. Draw a block diagram of your new design.
3. Create a block diagram for an automobile braking system.
4. Create a block diagram describing a typical medical checkup in a doctor's office.

1.5 Summary

By now, it must be clear to you that engineers are creating our world of tomorrow with their know-how and ingenuity. This is what engineers have done over many centuries. Each generation has had new challenges to face and has developed new technology to deal with them. Today what we take for granted as normal in terms of living conditions—fresh food, travel, and entertainment—would have been beyond the dreams of the elite aristocracy 500 years ago. In the same way, the technology we create today that seems so amazing to us will be ordinary in the future as new technology is created.

During the remainder of this book, you will have the unique opportunity to participate in this remarkable endeavor by creating and inventing new technologies. You will be taught the skills and knowledge necessary to take your dreams and turn them into tomorrow's reality. As you move forward in this book, please remember the following statement made by Theodore Van Karman:

"Scientists explore what is; engineers create what never has been."

Master Design Problem

Now that we have learned about the world of engineering, it is time to apply our knowledge to a futuristic design problem. Most innovations begin with someone asking the question, "Is it possible to . . . ?" The engineer must determine not only whether the concept is feasible, but also how to actually create the design and then build and test the product.

In each chapter of this book, you will have the opportunity to assume the role of an engineer working on a Master Design Problem. While there is still much to learn, we can gain insight into the creative side of engineering by stepping through the design process for a futuristic design problem: **Create a digital system that can produce award-winning movies from scratch by simply using a few suggestive keywords typed in by a user.**

- *System Use:* An ideal design is easy to use. For example, selecting "Western, romance, comedy" would cause the system to automatically create the characters, plot, scenes, speech patterns of the digital actors, soundtracks, and any and all other components of this hypothetical Western romantic comedy.
- *First Step—Product Evaluation:* The first question that needs to be addressed when developing a new technology is simply, *Would anybody want this? Does it make any economical sense to even attempt to develop such a system?* This is not always an easy question to answer. Times change and with it, people's interests change. As an example, video rental stores continue to be popular in the face of other movie distribution alternatives, because people enjoy the communal experience of wandering the aisles of the store. Will this ever change? Only time will tell. Will our home movie production system be so desirable that it will change people's behavior? Before we get started, it is important that we attempt to answer these questions.
- *Engineering Design:* If your analysis shows that our new proposed technology to create individualized movies is wanted and needed, then we are off to begin the challenging and fun work of designing and building this system. Taking a good idea and turning it into a working product is an important dimension of being a successful prac-

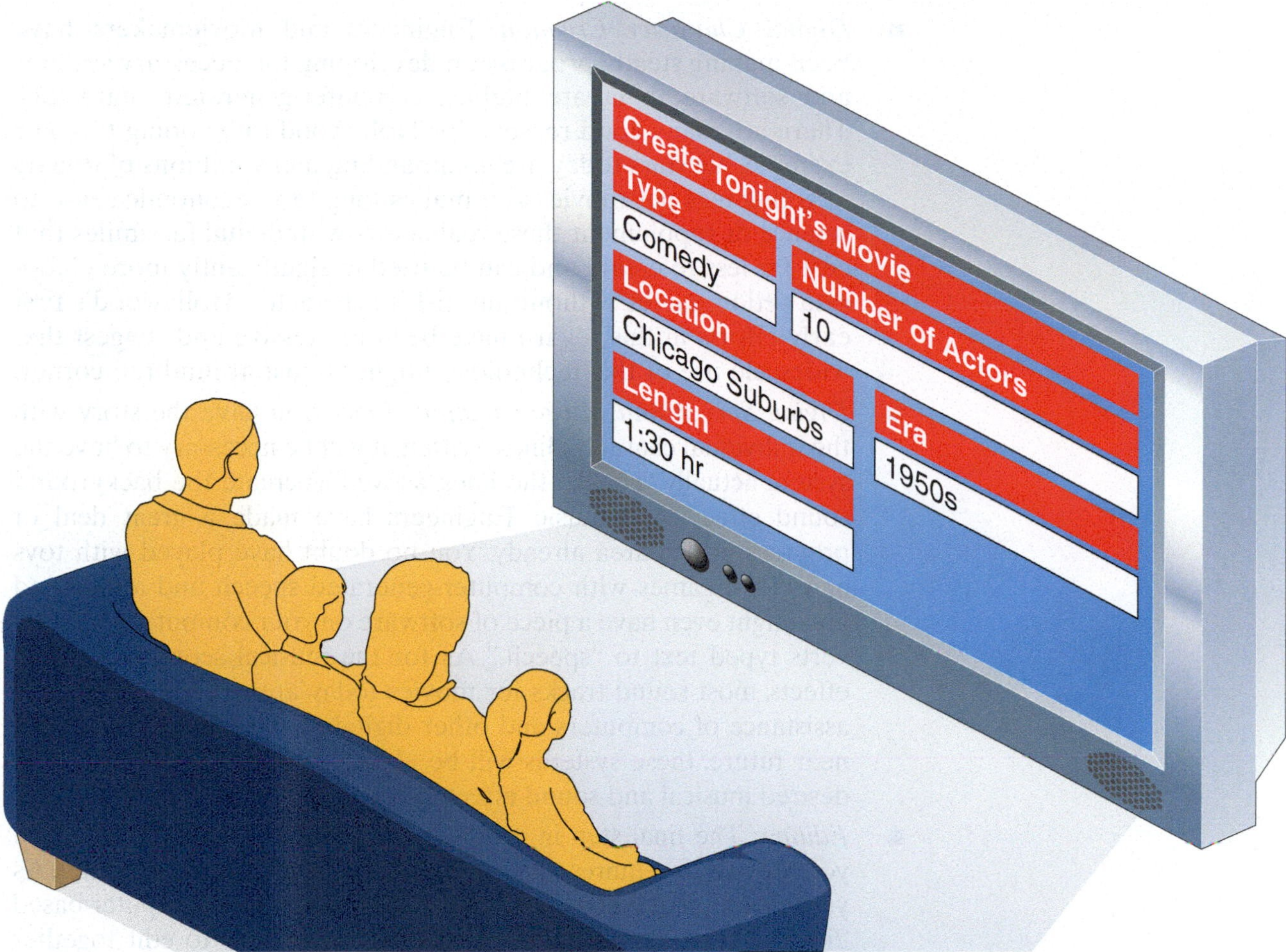

Figure 1.12 Family night in front of the "home movie production system."

ticing engineer. To accomplish this goal, we must determine whether it is even possible to create our proposed concept—and if it is not possible today, to assess the state of technology and predict the year when it will be possible. Let's consider a few of the key components of our futuristic design and attempt to estimate when the necessary technology will be mature enough for us to use to build this system.

- *Automated Story Creation:* As a critical component of the overall design, we need to develop a computer system that takes in a few suggestive keywords and then creates a text version of the complete plot, with a detailed description of all of the scenes and specific character lines. At first, this seems like a very difficult challenge. However, one simple, yet tedious, way to create such a system is to hire a large team of writers that would put together many, if not thousands, of small generic pieces of stories that could then be linked and edited together by a computer to produce a wide range of plots. Practically speaking, and in most cases, these stories would generally meet the interests of the users of the system. How often have you seen a movie with a misleading description in the newspaper that you still ended up enjoying? So long as the resulting movie is entertaining and loosely matches the keywords, the end users will be satisfied with the system. This approach, while somewhat primitive, might be an appropriate first version of the system that could be built to test the market interest. Future versions could involve more sophisticated approaches that will be available in future years.

- *Digital Character Creation:* Engineers and moviemakers have been making steady progress on developing the necessary systems and software to create lifelike, computer-generated characters. There are many good reasons for Hollywood to be doing this. For example, big stars today are commanding many millions of dollars in acting fees per movie, so it makes long-term economic sense to replace or supplement these real actors with digital facsimiles that cost far less to create and can be used in significantly more elaborate action scenes without any risk to the actor. Hollywood's first early efforts in this regard have been impressive and suggest that widescale use of this technology might be just around the corner.
- *Digital Speech and Music Creation:* Once you have the story with the individual character lines written, it will be necessary to have the system actually "speak" the lines, as well as create the background sound effects and music. Engineers have made a great deal of progress in this area already. You no doubt have played with toys and video games with computer-generated speech and audio, and you might even have a piece of software on your computer that converts typed text to "speech." As for the musical score and sound effects, most sound tracks for movies today are generated with the assistance of computers and other digital technologies. In the very near future, these systems will be able to automatically create the desired musical and sound effects without a human in the loop.
- *Editing:* The final step in producing a high-quality movie based on your keywords requires that we edit all these components together. As you may already know, nearly all movies today use computer-based digital editing systems. Extending their capabilities to edit together our computer-generated components will require some engineering, but it is conceivable that, with the right people working on the problem, engineers can achieve this goal within the next decade.

So, will people want to have digital systems create movies for them? Ask your friends and family. Can engineers do it? Yes, with enough creativity, determination, and time, this form of entertainment can be part of our future!

Is This Art?

Will these movies be any less works of art because they are created by engineers and computer scientists working with artists and psychologists rather than by producers, directors, and actors? This will be up to you and the marketplace to decide.

Big Ideas

Math and Science Concepts Learned

In this chapter, we discussed what makes engineers special, and we learned the basic processes by which engineers create the new technologies and inventions around us.

- The modern digital world began with the invention of the transistor in 1947.
- The mathematical foundation for the digital revolution is the engineer's ability to convert the physical, or analog, world into numbers to be collected, processed, and stored on digital technologies.
- High-speed, low-cost digital technologies were made possible by Jack Kilby's invention of the integrated circuit (IC) in 1958.
- Digital technologies, including computers, are accelerating in capability and performance at the rate predicted by Moore's law, doubling every two years.
- Bits and simple binary, or base-2, arithmetic are fundamental to the mathematics of all new digital technology.
- Engineers rely on block diagrams to visually describe processes and systems.

In the following chapters, we will apply these concepts to a variety of exciting design projects, including music, pictures, movies, and the Internet.

Important Equations

It is important for us to be able to predict the rate at which digital technology and engineering is accelerating. Moore's law allows us to predict the number of transistors used on future computer chips and integrated circuits (N_2), based on the number of transistors in computer chips today (N_1.)

Moore's law states the following: If you know that there are N_1 transistors on a computer chip in year Y_1, then in year Y_2 there will be N_2 transistors, where

$$N_2 = 2^{\left(\frac{Y_2 - Y_1}{2}\right)} \times N_1$$

Building Your Knowledge Library

Pearson, Greg, and A. Thomas Young, Editors, *Technically Speaking: Why All Americans Need to Know More about Technology*, National Academy of Engineering Press, 2002.

Technological literacy—a broad understanding of the human-designed world and our place in it—is an essential quality for all people who live in the increasingly technology-driven 21st century. This book explains what technological literacy is, why it's important, and what's being done to improve it.

Ambrose, Susan A., Kristin L. Dunkle, Barbara B. Lazarus, Indira Nair, and Deborah A. Harkus, *Journeys of Women in Science and Engineering: No Universal Constants*, Temple University Press, 1997.

Features short bios of 88 research scientists and engineers in areas from biochemistry to mathematics, from neuroscience to computer science, and from animal science to civil engineering. Includes those who have made careers in public service, such as Dr. Jocelyn Elders and Rhea L. Graham, as well as Nobel Prize winners, beginning assistant professors, division directors of corporations, and an engineering school dean. Each woman talks candidly about how she got into science or engineering, her work environment, and discrimination she may have encountered.

Uwe, Erb, and Harald Keller, *Scientific and Technical Acronyms, Symbols and Abbreviations*, John Wiley & Sons, Inc., 2001.

Never heard of a particular engineering and technology term before? This book has them all.

Creating Digital Music

chapter 2

OUTLINE

- Introduction
- Music, Sound, and Signals
- Making Music from Sines and Cosines
- Improving the Design—Making Different Instruments

Technology has changed nearly all aspects of the arts, especially the art of music. Nearly everyone listens to vocal and instrumental music on the radio, at concerts, or as part of movies and theater performances. Technology has made dramatic changes in our ability to create and enjoy a wide selection of music. Modern recording and communication methods make it possible for us to hear music performed long ago or in far-off places, using relatively inexpensive and widely available devices, such as DVD players and personal audio players.

Artists who compose and perform music now can use digital technology to enhance the sound they create by adding instruments and sound effects or even correcting performance errors. The sound of a new instrument can be created without actually constructing a new physical instrument. "Electronic compositions" are often used for the background audio of movie soundtracks. These digital techniques help make the pop stars of today sound as good as they do. Who knows—we may not be too far away from creating a "virtual singer" whose entire band shares her or his space inside of a digital device.

Because of the nature of sound, music is tied very closely to mathematics. We can use mathematics to create sounds that imitate traditional instruments or to create completely new sounds for instruments that have never even existed. Mathematics can be used to mimic the physical behavior of both traditional and new instruments. Mathematics also can be used to store music in digital form as a list of numbers that later can be converted back into music. A piece of music can be created and stored entirely inside a computing device, using numbers and equations, before it is turned into sound for everyone else to hear.

Design Objective for Creating Digital Music: A "Digital Band"

INTERESTING FACT:

In 1957, Max Mathews from Bell Laboratories created the first music synthesis program. An improved system named GROOVE was designed in 1970 to allow music to be created from a simple description so that the user could influence the sound as it was created. The electronics for this system was so large it would fill a walk-in closet.

Previous audio technologies, such as the compact disc and, more recently, DVD audio discs, have focused primarily on playing existing works. Let's use our engineering know-how to go beyond this task to design and build a new system that can create new music of any style, with any collection of instruments and performers at any time we want to hear it—our own "digital band."

In this book, we use the engineering design process or algorithm to help us understand and solve our problems. This algorithm begins with a series of questions to help us organize things:

- **What problem are we trying to solve?** It is always important to clearly define the problem we want to solve so that we use our time and resources effectively. In this chapter, we want to design a new device—a digital band—that will allow us to create a wide range of music without requiring us to have either extensive music training or a complete music library.
- **How do we formulate the underlying engineering design problem?** All design problems include a set of *specifications*, or features, that describe what we want our final design to do. Different approaches to the design are evaluated by how well they meet the specifications. Some of the most important capabilities of our digital band include the following:
  1. It should be able to re-create any combination of instrument sounds desired. For example, it should allow us to select known instruments or even to create the sound of new ones.
  2. It should be able to combine newly created sounds with existing recordings or modify previously recorded sounds to change the sounds' style or the type of instruments being played.
  3. It should be able to read some form of notation or score and convert performances into notation by itself without any special training.
  4. It should be able to imitate the environment in which we would like to hear the music performed so that it can sound as if the music comes from a concert hall, a recording studio, or even our echo-filled bathroom.
  5. It should be able to create music for a long time, so that we can hear a complete concert or listen to hours of background music.
  6. It should allow us to share the music we create easily with others or to take music our friends have created and change it to make it sound better to us.
  7. It should be small and lightweight so that we can take it wherever we go.

8. It should be durable and built to last a long time.
9. It should be as inexpensive and as easy to make as possible.

- **What will we achieve if our design meets our goals?** If we do a good job in the design of our digital band, we would achieve a number of important benefits for ourselves and our friends:
  1. We would have a very flexible device that would allow more people to enjoy their own music whenever and wherever they want to hear it.
  2. We would have a new creative device that would allow practicing musicians to make music that is not possible with existing instruments and devices.
  3. We could sell the system to others and make money.
  4. By using this system, we could come up with newer and better ways to make music—perhaps leading to an even better design for the system.
- **How will we test our design?** The quality of sound is always subjective, and the music from our digital band must sound good to the people who will use it. In order to achieve this goal, however, the design must be tested and improved at many stages of the system's development. Specific performance measures define the device's capabilities so that users will know what they can expect to do with it, and these specifications will be tested to make sure our system will function as advertised. Some of these tests might address the following inquiries:
  1. What is the range of notes that the system can create? How is this range of sounds related to what we humans can hear?
  2. How many different types of instruments or voices can be made? How many can be combined at one time?
  3. How long can new compositions or combined recordings be?
  4. How large is the device? How heavy is it? How much power does it need in order to operate? Is it a portable device, or does it need to be installed in a permanent location?
  5. What devices will be able to reproduce or play the music created by the system? If the devices don't already exist, how easy will it be to make them?
  6. How do listeners react to the quality of music from our system as opposed to music made by instruments that already exist? Do they enjoy what they hear when listening to our system?

2.1 Introduction

First Steps

When designing and building any device for the first time, we must make choices about the components, materials, and methods that we will use. What materials should we use to create our digital band? And what basic technology and components should it use to make music?

We begin our design task by drawing a block diagram of the system. As we learned in Chapter 1, each block has inputs, outputs, and a well-defined task. Once we have set the big picture in place with our high-level block diagram, we can then divide each of these major blocks into

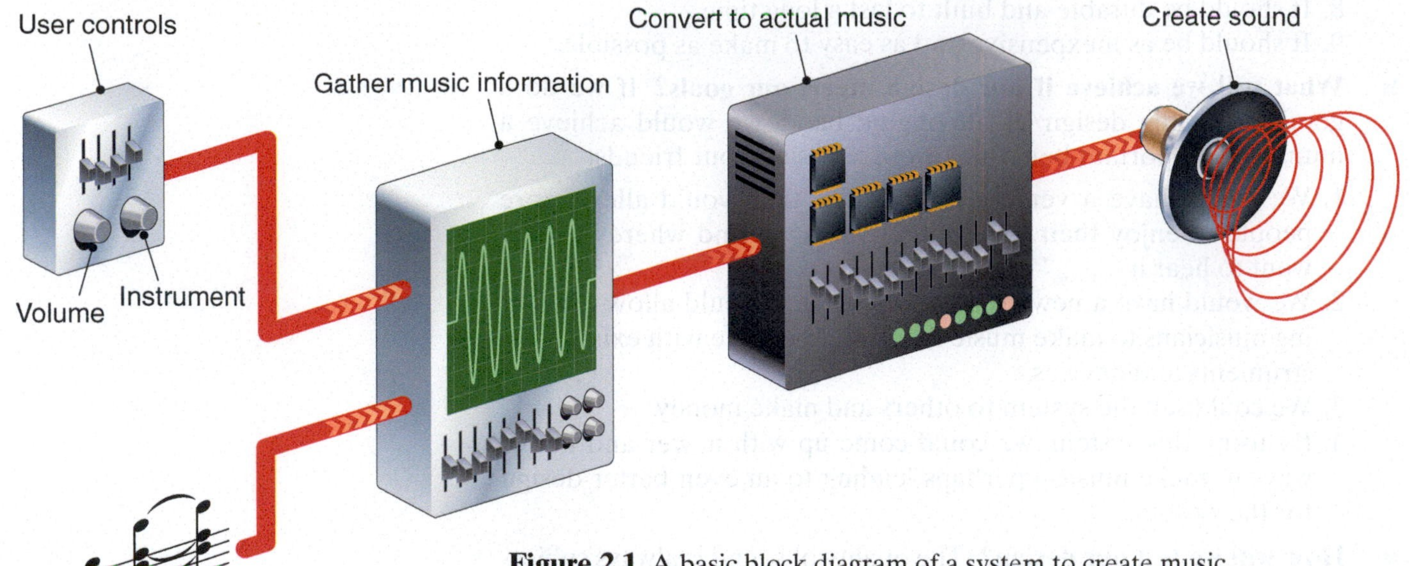

Figure 2.1 A basic block diagram of a system to create music.

smaller, more manageable, blocks for our design. After we have defined a good block diagram that includes all of our device's features, we can decide what technology to use to build each of the blocks. The choice of technology normally is based on what is currently available or expected in the near future and our estimate of the cost for the chosen technology.

A basic block diagram for our system is shown in Figure 2.1. The first block, labeled "Gather music information," has two general types of inputs. One input is the music source that describes the music we want to hear. Normally, this source would take the form of sheet music used by musicians to describe a piece of music. We will call the second type of input "user controls." These controls allow you to make changes while you are listening to the sounds of your device, such as by selecting the music from the source or adjusting the loudness with a volume control. The second block in Figure 2.1, labeled "Convert to actual music," takes the description of the music and the user controls and processes them to make an electrical form of the music before it is converted into sound waves. The third block, labeled "Create sound," translates the electrical form of the music into sound waves so that people can hear the music.

This block diagram is quite general. It can describe both simple and sophisticated music-generating systems. The details of the design of the blocks and the implementation technology will determine the capabilities of the system, its flexibility, its cost, and how well it meets our described design goals.

Engineers are always borrowing from the past and building on existing techniques and methods. They add new features and take advantage of current technology to create newer and better products. Before beginning a design, it is wise to review the development of the related products in order to understand what can be reused and to see where improvements are needed. Today, most of us listen to music from CDs or DVDs played on portable or home music systems. Because engineers are so effective at using what they already know, you might not find it surprising that CD and DVD players spin their discs in a way that is not much different from the method used by the vinyl record players that preceded them.

Ways to Make Music

Figure 2.2(a) shows a person, a pianist, making music on a piano. The pianist reads the notes to be played from sheet music that contains the instructions for playing a song. The conversion of the sheet music's description into sound is done by the fingers of the pianist striking the piano keys. The pianist can control the quality of the music by changing the way in which the keys are struck and by changing the position of the pedals near the base of the piano. Finally, the vibrating strings and sounding board of the piano make the sounds we hear. So, a person playing a piano is an example of our general block diagram of Figure 2.1.

Music can be played automatically using mechanical methods to describe and create the music. For example, mechanical windup musical toys use a small cylinder with spokes inside to play a short melody once the toy is wound up. The way these windup toys work is similar to that of the old-style music boxes that were first designed by Swiss watchmakers in the late 1700s. The player pianos of the late 1800s and early 1900s worked on similar concepts. (See Figure 2.3.)

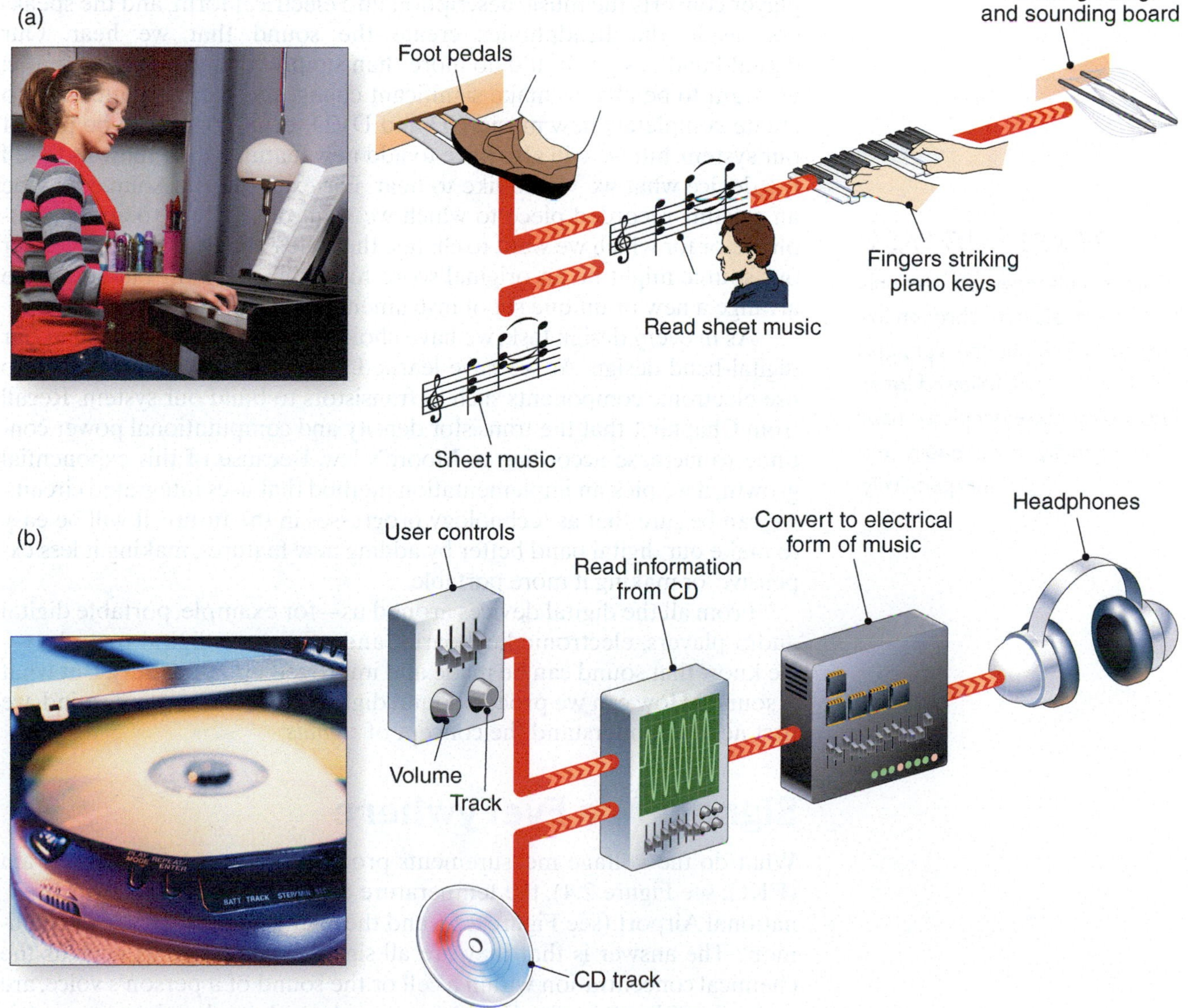

Figure 2.2 Two systems for creating music and their corresponding block diagrams. (a) A person playing a piano while reading sheet music. (b) A portable CD player.

Figure 2.3 Player piano.

Another way to enjoy music is by listening to a past performance that has been recorded. The first instrument to reliably record musical sound was the phonograph, invented by the American inventor and engineer Thomas Edison. Edison's phonograph was a mechanical device that was later improved by adding an electronic amplifier. It brought much longer and more complex musical performances to people who rarely had the opportunity to listen to the great singers and symphonies of the world. Engineers working in the 1970s further revolutionized the field of recorded music by developing the compact disc (CD) and other digital technologies. This invention removed most of the mechanical components of music reproduction and dramatically improved the quality of recorded music. Improvements in manufacturing methods allowed music distribution companies to make CDs quickly and inexpensively, and the size of the discs is small enough so that they may be carried around in a backpack. All of these developments resulted in a much wider availability of systems to play CDs, along with a much wider selection of recorded styles and performers.

Figure 2.2(b) shows a portable CD player and its associated block diagram. In this system, the CD provides the music description, the CD player converts the music description into electrical form, and the speakers inside the headphones create the sound that we hear. Our digital-band design should do more than simply replay a piece of music; we want to be able to make significant changes to recorded music or to create completely new music. CD and DVD technologies can be part of our system, but we will also have to add new features that allow us to tell the device what we would like to hear. For example, our source may be an existing recorded piece to which we want our system to add a saxophone or for which we want to change the style by making it go faster. Or our source might be an original score for which we want our system to arrange a new or unique set of instruments to perform the piece.

INTERESTING FACT:

Before Edison's phonograph, the only way to hear music was through live performances. The phonograph, and the CDs and DVDs that followed later, enabled many more people to hear high-quality music easily and inexpensively.

As in every design task, we have choices as to how we implement our digital-band design. As we have learned, we have some good reasons to use electronic components such as transistors to build our system. Recall from Chapter 1 that the transistor density and computational power continue to increase according to Moore's law. Because of this exponential growth, if we pick an implementation method that uses integrated circuits, we can be sure that as technology progresses in the future, it will be easy to make our digital band better by adding new features, making it less expensive, or making it more portable.

From all the digital devices around us—for example, portable digital audio players, electronic keyboards, and miniature digital recorders—we know that sound can be made and improved upon digitally. But what is sound? How can we process sound digitally? To understand sound, we first need to understand the concept of *signals*.

Signals Are Everywhere

What do the voltage measurements produced by an electrocardiogram (EKG; see Figure 2.4), the temperature at the Dallas–Fort Worth International Airport (see Figure 2.5), and the sound of a guitar have in common? The answer is that they are all signals! Some signals, such as the chemical concentration within a cell or the sound of a person's voice, are created inside of our bodies. Other signals, such as the electromagnetic waves produced by a radio transmitter or cellular telephone, the sound

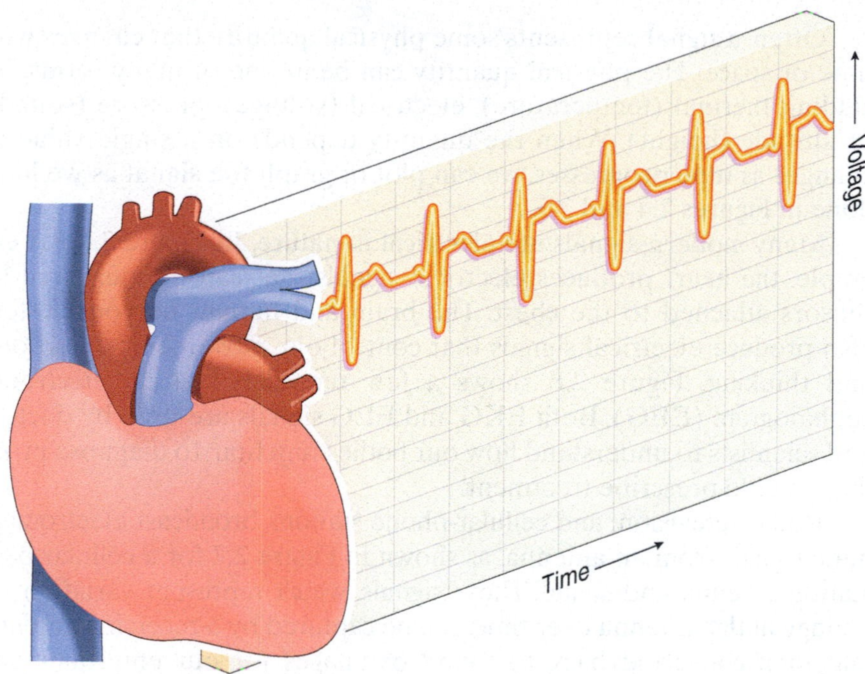

Figure 2.4 An EKG of a human heart.

of thunder, the vibration of an earthquake, or the light from the sun, are created in our surrounding environment.

What exactly is a signal? A **signal** is a pattern of variation over time that contains information. To understand this concept, let's return to the example of the temperature at Dallas–Fort Worth International Airport. What information can we get from this signal? You can see from the plot in Figure 2.5 that the temperature does in fact vary at the airport over the course of a day. The coolest time of day is 6:00 AM, and the hottest time of day is 3:00 PM.

Signal: A pattern or variation that contains information, usually denoted as $s(t)$.

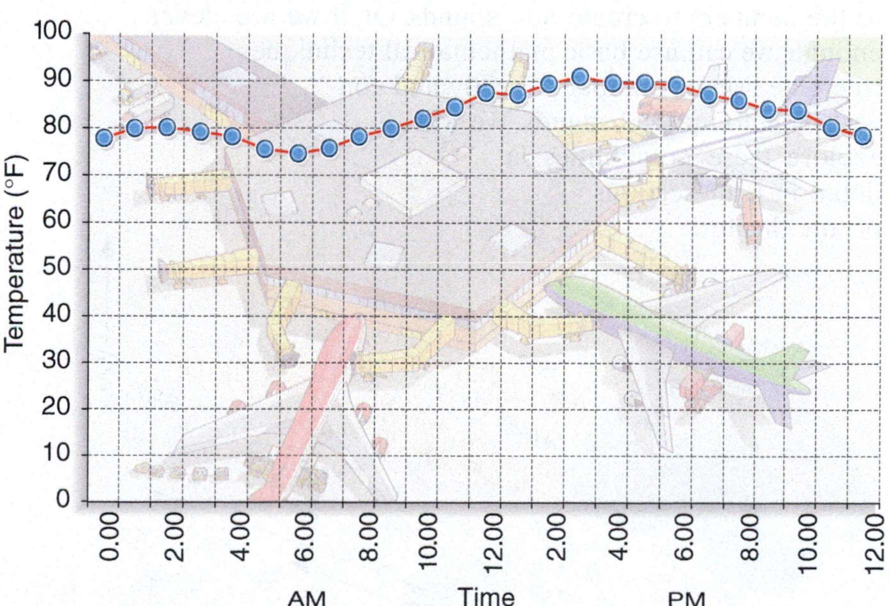

Figure 2.5 Temperatures recorded over a 24-hour period at the Dallas–Fort Worth International Airport. The range of the temperatures, the difference between the lowest and the highest, is less than 20°F.

Often, a signal represents some physical quantity that changes with time or space. The physical quantity can be in one of many forms; including thermal (temperature), electrical (voltage), pressure (sound), or altitude (height). When the quantity depends on a single value or changes as time progresses, we can plot or graph the signal as we have done in Figures 2.4 and 2.5.

Many modern signals are electrical in nature. In Figure 2.4, for example, the heart produces electrical signals that can be measured by sensors attached to the chest. The brain and human nervous system also produce electrical signals that control our movement, sensations, and thinking. Figure 2.6 shows a few recordings of an electroencephalogram (EEG). Both EKG and EEG signals are used by doctors and scientists to understand how our bodies function, to diagnose problems, and to prescribe treatment.

Radio, television, and cellular-phone stations broadcast electromagnetic signals from an antenna, as shown in Figure 2.7 for a cellular base station antenna and signal. These signals, which represent variations in voltage at the antenna over time, can be captured by a receiving antenna and then converted back to sound or images for our entertainment. These same signals also can be viewed directly by technicians or engineers on instruments that graph the signals as a function of time.

Music is also a signal. By capturing music with a microphone, music can be placed in an electronic form just like an EKG or a radio signal, and its variations can be plotted as a graph as in the previous examples. Like any graph, the signal can also be saved as a list of numbers and re-created later as was the case in Figure 2.4. In Figure 2.8, we see a group of musicians making music sounds. Each instrument is creating its own characteristic signal. These signals are sounds that combine in the air so that we hear them all at once. When each of these signals is converted to a list of numbers, we can use modern digital technology to re-create the sound. More importantly, we can also make changes to the numbers to create new sounds. Or, if we are clever enough, we can use basic mathematical techniques to create lists of numbers directly and convert them to musical sounds. We will explore these possibilities in detail in later sections of this chapter.

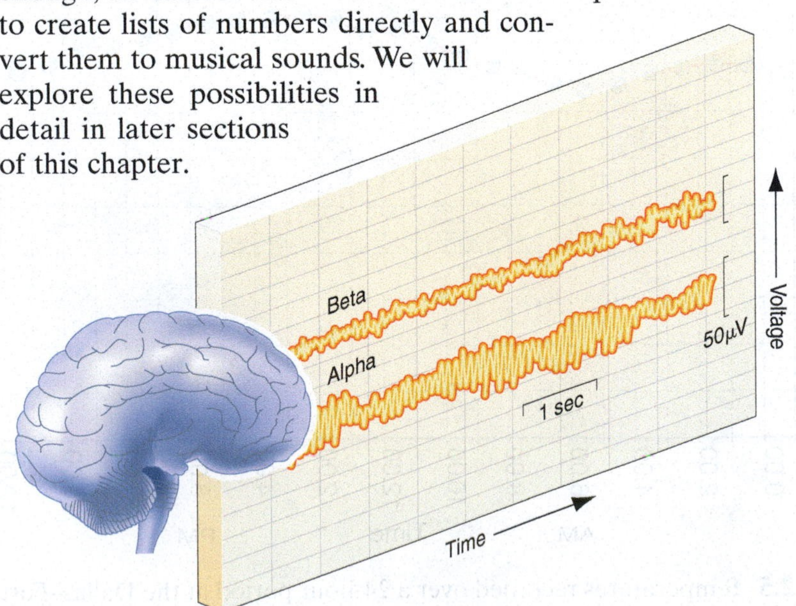

Figure 2.6 EEG signals from the human brain.

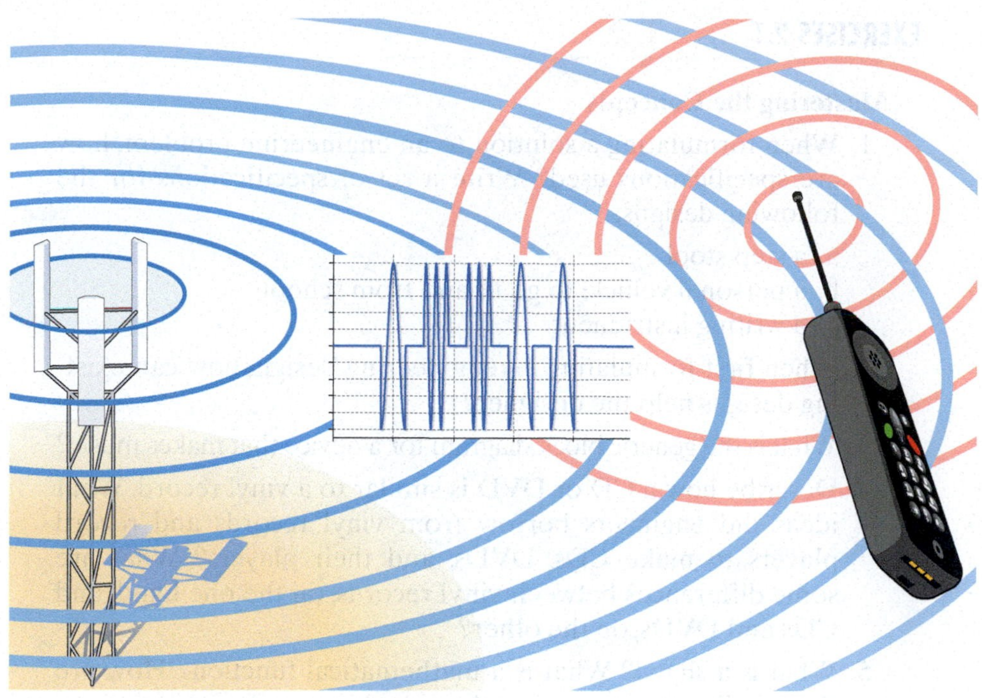

Figure 2.7 Cellular telephone signal.

Figure 2.8 A group of musicians and the signals they create.

EXERCISES 2.1

Mastering the Concepts

1. When formulating a solution to an engineering problem, how are specifications used? Write a set of specifications for the following designs:
   a. a step stool
   b. a personal vehicle to go to and from school
   c. a writing instrument
2. When first formulating an engineering design, how can existing designs help the designer?
3. What is the generic block diagram for a device that makes music?
4. Describe how a CD or DVD is similar to a vinyl record. What ideas did engineers borrow from vinyl records and record players to make CDs, DVDs, and their players? What are some differences between vinyl records, on the one hand, and CDs and DVDs, on the other?
5. What is a signal? What is a mathematical function? How are they similar? How are they different?

Back of the Envelope

6. Plot examples of five different signals that come from the real world. These signals can be from anywhere, but they must be able to be represented by a plot. Be sure to label both axes. How are these signals used in a particular application?
7. Find pictures of systems or devices that process signals. The sources of the pictures can be printed media such as magazines, newspapers, or catalogs, or you can find and print out pictures from the Web. For each device, do some research to find out the following:
   a. what technology the device uses
   b. whether the device contains or processes digital representations of signals

2.2 Music, Sound, and Signals

What Is Sound?

Before designing a new product or system, an engineer must understand how people will use it and understand the related science and technology that can be used in the design process. So, before we can make music, we have to understand what makes sound and how we hear sound.

INTERESTING FACT:

Sound causes air molecules to move back and forth. Wind causes air molecules to move forward in only one direction.

We can't see sound, but we can learn about sound through a visual analogy. We are all familiar with waves on the surface of water, such as those that move across a lake or crash on a beach. These waves are caused by wind or some other disturbance. Like waves on water, sound is a wave that travels through a physical medium, such as the air. Sound waves can travel only when there is some physical material through which they may be carried, so they can't travel in the near-vacuum of outer space. Although sound waves can travel in gasses, liquids, and solids, because of our interest in music we'll focus on sound waves in air.

Sound in air is created when a small disturbance causes the air molecules to move back and forth quickly. These disturbances come from mechanical vibrations such as the motion of our vocal cords when we talk or the vibrations of a plucked guitar string. These small "ripples" of motion actually move from one place to the next in a coordinated pattern called a "traveling wave," similar to the ocean waves we see at the beach.

INTERESTING FACT:

Two astronauts cannot talk to each other through the vacuum of outer space without some assistance. They use wireless radios to convert their voices to electromagnetic waves and back to sound again.

The Speed of Sound

Anything that moves has a number or value for its speed, and sound is no different. The speed of sound in air depends on several things, such as temperature, elevation, and humidity. The speed of sound at sea level in the Earth's atmosphere is about 340.4 meters per second (m/s) when the air temperature is 15° Centigrade (C). We can convert this number to miles per hour (miles/hour) as follows:

$$(340.4\ \text{m/s}) \times (60\ \text{s/min}) \times (60\ \text{min/hour}) \times (1\ \text{mile}/1609\ \text{m}) = 761.6\ \text{miles/hour}$$

This speed is faster than most normal passenger jets fly.

Since 761.6 miles per hour is faster than most of us have ever traveled, let's see if we can put this speed into a more understandable context. If sound travels 761.6 miles every hour, then it will take 1/761.6 hours, or 0.001313 hour, for sound to travel 1 mile. We can convert this number to seconds per mile as follows:

$$(0.001313\ \text{hour/mile}) \times (60\ \text{min/hour}) \times (60\ \text{s/min}) = 4.727\ \text{s/mile}$$

So, it takes sound about 5 s for sound to travel 1 mile. It takes us about 65 s to travel 1 mile in a car at highway speeds. However, the speed of sound is very slow when compared with the speed of light and electricity, which is about 186,000 miles per second! Light and electricity take only 0.0000054 s $= 5.4\ \mu$s to travel 1 mile—a time frame that is almost instantaneous compared with that of sound. The difference between the speeds of sound and light can be used to figure out how far away lightning is in an approaching thunder storm.

Figure 2.9 A group of fans doing "the wave."

INTERESTING FACT:

The way sound moves is similar to the way that "the wave" in Figure 2.9 is created by people as it moves through a stadium at an athletic event. In this analogy, the people are likened to air molecules, and their movements are likened to the sound waves moving through the air.

Sounds and Signals

If we had the ability to view air molecules and were to put ourselves in the middle of a sound wave, we would see that the density of the air molecules surrounding us would be changing all the time. We call this pattern of variation a **sound signal**. It is a signal that our ears or a microphone respond to and is different for each type of sound that we hear. This variation in density or air pressure causes our eardrums to move with the same pattern of variation. We perceive this movement as sound because our inner ears translate these small movements of the eardrum into nerve impulses that our brain interprets as sound.

Sound Signal: A pattern or variation in the motion of air molecules that a sound makes.

What is so amazing is that the sound waves from many voices or musical instruments combine at each of your ears into a single sound wave. This fact means that we can use simple devices to sense and replay such signals. A **microphone** is a device much like our eardrums in that it converts sound energy into electrical energy allowing the energy

Microphone: A device that converts sound energy into electrical energy.

Infinity Project Experiment: Plots of Speech

Like music, speech can be represented by a signal that can be plotted easily on a computer screen. Examine the shape of your own voice as you talk into a microphone. How does it look over several seconds? What happens if you zoom in on your speech signal over a fraction of a second? Try making different speech sounds by singing into the microphone or by talking in a different language, if you can, and look at the resulting signals. How do the signals change? How does the signal change if you whistle into the microphone?

Loudspeaker: A device that turns electrical energy into sound energy.

to be stored, changed, or displayed inside of an electronic device. Similarly, a **loudspeaker** is a device that converts electrical energy into sound energy so that it can be heard.

If we were to plot a sound signal, it might look like the plot shown in Figure 2.10, which depicts the signal picked up by a microphone when someone was saying "Hello, my name is. . . ." This plot of the sound signal mathematically represents what our ears respond to when we hear the sound of this particular person's voice. Mathematicians would call these signals *functions*, but we'll use the name "signal" because it is more common in engineering.

INTERESTING FACT:

The next time you are in a thunderstorm, watch for any bright flashes of lightning. As soon as you see one, count the number of seconds before you hear the sharp clap of thunder. By dividing the number of seconds you have counted by five, you can roughly calculate how many miles away the lightning is from you. Try it!

Most music is a combination of sounds created by several musicians playing instruments or singing together, as shown in Figure 2.8. Each instrument sounds different because the sound signal each creates is different. The characteristics of the sound we hear from a single instrument will depend on what instrument is used, what notes are being played, and the capabilities of the instrument player.

Figure 2.11 illustrates three sound sources—a tuning fork, a guitar, and a bird. Part of the signal recorded by a microphone for each is shown on the right. The signal produced by each source is different because the sound that each makes is different.

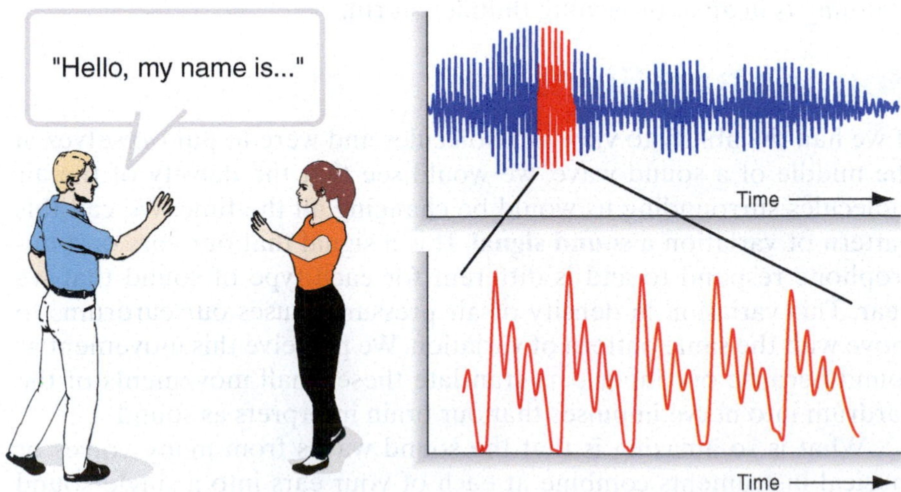

Figure 2.10 Person saying "Hello, my name is . . . ," and sound signals of the person's voice plotted on two different scales. The upper plot shows the whole signal, while the lower plot shows the time interval highlighted in red.

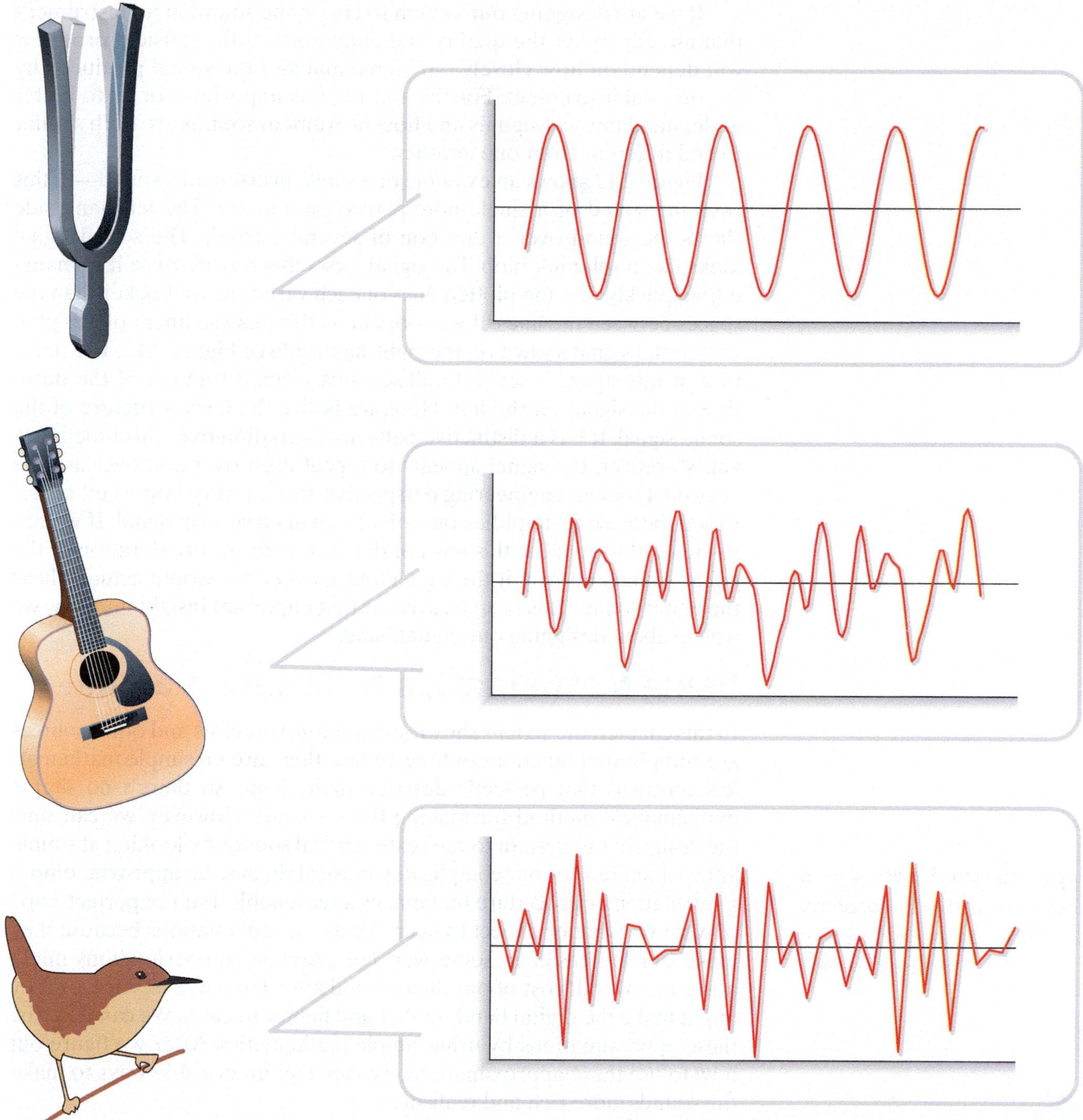

Figure 2.11 Three sound sources—a tuning fork, a guitar, and a bird—and the sound signals they create.

The design objective we've chosen for this chapter is to make a device that can create combinations of sounds from several instruments. Since sounds from instruments combine in the air, we can first design and test a device to re-create the sound of a single instrument. If we can figure out how to make the signal associated with any particular sound, such as those shown in Figures 2.8, 2.10, and 2.11, then we only have to convert those sound signals into vibrations in the air in order to re-create the sound itself. Once we have solved this problem for one instrument, we can explore methods for creating sounds from several instruments at the same time.

If we are designing our system to create the sound of an instrument that already exists, the quality and enjoyment of the music it produces will depend on how closely our signal matches the signal produced by the original instrument. For this reason, it is important for us to better understand musical signals and how instrument sounds are both similar to and different from one another.

Figure 2.12 shows an example of a single instrument's sound—in this case, the sound of a single note played on a guitar. The left-hand side shows the sound over a duration of several seconds. The sound signal looks like a solid ink blob. The signal looks this way because it is changing so quickly that the plotted lines of each variation are thicker than the spaces between the lines. If we zoom in on the small segment of this plot, we see the signal shown on the right-hand side of Figure 2.12. The duration of this signal is several milliseconds, a small fraction of the duration of the signal on the left. Here, we notice the basic structure of the sound signal. It has a distinctive pattern of variation over this time interval. Moreover, the signal appears to repeat itself over and over as time goes on. From an engineering perspective, this is a very important observation, because it simplifies our job of making a similar signal. If we can re-create this signal at this level of detail over the entire duration of the guitar sound and play it through a loudspeaker, we would actually hear the guitar sound. This fact gives us our first important insight into how we will go about designing our digital band.

Using Mathematics to Create a Signal

In this chapter, the signals shown so far for instruments and other sources are complicated functions of time. In fact, there are no simple mathematical formulas that perfectly describe their shape, so there's no simple mathematical method for making these sounds. However, we can start the design of our system to re-create musical sounds by looking at simple approximations for our complicated musical signals. An **approximation** is a calculation or procedure that makes a reasonable, but not perfect, copy of what we want or expect to hear. We use approximations because they make our life easier in some way. For example, approximations might make the overall cost of our digital band affordable to everyone, or they might make the digital band smaller and lighter to carry. We can describe these approximations by using simple mathematics. After we figure out how to use these approximations, we can explore possible ways to make the sounds more rich and realistic.

Approximation: A calculation or procedure that makes a reasonable, but not perfect, copy of what we want or expect to hear.

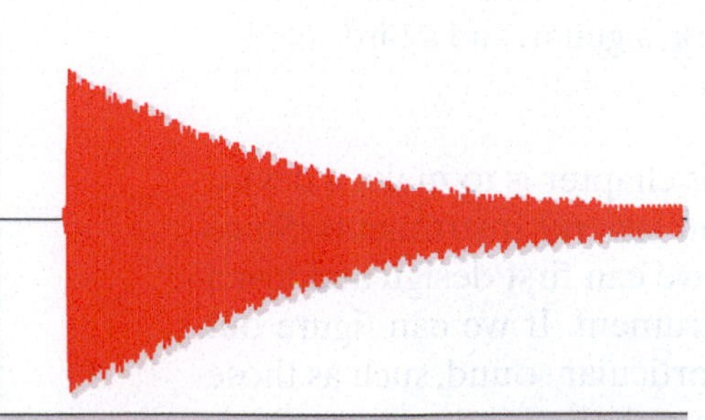

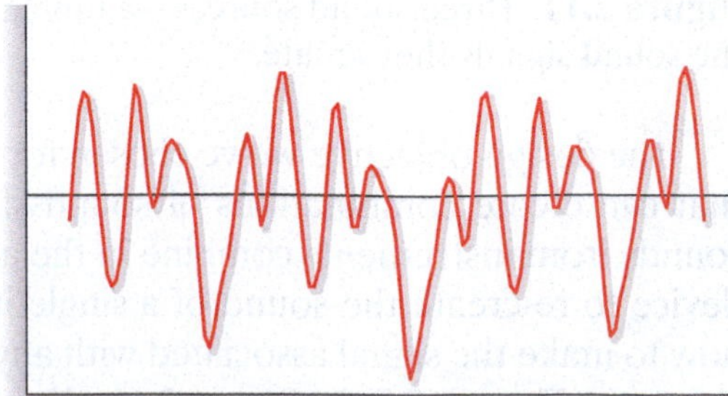

Figure 2.12 The guitar signal in Figure 2.11 shown over two different time scales. The plot on the left shows several seconds of the signal, while the plot on the right shows several milliseconds of the signal.

To proceed with our design, we will first need to figure out how to plot functions from their mathematical descriptions, so that we can compare the plot of a sound's approximation to that of the actual sound signal. If we have a formula for a signal, how do we plot it? Any signal that is a function of time is written as $s(t)$, where t has units of time. The value of $s(t)$ at any given time instant t is called the signal's **amplitude**, which is the height of the signal as measured from the time axis. When we plot such signals, if $s(t)$ is positive the value of $s(t)$ is the height of the curve above the t-axis, or time axis; if $s(t)$ is negative, it will be plotted below the time axis. Figure 2.13 shows three examples of simple functions and their corresponding plots.

Amplitude: The height of a signal, *s*, at time, *t*.

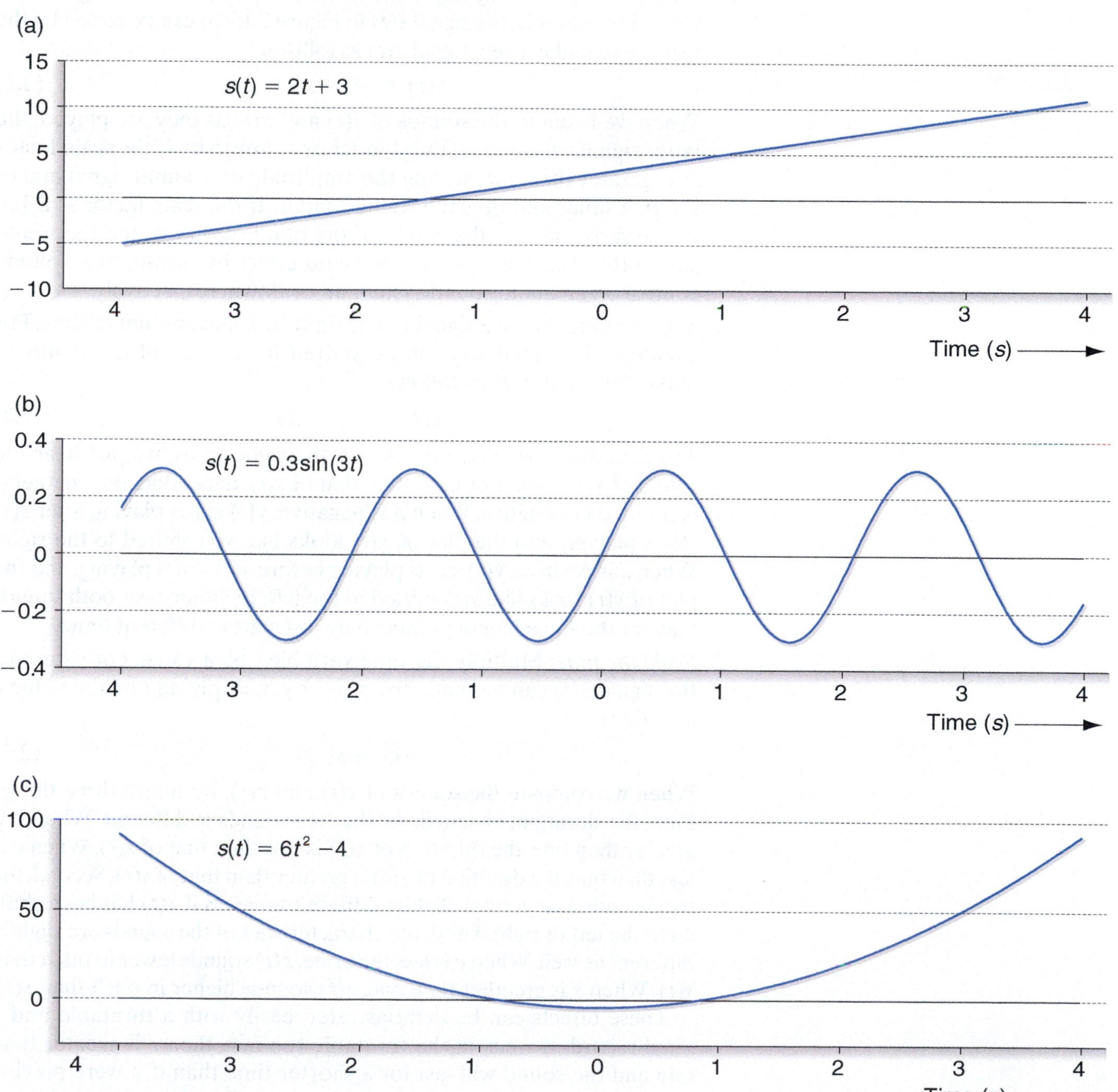

Figure 2.13 Plots of three different equations for $s(t)$ for values of t ranging from −4 to +4 seconds.

If we compare the plot of the function in Figure 2.13(b) with the plot in Figure 2.12(b), we see that they look very similar, which means that when they are converted to sound by a speaker, they will sound similar. This realization gives us the hope that we might be able to approximate a musical instrument's sound with some simple signals that we can compute easily. To produce realistic-sounding music from mathematics, our mathematical function should look as much as possible like the instrument's signal. Therefore, we will need simple ways to change the mathematical function to make it look more like an instrument signal. There are three basic changes we can implement:

1. *Scale the amplitude:* Multiply the height, or amplitude, of a signal by a value. For example, the signal $s(t)$ in Figure 2.13(b) can be scaled by the value A to make a new signal $x(t)$ as follows:

$$x(t) = A \times s(t) \tag{2.1}$$

When we listen to the sounds of $s(t)$ and $x(t)$ as they are played, the latter signal will sound either louder or softer to us. If the scale factor A is greater than one, scaling the amplitude of a sound signal makes the plot taller and thus the sound louder. If the scale factor A is between zero and one, the height of the plot is reduced, and the sound gets softer. We could create the same effect by turning the volume control on an audio sound system up or down, respectively.

2. *Shift the time:* Move a signal left or right by some amount of time. For example, the signal $s(t)$ can be shifted by a value of d seconds to make the signal $y(t)$ as follows:

$$y(t) = s(t + d) \tag{2.2}$$

The sounds of $s(t)$ and $y(t)$ will be the same to us; we just hear the sound of $y(t)$ earlier or later than that of $s(t)$, depending on whether d is positive or negative. When d is negative, $y(t)$ starts playing after $s(t)$ starts playing, and the plot of $y(t)$ looks like $s(t)$ shifted to the right. When d is positive, $y(t)$ starts playing before $s(t)$ starts playing, and the plot of $y(t)$ looks like $s(t)$ shifted to the left. In either case, both sounds play for the same length of time; they just start at different times.

3. *Scale the time:* Multiply the time variable t by a value. For example, the signal $z(t)$ can be made from $s(t)$ by multiplying t by the value c as follows:

$$z(t) = s(ct) \tag{2.3}$$

When we compare the sounds of $s(t)$ and $z(t)$, we notice three things. First, the duration, or length, of the two sounds is different. When c is greater than one, the duration of $z(t)$ is less than that of $s(t)$. When c is less than one, the duration of $z(t)$ is greater than that of $s(t)$. Second, the two sounds may start at slightly different times, as if $z(t)$ has been shifted to the left or right. Third, the characteristics of the sounds are slightly different as well. When c is less than one, $z(t)$ sounds lower in pitch than $s(t)$. When c is greather than one, $z(t)$ sounds higher in pitch than $s(t)$.

These effects can be demonstrated easily with a turntable and a vinyl record. If we spin the turntable too fast, then c is greater than one and the sound will last for a shorter time than if it were playing at a normal speed. In addition, the sound will seem to have a higher pitch. For example, a man's voice may sound more like a child's voice. If we spin the turntable too slowly, the sound takes longer to play and is much lower in pitch. These notions of lower and higher pitch are directly connected to how we perceive musical information. We'll learn more about pitch shortly.

EXAMPLE 2.1 Plotting Signals

Suppose we have the signal $s(t) = 2t + 4$, which is similar to the signal shown in Figure 2.13(a). Three different signals $s_2(t)$, $s_3(t)$, and $s_4(t)$ are created from $s(t)$ as follows:

$$s_2(t) = 3s(t)$$
$$s_3(t) = s(t - 1)$$
$$s_4(t) = s(2t)$$

Plot each of these new signals on its own set of axes, and then plot $s(t)$ on each set of axes. Verify that each signal is a line, and identify the slope and the vertical and horizontal intercepts of each signal.

Solution

When plotting equations or signals, it helps to get an idea of what to expect in the plot. For example, we could say that $s(t) = 2t + 4$ is a line that has a slope of 2 and that, when $t = 0$, it crosses the s-axis at the value $s = 4$. When $t = -2$, $s(t) = 0$.

(a) The signal $s_2(t) = 3s(t)$ changes the amplitude of $s(t)$ by a factor of three, or

$$s_2(t) = 3s(t) = 3(2t + 4) = 6t + 12$$

This signal is a line that has a slope of 6 and crosses the s-axis at the value $s = 12$. Because the amplitude of $s(t)$ is scaled by a factor of three, both the slope and the vertical intercept are increased by a factor of three. However, the horizontal intercept (where $s = 0$) does not change. Plots of both $s_1(t)$ and $s(t)$ are shown in Figure 2.14(a).

(b) The signal $s_3(t) = s(t - 1)$ is a delayed version of $s(t)$. The line is shifted to the right by 1 second, because

$$s_3(t) = s(t - 1) = 2(t - 1) + 4 = 2t + 2$$

The slope is still 2, because we only shifted the signal; we did not change its amplitude. However, the shift changes both intercepts. The vertical intercept is changed from 4 to 2, because $s_3(t)$ is shifted to the right of $s(t)$. And now $s_3(t) = 0$ at $t = -1$, which is one time unit later than when $s(t) = 0$. This plot is shown in Figure 2.14(b).

(c) The signal $s_4(t) = s(2t)$ is a time-scaled version of $s(t)$. This operation compresses time by a factor of two, such that

$$s_4(t) = s(2t) = 2(2t) + 4 = 4t + 4$$

Because time has been compressed, the slope of the original signal increases by a factor of two and the horizontal intercept is reduced by a factor of two. Now $s_4(t) = 0$ when $t = -1$ instead of -2. The vertical intercept is not changed. This plot is shown in Figure 2.14(c).

This example illustrates the three basic operations we can use to change simple mathematical functions to better match real instrument signals.

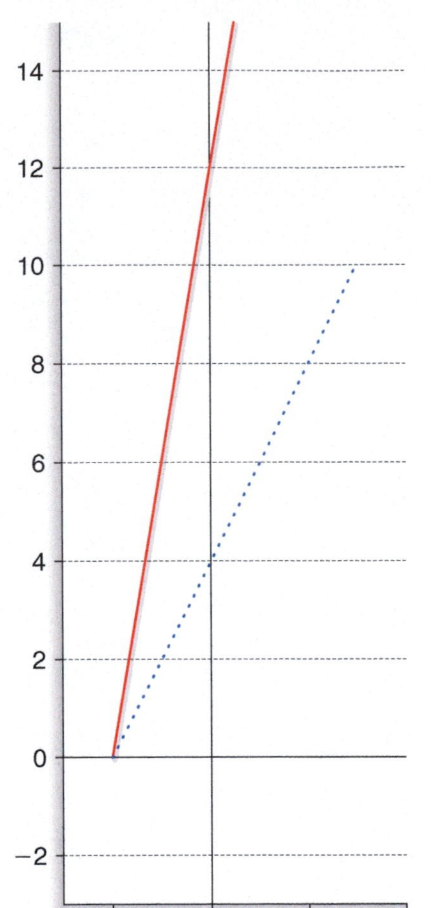

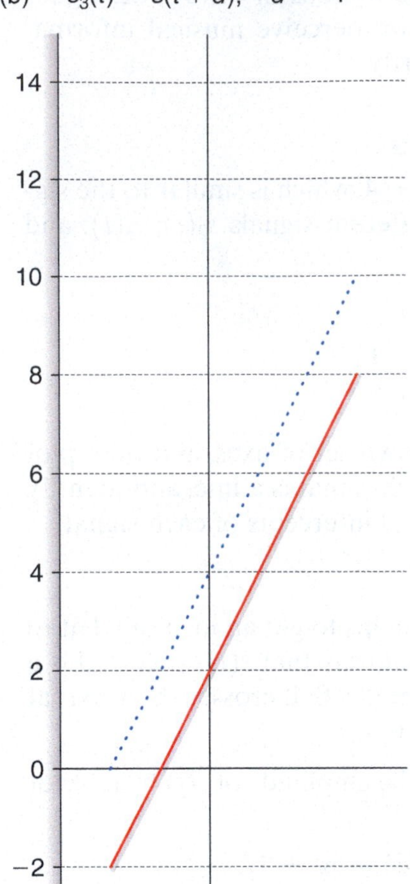

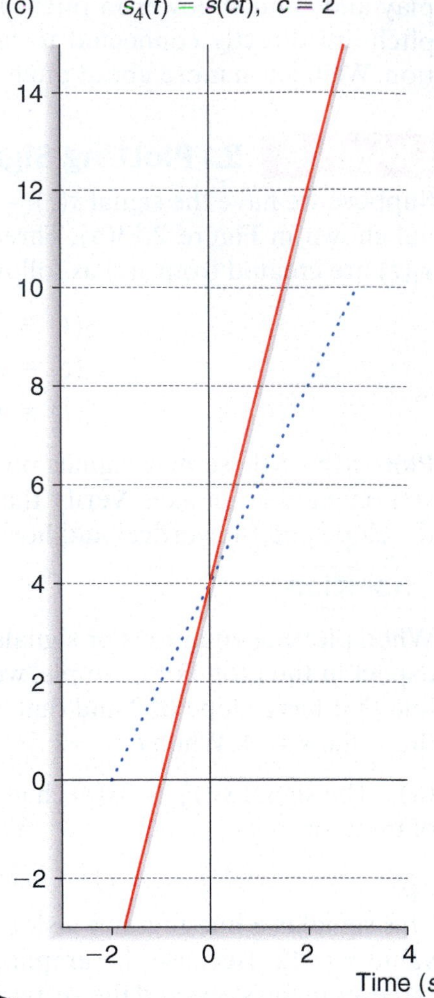

Figure 2.14 Plots of a signal $s(t)$ for t between −2 and 2 after (a) scaling its amplitude, (b) shifting in time, and (c) scaling the time variable. The original $s(t)$ is plotted as a dotted blue line for reference.

Pitch, Frequency, and Periodic Signals

To make music, we need to construct signals that have the special characteristics of signals produced by instruments. What are these special characteristics? We all know that music sounds different from other sounds, such as the roar of an engine. What makes music different? The answer lies in how instruments make sound.

Notes: The signs with which music is written.

Pitch: The perceived frequency of the sound of a note when we hear it.

When a musician plays an instrument—by striking an individual key of a piano, for example—each sound that is played is called a **note**. Some notes sound high, whereas others sound low. This perceptual difference is called **pitch**. Pitch is tied closely to a particular characteristic of a musical instrument's sound signal: the rate at which the sound *oscillates*, or how fast the amplitude of the signal moves up and down over time. We will need to learn how our perception of pitch is related to the signal of our device so that our system will to be able to create different notes.

Figure 2.15 shows the signals from two different notes that have been played on a piano keyboard. On the left, both signals are displayed over

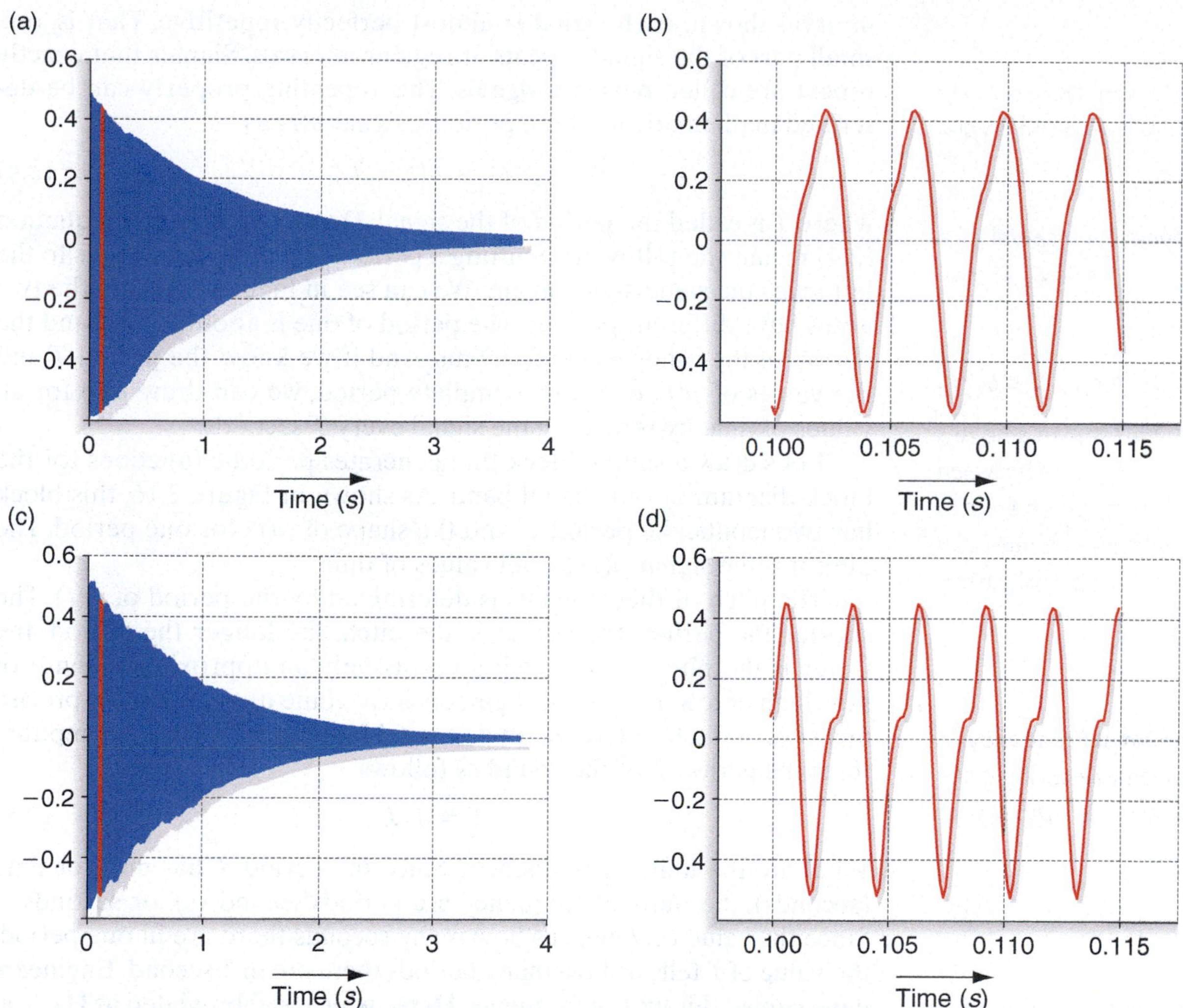

Figure 2.15 Plots of two piano notes are shown on the left for a time interval of 4 seconds each. The 15-ms interval highlighted in red in (a) is shown in (b). The 15-ms interval highlighted in red in (c) is shown in (d). The note shown in (c) has a higher pitch than the one shown in (a), so the highlighted part plotted in (d) looks compressed horizontally compared with (b).

4 seconds so that we can see the complete sound from when it starts to when it ends. On the right, we've zoomed in on a short portion of each sound over a time interval of 0.015 seconds (s), or 15 milliseconds (ms). This small 15-ms interval is highlighted in red on the signal plots on the left. The plots on the right show the characteristics of the sound created by the piano and of the specific note being played. Both signals on the right have a similar shape because they were both made by a piano as opposed to some other instrument. One of the signals, however, is compressed in time relative to the other, so it oscillates up and down at a different rate. This difference in time scale is what causes us to hear one note at a higher pitch than the other. Our observation gives us a key idea for making musical sound mathematically: Once we know the characteristic signal of an instrument for one note over a short interval, we can approximate other notes from the same instrument by scaling the time variable, thus expanding or compressing the signal in time.

Like the guitar sound in Figure 2.12, each signal on the right side of Figure 2.15 has an oscillating, or vibrating, quality to it. Over the time

Periodic Signal: A signal that exactly repeats at regular intervals.

Period: The repeating interval of a periodic signal.

INTERESTING FACT:

High-fidelity music systems are designed to produce sounds between 20 Hz and 20 kHz because that is the range of pitches humans can normally hear.

Fundamental Frequency: A mathematical quantity describing the repetition rate of a periodic signal.

Hertz: The units of frequency, represented in periods/second, or seconds^{-1}.

interval shown, each signal is almost perfectly repetitive. That is, one small part of the signal repeats at regular intervals. Signals that exactly repeat are called **periodic signals**. This repeating property can be described mathematically for a periodic signal $p(t)$ as

$$p(t) = p(t + T) \tag{2.4}$$

where T is called the **period** of the signal. Described in words, Equation (2.4) means the following: Shifting a periodic signal by T seconds to the left gives the same signal again. We can see in Figure 2.15 that the two notes have different periods. The period of one is about 3.8 ms, and the period of the other is about 2.9 ms. And if we know the period T and the values of $p(t)$ over one complete period, we can draw $p(t)$ for all values of time by repeating the signal every T seconds.

Let's draw a simple block that generates periodic functions for the block diagram of our digital band. As shown in Figure 2.16, this block has two inputs: the period T, and the shape of $p(t)$ for one period. The output is the signal $p(t)$ for all values of time.

The pitch of these signals is determined by the period of $p(t)$: The shorter the period, the higher is the pitch; the longer the period, the lower is the pitch. Since pitch gives us only an approximate sense of how high or low a sound is, engineers have come up with a more precise measure of pitch called the **fundamental frequency**. It can be computed from the period T of the sound as follows:

$$f = 1/T \tag{2.5}$$

What are the units of frequency? Since the period T has units of time (seconds), the units of frequency are periods/second (s), or seconds^{-1}. Since the value of T tells us how many seconds there are in one period, the value of f tells us how many periods there are in 1 second. Engineers have named this unit of frequency **Hertz**, which is abbreviated as Hz. You may have seen the units of Hz before when reading about high-quality audio equipment that can produce frequencies as low as 20 Hz and as high as 20 kHz. A 20-Hz sound has a period of $\frac{1}{20}$ s = 0.05 s, or 50 ms, whereas a sound with a frequency of 20 kHz = 20,000 Hz has a period of $\frac{1}{20{,}000}$ s = 0.00005 s, or 0.05 ms. This is the range of frequencies that a young man or woman can normally hear.

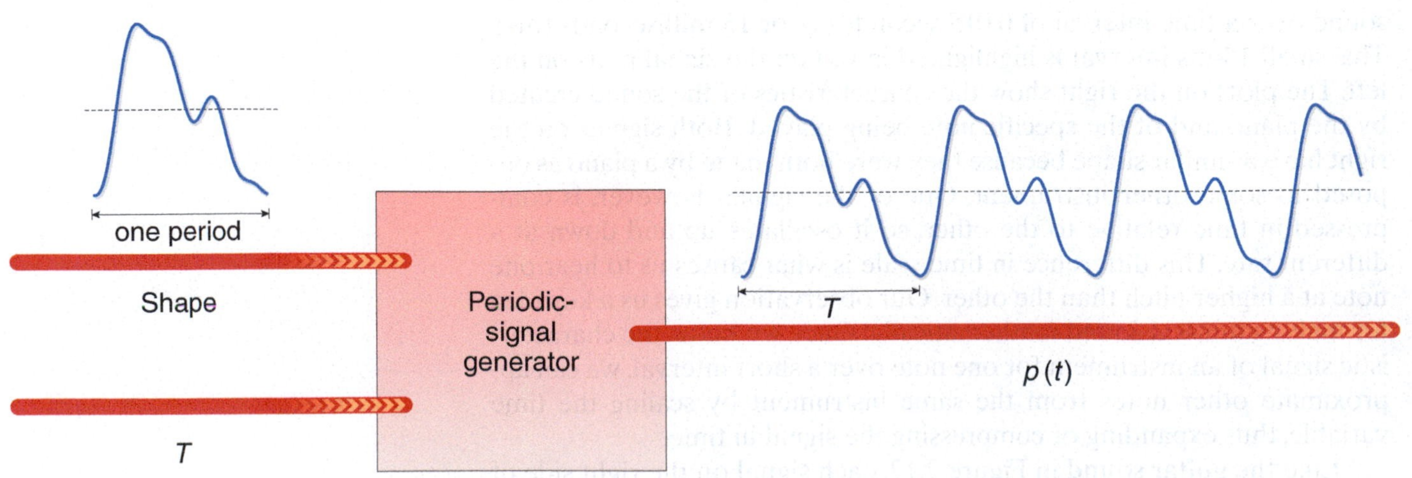

Figure 2.16 Block diagram of the periodic-signal generator.

When an instrument is making a sound, it is producing a periodic signal that has a fundamental frequency. Our perception of pitch comes from this fundamental frequency of a sound. When we hear periodic signals, we hear differences in the pitches of notes because the different notes have different fundamental frequencies.

The center of Figure 2.17 shows the names of notes assigned to eight of the keys in the middle of a piano keyboard, as well as the fundamental frequencies of the keys. The leftmost highlighted key in this cluster of eight shaded keys is called middle C. This key is often used as a reference key for piano music. If we compare the F-above-middle-C note, with a frequency of 349.23 Hz, with the middle-C note, which has a frequency of 261.63 Hz, we find that the F note sounds higher, because it has a higher frequency. We also say that its pitch is higher.

There are only seven unique note names, from the letter A to the letter G, in Western music. For notes that are above and below those on the piano highlighted in Figure 2.17, the letters A through G are used over and over again to specify the frequencies we hear. Every time a note name is reused above this range, the note's frequency is twice that of the identically named note below. Likewise, every time a note name is reused below this range, the note's frequency is half that of the identically named note above. In the figure, the frequencies of all the C notes are shown. As we move to the right, the frequency of each C note is double the frequency of the previous C note.

INTERESTING FACT:

Heinrich Hertz (1857–1894) was a German physicist. An experimentalist, he demonstrated that electromagnetic waves exist and showed how they could be made. His discoveries paved the way for all modern wireless communications technologies, including radio, television, cell phones, and radio astronomy.

EXAMPLE 2.2 Find the Frequency of a Periodic Signal from its Plot

From the signal plots in Figure 2.15 and the definition of the frequencies for piano keys in Figure 2.17, determine which notes were played when these signals were recorded.

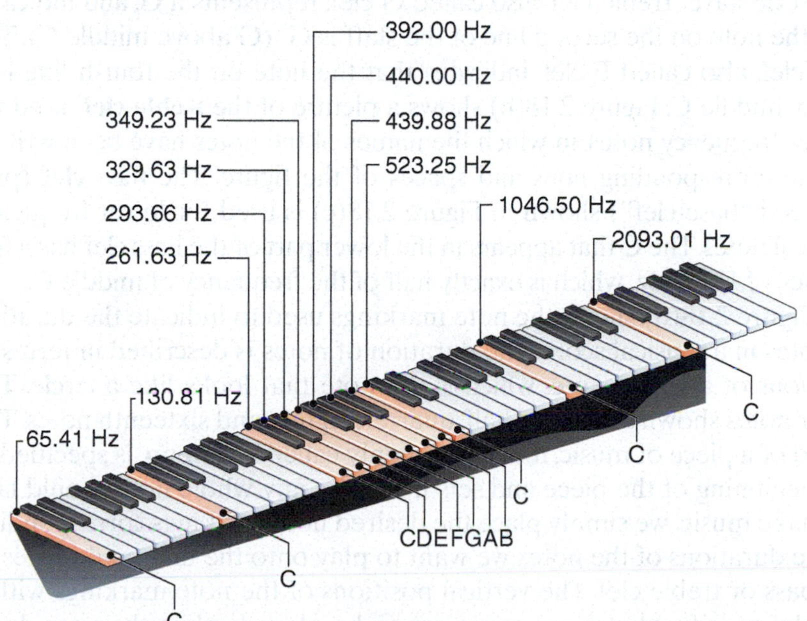

Figure 2.17 Frequencies for notes on the piano keyboard. The frequency labels shown in the upper portion of the figure correspond to the white keys labeled with the names of the respective notes in the lower portion. The frequencies of all notes named C increase by a factor of two moving from left to right.

T = Period from Plots	$f = 1/T$	Closest Piano Key
2.9 ms	3.4×10^2	F at 349 Hz
3.8 ms	2.6×10^2	C at 262 Hz

Solution

From the plots, we already estimated that the periods of the two signals were 2.9 ms and 3.8 ms. Using Equation (2.5), we can compute the corresponding fundamental frequencies from the period values. The frequencies are shown in the center column of the table to the left. Our measurements from the plot are not accurate enough to compute the frequencies to the numerical accuracy shown in Figure 2.17, but since the notes came from a piano, we can choose the piano key that has the frequency closest to our result. From this calculation, we see that the first note corresponds to middle C and the second note corresponds to F above middle C.

Melodies and Notes

Melody: A sequence of notes that make up a piece of music.

Score: Notation showing all parts or instruments.

Clef: The sign written at the beginning of a staff to indicate pitch.

Treble Clef: Also called G clef. Indicates that the note on the second line of the staff is G' (G above middle C).

Bass Clef: Also called F clef. Indicates that the note on the fourth line is F below middle C.

Tempo: The speed of a piece of music.

How interesting a song is to us depends on many factors, but perhaps the most important factor is the song's melody. A **melody** is a sequence of notes that make up a piece of music. When you whistle a song, you are whistling a melody. Notes are the "letters" that make up the "language" of the melodies that we hear in music. To describe melodies so that others can reproduce them with instruments, we need a formal method to list the sequence of notes in the music. The description we use must specify the frequency of each note as well as its time duration. The traditional method by which music is specified uses the notation and graphical presentation of music scores.

A complex composition such as *Don Giovanni,* by Wolfgang Amadeus Mozart, can be performed from its sheet music, or score. A small part of this score is shown in Figure 2.18(a). The **score**, which is a kind of recipe for creating music, represents in a graphical manner what notes should be played and when they should be played. The system of lines and spaces is known as a staff or stave. Treble clef, also called G clef, represents a G, and indicates that the note on the second line of the staff is G' (G above middle C). The bass clef, also called F clef, indicates that the note on the fourth line is F below middle C. Figure 2.18(b) shows a picture of the **treble clef**, used for higher frequency notes, in which the names of the notes have been written on the corresponding lines and spaces of the figure. The **bass clef** (pronounced "base clef") shown in Figure 2.18(c) is used for lower frequency musical notes. The C that appears in the lower part of the bass clef has a frequency of 130.8 Hz, which is exactly half of the frequency of middle C.

Figure 2.18(d) shows the note markings used to indicate the duration of notes in a musical score. The duration of notes is described in terms of fractions of a whole note, which is the note that looks like a circle. The other notes shown are called half, quarter, eighth, and sixteenth notes. The speed of a piece of music, more commonly called its **tempo**, is specified at the beginning of the piece and sets how long any whole note should last. To make music, we simply place the desired note markings corresponding to the durations of the notes we want to play onto the lines and spaces of the bass or treble clef. The vertical positions of the note markings within the clef specify what frequencies are to be played, while the note durations are specified by the shape of the note. The order of the notes proceeds from left to right on the page just like this text.

If we want to re-create a piece of music from a score, we first identify each note and its frequency and then create a corresponding periodic sig-

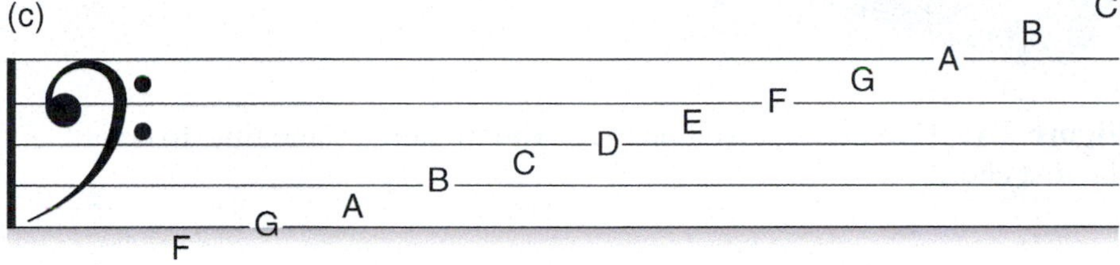

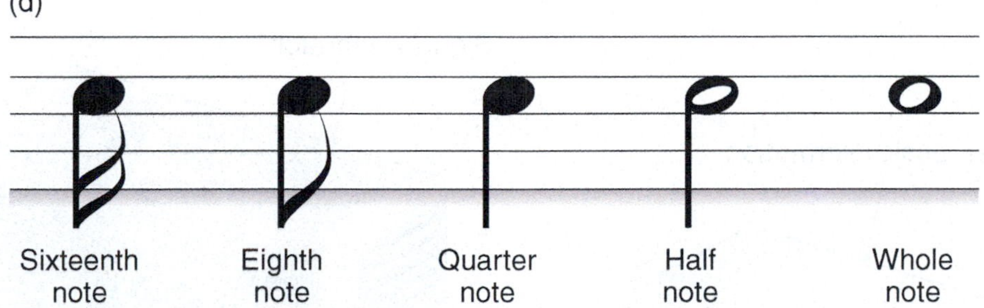

Figure 2.18 Music notation.

INTERESTING FACT:

The sheet music we use today employs a form of musical notation that dates from 1600 A.D. in Europe. It was used mainly in churches and cathedrals where music was performed for ceremonies. Music notation was invented as a way to communicate ideas, much like a language.

nal with the fundamental frequency of each note to be played. The shape of each note on the score tells us how long to play its corresponding signal, and the sequence of notes tells us the order in which they should be played. When two or more notes are aligned vertically in a musical score, they are played at the same time. This structure is called a **chord**. An example of a chord is shown in the highlighted part of the score shown in Figure 2.19 (repeated from Figure 2.18(a)). In the figure, each note in the chord is highlighted on the keyboard shown below the score.

Chord: A collection of simultaneously played tones.

Refining Our Design of the Digital Band

After our study of musical sound and signals, we are in a better position to specify the blocks within our digital-band design in Figure 2.1. For the moment, we shall consider a design that plays only one note at a time. While this design may not produce the high quality of music that we ultimately want, it will allow us to test our approach in a controlled

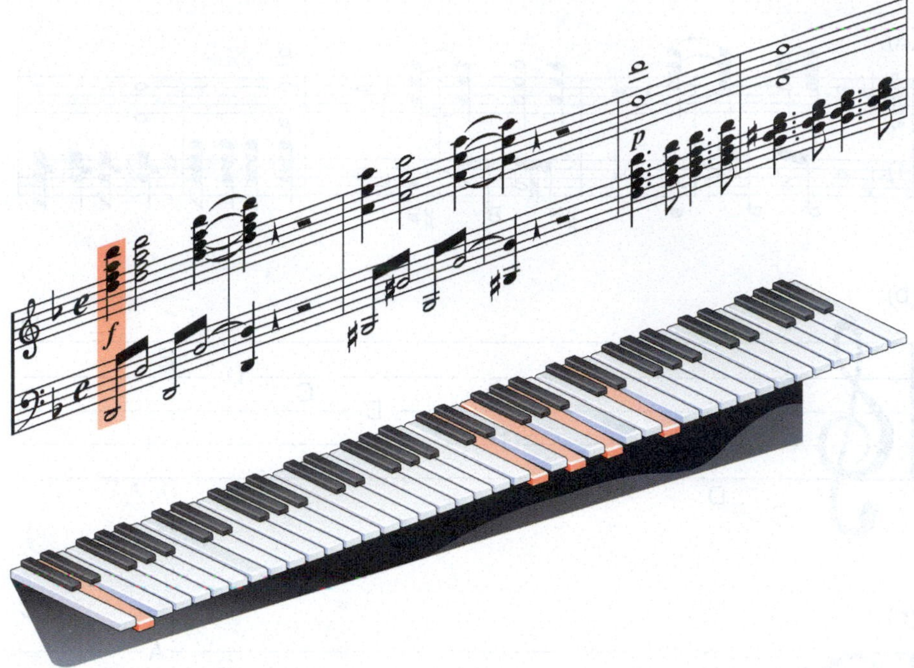

Figure 2.19 Five notes played on the keyboard at the same time to create the first chord.

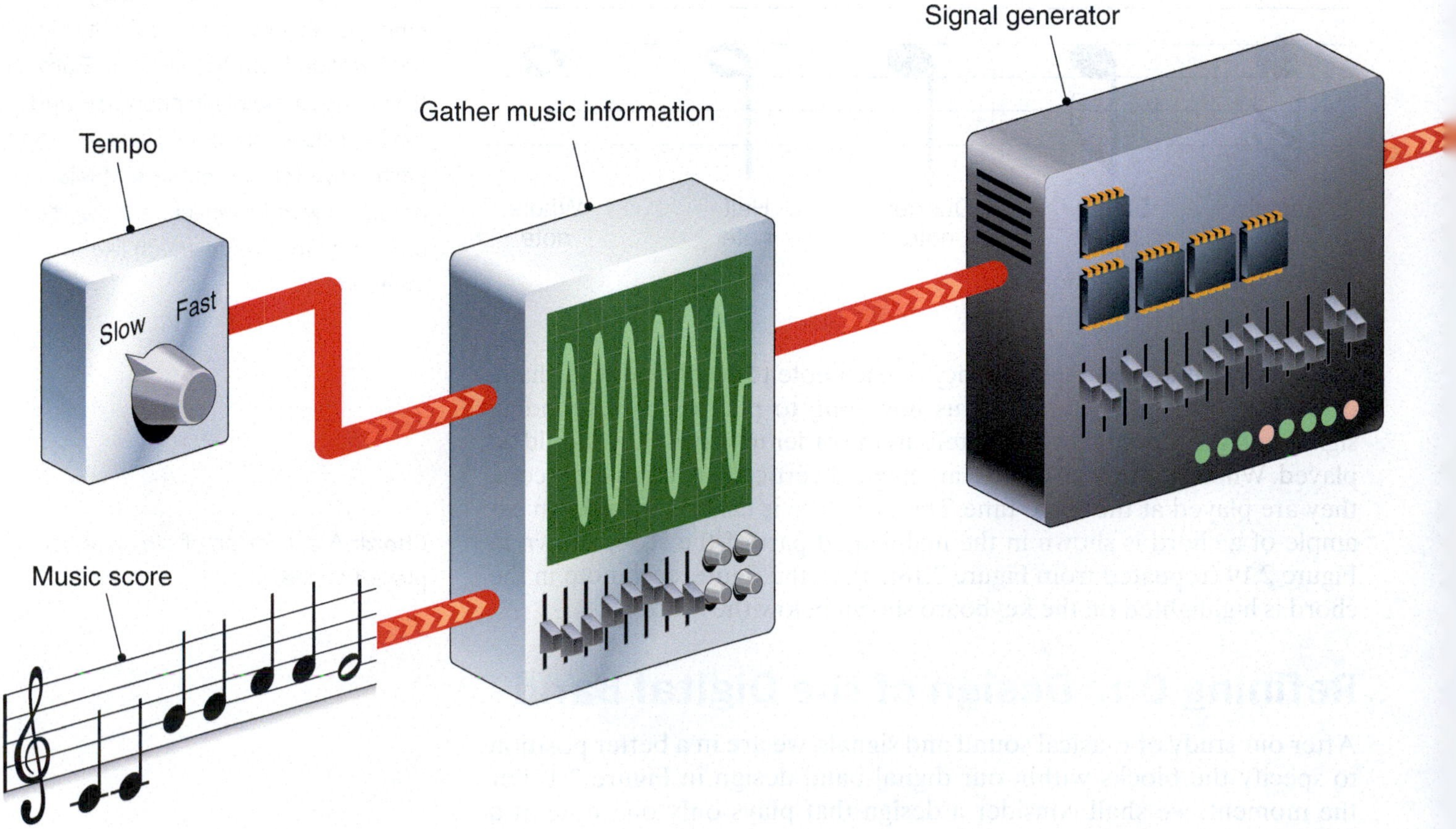

manner. Our improved design is shown in Figure 2.20. The input to our system is the musical score that specifies the notes of the song we want to hear. The frequencies of these notes are used by the sound-generation block to create signals for the specified notes. The sound-generation block also uses the shape of a periodic signal to define the signal's sound. We will use $T = 1/f$ to compute the period from the frequency f. The output of this signal generator is sent to a sound-creation device, which usually contains a loudspeaker and amplifier to boost the amplitude of the signal so that we can hear it at our desired volume.

As in all engineering designs, specifying the system can be easy when it is broken down into its parts or individual blocks. But how do all of these parts work? And how can we increase the capability of our system to enable it to play the sounds of different instruments or more complex sounds and music? In the next section, we'll address these questions, and, in the process, we will improve our design significantly.

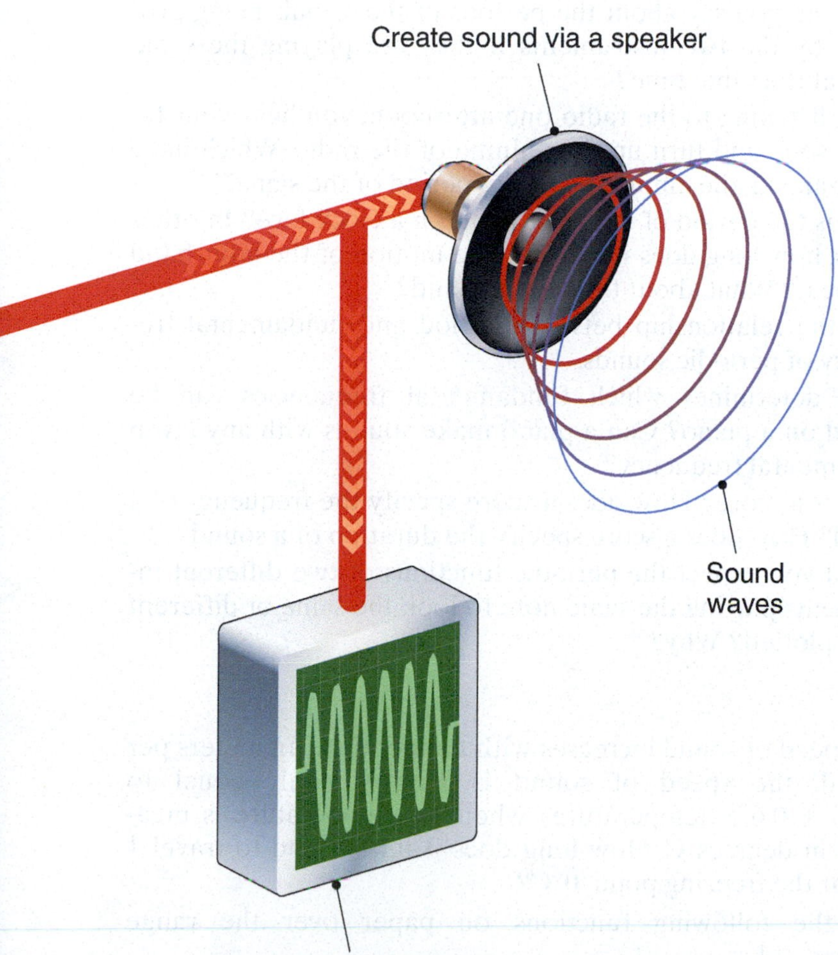

Figure 2.20 Block diagram for a system that can create the sound of a single instrument. The signal-generator block in the center is the periodic-signal generator.

EXERCISES 2.2

Mastering the Concepts

1. One of your friends says that loud sounds move air molecules like a race car zooming down the highway. Another friend says that loud sounds move air molecules like bumper cars bouncing back and forth in an amusement park ride. Who is right?
2. Can really complicated sounds, such as the sound of a 100-instrument orchestra, be represented by a single signal?
3. Name three ways that you can modify a sound signal.
4. Delaying a signal amounts to shifting the signal to the right on a plot. If $s(t)$ is the signal to be delayed and T is a positive number indicating the amount of time delay, is the delayed signal represented by $s(t + T)$ or $s(t - T)$? Draw an example so that you may be sure of your answer.
5. If the amplitude scaling factor A of a signal is negative, how will $s(t)$ be changed by scaling?
6. How are pitch and frequency related?
7. You hear two musical instruments playing the same tune. What can you say about the periods of the signals being produced by the two instruments if they are playing the same notes at the same time?
8. While listening to the radio one afternoon, you hear your favorite song and turn up the volume of the radio. Which have you changed, the amplitude or the period of the signal?
9. What is the period of the hour hand on a clock face? In other words, how long does it take for the motion of the hour hand to repeat? What about the minute hand?
10. Give the relationship between period and fundamental frequency of periodic sounds.
11. What determines which fundamental frequencies can be played on a piano? Can a piano make sounds with any given fundamental frequency?
12. What is a score? How does a score specify the frequency of a sound? How does a score specify the duration of a sound?
13. Would you expect the periodic functions of two different instruments playing the same note to look the same or different when plotted? Why?

Try This

14. The speed of sound increases with temperature. In meters per second, the speed of sound is approximately equal to $(331.4 + 0.6 \times \text{temperature})$ when the temperature is measured in degrees C. How long does it take sound to travel 1 mile at the freezing point, 0°C?
15. Plot the following functions on paper over the range $0 < t < 0.1$ s:

    a. $s_1(t) = 0.7 \sin(2\pi 45t)$
    b. $s_2(t) = 11\,t - 0.07$
    c. $s_3(t) = t - 0.25\,t^3$

16. Using the functions from Exercise 2.2.15, plot the following functions on paper over the range $0 < t < 0.1$ s:
    a. $s_4(t) = s_1(t - 2)$
    b. $s_5(t) = s_2(t + 3)$
    c. $s_6(t) = s_3(t + 3)$
17. Plot $s_1(t)$ for the following signals:

$$s_1(t) = -t - 2$$

$$s_1(t) = |t| = \begin{cases} t, & \text{if } t \geq 0 \\ -t, & \text{if } t < 0 \end{cases}$$

$$s_1(t) = 3 - 4t + t^2$$

    In each case, plot the corresponding signals $s_2(t)$, $s_3(t)$, and $s_4(t)$ as defined in Example 2.1.
18. Plot the following signals either by hand or on your calculator:
    a. $s_1(t) = 0.5\cos(0.3t + 0.2)$
    b. $s_2(t) = 2\,s_1(t)$
    c. $s_3(t) = s_1(t - 2)$
19. What is the fundamental frequency of the periodic signal in Figure 2.13(b)?

In the Laboratory

20. Have a friend bring a musical instrument to the lab. Use laboratory equipment from the "Plots of Speech" Infinity Project Experiment on page 44 to display the instrument's sound. Measure the signal produced by the instrument for several notes across the frequency range of the instrument and player. How do these signals differ? How are they similar?
21. When you whistle, you make a nearly periodic sound. Take a bottle, blow across the top of it, and measure its sound signal. How periodic is the signal? Can you measure its fundamental frequency?

2.3 Making Music from Sines and Cosines

The knowledge that we have gained about sound and music has enabled us to add many important and specific details to the design of our digital band in Figure 2.1. We are much closer to our design objective, but we need to provide more details than in our current design in Figure 2.20 if we want our system to make music. To do so, we will study each block, figure out exactly what it will do, and determine its inputs and outputs. Some important questions to help us in this task are as follows:

- How do we get our information into the device?
- How will this information be converted into a sequence of notes?
- What signals will we use to convert the notes into sound?

Using MIDI to Specify Information

Let's start by specifying the first block of our digital band in Figure 2.20. We've learned that music is described by a sequence of notes that define the melody we want to hear. Since the sound of each note can be made by a periodic function, this first block has to specify the fundamental frequency of each note. It also has to specify the order in which the notes are played and how long each note will last.

There are many ways to specify information of this sort. Sheet music contains a graphical presentation of this information that has been used by live performers for centuries. The following are some advantages of sheet music:

- The graphic layout of sheet music can be learned easily by most people, to the point that musicians can read it as fast as they can play.
- The use of pictures for the notes does not depend on any particular spoken language. Thus, anyone in any part of the world can learn to read and make music for others to hear and play. Sheet music represents a **standard** that allows musicians from anywhere in the world to play together.
- The paper on which sheet music is written is durable, portable, easy to make, and easy to copy.

Standard: A description for a method or process for using or building something that a group of people has agreed to use. By establishing a standard, people can use, enjoy, and even build on other people's work to make it better. Standards are important in music, because they allow us to make, share, and enjoy more music together.

Even with these advantages, sheet music is probably not the best solution for the first block of our digital band. If we were to use sheet music as our musical input, we would have to design a system to read the notes from the page. We could build such a system, but the cost and size of the scanning device would probably be too great for us to meet the other design goals for our digital band. Is there a better way if we start from scratch?

Let's think again about a piece of music and how it is specified. We know that each note corresponds to a frequency and lasts a specific length of time. Why not just list the frequencies and lengths of time in the order that they are played? We could create a document or computer file that has this information. Then, an electronic device could read the file and play the piece from the instructions contained in the file. The advantages of such a solution are many and include the following:

- Since the piece of music is in electronic form, it can be manipulated easily by a computer, saved on a CD or hard disk, and transmitted by electronic means such as the Internet.
- As with sheet music, we could convince others to use our description, allowing us to share the piece. In this way, we will have created a standard that allows us to communicate and share music more easily.
- Designers of electronic instruments can use our description to communicate information between other electronic devices and instruments, such as from a computer to an electronic piano, saxophone, or trumpet, and back again.

The biggest drawback to our method is that we would have to convert existing pieces of music to our new format—but given the ease with which we can communicate electronically, we'd have to convert each piece only once.

Our method of specifying a piece of music would allow us to create simple melodies, like "Mary Had a Little Lamb." There are a couple of problems with our methods, though:

1. It wouldn't enable us to make more complicated pieces of music that have periods of silence, called *rests*.
2. Most pieces of music have more than one note playing at a time, and there's no way for us to specify when more than one note should be played.

We need to change our method to add this information.

Fortunately, designers of electronic instruments have come up with a similar system to represent music electronically that overcomes the two problems we've identified. This system is called Musical Instrument Digital Interface, or **MIDI** for short. Music in this form is stored in MIDI files on a computer, often denoted by the extension ".mid". These files can contain many types of information, but the most critical information for our digital band is the note information. Music in a MIDI file is stored as a list of instructions to turn notes on and off. That way, two notes can be turned on at the same time, and we can turn all the notes off at any time as well. MIDI is also a great choice for our digital band, because it is already a widely accepted standard among companies that make and sell systems for creating and manipulating music.

MIDI (Musical Instrument Digital Interface): A specification for how information is communicated between electronic instruments. It was developed in the 1980s by engineers from Sequential Circuits, Roland Corporation, and Oberheim Electronics—three electronic-instrument companies. The specification was first published in August 1983 and has since been modified to include certain types of non-musical-performance information.

Figure 2.21 shows a simple score on the left and a corresponding pseudo-MIDI-file list of instructions on the right. On the left of the MIDI list is a column labeled "timestamp" that describes when an instruction occurs. For each timestamp value, we see a note value in the form of a letter and number, such as C4 and G4, and an "event" on the same line. The letter is the note name, and the number specifies the exact note with that name. In Figure 2.17, the note designated as C4 (middle C) has a frequency of 261.63 Hz. In our simple example, the events are either "On" or "Off," corresponding to whether we turn the note on or off, respectively. Other information that might be contained in a MIDI file includes the channel of the instrument being played, the loudness of the note, and other performance information. A more complicated example might show such information as well as multiple notes being played simultaneously.

For fun, play the notes in Figure 2.21 on a piano or some other instrument and see if you can identify this familiar children's song.

Making Signals

Now that we have a way of getting music into our digital band using a widely accepted standard, we shall turn our attention to the next block of the block diagram in Figure 2.20. This block takes the information contained in the MIDI file and turns it into signals. We will first explore simple and well-understood ways for making these signals.

The periodic signals shown in Figures 2.12 and 2.15 each have distinctive plots that look like they might be hard to describe mathematically. Moreover, we don't have a simple way to calculate them. Have you ever seen a saxophone or tuba key on a calculator?

Sinusoid: A simple oscillating waveform created from the sine and cosine function.

A mathematically simple periodic function is one that you might already have seen plotted in a mathematics textbook: the **sinusoid**.

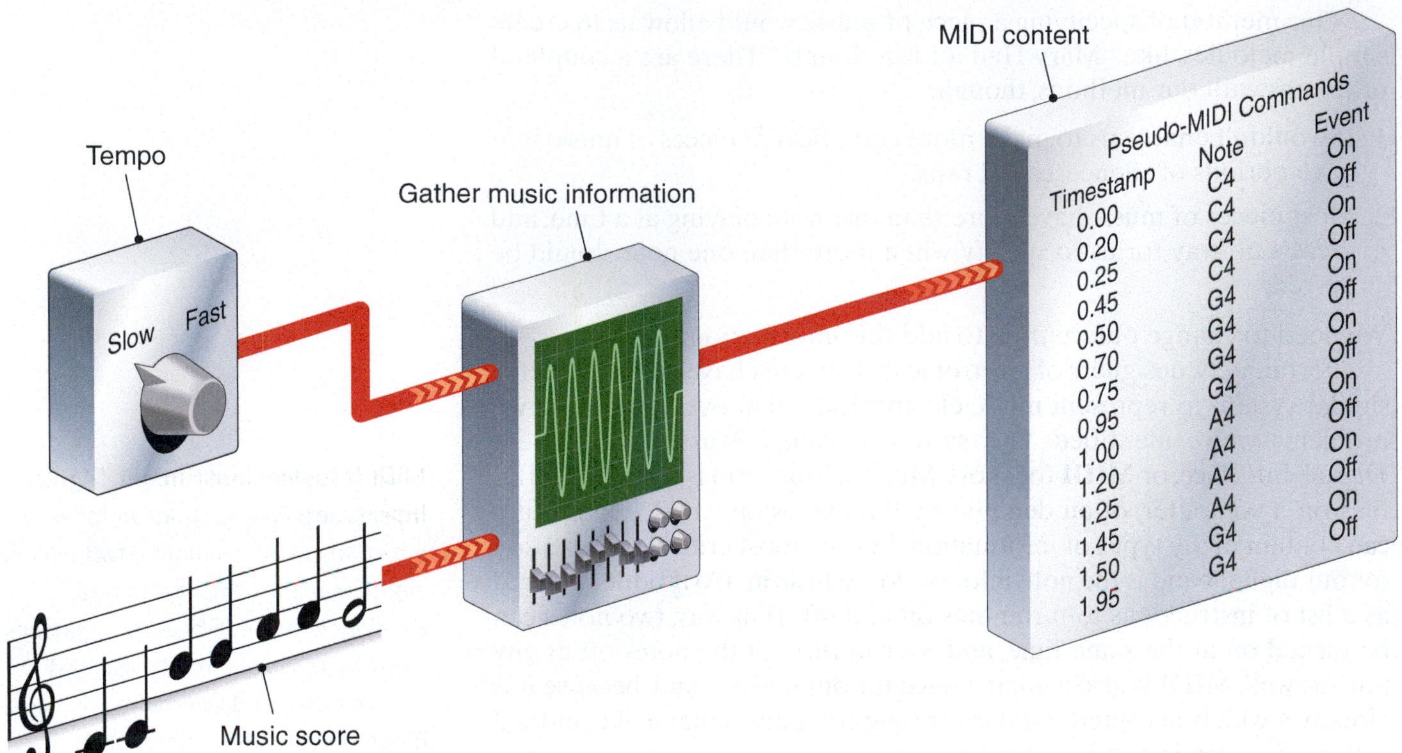

Figure 2.21 An example of how Musical Instrument Digital Interface (MIDI) can be used to specify information for a piece of music.

Figure 2.22 shows a sinusoidal signal being played by a MIDI file. Sinusoids are very common signals, and all of them have the same simple shape. The tuning fork in Figure 2.11 produces a change in acoustic pressure that is a nearly perfect sinusoid. A guitar string produces a signal that is approximately a sinusoid and becomes more and more like a sinusoidal signal as the sound decays away.

The Sine and Cosine Functions Sinusoids are a type of mathematical function that includes both the cosine and sine functions, using standard definitions from right triangles. Figure 2.23(a) shows a circle with radius one. Any point on the circle can be said to have coordinates (x, y), which are at a distance of one from the origin of the coordinate system, because the radius of the circle is one. The specific location on the circle can be described by the angle measured from the horizontal axis. The symbol for this angle is the Greek letter θ. When the angle θ is known, the Cartesian (x, y) coordinates can be computed from the sine and cosine of the angle. The x value is $\cos(\theta)/1$, or $\cos(\theta)$, and the y value is $\sin(\theta)/1$, or $\sin(\theta)$, where the first three letters of "sine" and "cosine" are used as abbreviations.

We can get a good idea of what $\cos(\theta)$ and $\sin(\theta)$ look like as θ increases by studying the diagram in Figure 2.23. When $\theta = 0$, we are at the rightmost part of the circle, where $x = \cos(0) = 1$ and $y = \sin(0) = 0$. Similarly, at the top of the circle, $\theta = 90°$, so $x = \cos(90°) = 0$ and $y = \sin(90°) = 1$. Continuing around the circle to the leftmost point, we

INTERESTING FACT:

A tuning fork makes a sound that is almost a perfect sinusoidal signal. The signals from the sound of a flute and a person whistling are also very close to sinusoidal.

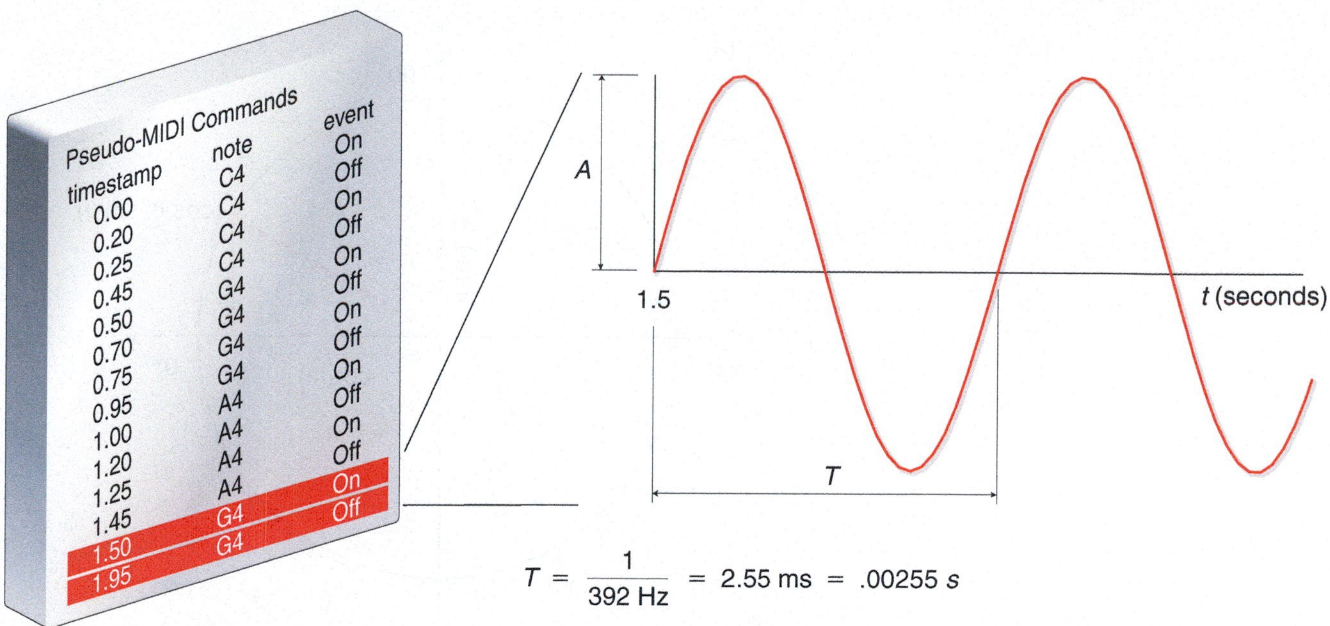

Figure 2.22 MIDI file producing a sinusoidal signal for the note G4 with a frequency of 392 Hz. The amplitude A and period T of the signal are shown.

have $x = \cos(180°) = -1$ and $y = \sin(180°) = 0$. At the bottom of the circle, $x = \cos(270°) = 0$ and $y = \sin(270°) = -1$. When we complete our trip around the circle to 360°, we find ourselves back where we started, and the same sequence of values for sine and cosine will be produced as the angle continues to increase. The values for some intermediate angles can be measured from the coordinate values on the circle, since $x = \cos(\theta)$ and $y = \sin(\theta)$. These values are summarized in Table 2.1.

We can compute the Cartesian coordinates of any point on this circle by using the cos() and sin() functions on a scientific calculator.

EXAMPLE 2.2 Computing Cartesian Coordinates

Using your calculator, determine the x and y coordinates of the point A on the plot in Figure 2.23(b).

Solution

After making sure that your calculator is set to angular measurements in degrees, the values of x and y can be computed to three decimal places as

$$x = \cos(72) = 0.309 \quad \text{and} \quad y = \sin(72) = 0.951$$

These numbers seem to agree with the points along the x and y axes of the plot. If we measure $\cos(\theta)$ and $\sin(\theta)$ for all values of θ between 0 and 360°, we can plot $\cos(\theta)$ and $\sin(\theta)$ as a function of θ, as shown in

INTERESTING FACT:

MIDI can also be used to transmit information from one location to another, such as through a cable from a piano keyboard to a separate sound module or even through a connection over the Internet. With MIDI, several musicians can play music together without being in the same room, participating in a "virtual jam session."

INTERESTING FACT:

Mathematicians often use Greek letters for symbols so they will not be confused with text. The most common symbol used for an angle is the Greek letter theta, θ.

Table 2.1 Values of $x = \cos(\theta)$ and $y = \sin(\theta)$ for different angles in Figure 2.23(a)

θ in degrees	0°	45°	90°	135°	180°	225°	270°	315°	360°
θ in radians	0	$\pi/4$	$2\pi/4$	$3\pi/4$	π	$5\pi/4$	$6\pi/4$	$7\pi/4$	2π
$x = \cos(\theta)$	1	0.707	0	−0.707	−1	−0.707	0	0.707	1
$y = \sin(\theta)$	0	0.707	1	0.707	0	−0.707	−1	−0.707	0

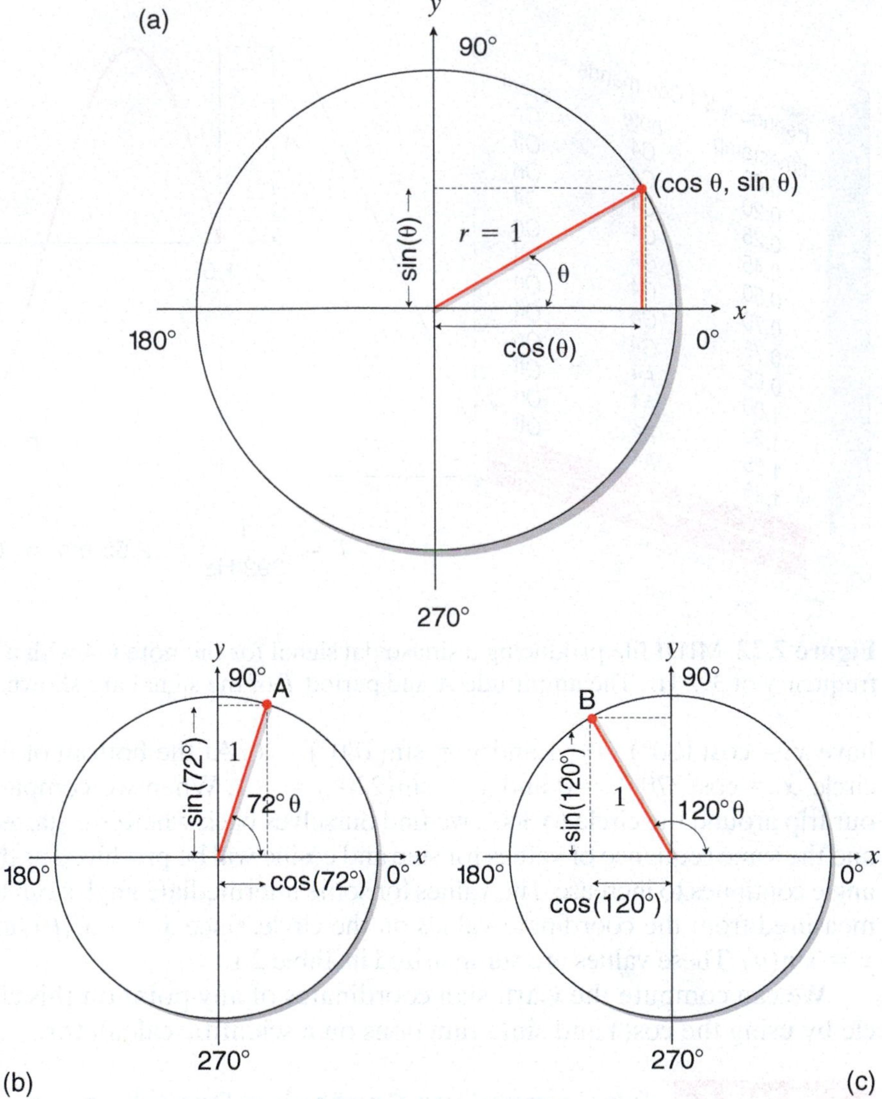

Figure 2.23 A circle with a radius of one, showing the sine and cosine of an angle.

Figure 2.24. You can verify that the plots and the table have the same values of $\cos(\theta)$ and $\sin(\theta)$ for the angles listed in Table 2.1. From this figure, we clearly see that $\cos(\theta)$ is the same as $\sin(\theta)$ shifted to the left by 90°. This relationship is described mathematically as follows:

$$\begin{aligned} \cos(\theta) &= \sin(\theta + 90°) \\ \sin(\theta) &= \cos(\theta - 90°) \end{aligned} \tag{2.6}$$

The effect of adding or subtracting 90° from the angle θ is the same as that of shifting a signal in time—the plots move to the left or right, respectively.

The sine and cosine functions are periodic functions with a period of 360°. Since we know the period and we know the value of the functions for one full period, we can extend the plots to as large a value of θ as we wish by repeating the values of the function every 360°:

$$\begin{aligned} \sin(\theta) &= \sin(\theta + 360°) \\ \cos(\theta) &= \cos(\theta + 360°) \end{aligned} \tag{2.7}$$

INTERESTING FACT:

The human hearing system is not very sensitive to time shifts in sinusoids or in any other periodic signals. A human listener cannot hear the difference between a cosine and a sine function when played through a loudspeaker.

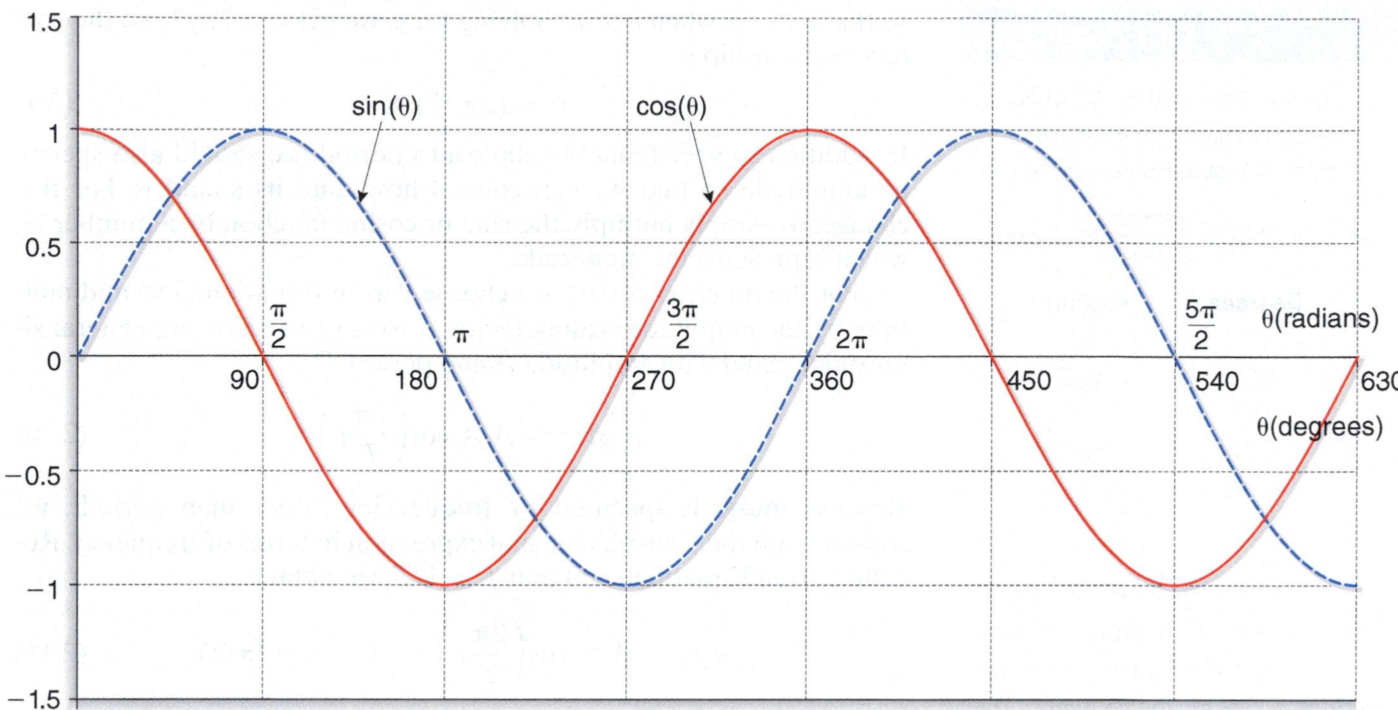

Figure 2.24 Sine and cosine as a function of angle. The angles are shown in both degrees and radians.

Infinity Project Experiment: Generating Sine and Cosine Signals

Sine and cosine signals are directly related to how a wheel rotates. From the spinning "dot" on your computer, can you see how a sinusoid is created? How about a cosine signal? Try changing the speed of rotation. What happens to the signal produced?

Using Radians to Measure Angles We use degrees for the units of an angle when we first learn about sines and cosines, but mathematicians and engineers often use a different unit of measure called a **radian**. For angles, 2π radians means the same thing as 360°. It turns out that using radians makes many of our equations simpler.

Radian: An alternative way to measure angles that is often used by engineers and mathematicians. One radian is about 57.3°, and 360° is 2π radians.

Cosines and Sines as Sound Signals Now that we have the cosine and sine functions as possible candidates for making signals, we would like to turn one of these functions into a sound signal that is a function of time. The simplest way to do this is to make θ equal to some constant multiplied by time so that the angle will increase as time increases. In other words,

$$\theta = c \times t \tag{2.8}$$

where c is our scaling factor and is constant for each note. We want the signal to be periodic with period T seconds, but sinusoids are periodic with period 2π radians (360°). So, we should choose the scaling factor

KEY CONCEPT

To convert degrees to radians, multiply the angle in degrees by $(2\pi/360)$ radians per degree, or

$$\text{radians} = (2\pi/360)\frac{\text{radians}}{\text{degree}} \times \text{degrees}$$

Degrees	Radians
0°	0
90°	$\pi/2$
180°	π
270°	$3\pi/2$

INTERESTING FACT:

Besides amplitude and frequency, there is a third parameter that defines a sine or cosine signal. The *phase* of a sine function is the amount in radians that the sinusoid is offset with respect to the origin. When the angle of a sine function is made proportional to time, changing the phase of the sine function is equivalent to time-shifting the sinusoidal signal by some amount. Phase is important for specifying the exact forms of sinusoidal signals and their combinations when matching a plotted signal. It is also very important for communications systems. Phase is less important when making music, because our ears are not at all sensitive to small changes in the phase of a sinusoid.

so that $\theta = 2\pi$ when $t = T$. Solving for c, we get $c = 2\pi/T$, so the correct relationship is

$$\theta = (2\pi/T) \times t \tag{2.9}$$

In addition to specifying the sinusoid's period, we should also specify its amplitude so that we can control how loud its sound is. For this change, we simply multiply the sine or cosine function by a number A, which represents the amplitude.

For the function $\cos(\theta)$, we choose θ as in Equation (2.9) and multiply by the amplitude scaling factor A to get the following general sinusoidal signal with amplitude A and period T.

$$s(t) = A \times \cos\left(\frac{2\pi}{T}t\right) \tag{2.10}$$

Because music is specified by frequencies rather than periods, we should write the sinusoidal-signal expression in terms of frequency. Recalling from Equation (2.5) that $f = 1/T$, we obtain

$$s(t) = A \times \cos\left(\frac{2\pi}{T}t\right) = A \times \cos(2\pi f t) \tag{2.11}$$

Figure 2.25 shows a plot of two cosine functions, one with a frequency of 625 Hz and one with a frequency of 250 Hz. The period T and amplitude A of each is marked on the plot. We can now see the meaning of both parameters: A corresponds to the maximum height of the sinusoid above or below from zero, and T is the repeating interval. The frequency of the sinusoid corresponds to the number of times the function repeats itself every second. We can think of the frequency f as the number of periods per second, and the period T as the number of seconds per period.

A third way to change a signal $s(t)$ is by shifting it in time, as shown in Equation (2.12). Shifting a cosine 90°, or $\pi/2$ radians, to the right produces a plot that is exactly the same as a sine. We can shift a sine or cosine signal that is a function of time by any amount. When we do, it will not start or end exactly at a value of zero (as does a sine function) or a value of one (as does a cosine function). Equation (2.12) shows the most general expression for a cosine signal that has been modified using amplitude scaling, time scaling, and time shifting:

$$s(t) = A \times \cos(2\pi f(t + d)) \tag{2.12}$$

EXAMPLE 2.3 Plotting Sines and Cosines

Plot the sine and cosine functions from $t = 0.0$ ms to $t = 6.0$ ms if each has a frequency of 400 Hz and an amplitude of 3.2.

Solution

We first specify the axes for our plot before drawing the functions. The horizontal axis goes from 0 to 6 ms. Since the amplitude of the sinusoid is 3.2, we know that the final scaled sinusoid will always be between −3.2 and 3.2 on the vertical axis. We next convert the frequency of the sinusoid into its period, using $T = 1/f$:

$$T = 1/400 \text{ Hz} = 0.0025 \text{ s}$$

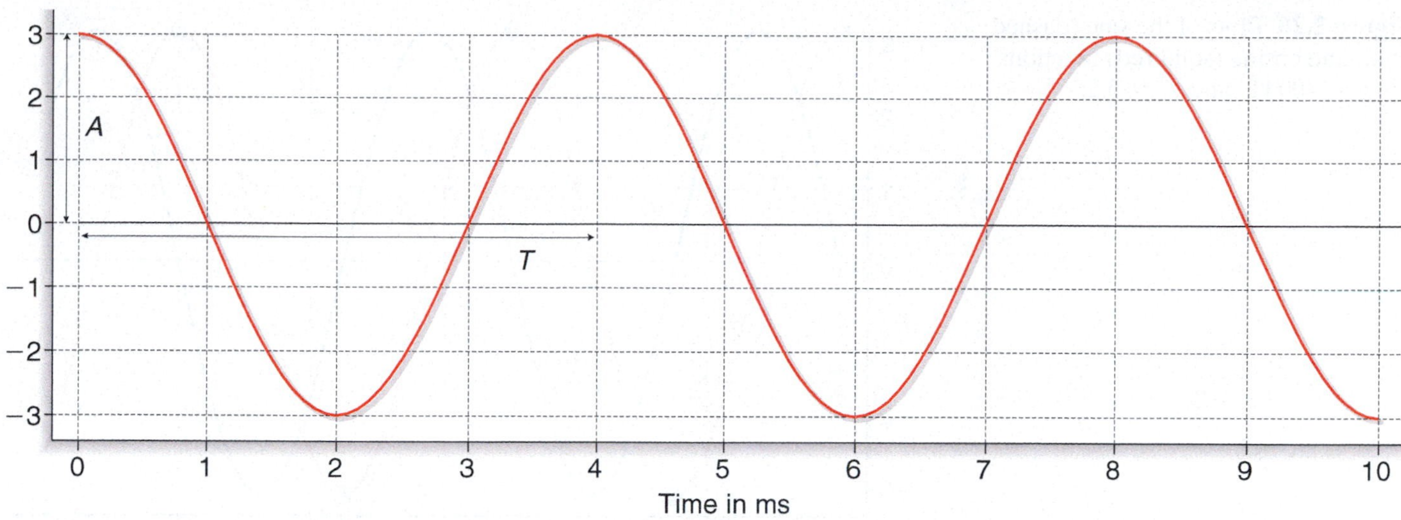

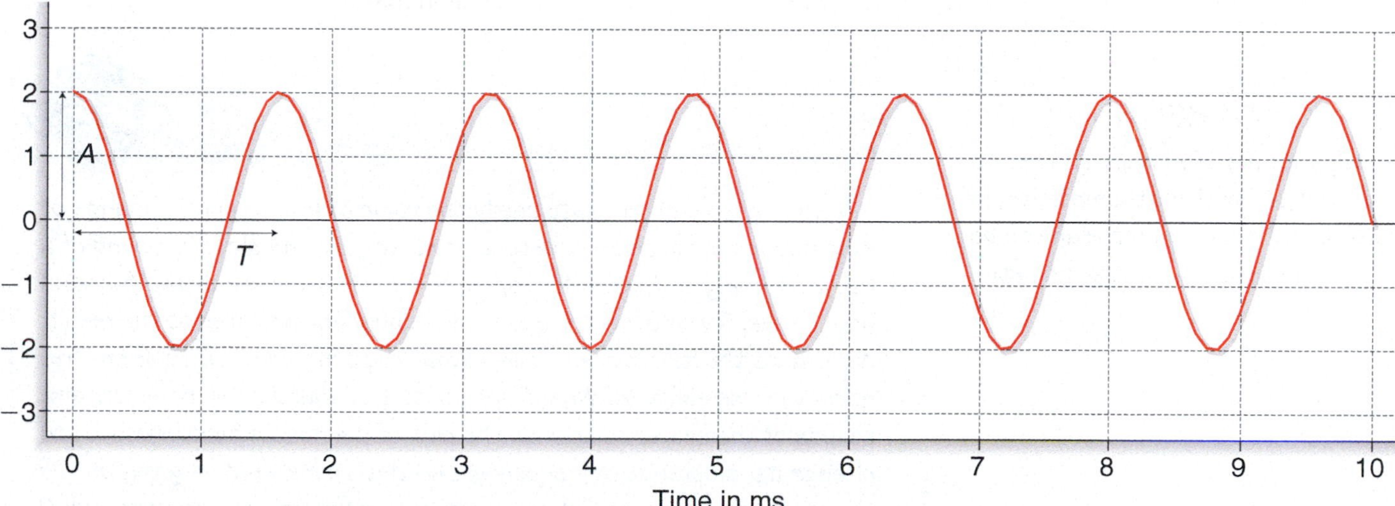

Figure 2.25 Plots of two cosine functions for 10 ms. In both plots, the amplitude A is shown by the vertical arrow and the period T is shown by the horizontal arrow. In the upper plot, the amplitude is 3 and the frequency is 250 Hz, so the period is 4 ms. In the lower plot, the amplitude is 2 and the frequency is 625 Hz, so the period is 1.6 ms.

There are 1000 ms in 1 second (s), so $T = 0.0025\text{ s} \times 1000\text{ ms/s} = 2.5\text{ ms}$. The cosine will have its maximum value at $t = 0$, $t = T$, $t = 2T$, and so on. Let's mark the values of $s(t) = 3.2$ at times $t = 0.0$, 2.5 ms, and 5 ms on the graph. These values are shown in green in Figure 2.26. The cosine will have its minimum value at $t = T/2$, $3T/2$, $5T/2$, and so on. The values of $s(t) = -3.2$ at $t = 1.25$ ms and 3.75 ms are marked in blue on the plot. Finally, we can sketch the function $A \times \cos(2\pi ft)$, where $A = 3.2$ and $f = 400$ Hz, using our knowledge of what a cosine function looks like. This function is shown as the solid red line in Figure 2.26. We can repeat this process for the sine function, shown with a dotted blue line in Figure 2.26, by simply shifting all the points $T/4 = 0.625$ ms, or one-quarter period, to the right.

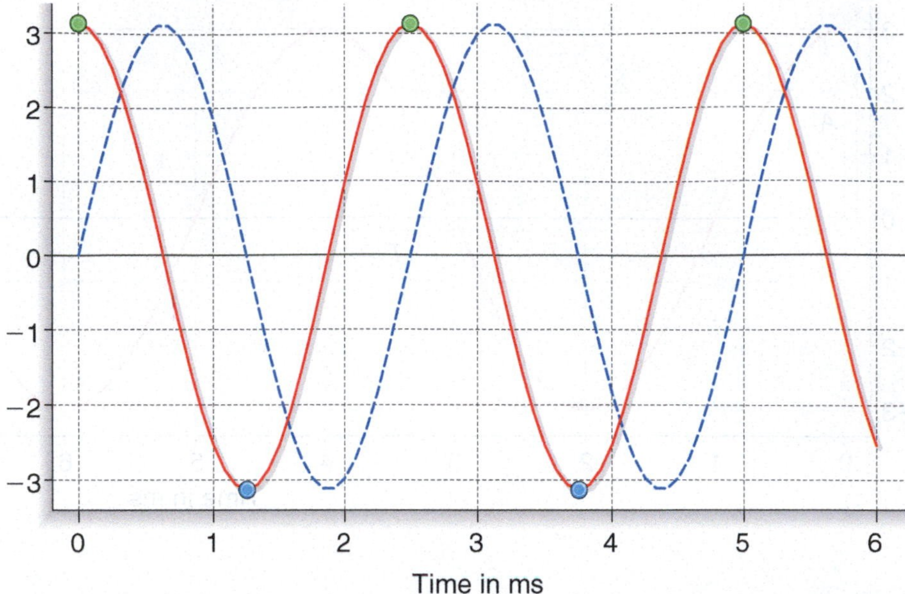

Figure 2.26 Plots of the sine (dashed blue) and cosine (solid red) functions for $f = 400$ Hz and $A = 3.2$.

INTERESTING FACT:

Hipparchus of Rhodes (190 BC to 120 BC), a Greek mathematician, is considered by some historians to be the inventor of trigonometry.

Infinity Project Experiment: Listening to Sines and Cosines

Digital devices can calculate easily the value of a function thousands or even millions of times every second. If we amplify and play the sound of a function using a device's loudspeaker, we can hear what the function sounds like. Try creating the sound of a sinusoid, and listen to the result. What does the function sound like? What happens when you raise the frequency of the sinusoid? What happens to the sound if you raise the amplitude of the sinusoid? Look at the plot of a sinusoid function while you change its amplitude or frequency, and determine what is going on. Do these plots make sense? What is the highest frequency you can make? What is the highest frequency you can hear? What is the lowest frequency you can make? What is the lowest frequency you can hear?

EXAMPLE 2.4 Writing the Equation of a Sine or Cosine Function from Its Plot

From the plot of the sinusoid in Figure 2.27, determine the values of the amplitude A and the frequency f, and write the equation for the sinusoid, using either the sine or cosine function.

Solution

By finding the highest and lowest values of the signal, we see that $A = 3.0$. We next need to find the period T by finding the interval between repetitions of the signal pattern. The maximum values occur at $t = 0.2$ s, $t = 1.0$ s, and $t = 1.8$ s. We can calculate the period as the time interval between maximum values, so $T = 1.0 - 0.2$ s $= 0.8$ s. Using Equation (2.5), we can compute the frequency $f = 1/T = 1/0.8$ s $= 1.25$ Hz. We could have picked any point on the curve to measure the period, but choosing two maximum values or two minimum val-

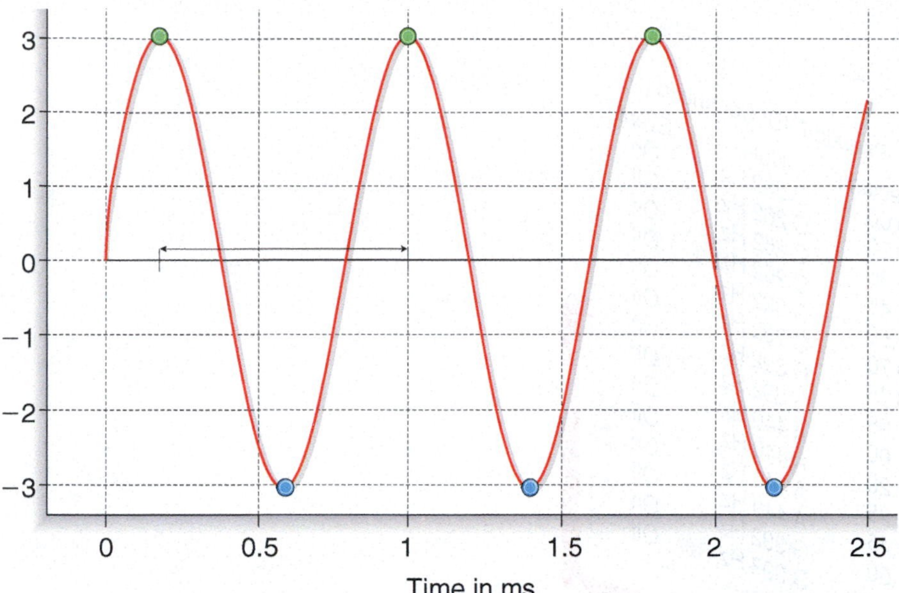

Figure 2.27 Plot of a sinusoidal function.

Infinity Project Experiment: Measuring a Tuning Fork

A tuning fork makes a sound that is nearly sinusoidal. The frequency of this sinusoid is determined by the tuning fork's size and physical makeup. By measuring the signal of a particular tuning fork's sound, we can determine the period of that sound. What is the frequency of the tuning fork? How close to sinusoidal is its signal? What is the signal's amplitude? If we know both the frequency and amplitude of the tuning fork's signal, we can re-create the signal by using a sinusoidal-function generator. Can you use such a generator to produce a sinusoid that sounds like the one produced by the tuning fork?

ues makes it easy to find the same point in the next period. The signal starts at 0 when $t = 0$ and has its first maximum value at $T/4 = 0.2$ s. Comparing this signal to the values in Table 2.3, we see that it has the pattern of a sine. Our final expression is

$$s(t) = A \sin(2\pi f t) = 3 \sin(2\pi \times 1.25t)$$

Making Melodies with Sinusoids

A melody is a sequence of notes. Our first digital-band design will use simple melodies with only one note played at any one time. Each individual note can be played by making a periodic signal whose fundamental frequency corresponds to the pitch of the note. Therefore, all we have to do to make music is to play these periodic signals, one after the other, in correspondence with the notes to be played. Let's consider a simple example.

Figure 2.28 shows a block diagram for the creation of the first few notes from a familiar children's song, starting with the input of the information and ending with the production of sound. Suppose this piece of music has already been converted into a MIDI file and saved. The MIDI-file information shows the notes being played as well as their frequencies, which can be figured out using Figure 2.17.

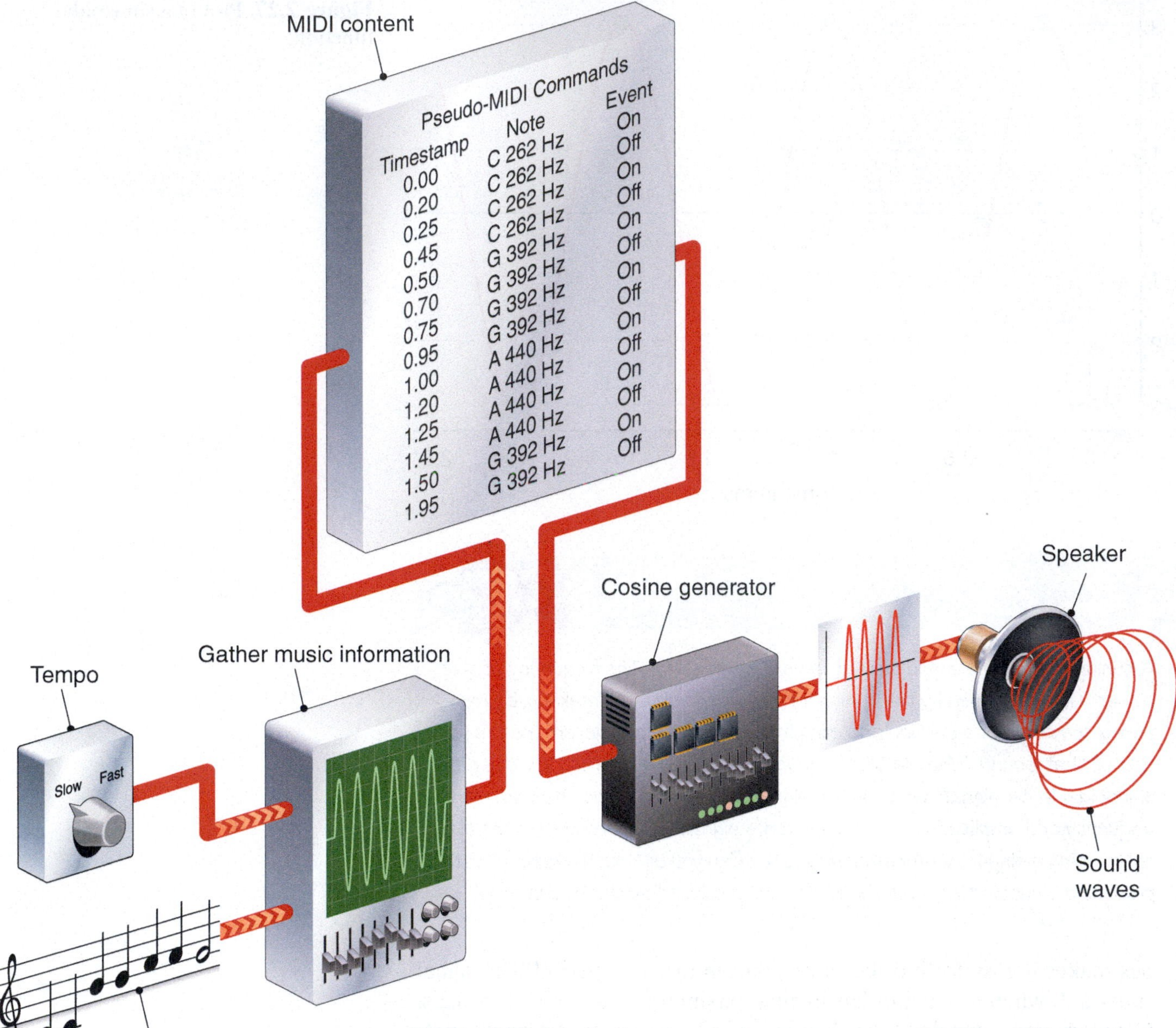

Figure 2.28 Block diagram of a system to create a familiar children's song from a MIDI file.

The cosine generator will compute the values of a cosine function at the specified frequency and will turn each cosine function on and off according to the time information in the MIDI file. Each cosine has the form of Equation (2.11), with amplitude $A = 1$ and frequency f determined by the MIDI file. A very short segment typical of part of the music signal is also shown in Figure 2.28. The sequence of cosine waves generated at the selected frequencies and times is our music signal that will be connected as the input to the "Convert to sound" block. This last block makes sounds from the signal, using a device that you probably already know about—a loudspeaker.

The short melody in Figure 2.28 should last for almost 2 seconds (s). Since the periods of our sinusoidal notes are between 2.3 ms and 3.8 ms, we cannot look at the entire music signal in one plot and expect to see the detail of the individual periods of the sinusoids. In Figure 2.29, we have

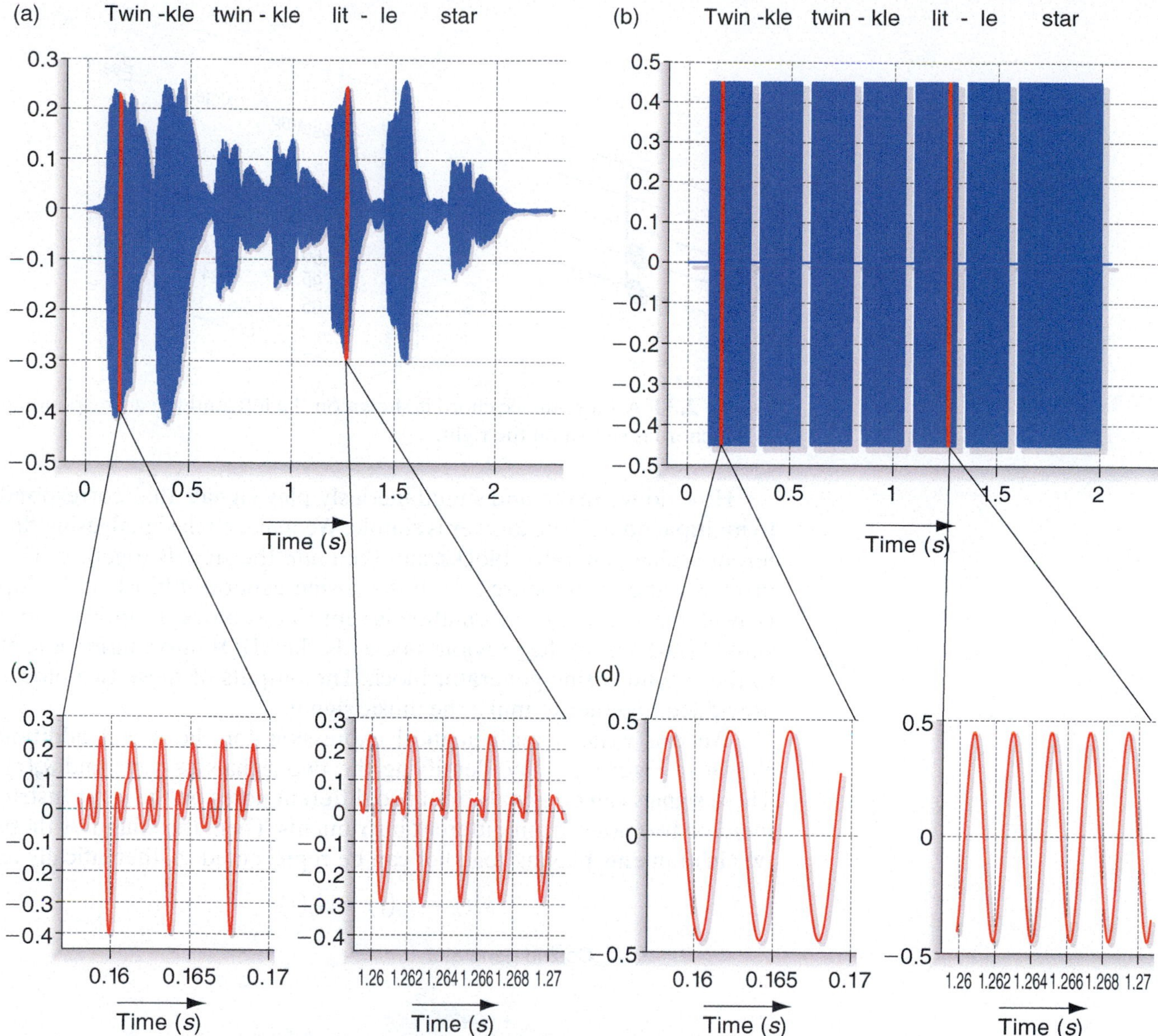

Figure 2.29 The plot of the signal for a saxophone playing a popular children's song in (a) can be compared with the version from our sinusoidal-signal generator in (b). Plots of short intervals of two notes from each (highlighted in red) are shown in (c) and (d), respectively.

taken a few periods out of each note and plotted the notes in sequence to give you an idea of what the music signal looks like. If the whole signal were plotted at a scale of 20 ms per inch, it would be over 8 feet long.

Making Music with More than One Note at a Time

Most interesting music is made up of more than just simple melodies. Usually, two or more notes should be played together, often by more than one instrument at a time, as in the chords of Figure 2.18. Figure 2.30 is an example of a simple chord and its MIDI file description.

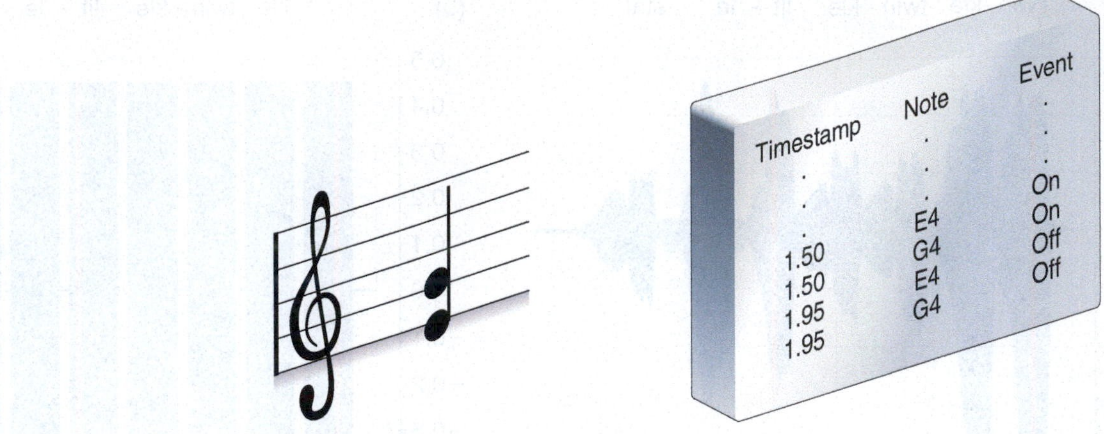

Figure 2.30 A sheet-music chord is shown on the left, and the corresponding MIDI chord is shown on the right.

How do we make and simultaneously play signals that correspond to multiple notes? The answer is simple: We make each signal, using different cosine-generator blocks, and then *add* the signals together. This process is shown in Figure 2.31. If the cosine-generator block at the top is available, the MIDI information is sent to this block. If another note should start before the previous one ends, the MIDI information is sent to the second cosine-generator block. The outputs of these two blocks are added together to make the music signal.

We can write a mathematical expression for the signal addition shown in Figure 2.31 by identifying the two signals as $s_a(t)$ and $s_b(t)$. These signals can be signals from two different notes on the same instrument or two notes from different instruments. Then, the sound of these two instruments playing together can be represented mathematically as

$$s(t) = s_a(t) + s_b(t)$$

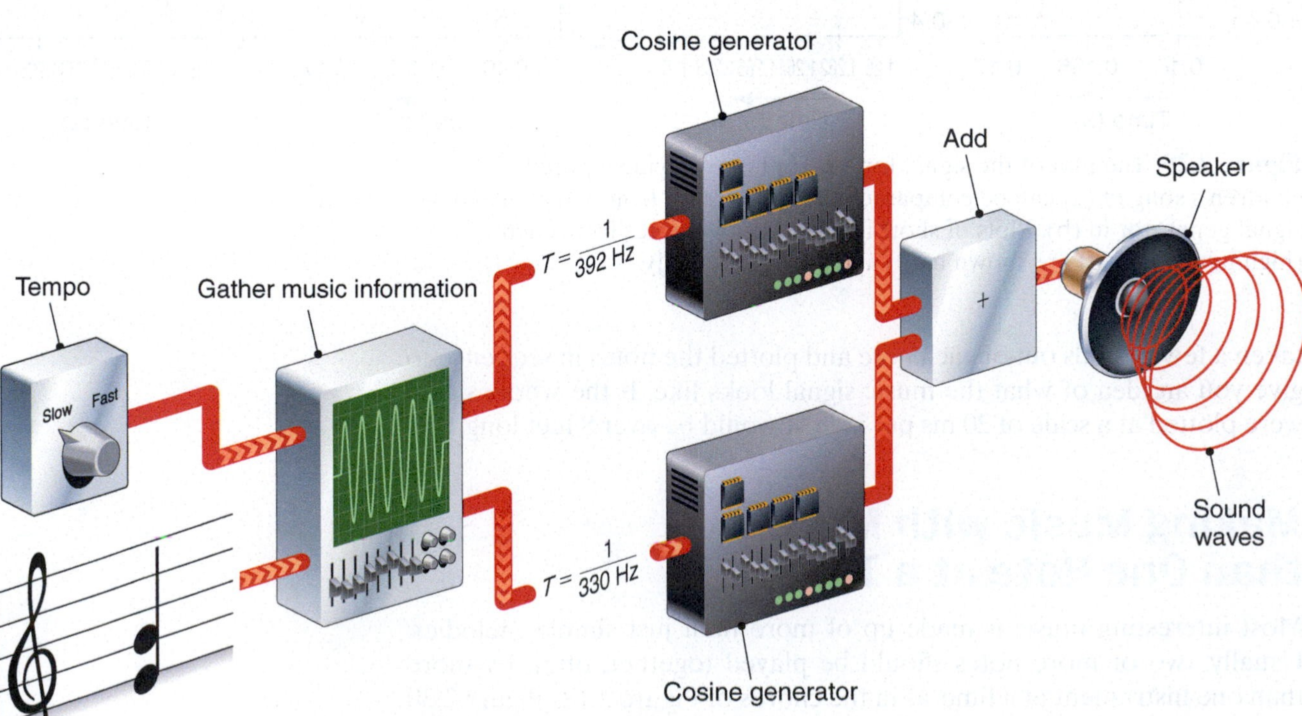

Figure 2.31 Adding two signals to make a musical chord.

The addition is done by finding the value of each signal at a specific time instant $t = t_0$ and adding the values of the signals together. For example, suppose a guitar sound and a singing voice are added together. If the value of the guitar sound signal at time $t = 1.5$ s is found to be $s_a(1.5) = 0.7$ and the value of the voice signal at time $t = 1.5$ s is found to be $s_b(1.5) = -0.4$, then the value of the combined signal $s(t)$ at time $t = 1.5$ s is

$$s(1.5) = s_a(1.5) + s_b(1.5) = 0.7 + (-0.4) = 0.3$$

To compute the signal $s(t)$ for any other time instant, we would do a similar calculation using the corresponding numbers. When this operation is done for all time values, we have the complete signal $s(t)$.

Adding two signals over a time interval to create a plot can be done by hand for simple signals. In such cases, significant points, such as when the signal is at a maximum, a minimum, or zero, are identified for each of the signals to be added. Then the values of both signals at each of these times are determined so that they can be added. When a signal has a sharp change, then the value of the sum can be computed just before and just after the change.

EXAMPLE 2.5 Adding Two Signals Together

Add the two signals in Figure 2.32(a) and (c) together, and draw the resulting signal.

Solution

We will solve this problem in two ways. The first method takes advantage of the particularly simple form of $s_1(t)$. The second way, using a table, will work for all sums of signals. The two signals to be added are plotted on the same axes in Figure 2.32(b), and the sum is plotted in (d). If one signal represented a clarinet and the other a flute, the sum would represent the signal for both instruments played together.

Method 1:

Since $s_1(t)$ is always $+1$ or -1, we will divide the time axis into segments where $s_1(t)$ is not changing and then plot over those segments.

Time Interval	$s_1(t)$	$s_2(t)$	For $s(t) = s_1(t) + s_2(t)$, Plot
$t = 0$ to $t = 0.5$	1	$s_2(t)$	$s(t) = 1 + s_2(t)$
$t = 0.5$ to $t = 1.0$	−1	$s_2(t)$	$s(t) = -1 + s_2(t)$
$t = 1.0$ to $t = 1.5$	1	$s_2(t)$	$s(t) = 1 + s_2(t)$
$t = 1.5$ to $t = 2.0$	−1	$s_2(t)$	$s(t) = -1 + s_2(t)$

Method 2:

This method works for all types of signals. To start, make a table of values for the two functions, and add corresponding pairs of values.

t	0.05	0.25	0.45	0.55	0.75	0.95	1.05	1.25	1.45	1.55	1.75	1.95
$s_1(t)$	1.0	1.0	1.0	−1.0	−1.0	−1.0	1.0	1.0	1.0	−1.0	−1.0	−1.0
$s_2(t)$	−1.8	−1.0	−0.2	0.2	1.0	1.8	1.8	1.0	0.2	−0.2	−1.0	−1.8
$s(t) = s_1(t) + s_2(t)$	−0.8	0.0	0.8	−0.8	0.0	0.8	2.8	2.0	1.2	−1.2	−2.0	−2.8

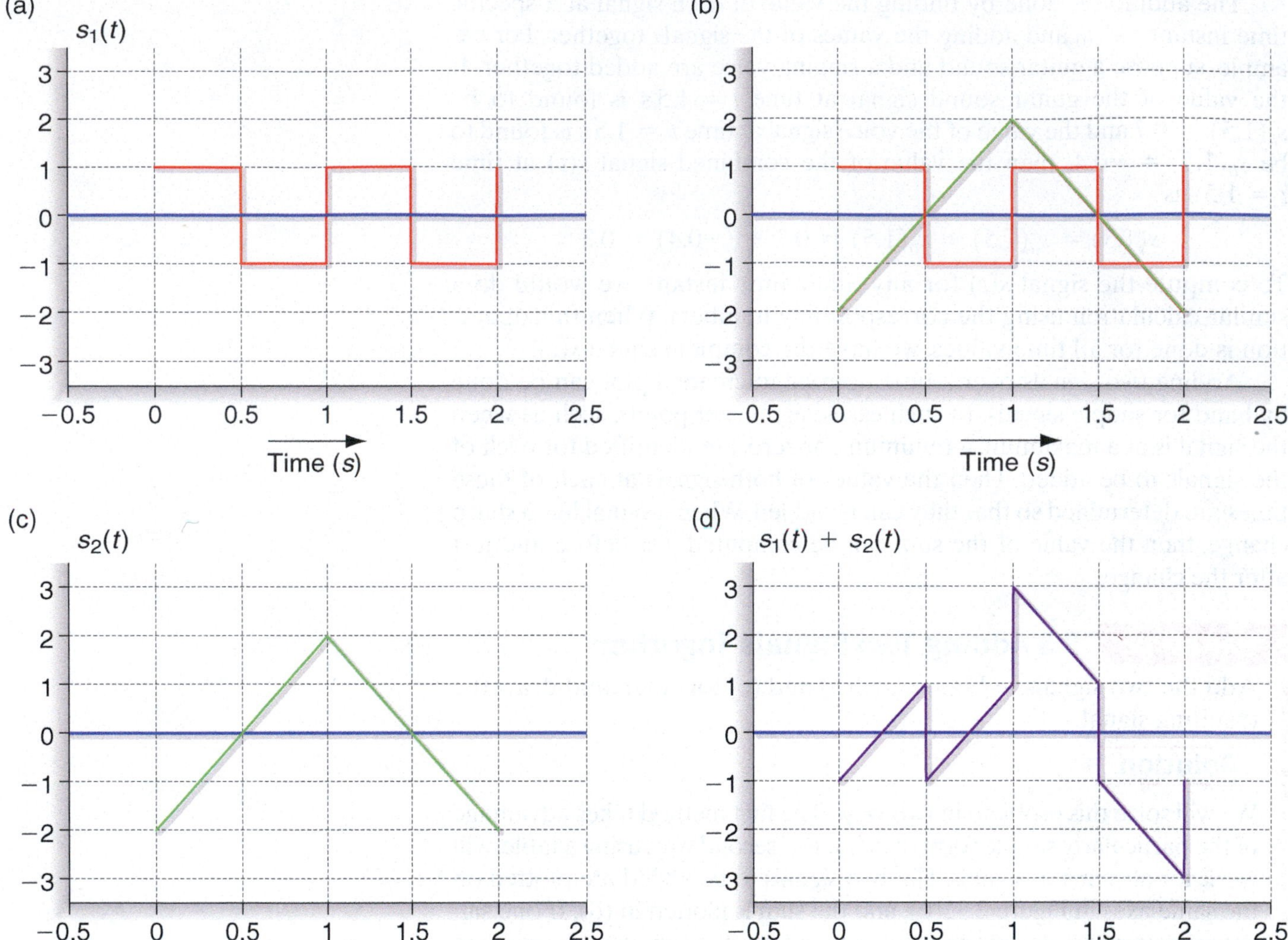

Figure 2.32 Sum of two plots of signals. The individual signals are shown in (a) and (c), the two signals are plotted on the same axis in (b), and the sum is shown in (d).

Once we have the sums of the pairs, we can sketch out the desired signal. The points used in the table are shown in both the individual signals and the sum in Figure 2.32(d).

The results using Method 1 and Method 2 are exactly the same.

Adding Two Sinusoids Together The first design of our digital band used sinusoids for the periodic functions. If we want the band to play two different notes together, we sum two sinusoids together, where each sinusoid corresponds to each note. Consider the sum of two cosines with the same amplitude $A = 1$, but different fundamental frequencies f_1 and f_2, so that

$$s(t) = \cos(2\pi f_1 t) + \cos(2\pi f_2 t)$$

In this sum, the values of f_1 and f_2 could correspond to two notes that are being played by an instrument, such as a note we'll call concert A (440 Hz) and another note we'll call the F below concert A (349.23 Hz).

These two cosine functions could be added together using the table method from the previous example, although this operation is more easily done with a calculator. Figure 2.33 shows the two individual signals and the resulting sum $s(t)$. In general, it is hard to guess the appearance of the plot of the sum of two sinusoids. In most cases, the sum is not periodic, because the individual signals will never start over again at the same time. But in this case, the ratio of the frequencies is approximately 4 to 5, and the sum will repeat almost perfectly after only a few periods of each signal. When we hear sounds with this characteristic, they usually sound good to us.

INTERESTING FACT:

Why do the sounds of certain pairs of notes sound good when they are played together? The ratio of the periods of the sounds is the key to the answer. If this ratio contains two small integers, such as $\frac{3}{2}$, $\frac{4}{3}$, and $\frac{5}{3}$, the sound signal produced is periodic and has a pleasing sound. This chordal or vertical structure of music (as opposed to melodic or horizontal structure) is known as harmony .

Reverse Engineering the Score

Up to this point, we've been focused on building a system that takes a description of music and turns it into sound. Suppose we already have a piece recorded by somebody else that we want our digital band to play.

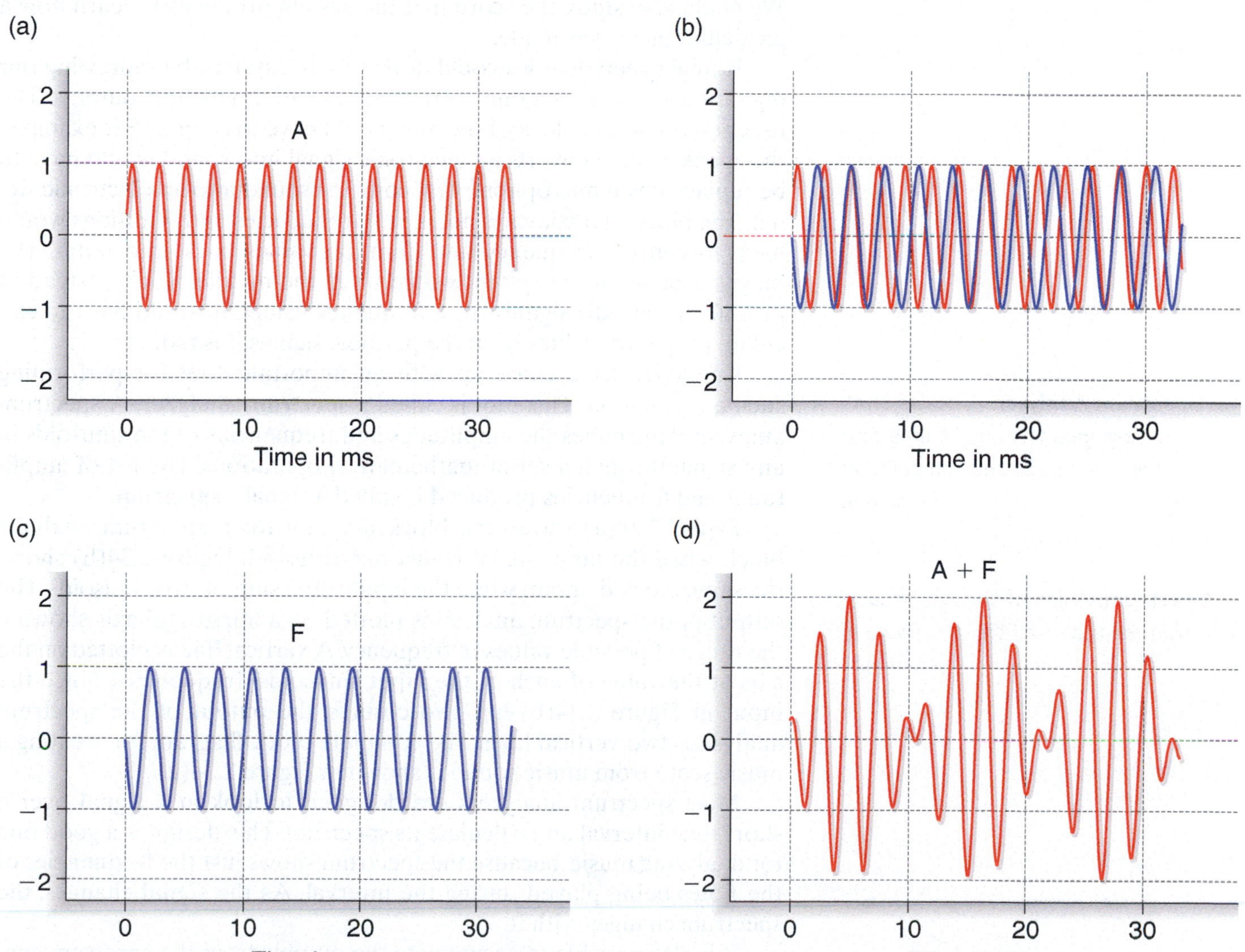

Figure 2.33 Sum of two sinusoidal signals at the frequencies of A and F. The individual signals are shown in (a) and (c), the two signals are plotted on the same axis in (b), and the sum is shown in (d). The sum is not a sinusoidal signal.

Infinity Project Experiment: Building the Sinusoidal MIDI Player

This experiment allows you to construct your own virtual keyboard inside of a computer, using sinusoidal-function blocks, adder blocks, and a MIDI message-reader block. Once you've built your synthesizer, go to the Web and get a MIDI file of your favorite piece of music by your favorite artist and listen to it. How does your favorite piece of music sound like when played using sinusoids? What do the signals look like as the piece is playing?

Is there some way to create the score or the MIDI file from the sound of the music itself? If so, there would be many things that we could do with such a system. For example, we could use the system to translate all the pieces we like into MIDI files automatically without our help. We could also study the score that the system produces to learn how a particular piece was made.

It might seem like we could design such a system by redrawing our digital-band block diagram to reverse all the inputs and outputs. The functions of all the blocks, however, would have to change. For example, the speaker that converts an electronic signal into sound would have to be replaced by a microphone that converts sound into an electronic signal. The block that converts notes into frequencies and durations would have to convert frequency and duration values back into notes. The biggest change in the system would be in the middle block. Instead of producing periodic signals from frequency values, it would have to calculate frequency values from the periodic signals it is fed.

Spectrum Analyzer: A device that determines the amplitudes and frequencies of the sinusoids in any signal.

Spectrum: A list of the amplitudes and frequencies of all the sinusoids in a signal.

Engineers have come up with an important tool for performing such a calculation. This tool is called a **spectrum analyzer**. A spectrum analyzer determines the amplitudes and frequencies of the sinusoids in any signal through a set of mathematical operations. The list of amplitudes and frequencies produced is called a signal's **spectrum**.

Figure 2.34(a) shows the block diagram for a spectrum analyzer block when the input signal is just one sinusoid. Figure 2.34(b) shows the same block diagram when the input is the sum of two sinusoids. The output of the spectrum analysis is plotted on a horizontal axis showing the range of possible values of frequency. A vertical line is plotted on the axis at the value of each of the input sinusoids' frequencies. Since the input in Figure 2.34(b) has two cosines, the output of the spectrum analysis is two vertical lines. The complete block diagram for creating a music score from music sound is shown in Figure 2.34(c).

Most spectrum analyzers are designed to look at a signal over a short time interval and calculate its spectrum. This design is a good one for analyzing music, because the spectrum shows just the frequencies of the notes being played during the interval. As the signal changes, the spectrum changes with it.

It is also possible to accumulate the outputs from the spectrum analyzer and make a graphic display that is very similar to a score. The axis of possible frequencies is horizontal for the spectrum analyzer and vertical for the score's treble and bass clefs, so we will have to rotate the spectrum and make it a vertical strip to match the score. When we accumulate

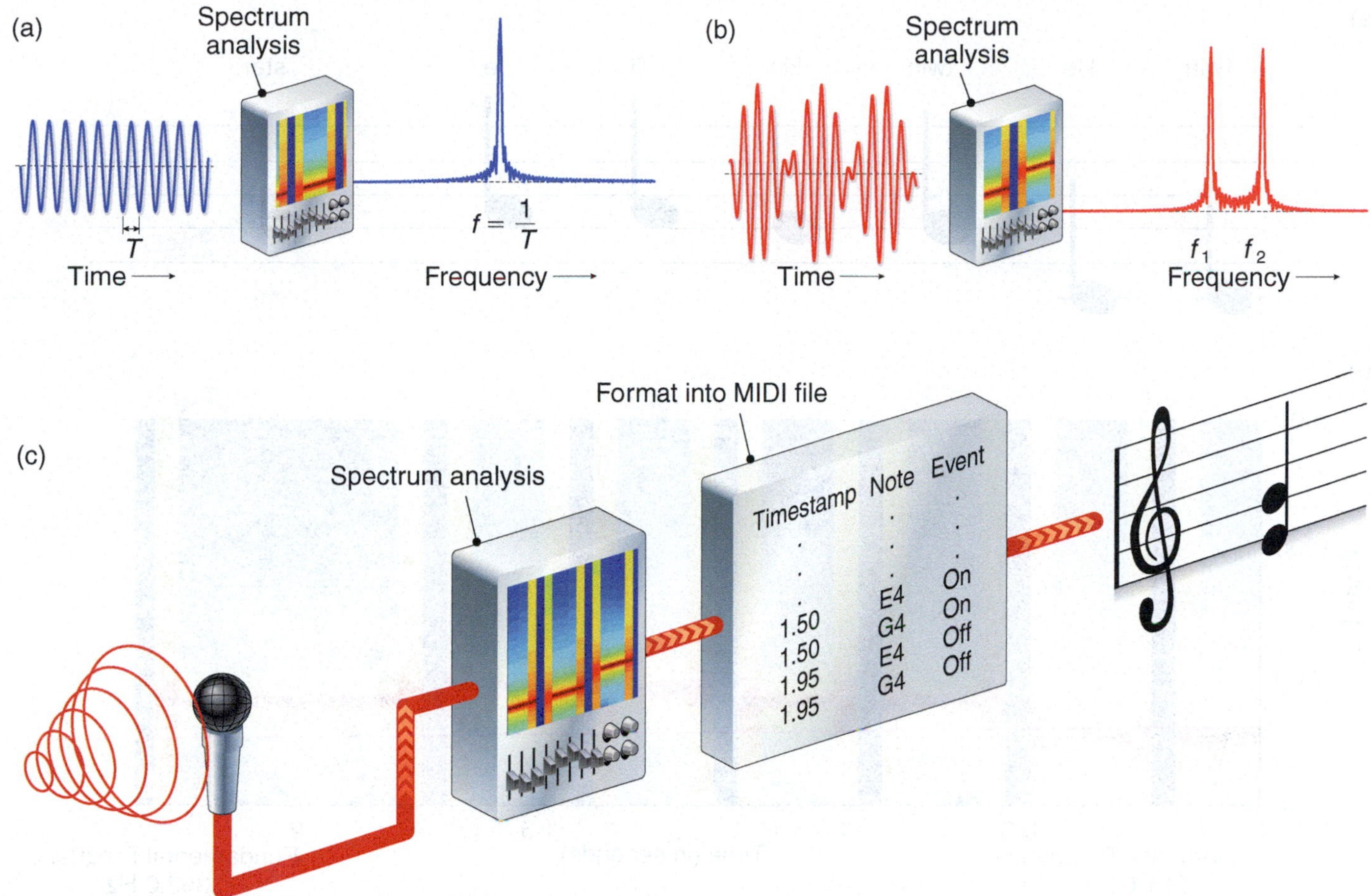

Figure 2.34 Block diagram showing conversion of music sound to MIDI information, using a spectrum-analyzer block. In (a) a signal with a single sinusoid is identified as one note. In (b) a signal with two sinusoids is identified as two notes and the frequency of each is shown.

these strips and place them side by side, we will have lines that show us what notes were being played. This graphic display is called a **spectrogram**. It will not be as stylized as a score, but it will show the frequencies of the notes as they are played and when each began and ended. In Figure 2.35(a), the score for the children's song is shown again for reference. In Figure 2.35(b), we can see the output of the spectrogram of the music signal created by our device, using sinusoidal signals. The dark red horizontal bars show the frequency and duration of each of the notes. Color is used to show the amplitude of each sinusoid with dark red for the highest or loudest values and dark blue for the lowest values. The vertical bars that are completely dark blue represent the quiet time between notes, and the vertical yellow bars show when a note starts and stops.

Spectrogram: A two-dimensional image describing the spectrum of a sound over time.

In Figure 2.35(c) we see the spectrogram of the same sequence of notes played by an instrument with a more complex sound. The periodic function generated by this instrument is the sum of a sinusoidal signal at the fundamental frequency of the note and several other sinusoids which are called harmonics.

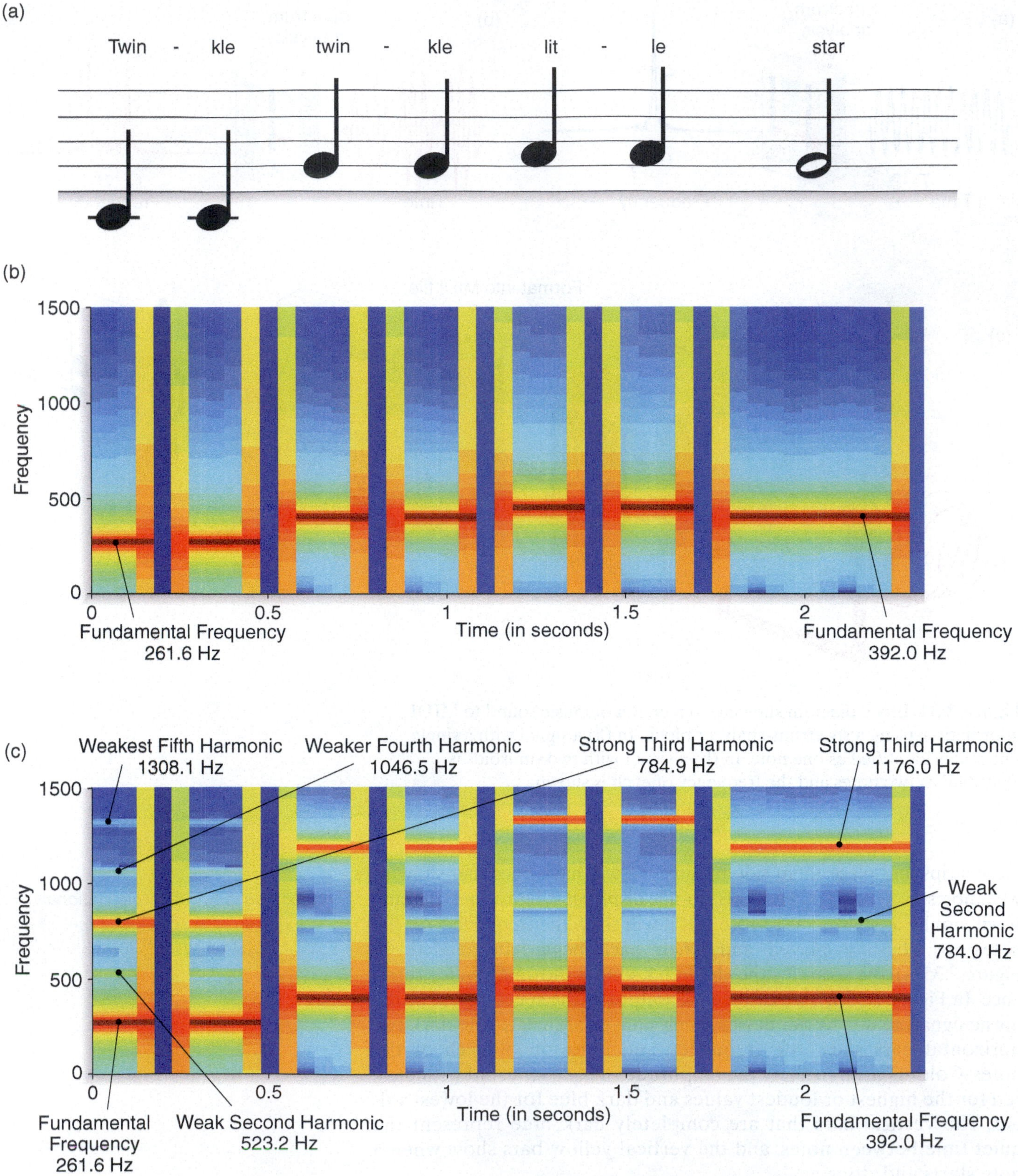

(a)
Twin - kle twin - kle lit - le star
(b)
1500
1000
500
0
Frequency
0
0.5
1
1.5
2
Time (in seconds)
Fundamental Frequency
261.6 Hz
Fundamental Frequency
392.0 Hz
(c)
Weakest Fifth Harmonic
1308.1 Hz
Weaker Fourth Harmonic
1046.5 Hz
Strong Third Harmonic
784.9 Hz
Strong Third Harmonic
1176.0 Hz
Weak
Second
Harmonic
784.0 Hz
1500
1000
500
0
Frequency
0
0.5
1
1.5
2
Time (in seconds)
Fundamental
Frequency
261.6 Hz
Weak Second Harmonic
523.2 Hz
Fundamental Frequency
392.0 Hz

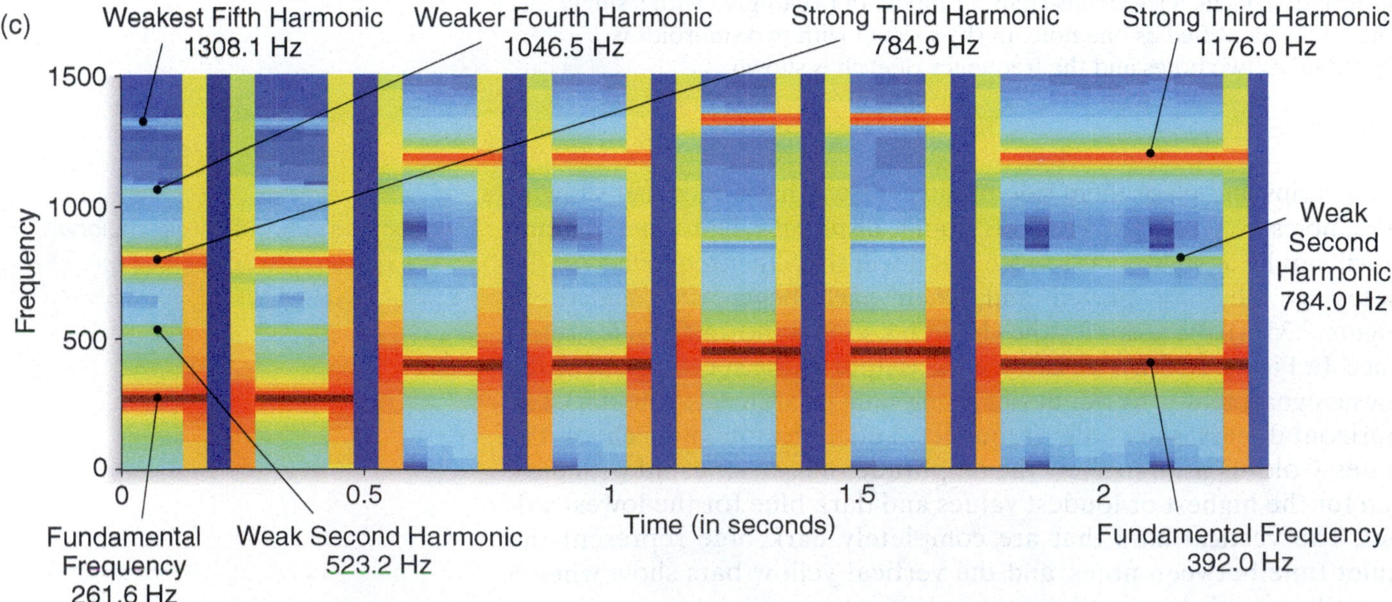

Figure 2.35 The music shown in (a) produces the spectrogram in (b) when played by our sinusoidal music generator. The dark-red horizontal bars show the frequency of the note played at each time, and the blue background corresponds to the quiet intervals between notes. The spectrogram also displays a yellow vertical stripe at the beginning and end of each note. An instrument with a more complex sound produces the spectrogram in (c). It has a weak second harmonic, a stronger third harmonic, and much weaker fourth and fifth harmonics that can be seen only for the first two notes.

Infinity Project Experiment: The Spectrogram

The spectrogram is a useful tool for understanding the frequency structure of any sound—not just music. The spectrogram can be applied to any signal, not just music. Examine the spectra of several simple sounds, and write down the amplitudes, frequencies, and time instants of the sinusoids that are contained in them. What does the spectrogram of your voice look like? Does it have a simple description? How about that of a recording from your favorite performer?

EXERCISES 2.3

Mastering the Concepts

1. Suppose you want to create the sound of a piano digitally. Can you use a MIDI file to create an accurate piano sound? Why or why not?
2. How are songs described using MIDI information? If you were to open up a MIDI file and read its instructions, what would you see?
3. What is a sinusoid? Sketch out an example of a sinusoid, and label all the important quantities of the sinusoid on your sketch.
4. How is the sine function related to a circle with a radius of one? How is the cosine function related to this circle?
5. What's the difference between radians and degrees? Which measure of angle is most often used by engineers and mathematicians?
6. What is the frequency of a sinusoid? How do we change a sinusoid's frequency?
7. What is the amplitude of a sinusoid? How do we hear the effect of changing the amplitude of a sinusoid?
8. Suppose we want to play the sounds of several sinusoids at the same time. How do we do that?
9. Engineers have come up with a way to calculate the frequencies of the sinusoids in a piece of music and display them in a plot. What is this type of plot called? Describe how the frequency information is contained in the plot.

Try This

10. A piece of music contains the following note sequence: E4, D4, C4, D4, E4, E4, E4. Each note lasts 0.25 second, and no two notes are played at the same time. Write the list of instructions that would be contained in a MIDI file that describes this piece of music.

11. Convert the given angle values to Cartesian coordinates, using $(x, y) = (\cos(\theta), \sin(\theta))$. Then sketch the (x, y) points on a graph.
    a. $\theta = 15°$
    b. $\theta = \pi/3$ radians
    c. $\theta = 4\pi/5$ radians
    d. $\theta = 4\pi$ radians
12. Plot the following sine and cosine signals on a graph over 7 ms.
    a. $s(t) = 2\cos(2\pi \times 675t)$
    b. $s(t) = 3\sin(2\pi \times 432t)$
    c. $s(t) = 6\cos(2\pi \times 500(t - 0.00025))$
13. Look at the plots of the sinusoids you drew in Exercise 2.3.12. From those plots, calculate following:
    a. the amplitude of each sinusoid
    b. the period of each sinusoid
    c. the frequency of each sinusoid
14. Plot the sum of $s_1(t)$ from Figure 2.32 and $s_2(t) = 2\sin(\pi t)$ for 2 seconds (s).
15. Sketch out a picture corresponding to the spectrogram for the piece of music in Exercise 2.3.10 if the frequencies of C4, D4, and E4 are 261.6 Hz, 293.7 Hz, and 311.1 Hz, respectively. Carefully label the axes of your plots.
16. Sketch the outputs of a spectrum analyzer that is analyzing the following signals:
    a. $s(t) = 2\cos(2\pi \times 675t) + 3\cos(2\pi \times 544t)$
    b. $s(t) = 1.3\sin(2\pi \times 182t) + 2.4\cos(2\pi \times 215t)$

In the Laboratory

17. A tuning fork makes a signal that is almost exactly a sinusoid. How close are other signals to sinusoids? Have a musician come into the lab and play her or his instrument for you while you measure the signal produced by the sinusoid. Try to make a sinusoid that has the same frequency as the note the musician plays. How similar do the two signals sound? How different do the two signals look?

Back of the Envelope

18. Would it make sense for a MIDI file to have out-of-order timestamps for its notes? Why or why not?
19. Figure out how many note entries a typical MIDI file might have. To do so, suppose a MIDI file describes a 3-minute piece of music and that no more than five instruments are being played at any one time. Also, suppose each instrument plays an average of five different notes each second. What is the largest number of note entries that the MIDI file will have? Remember, it takes two entries to create each note (one to turn the note on and one to turn the note off).

2.4 Improving the Design—Making Different Instruments

Instrument Synthesis

We like variety in music. Just look at all of the different instruments used in making music today—trumpets, violins, snare drums, piccolos, bassoons, guitars, etc.—and we haven't even considered the different styles of music, such as rock, R&B, jazz, country, and so on. The lists seem endless. At this point, our digital band can play only sinusoidal signals. With these signals, we can hear the melody and chords very clearly, but the sound it creates is often described as "hollow," or worse yet, boring. If this type of music were all we listened to, the music world wouldn't be nearly as rich and textured as it actually is. We need to improve our making-music block in our digital band so that it produces real-world instrument sounds, as well as the sounds of new instruments that have never existed. But how can we do this?

To help us figure out a better making-music block, we will first study what makes instrument sounds different from sines and cosines. You will recall that we chose sine and cosine signals because they were simple. Unfortunately, very few instruments make a signal that is an exact sinusoid, even if we choose the period of the sinusoid to be the same as that of the instrument's sound. For example, the guitar sound in Figure 2.12 is not exactly sinusoidal, nor are the two piano sounds shown in Figure 2.15. Another important feature of an instrument's sound is the way its loudness changes over time. For example, each of the two piano sounds at the top of Figure 2.15 has a large amplitude at the start of the note, and the sound gradually gets softer over time. We will have to take these signal properties into account if we want to make high-quality instrument sounds.

There are several methods we might explore to create realistic instrument sounds. One simple approach might be to record all the instruments we want to use. We would have to record each instrument playing every possible note. Then we would combine signals we've recorded from each instrument to make the music we want to hear. From an engineering perspective, this approach has several problems. First, it would require storing all of those sounds and then finding them when they were needed. This method also wouldn't enable us to make new sounds that weren't combinations of our saved sounds. A better approach is to find a way to actually create these sounds as we need them.

Synthesis is the creation of useful and complicated items from more basic ones. **Sound synthesis**, then, is the creation of useful and complicated sounds from more basic sounds. In the next section, we will focus on two of the most popular techniques for creating periodic waveforms that do a very good job of matching the characteristics of instrument sounds. We will also figure out ways to control how these sounds turn on and off so that they more closely mimic the sounds of instruments. Combining these two ideas will help us make a much better digital band.

Synthesis: The creation of useful and complicated items from more basic ones.

Sound Synthesis: The creation of useful and complicated sounds from more basic sounds.

Waveform Synthesis: A synthesis technique that stores a characteristic signal of an instrument for one specific note and then shrinks or stretches the signal in time to play several different notes with different fundamental frequencies.

Additive Synthesis: A synthesis technique that creates a periodic signal by adding sinusoids together.

INTERESTING FACT:

Besides wavetable and additive synthesis, engineers have a third way of making a realistic instrument sound. Called *physical modeling synthesis*, the technique uses a mathematical model of the physics of an instrument to compute the instrument's sound. This synthesis method is the most sophisticated and the newest. When it is properly implemented, this method can provide extremely realistic sounds.

Making Periodic Signals

The two most common digital techniques for making periodic signals with a specific shape are called **waveform synthesis** and **additive synthesis**:

1. *Waveform synthesis,* or *wavetable synthesis,* stores a characterisctic sound signal of one specific note for the actual instrument being re-created and then shrinks or stretches it in time to play different notes with different fundamental frequencies. While this synthesis method is the more straightforward of the two methods, it does use significant amounts of memory to store the characteristic sound of each instrument.
2. *Additive synthesis* (also known as *Fourier synthesis*) uses the sinusoidal functions we've put in our digital band as building blocks to create sounds. Each characteristic sound is created by starting with a cosine signal whose period matches that of the note we want to play. We then improve upon this signal by adding more sinusoids at higher frequencies so that the shape of the signal follows that of the desired sound more closely. This synthesis method was used in most early music synthesizers, although it is less popular in modern-day synthesizers.

Waveform Synthesis Suppose that we know the characteristic signal $p_A(t)$ of an instrument for one note with a fundamental frequency f_A and a corresponding period $T_A = 1/f_A$. We would like to create that same instrument sound for a new note with a fundamental frequency f_{new}.

One way to change the period T in a periodic signal is by time scaling the signal as described in Equation (2.3). With time scaling, a signal is "stretched" or "compressed" by expanding or shrinking, respectively, the period of the sound without changing the signal's overall shape. Then the new signal $p_{\text{new}}(t)$, with new period T_{new}, can be created from the original periodic signal as follows:

$$p_{\text{new}}(t) = p_A\left(\frac{T_A}{T_{\text{new}}}t\right) = p_A\left(\frac{f_{\text{new}}}{f_A}t\right) \tag{2.13}$$

Either the ratio of the periods or the ratio of the frequencies can be used to scale the time variable in this calculation. A block diagram for waveform synthesis using time scaling is shown in Figure 2.36.

Figure 2.36 Block diagram for time scaling of signals.

EXAMPLE 2.6 Using Time Scaling to Create Mathematical Expressions

Figure 2.37(a) shows the characteristic signal $p_{250}(t)$ for a new instrument being played at a fundamental frequency of 250 Hz over 8 ms. Use time scaling to create mathematical expressions for signals from this instrument when it plays notes with fundamental frequencies of 300 Hz and 400 Hz, respectively. Plot the signals for both notes over 8 ms, and plot the 250-Hz signal on each plot. Also, plot the 300-Hz signal over 6.6 ms, and plot the 400-Hz signal over 5 ms. How do these last two plots compare with that of $p_{250}(t)$?

Solution

Using Equation (2.13), we can write the following expressions for the new periodic functions:

$$p_{300}(t) = p_{250}((300/250)t) = p_{250}(1.2t)$$
$$p_{400}(t) = p_{250}((400/250)t) = p_{250}(1.6t)$$

We can check that these two answers are correct by first noting that $p_{250}(4\text{ ms}) = p_{250}(0)$, because the period of the original signal is $\frac{1}{250}\text{ Hz} = 4\text{ ms}$. Since the 400-Hz signal has a period of 2.5 ms, we must have $p_{400}(2.5)\text{ms} = p_{400}(0)$. Using the expression above, $p_{400}(2.5\text{ ms}) = p_{250}(1.6 \times 2.5\text{ ms}) = p_{250}(4\text{ ms}) = p_{250}(0) = p_{400}(0)$. Therefore, our expression for $p_{400}(t)$ is periodic with a period of 2.5 ms. A similar check can be made for $p_{300}(t)$.

All of the requested plots are shown in Figure 2.37 (b–e). When plotted over the same time scale as that for the original signal, $p_{250}(t)$, the new signals clearly have different periods. All of the signals look similar, however, when plotted over time intervals that are scaled according to their periods in (c) and (e).

Using waveform synthesis to make sound requires a technique for specifying the function $p_A(t)$ with a fundamental frequency f_A. We could develop a database of recorded waveforms and load each one in when it is needed. An even simpler way would be to draw the waveform on the computer, given a hard-copy picture of it. The next Infinity Project Experiment allows us to try our hand at using waveform synthesis to make sounds and to hear the music that it creates.

Additive Synthesis Engineers are always looking for new and useful ways to do things. Waveform synthesis is just one way to realistically re-create the sounds of instruments. Are there other ways? In particular, is there any way we can use the sinusoid as a simple building block for creating interesting sounds?

A mathematical technique called *additive synthesis* is another way to create periodic signals that closely match those of instruments. Additive synthesis is built on the following fundamental concept:

The shape of any periodic signal can be approximated to any desired accuracy by adding together enough different sines and cosines with the right amplitudes, frequencies, and time delays.

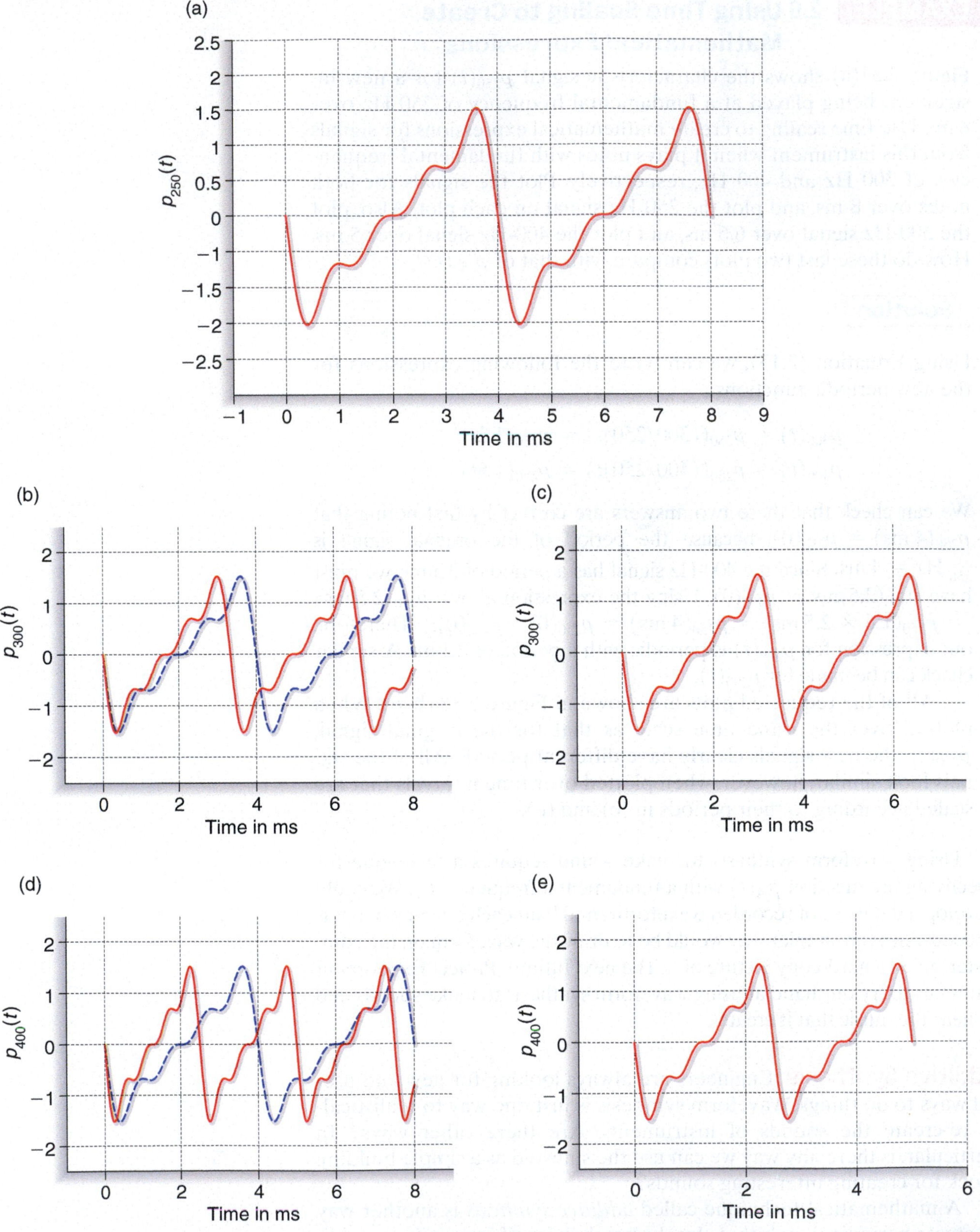

Figure 2.37 Two new periodic signals, $p_{300}(t)$ and $p_{400}(t)$, are created from $p_{250}(t)$ by scaling the time variable t. In (b) and (d), the two new signals are plotted for 8 ms on the same axes as the original $p_{250}(t)$, which is shown by the dashed blue plot. The effects of the time scaling are easily noticed. In (c) and (e), each of the new signals is plotted for two periods, which are 6.6 ms and 5.0 ms, respectively. Since just the time scales are different, the new signals look exactly like the original signal $p_{250}(t)$ plotted for two periods.

Infinity Project Experiment: SketchWave with MIDI

Sketch Wave is a system that allows you to sketch one period of the function $p(t)$ and hear the resulting sound $s(t)$ while it is being created. Try sketching different periodic parts and listening to the music as the periods of the signals are being changed by the MIDI block. Then try to imitate the waveforms shown in Figures 2.12, and 2.15. Can you re-create the sounds of these instruments to make the music you want to hear?

This idea allows us to construct a periodic signal such as that produced by an instrument, using the same building blocks that we used to make sums of sinusoids in the last section. Instead of using the frequencies, amplitudes, and note on–off commands defined by the score, we can use the frequencies, amplitudes, and time delays to re-create the periodic signal of any instrument. The values of the frequencies and amplitudes for a periodic signal can be calculated using a spectrum analyzer. An example shows how this process works.

EXAMPLE 2.7 Finding Amplitudes and Frequencies of Sinusoids in Periodic Signals

Figure 2.38 shows four simple periodic signals on the left and their corresponding spectra on the right. For each periodic signal, write the list of the amplitudes and frequencies of its sinusoids. Find the period of each signal and identify its fundamental frequency.

Solution

(a) The spectrum analyzer shows an amplitude of 1 at both 200 Hz and 400 Hz. The period is $T = 5$ ms $= 0.005$ s, so the fundamental frequency is 200 Hz.

(b) The spectrum analyzer shows an amplitude of 1 at 400 Hz and an amplitude of 0.5 at 1200 Hz. The period is $T = 2.5$ ms $= 0.0025$ s, so the fundamental frequency is 400 Hz.

(c) The spectrum analyzer shows an amplitude of 1 at 300 Hz, 900 Hz, and 1500 Hz. Three periods take 10 ms, so the period is $T = 3.33$ ms $= 0.00333$ s. The fundamental frequency is 300 Hz.

(d) The spectrum analyzer shows an amplitude of 1 at both 200 Hz and 300 Hz. The period is $T = 10$ ms $= 0.1$ s, so the fundamental frequency is 100 Hz.

Figure 2.39 shows how additive synthesis works. Part (a) shows a plot of a saxophone sound in blue. The signal is periodic with period 3.8 ms and a sinusoid with that period is shown with a dashed red plot. In (b)–(d), we have attempted to reconstruct the shape of the saxophone sound, using sums of 2, 4, and 11 sinusoids, respectively. The sum of sinusoids is shown in blue and the dashed red plot of the true signal is also shown for reference. As we add more sinusoids together, we see that the signal produced by this sum begins to take on the shape of the original

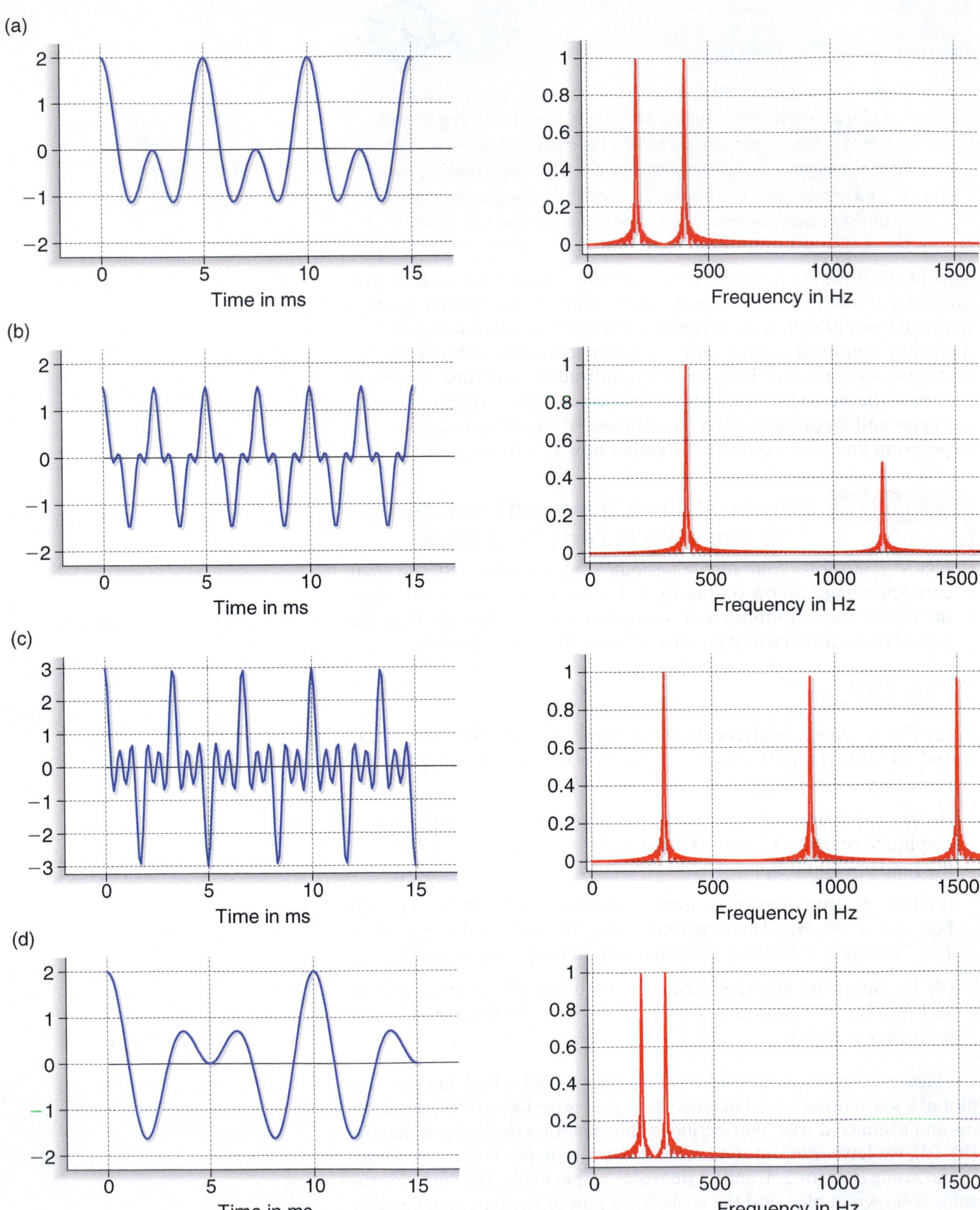

Figure 2.38 Spectrum-analyzer information is shown on the right for the four different periodic signals shown on the left.

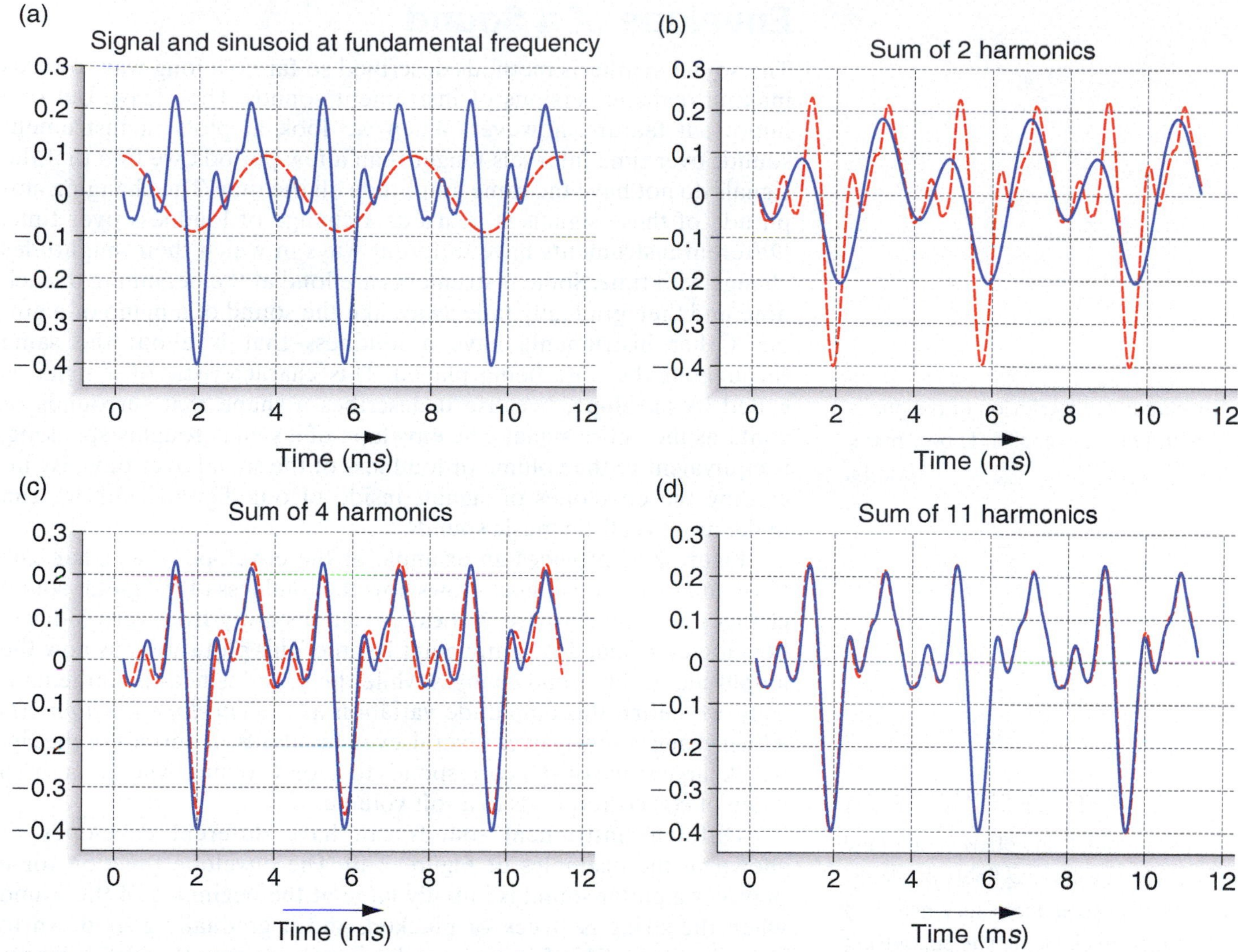

Figure 2.39 In (a), three cycles of a saxophone waveform are shown in blue, with the sinusoid at the fundamental frequency shown in red. In (b), the sum of the first 2 harmonics is shown. The sum of the first 4 harmonics in (c) is much closer to the correct signal shape, and the sum of 11 harmonics in (d) is almost an exact match.

saxophone sound. By adding enough sinusoids together, the synthesized sound becomes so much like the original one that our ears can't tell the difference.

Additive synthesis is a great way to make music because it uses simple blocks—sine generators and adders—in a simple way. Our ability to make sound by using additive synthesis is always getting better, because the number of sine generators and adders we can implement in a single device is always increasing, due to Moore's law. Despite these apparent advantages, additive synthesis is not as popular a method for re-creating audio as waveform synthesis. The reason is that the amplitudes, frequencies, and time delays needed to make really good instrument sounds change from time instant to time instant. Specifying and controlling the values of all of these mathematical quantities is hard. For this reason alone, additive synthesis is not often used in modern electronic musical synthesizers today.

Envelope of a Sound

The sound-synthesis methods described so far go a long way towards making realistic versions of instrument sounds. They leave out one important feature, however. When we look at plots of instrument sounds over time intervals longer than a few periods, we find that the signals do not have the same amplitude all the time. The changing amplitude of these signals is heard as a change of loudness over time. Different instruments have different ways in which their amplitudes change with time. Some instruments are loud at the beginning of each note and then gradually fade away, like the sound of a piano or a guitar. Other instruments have a loudness that is about the same throughout the note being played. This characteristic of a signal is called its **envelope**, because it describes a shape that surrounds or contains the entire signal. The envelope of a signal, roughly speaking, is equivalent to the volume or loudness of the signal over time. By including the envelopes of signals inside of our digital band, we can make more realistic music sounds.

Envelope: A description of a signal's general size or amplitude over many periods.

Figure 2.12 provides an example of the envelope of a signal. The left-hand side of this signal shows how the loudness of the guitar sound changes with time, where vertical deviations away from zero indicate how loud the sound is at any point in time. This plot shows us how the amplitude of the sound changes while the sound is heard. Mathematically, we define this amplitude variation as the *envelope function* $e(t)$ which changes slowly over several, even hundreds, of periods of the signal. A large value of $e(t)$ corresponds to a loud volume, whereas a small value of $e(t)$ corresponds to a soft volume.

INTERESTING FACT:

Exponential functions show up in many different problems and fields of study, from biology to economics. When food and space are plentiful, the population of a living species in a particular environment grows exponentially. If you put your money in a savings account, it also grows exponentially. Unfortunately so does your debt if you borrow money from the bank.

Different instrument sounds can have different envelopes, as shown in the examples of Figure 2.40. The envelope function for a piano or a guitar sound is initially large at the beginning of the sound when the string is struck or plucked, and it gradually goes down to zero over time. This function can be written as $\exp(t) = 10^{-at}$, where the value of a controls how fast the sound fades away. In Figure 2.40(a) and (b), we used the values of $a = 1$ and $a = 2$, respectively. The envelope of a trumpet or trombone sound typically looks fairly constant while the note is being played. The envelope of a violin or cello sound could start with a gradual increase to a particular point and then decrease steadily back down to zero. The envelope of a clarinet sound might start out large and then get softer as the clarinetist plays. All of these examples are approximate envelopes for these instruments, but we could measure more accurate ones from the actual sounds if we wanted to.

In our digital-band design, we would like to re-create the sounds of real instruments by applying the envelope function $e(t)$ to the periodic function $p(t)$. This process is relatively straightforward if we recall that $e(t)$ is nothing more than an amplitude that changes with time. All we have to do is scale the periodic function by the envelope function. A very good approximation to a musical signal is given as

$$s(t) = e(t) \times p(t) \tag{2.14}$$

where $e(t)$ is the time-varying description of the signal's amplitude and $p(t)$ contains the underlying periodic structure of the signal. In this

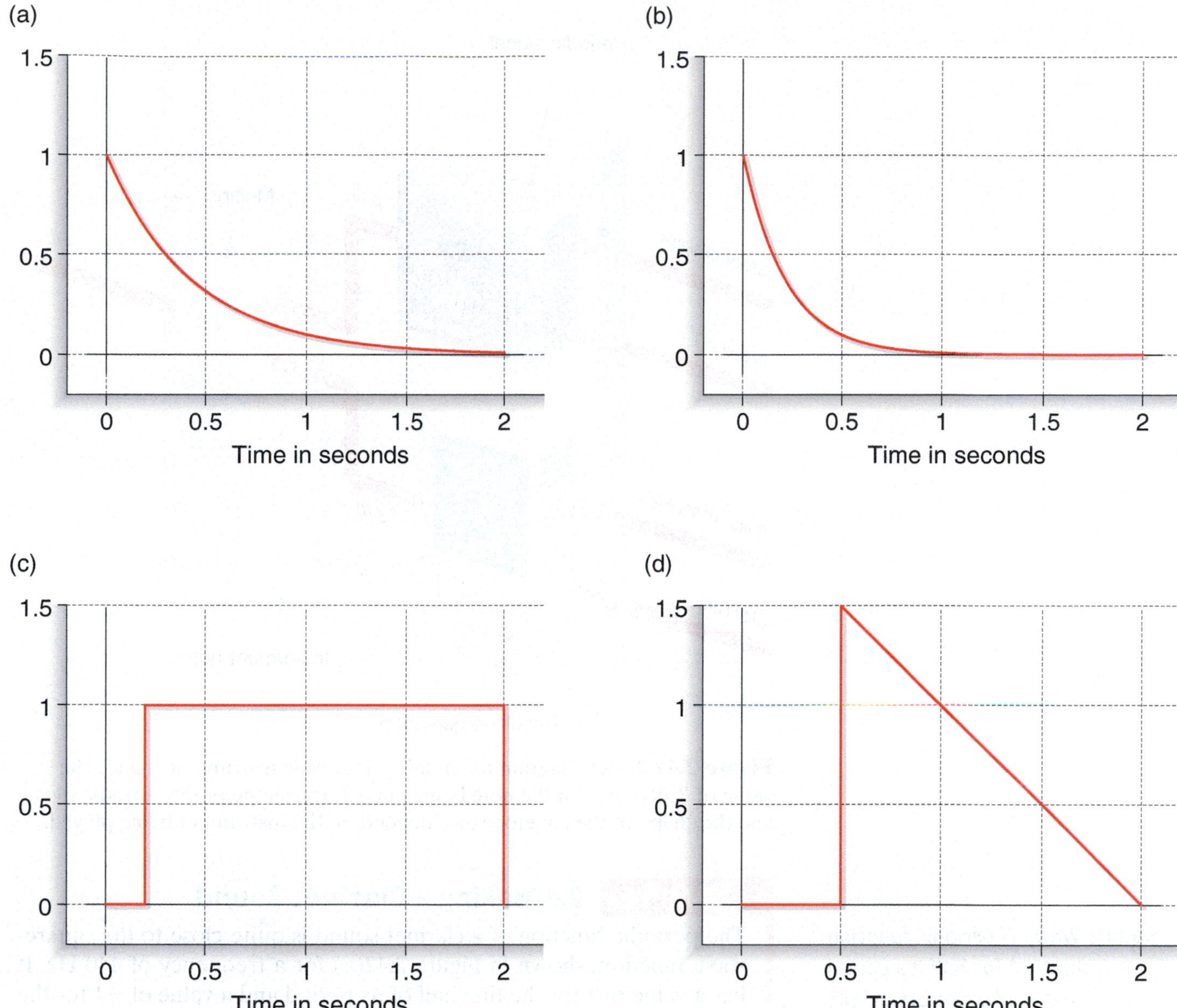

Figure 2.40 Four envelope signals. In (a) and (b), the exponentially decaying envelopes are described by 10^{-t} and $10 - 2t$, respectively. These envelopes are good models for stringed instruments. In (c), the envelope of a trumpet is modeled as a constant amplitude for 1.8 s. In (d), a clarinet envelope is modeled as a constant decrease in amplitude for 1.5 s.

case, the product of the two signals $e(t)$ and $p(t)$ is taken at each time instant. For example, if at time $t = 1.5$ s the values of $e(t)$ and $p(t)$ are $e(1.5) = 0.4$ and $p(1.5) = -0.3$, then the value of $s(t)$ at $t = 1.5$ s is

$$\begin{aligned} s(1.5) &= e(1.5) \times p(1.5) \\ &= 0.4 \times (-0.3) \\ &= -0.12 \end{aligned}$$

Figure 2.41 shows a block diagram for the creation of a realistic note sound, using both the envelope and the periodic signal of a instrument sound. Now that we understand this process, we can create realistic sounds for several musical instruments.

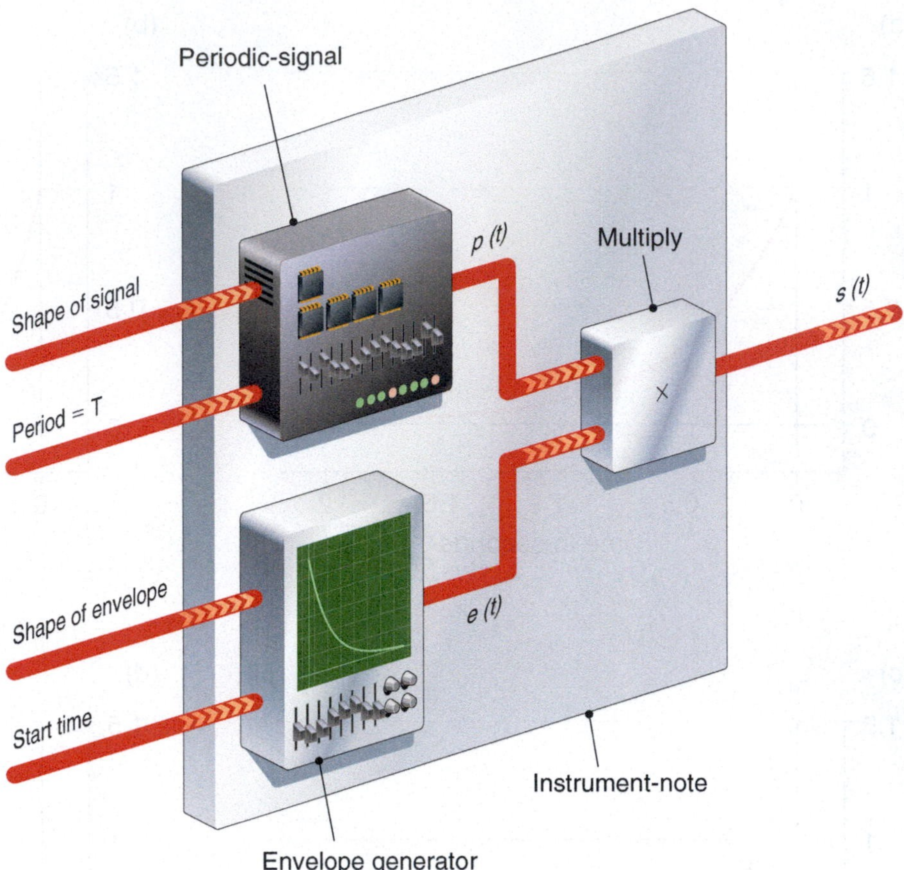

Figure 2.41 Block diagram for creating a realistic instrument sound. The value of T depends on the note being played. The shape of the periodic signal and the shape of the envelope also depend on the instrument being played.

EXAMPLE 2.8 Making a Clarinet Sound

The periodic function of a clarinet sound is quite close to the **square-wave** function, shown in Figure 2.42(a) for a frequency of 440 Hz. It has a value of 1 for the first half of its period and a value of −1 for the second half of its period. The envelope of a clarinet sound depends on how the clarinetist plays the instrument and, more specifically, how much air the clarinetist blows through the instrument. Suppose that a particular clarinetist creates a sound with an envelope given by

$$e(t) = \begin{cases} 0, & t < 0.5\,\text{s} \\ 2 - t, & 0.5\,\text{s} < t < 2\,\text{s} \\ 0, & t > 2\,\text{s} \end{cases}$$

The value of $e(t)$ is plotted in Figure 2.40(d). The combined signal $s(t)$ is computed using Equation (2.14) and is plotted in Figure 2.42(c). Over two seconds, the function $p(t)$ repeats 880 times, so we cannot see the individual periods of $p(t)$ in this plot. If we were to zoom in on any small region of the plot, $p(t)$ would look very much like Figure 2.42(a) with a different amplitude, because $e(t)$ changes very little over a few periods. To get a better sense of what is going on, Figure 2.42(b) and (d) show the same set of plots for a $p(t)$ signal with a fundamental frequency of 2 Hz. We could not hear such a low frequency, but the plots show how the envelope changes the size of the periodic signal.

Square Wave: A periodic function that is 1 for half its period and −1 for the other half.

INTERESTING FACT:

Our perception of the sound of an instrument is affected by its initial *attack*, or the way the note first begins. Instruments that are excited using blown air, such as flutes, clarinets, saxophones, trumpets, and the like, generally have more gradual attacks than pianos or guitars. For this reason, $e(t)$ increases gradually at first for these instruments.

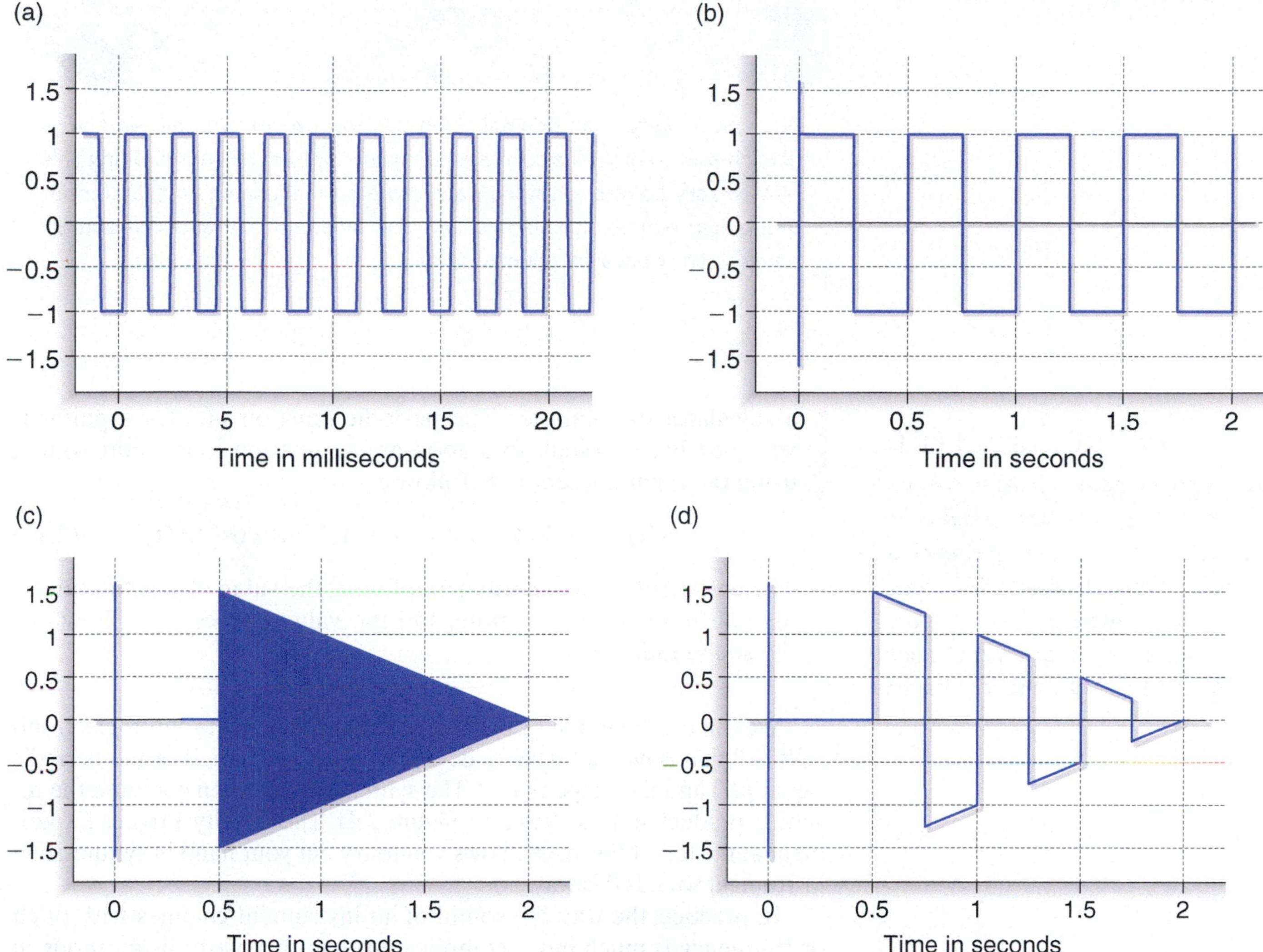

Figure 2.42 Plots of clarinet note with periodic function and envelope. In (a), the square wave at a fundamental frequency of 440 Hz is shown for 25 ms. In (c), the square wave is multiplied by the envelope function from Figure 2.40(d) and displayed for 2 s. In (b), a square wave with a fundamental frequency of 2 Hz is shown, and in (d) the same wave is shown multiplied by the same envelope function.

EXAMPLE 2.9 Making a Guitar Sound

When you listen to someone pluck a guitar string, you can hear the sharp attack of the sound followed by a slow decay to silence. This envelope is very well modeled by the **exponential function**

$$e(t) = B^{-at}$$

for $t \geq 0$, where B can be any constant value. The positive constant a describes the **rate of decay** of the function over time. This type of envelope is shown in Figure 2.40(a) and (b) for $B = 1.0$ and $a = 1$ and 2.

A larger value of a will cause a faster decay, making the guitar sound more like a ukelele or banjo. A smaller value of a will cause a slower decay, making a sound more like that of a bass guitar. Exponential functions show up in many real-world problems in engineering, physics, and mathematics, and they can be easily computed with

Exponential Function: A waveform that grows or decays at a constant rate.

Rate of decay: A number that describes how fast an exponential function decreases over time.

Infinity Project Experiment: SketchWave with Envelope Functions

We can modify our original SketchWave experiment by adding a sketch-pad to draw the corresponding envelope of the sound. Using this system, try to re-create the waveforms corresponding to the clarinet and guitar sounds of this section. How close do the resulting sounds come to their physical counterparts?

INTERESTING FACT:

The small changes in a instrument's signal shape over time are due to many factors. For example, when a saxophone is played, slight differences in the lip pressure and air pressure being exerted at the mouthpiece cause its sound signal to change significantly.

a calculator or computer. The periodic function $p(t)$ for a guitar is very nearly sinusoidal, so a good approximation to a guitar sound, using three parameters, is as follows:

$$s(t) = e(t) \times p(t) = A \times 10^{-at} \times \cos(2\pi f t) \qquad (2.15)$$

The parameter A is the initial amplitude, the value of f is the fundamental frequency of the note, and the value of a controls how fast the sound fades away.

We can use Equation (2.14) to synthesize instrument sounds electronically. All that is needed is some device to input one period of the periodic signal $p(t)$ and the shape of $e(t)$. The synthesizer can then compute the resulting product $s(t)$, as shown in Figure 2.41. The Infinity Project Experiment at the top of this page allows you to try out your hand in synthesizing instrument sounds this way.

In practice, the way the sound of an instrument changes with pitch or frequency is much more complicated than the synthesis methods in this section will allow us to describe. Most signals from instruments have shapes that change a little for each note and within each note while it is playing. Even so, the signals we have created sound quite similar to those of instruments, and they are an improvement over the sinusoids we were using before. More importantly, we now have ways to tailor the sound of our digital band to make more interesting-sounding music. We can now make reasonable approximations of many instrument sounds. We can then create these sounds from different notes and instruments, using the note-generating blocks defined by Figure 2.41. Adding together all of the outputs of these blocks produces the music that we want our digital band to play.

Infinity Project Experiment: Echo Generator

This experiment allows you to see the effect of echo on both recorded music and your own voice. Experiment with different echo amplitudes and delays. What happens when you make the echo amplitude larger than one? Does this sound seem natural?

INTERESTING APPLICATION

Sound Effects

Making great-sounding music doesn't end with the signals that each instrument produces. The room in which music is played alters the music's sound in a subtle, but important, way. A rock guitarist trying to create a new sound might turn up the loudness setting on his electric guitar until it starts distorting. We use the generic term "sound effects" to describe all of the changes that we can make to a sound after it has been recorded or while it is being played. We create sound effects by mathematically manipulating the sound signal before it is played.

Examples of some sound effects are described as follows:

—*Reverberation*, or "reverb," as it is sometimes called, is caused by reflected sound that we hear when we are in a room or concert hall. Reverberation is what makes us sound so good when we sing in the shower. It also is used by professional recording artists to make their tracks sound as if they are played in a concert hall or other room with good acoustics..

—*Echo* is caused by reflected sound that returns after traveling a large distance. Echo can be simulated electronically using simple digital devices. It can provide some really cool effects to a singer's voice or an instrument's sound.

—*Flanging* is a sound effect that is created by a single echo whose delay changes with time. Flanging is often used by electric guitar players to make the sound of their instruments "fatter" or "more alive."

Audio engineers constantly are coming up with new and interesting ways to make great sound effects. Who knows—you might experience some great sound effect the next time you go see a movie, listen to a CD, or play music with your friends!

EXERCISES 2.4

Mastering the Concepts

1. What are the two most popular methods for synthesizing sound? How are they similar? How are they different?
2. How is time scaling used in waveform synthesis? What do you need to know in order to change the frequency of a waveform?
3. Is it possible to add different sinusoids together and still make a periodic signal? How can this idea be used to make audio signals?
4. Write down the expression for synthesizing sounds by using additive synthesis. Can you describe what this equation is doing, in simple words?
5. What is the envelope of a sound? Why is it important in making signals?
6. Give three different examples of envelope functions and how they are used to make signals.

Try This

7. Figure 2.43 shows the plots of two signals $s_1(t)$ and $s_2(t)$. What is the period of $s_1(t)$? If $s_2(t)$ has a period of 2 s, plot it for $0 \le t \le 6$ s.
8. Suppose we want to convert the frequency of a periodic signal corresponding to a D above middle C (293.7 Hz) to that of the B below middle C (246.9 Hz). What time-scaling factor is required?
9. Suppose a periodic signal with period $T = 0.001$ s is to be converted to a signal with a frequency of $f = 600$ Hz. What time-scaling factor is required?

10. Suppose the envelope of a particular sound starts out loud with a value of 1.0 at time $t = 0$ and then decreases linearly to 0 over 4 s. Plot the envelope of the sound. Also, assuming that the period of the periodic signal for this sound is much smaller than 1 s, sketch the overall shape of the signal over 4 s.
11. After getting home from school, you turn on the radio. Somebody else left the radio volume on high, so it blasts a loud sound for 2 s before you turn it down to half its original volume. You then listen to another 60 s of music before turning it off to go do something else. Draw the envelope of the sound of the radio, where $t = 0$ indicates the time that you turned on the radio.

In the Laboratory

12. Say any word in your normal voice that has an "eee" sound in it, such as "weave." Capture your sound, using the equipment in the laboratory, and measure the period of the "eee" sound. How does the period of your sound compare to that of other students in the class?

Back of the Envelope

13. Can the envelope of a sound be periodic as well? Give an example of a signal with a periodic envelope.
14. Suppose you are standing on an highway, and a single truck appears on the horizon traveling toward you. The sound of its large diesel engine gets louder and louder as it comes toward you, and after 30 s, it races by you. Thirty seconds later, it disappears over the horizon again. Sketch the envelope of the truck sound that you heard, labeling the time axis carefully. Take an educated guess as to the shape of the envelope over the first 30 s (you need to know more about the physics of sound and the way the truck is traveling to get more specific), but sketch carefully the envelope of the latter 30 s.

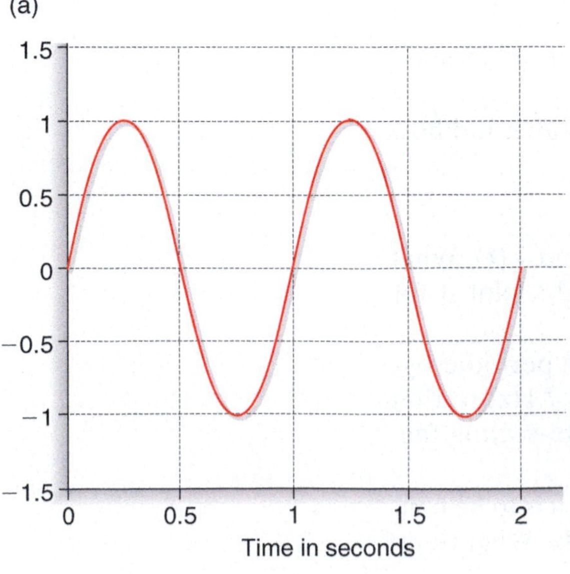

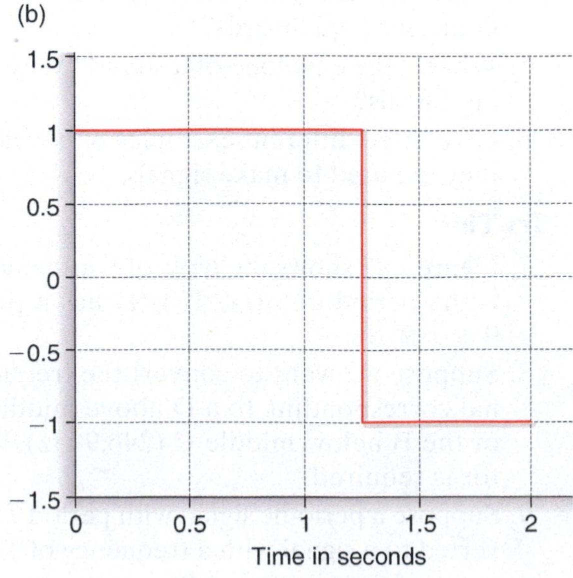

Figure 2.43 Plots of two signals.

2.5 Using Our Ability to Create Signals for New Applications

With the increasing computational power and the shrinking size of modern electronics, the closet-sized music synthesizer Max Mathews used in 1970 could now easily fit in the palm of your hand and still have far more capability. With this computational power and the methods for custom-designed signals we have developed for music creation, we can now explore how to use the same approach to create signals for new applications.

We have learned how to use a mathematically simple signal, a sinusoid, to create music sounds from any note of any instrument. We can also create notes that are not heard on traditional instruments and the sounds of new instruments that have never existed. All of this is possible with a few very simple mathematical operations. We can adjust the three parameters of the sinusoid, its amplitude, its period or frequency, and its time offset. We can add several different sinusoids together. We can turn the signals on and off with envelope functions.

These same methods and sinusoids can be used to create signals for a wide variety of other applications. Creating speech in any language from written text uses many of the same methods used for music synthesis, since for speech the human vocal tract is the instrument. The center directed by Max Mathews at Bell Laboratories also worked on speech processing. Sinusoids are also used to create signals for modern communications systems such as modems used by personal computers. Image creation and control in medical imaging systems is based on sinusoidal signals. In engineering, the techniques developed to solve a particular problem are often found to be useful in many other areas as well.

Master Design Problem

If you have gone to a nightspot or party place on a Friday or Saturday night, you might have participated in *karaoke*. A karaoke system allows you to add your singing voice to a wide selection of well-known music. Unfortunately, for many of us, the quality of the music that is produced is often not even close to what we might hope for. Perhaps we can design a new and improved karaoke machine—an "Ultimate Karaoke Machine"—that would improve the quality of any singing voice fed to it, with choices as to how the music and voice would sound. Then nightspots with the Ultimate Karaoke Machine would be more attractive to party goers than those without, because the music heard at these nightspots would sound better. Since advances in technology steadily give us more capabilities in a device at a lower cost and in a smaller size, we can also consider designing a portable or backpack version of our Ultimate Karaoke Machine for our own use wherever and whenever we want.

INTERESTING FACT:

"Karaoke" is a Japanese word formed from the combinations of "karappo," which means empty, and "oke," which is an abbreviated form of "okesutura," or orchestra.

We can start our design by thinking of all the features we might like to have in our Ultimate Karaoke Machine. This list of features may be different depending on how the machine is used, who will buy it, and who will use it. For example, a full-featured machine could be designed for nightspots, whereas a more streamlined and less expensive machine could be designed for home use. After creating a list of features, we can examine the list to see how much of the technology already exists and how much has to be created. For example, we might choose to use the digital band we have just designed, or perhaps just parts of it. We can also choose specific technologies on which to create new blocks for our machine. For example, we might use our spectrum analyzer within the design of a device that improves a piece of music by modifying its description.

The Ultimate Karaoke Machine for the Home

What should our more basic home version of the Ultimate Karaoke Machine do? The most important feature of this version would be the capability to help any singer sing the correct notes of a song at the correct times. Before we explore such features, it helps to figure out how singers would use the system. The system should be easy to use, yet have enough features to appeal to a wide range of singers with different abilities. We should include the following functions in our system:

User Control: The machine should provide some control over the amount of voice correction and enhancement that is applied to the singer's voice. Some singers would want to turn off all possible changes to their voice. Others might want to be able to vary their voice around the proper pitch for artistic expression and correct their voice only when their pitch is way off. Still others might want the system simply to hold their sung notes for the right amount of time. Many users would want the system to completely correct their voice so that they always sing in tune and with the right pitches. If designed properly, this device could be used for other purposes, such as the training of choirs to help their members improve their singing abilities.

User Input: As in all karaoke machines, the singer will need to hear the music that she or he will sing along with. If the singer's voice needs to be corrected, however, the machine will have to "listen" to the singer for a short time to see how to correct her or his voice. The system could become confused if it "hears" the music accompaniment instead of the singer's voice, so we will need to figure out a way to prevent this problem. And if our machine is used for training a singer, how would it let the singer know how to correct her or his singing?

Other Inputs: What other inputs should the system have? For example, would we have a ready-made library of music to sing along with?

Can we store and download the music from the Internet? Should the music come from our digital band? Can we use an audio recording from a CD or DVD?

Technology Design: Now that we have a big picture and a sense of the scope of our design, we next look at the technology we can use for improving a singer's voice. We know how to create a periodic signal of a instrument. In the Ultimate Karaoke Machine, the instrument is the voice-production system of the singer, made up of the vocal chords, throat, mouth, and head cavity. This system is always changing, depending on what words and syllables are being sung. Can we use a spectrum analyzer to help figure out how these sounds are being produced? We might use a spectrum analyzer to determine the pitch of the singer's voice and calculate how close it is to the correct pitch from the recorded music.

One possible design for the Ultimate Karaoke Machine could use our digital band to create a sound signal, with the singer's voice controlling the shape of the periodic signal and the music controlling the frequency and duration of each note. Such a system would correct every note sung. How would we change this design so that it controls only the duration of the notes? How would we let the singer know if the pitch is wrong and should be made higher or lower?

Exercise Make a complete block diagram for this basic system. Be sure to include all user inputs, all music inputs, and all outputs. For each of the internal blocks of the basic diagram, make a more detailed block diagram.

The Ultimate Karaoke System for Nightspots

For a business whose goal is to attract party goers, an Ultimate Karaoke Machine could be more expensive and somewhat larger than the home version. What features would we like to add to our home-system design for this version? Some possible enhancements to our existing design are given as follows:

- Play recorded music containing several singers as background, and let the user specify which of the singers' voices should be removed so that the user's voice may be added.
- Allow the option of automatic pitch correction to adjust the frequency of the user's voice to the nearest frequency shown on the keyboard in Figure 2.17 instead of to the frequency specified by the music.
- Modify the user's voice to sound more like a particular well-known singer such as Elvis Presley, Tina Turner, or whomever the user specifies.

- Modify the user's voice to sing in harmony with himself or herself.
- Modify the user's voice to sing an octave higher or lower than he or she is singing.
- Use the user's voice as the periodic signal for adding a new instrument to the total musical sound. The user's voice would replace the MIDI files or the music score to create music for the new instrument.

Are there anymore features that can be added to make the system more fun to use or easier to sell?

Now that we have a list of possible additional features, think of the cost and complexity of each, and decide which features should be in both systems and which should be features of the nightspot system only. You should include all of the features of the home system in the nightspot system. Some of the important issues to consider in making your choices include the following:

- This system could use any and all types of musical inputs that we have discussed so far, including MIDI files, prerecorded music, and live sound as captured by a microphone. Which of these inputs will you use? How many of each can be used at the same time?
- What kind of music libraries will you need to have for this design? If you wanted to enable the user to sound like Elvis Presley, Tina Turner, or any other well-known singer, you would need to know how to make the signals those singers' voices produced. How would you obtain this information?
- How would you interact with the system if it had all these new features? How would you tell it what instruments to create or how to modify your voice?
- How would you like to save the results of your music creation in order to hear later or send to others? Will you also be able to save the way you set your system up so that, at some future time, you can make new music in the same way?

After thinking about these issues, draw a basic block diagram for both the home and nightspot version of the Ultimate Karaoke Machine. Define what each block does, and break each block into smaller blocks to show how each one works. Since the nightspot version will be quite complex, it is best to have the members of a design team work on it together. After you have agreed on the highest-level block diagram, each individual team member can take responsibility for a different block and work on the detailed design of that block.

Big Ideas

Math and Science Concepts Learned

Any device or system for making music creates a complex sound that can be characterized by signals. A signal is a pattern of variation over time that contains information. Signals are everywhere. They can be used to describe physical phenomena, such as sound waves in the air, the temperature over a single day, and voltages inside of our bodies. We also communicate using radio signals that are transmitted through the air and electrical signals that are transmitted through wires.

Sound is a physical wave or disturbance that travels through air by moving air molecules back and forth. Sound doesn't travel as fast as light does; it takes sound about 3 s to travel 1 km, or 5 s to travel 1 mile. Sound can be converted into an electrical signal by using a microphone. We can also create sound by making an electronic signal and then converting it to sound waves, using a loudspeaker.

Signals are functions that can be plotted easily. Once described, signals can also be manipulated in simple ways. Some of the ways a signal can be manipulated include scaling the amplitude, shifting the time, and scaling the time. These manipulations are especially useful for making music.

When a musical instrument plays a single note, its corresponding sound signal has an inherent structure. The signal that it produces is periodic—repeating over and over—and this period is tied to the pitch that we hear. The numerical equivalent of pitch is the fundamental frequency of the sound. The fundamental frequency f and the period T are inversely related; that is, $f = 1/T$.

Simple melodies are made up of notes played one after the other. Sheet music describes the pitches and durations of the notes being played. Engineers have come up with a computer language—MIDI—to describe such information so that it can be easily stored, transmitted, and used electronically.

One of the simplest and most important periodic functions is the sinusoidal function. This function is generated using the sine and cosine functions from trigonometry by making the angle of these functions dependent on time. The units of angle can be given either in degrees or radians. A tuning fork makes a nearly sinusoidal sound. We can make simple melodies with sinusoids by changing the frequency of the sinusoidal function according to the frequencies in the melody. If we want to play more than one note at the same time, we simply add the signals corresponding to the individual notes being played. This simple technique produces recognizable music to our ears. We can also reverse engineer the score from music created in this way, using a computational tool called the spectrogram. A spectrum analyzer computes and plots the amplitudes and frequencies of the sinusoids within a short segment of a signal.

To make better music, we can employ more complicated sound-synthesis methods to create more interesting instrument sounds. These methods create the periodic shape of a instrument signal in different ways. Waveform synthesis uses a single period of a chosen periodic signal and time scales it so that its fundamental frequency matches that of the desired note. Additive synthesis uses the spectral content of the sound to create a version of the sound signal by summing sinusoidal signals together. The frequencies of these sinusoidal signals are multiples of the fundamental frequency and can be controlled by the fundamental-frequency value. Additive synthesis can be used to approximate any signal, and we have a choice as to how many sinusoidal functions to use in this approximation. Each of these periodic signals can then be combined with the envelope function, which describes the slowly varying amplitude of a sound from beginning to end.

Sound effects are simple manipulations of sound signals to make them sound more interesting or more realistic. Some common sound effects include reverberation, echo and flanging.

Important Equations

There are three basic ways to change a sound signal. The changes in mathematical form are directly related to our perception of the loudness of the signal, the time the signal starts, and the pitch of the signal.

Scaling the amplitude of a signal $s(t)$:

$$x(t) = A \times s(t)$$

Shifting the time of a signal $s(t)$:

$$y(t) = s(t + d)$$

Scaling the time axis of a signal $s(t)$:

$$z(t) = s(ct)$$

Most of the sound we hear when we listen to music or speech is made from periodic functions. Periodic functions also describe mechanical vibrations, planetary motion, visible light, and man-made communications signals. The most basic periodic functions are sinusoidal signals.

Property of a periodic signal where T = period of the signal in seconds:

$$p(t) = p(t + T)$$

Relationship between frequency and period, where f = frequency of the signal in Hz:

$$f = 1/T \quad \text{or} \quad T = 1/f$$

Relationship between sine and cosine:

$$\cos(\theta) = \sin(\theta + 90°)$$

Definition of a sinusoidal signal, where A = amplitude:

$$s(t) = A\cos\left(\frac{2\pi}{T}t\right) = A\cos(2\pi f t)$$

Using our methods of changing signals, we can use sinusoidal signals and simple periodic signal to make realistic music sounds.

Waveform synthesis (changing the period of a periodic signal):

$$p_{new} = p_A\left(\frac{T_A}{T_{new}}t\right) = p_A\left(\frac{f_{new}}{f_A}t\right)$$

Using envelope and periodic signals to synthesize sounds:

$$s(t) = e(t) \times p(t)$$

Simple extensions of these equations can be used to model the sound effects discussed briefly at the end of the chapter. One of the most basic effects is adding an echo to a sound signal. An echo signal can be created by scaling the amplitude of a signal and also shifting the signal's time variable. The scale factor, A, should be chosen to be less than 1, so the echo will not be as loud as the original signal. The time shift, d, should be less than 0 so the echo will happen after the original signal.

If $d = -T_{delay}$, then the equation for an echo of the signal $s(t)$ is

$$s_{echo}(t) = A \times s(t - T_{delay})$$

The total signal is the sum of the original signal and the echo which occurs after a delay of T_{delay}.

$$s_{tot}(t) = s(t) + s_{echo}(t) = s(t) + A \times s(t - T_{delay})$$

A signal with a double echo would be:

$$s_{tot}(t) = s(t) + A \times s(t - T_{delay}) + A^2 \times s(t - 2 \times T_{delay})$$

Another interesting sound effect can be created when the envelope function $e(t)$ is replaced by a sinusoidal function, and the periodic function $p(t)$ is replaced by any sound signal $s(t)$. This is called modulation. The most interesting sound effects are heard when f_m, the frequency of the sinusoidal modulation function, is no more than a few hundred Hertz, but for communications systems much higher frequencies are used.

Modulation:

$$s_{mod}(t) = \cos(2\pi f_m t) \times s(t)$$

Making Digital Images

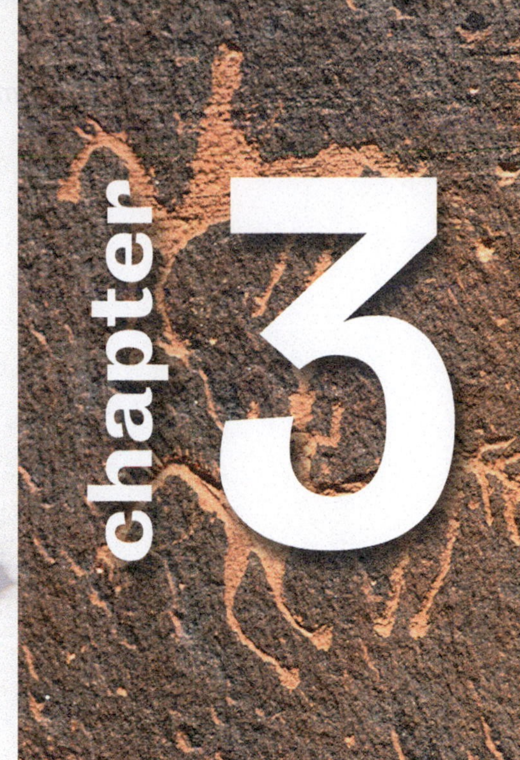

In today's world, most of our daily activities, such as learning, working, shopping, or entertainment, make use of digital technology. The two mediums of still photography and moving images (including video, DVD, film, and television) are key elements that help us interact with the world around us.

In the previous chapter, we learned about creating, recording, and changing sounds. Now we will learn to create, record, and improve images. We will study ways of digitally representing still and moving images. These digital representations allow us to design digital imaging systems for a wide variety of applications in fields ranging from personal entertainment to medical and scientific discovery. In this chapter, we will study the motivation for, benefits of, and mathematics behind digital imaging.

OUTLINE

- Introduction
- Digitizing Images
- Putting It Together
- Better Design within the Bit Budget

Design Objectives for Digital Imaging

Imagine this: You've got tickets to a popular rock concert. They are floor tickets, and you are desperately hoping that you will get close enough to <fill in the name of your favorite rock star!> to make eye contact and maybe even get a wave from him or her. The day arrives and you go to the concert. The stage starts moving around, and, wonder of all wonders, you *do* find yourself within 10 feet of the star and *do* get a wink in response to your hysteria. You come home walking on cloud nine and tell your friends all about it. The problem is that nobody believes you! They want some proof, like a photo! So you confidently tell them to wait until you get your film developed. As you look at your photos from the concert, however, your heart sinks. All the close-ups are too dark and blurred, because you moved or were bumped by the crowd. Some are just too fuzzy. Your friends remain as skeptical as ever. What could you have done differently?

INTERESTING FACT:

NASA's *Voyager I* and *Voyager II* missions were launched in 1977 with the goal of exploring the outer planets of our solar system. Over the past 25 years, these probes have continued to send spectacular images of space, millions of miles away from the Earth.

The National Aeronautics and Space Agency (NASA) has similar problems with its mission to explore Earth and the outer planets of our solar system. It has to design systems to create high-quality images of objects in space that can be hundreds of miles from the camera, which then must be returned to Earth from great distances, ranging upwards of millions of miles. In space, one doesn't get a second chance at doing something right; it has to be right the first time. Both the rock concert and space exploration examples illustrate an age-old problem:

How does one share a visual experience with other human beings?

An engineering approach to any problem begins with a clear definition of the problem and a specification of an ideal solution. At this stage, it is best to look at the problem as if you would be the person using your product or service, not as the person who will do the design. From this general description of the problem, a list of specific design objectives can be written. It is important that these two steps be completed before the actual design process begins, because without a clear set of objectives we might spend a lot of time, energy, and resources doing something that does not help us achieve our goals. This approach also helps us keep our minds open to different types of designs so that we will choose the best one, not the one we thought of first.

- **What problem are we trying to solve?** We need a system to capture visual experience and record it for later use. It must automatically produce images of high quality that can be easily reproduced and sent to other places.
- **What are the specific objectives?** Our new imaging system design should have the following characteristics:
  1. It should acquire images as quickly as possible, with as little operator interaction as possible.
  2. It should be "smart." That is, it should either adjust itself to get the best possible image or get an image that can be adjusted later to fix any mistakes or defects.
  3. The images should be easily duplicated and transmitted to other locations.
  4. It should be inexpensive to buy, use, and maintain.
  5. It should be easy to use.

- **What are the rewards of a successful design?** A successful design would allow us to have a high-quality visual record of events that are important to us so that we can share them with our friends now and perhaps with our grandchildren in the future. It would also allow us to see high-quality images from places that human photographers cannot easily access, such as outer space, the deep ocean floor, or inside a blood vessel. Many people with other imaging applications would find this system useful and perhaps would improve or customize it. Having such a system would also inspire related design efforts, for example, to improve image communications, image storage, or image display.

After the objectives have been clearly stated, the design process can begin. There are several important points to keep in mind:

- The big problem is big. It has to be broken into smaller subtasks that will fit together to make the whole project. This approach allows you to work on small, manageable pieces and allows teams of people to work together.
- The objectives are often competing, and you will have to use some judgment in balancing them. The cheapest solution will probably not produce the highest quality results.
- You should base your design on current and near-future technology, but you should also be aware of related historical developments. Technology moves forward by building on and branching out from existing technology and experience.
- At every point in the design, you should plan a test procedure. It is much easier and more reliable to test small parts and then use them than to try to test a large complex design all at once. When all the parts have been tested, then the full system or project can be tested for a specific application.

We will begin our design process of a better system for taking pictures at rock concerts and during space exploration by looking at ways that images have been created and preserved historically.

3.1 Introduction

Images throughout History

The goal of saving a permanent visual record is not just a desire for rock concert attendees and NASA space probes. Prehistoric humans faced the same objective when they tried to share their latest exploits of a woolly mammoth hunt with their fellow cave dwellers. They picked up some burned sticks and colored chalklike stones and started drawing pictures on the cave walls. The results, such as the cave painting shown in Figure 3.1(a), are there for all of us to see tens of thousands of years later. From these cave paintings, we can begin to understand the life experiences of the people who created them.

As human cultures develop and evolve, so do the methods of creating images and visual experience. For example, painting has progressed significantly in both style and media. Images can also be created using

INTERESTING FACT:

Cave paintings from the time of the Neanderthals can be found throughout central and western Europe. The oldest known site is in southern France and has been determined to be 30,000 years old.

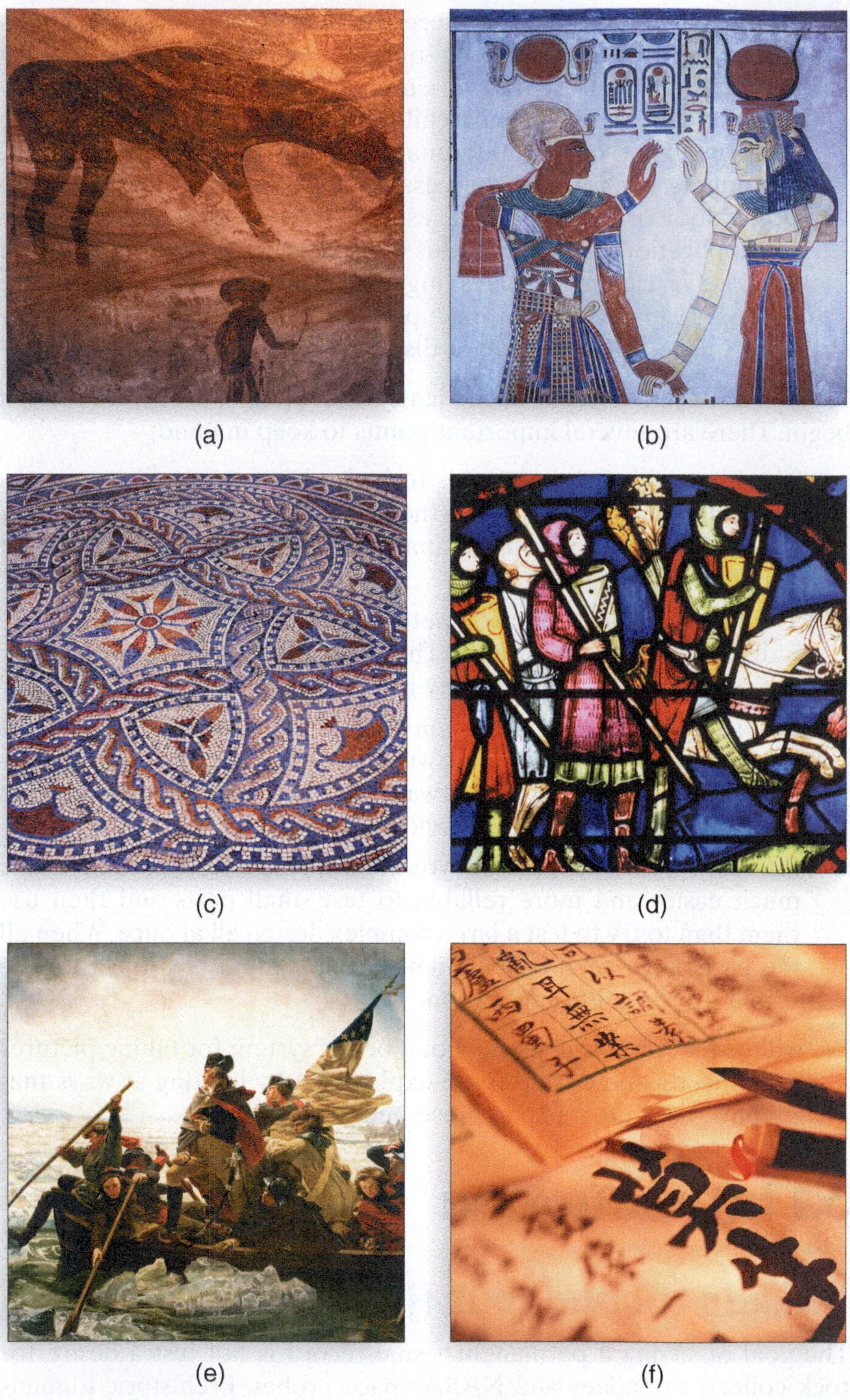

Figure 3.1 Examples of images recording events in different media throughout human history: (a) cave painting, (b) Egyptian painting, (c) mosaic, (d) stained-glass window, (e) painting of General Washington crossing the Delaware River, and (f) Chinese brush painting.

colored sands and tiles. In many cultures, mosaics created from colored tiles have been used to re-create scenes from daily life or from the artist's imagination. Stained-glass windows use colored glass to create both images from life and abstract designs. Figure 3.1(b)–(d) show examples of these image formation methods that convey events or ideas the artists wish to share. As painting techniques became more sophisticated, more detail could be communicated. In Figure 3.1(e), a historical event, General Washington crossing the Delaware River to capture Trenton during the American Revolutionary War, is depicted. It has a much more advanced painting style than the cave paintings, but it has been created for a similar purpose.

The focus throughout history on visual experience is understandable, since we derive so much information about the world through our visual senses. As a result, a large fraction of our brain activity—about half—is dedicated to gathering, processing, and storing visual information. This, in turn, has led to common expressions like "Seeing is believing" or "A picture is worth a thousand words." This accentuates the need to develop technologies for image sensing, processing, and storing in an efficient and flexible manner so that visual experience can be shared for the communication of information and personal enjoyment.

For thousands of years, the technology of image creation was basically the same as that used by our prehistoric cave-dwelling ancestors in that artists interpreted experience and then created a visual representation with paintings, drawings, mosaics, or other media. Some media, such as brush stroke painting, were analog, or continuously varying in physical form, but others, such as mosaic tiles and stained-glass pieces, were digital, because they used discrete units. These methods remained the primary means for storing and sharing our visual sensations right up to the middle of the 19th century. For example, we know how George Washington looked only through paintings made by artists, so we see him today through their eyes. Such paintings required many hours of work by one or more talented artists. Once the paintings were finished, they were hard to modify. Most art was never reproduced more than a few times, so works of art were hard or impossible to share and distribute.

Since the invention of photography in the early 19th century, we have seen an explosive growth in our ability to record, store, and distribute images. Photographs are relatively easy to reproduce in large quantity and then distribute. In addition, a photograph captures the world as it is, rather than as the artist interprets it, although artistic composition of a photograph and technical skill in taking the photograph make it more effective.

Figure 3.2 shows some examples of the wide variety of different types of images that we can expect to encounter in our daily life. These images can be ordinary photographs of natural scenes or of people, in either color or black and white. In Figure 3.3, we can see visuals from applications, such as X rays from medical imaging systems, thermal images, and scientific images produced by various instruments. These images are not the expression of creativity or the recording of events seen by an artist, but rather the recording and communication of information that is beyond our ordinary visual experience.

This explosion in producing single images has been matched and, indeed, exceeded by the growth in movie production. A movie is simply a sequence of single images shown at a specific rate so that our human visual system will interpret the sequence of images as continuous motion. The different images in the sequence are closely related to each other, with only small differences between subsequent images. Each of these individual images is referred to as a **frame**. Showing these closely related images quickly in time creates a sensation of movement in the image, hence the name "movie"! The rate at which the individual images are displayed is measured in images per second, which is also called frames per second. Single-frame images and movies together form an important component of the present-day "multimedia revolution."

INTERESTING FACT:

The term "photograph" literally means "record of light." The term has been ascribed to Sir John Herschel in 1839. During the American Civil War, Mathew Brady took 7000 photographs on glass plates, giving us the first photographically recorded historical event.

INTERESTING FACT:

The first motion pictures (movies) with sound were originally called "talkies."

Frame: An individual image in a sequence that, when displayed in time, constitutes a movie. If we push the pause button on a DVD player, the stationary image we see on the screen is a frame.

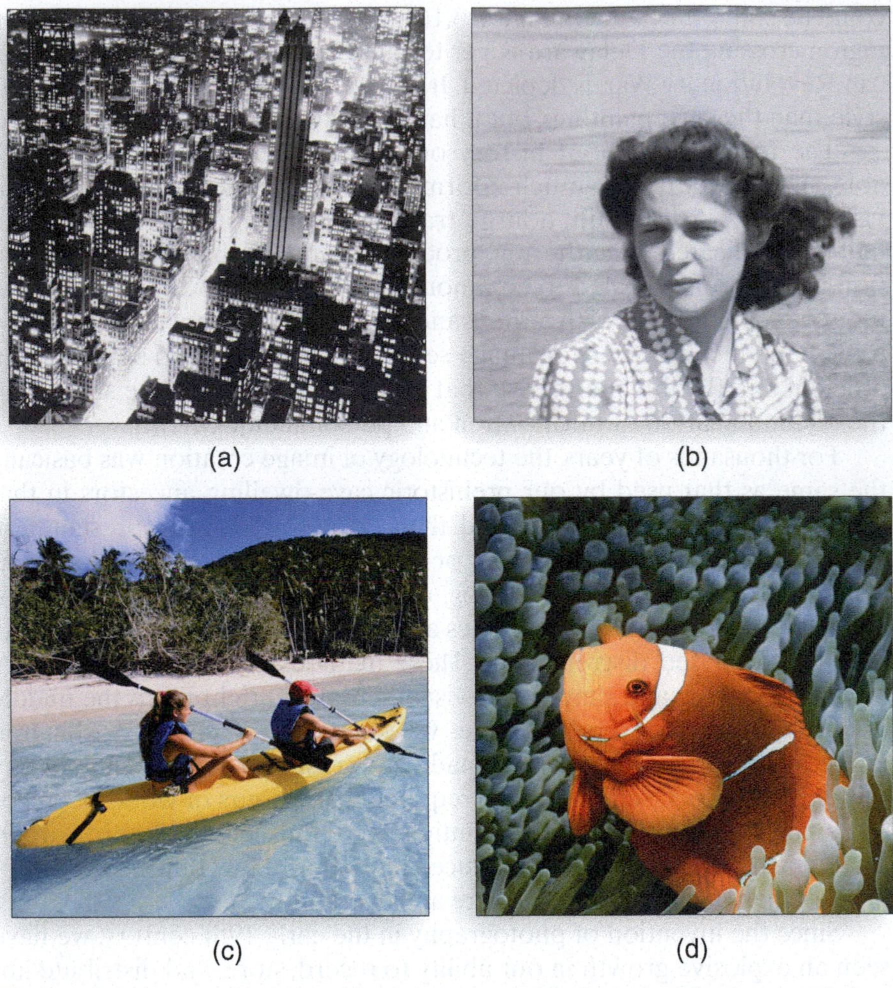

Figure 3.2 Types of images: (a) high-contrast black-and-white image of city lights, (b) black-and-white image of a woman's face, (c) color image of kayakers, and (d) color image of clownfish.

The tremendous advances in photographic science and technology still would not meet completely your need to impress your friends with your good fortune about being close to the rock star or NASA's objectives of transmitting images of Earth or other planets from its spacecraft. In the first case, having to wait for the film to be developed in order to determine the quality of the images resulted in disappointing photographs that could not be retaken. Immediate feedback about the image quality would have allowed you to make adjustments at the concert and create better images. In the case of space exploration, it is clearly not reasonable to expect the spacecraft to return to Earth from planetary exploration in order to deliver film for developing. The next step beyond film photography is needed to allow interactive control of photographs as they are taken, to allow delivery of images from distant sources such as outer space, and to allow images to be created where film cannot be used.

INTERESTING FACT:

The rate at which images are displayed to create movies is measured in frames per second, or images per second. In the early days of the motion picture industry, the frame rates were lower than present-day rates, which caused the images to flicker. This flickering led to the use of the slang "flicks" for movies.

The Basic Idea behind Digital Imaging

A verbal description of an image can be copied easily without error, but such descriptions are typically long, and words tend to be imprecise in describing all the details of the image's content. Images created from verbal

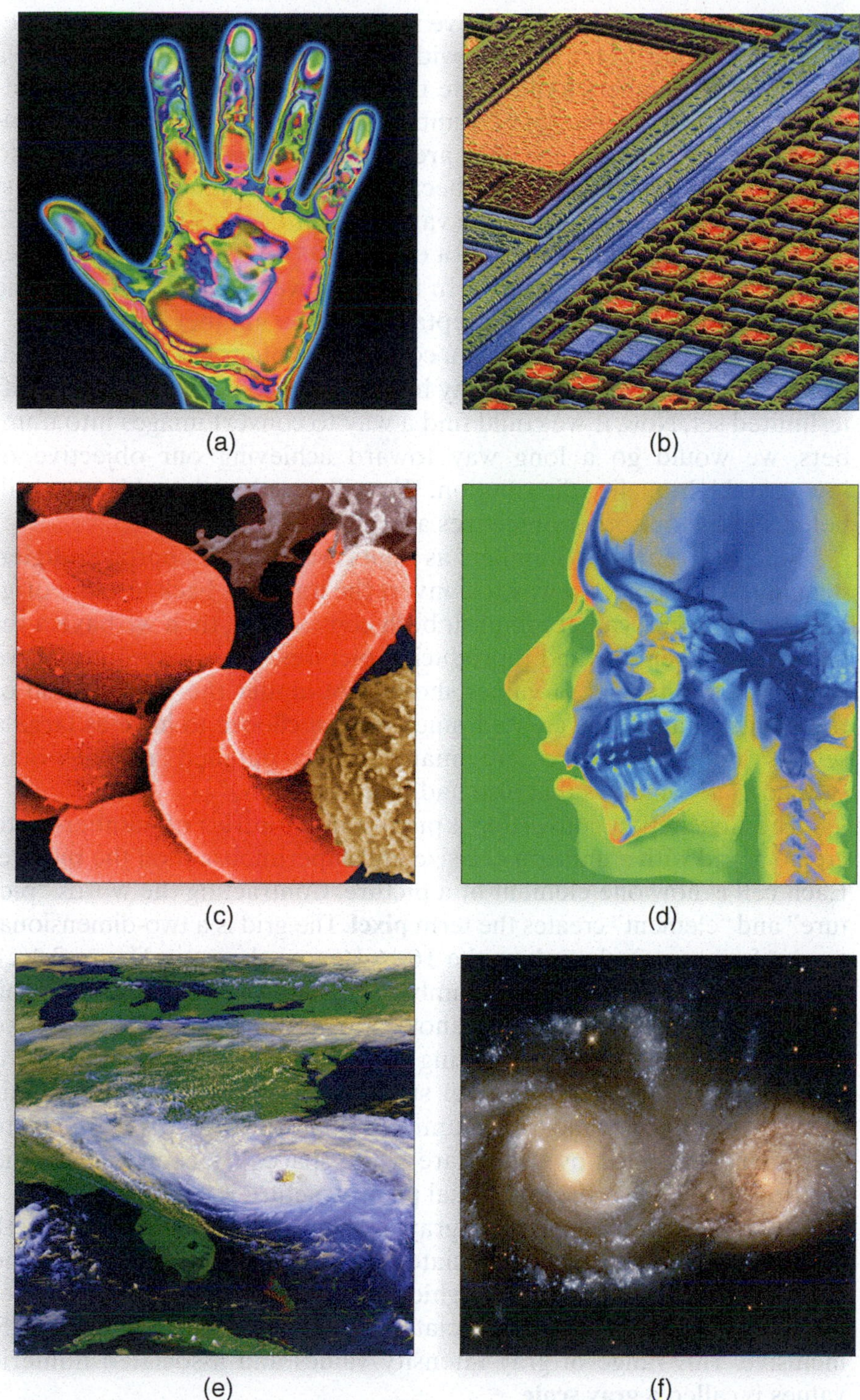

Figure 3.3 Imaging applications: (a) thermal image of a hand, (b) SEM image for an integrated circuit, (c) red blood cells as seen under a microscope, (d) color enhanced X ray image, (e) aerial view of a hurricane, and (f) view of colliding galaxies in space.

descriptions are often very different from the original image. When photographs are copied many times with an office photocopying machine, they lose their quality very quickly. Higher quality reproductions can be obtained only at higher cost from photographic processes.

We need to find an inexpensive method of reproducing images precisely if we want to make their wide distribution possible. Let's use a simple set of numbers to illustrate this concept. Numbers from a limited set (for example, the set of numbers 1, 2, 3, and 4) can be communicated very precisely. When they are duplicated, they are less likely to lose their essential information, because we know that each value must correspond to one of the specific values in our original limited set (1, 2, 3, and 4). If a copied number is not on the list (for example, the number 3.7), we know we have an error. In this case, we may choose to use the closest number that is in the acceptable set (the number 4).

In contrast, when we look at a copy of a photograph, we cannot tell if the copy has errors, because any image is possible. There is no specific, limited set. Now, if we could find a way to convert images into numbers, we would go a long way toward achieving our objective of low-cost, high-quality distribution. We will see that this objective will help us attain our other objectives as well.

Digital Imaging: The technique of representing images (or movies) as a sequence of numbers.

Since we represent numbers as digits, we call this approach **digital imaging**. We also need a way to convert the numbers back into the original picture. If we can accomplish both conversions, then we have a reliable way of sending and reproducing pictures. The constraint that we use only a limited set of values should not hinder the composition of the images that can be created, since we can choose to have as large a set as required. There are many finely detailed mosaics and stained-glass windows that also had this limitation.

Pixel: Contraction of the words "picture" and "element." A pixel corresponds to the smallest detail in a picture that one wants to preserve. This concept is equivalent to one square tile in a mosaic.

The first step in converting a picture into a series of numbers is to create a grid with a uniform cell size and superimpose it on the picture. Each cell is now one element of a picture. Contracting the words "picture" and "element" creates the term **pixel**. The grid is a two-dimensional array of square pixels such as the 16 × 16 area shown in Figure 3.4(c). The row number and column number of each pixel define its location within the image. Figure 3.4(a) shows a digital image of the Statue of Liberty. Any image with interesting detail will have many pixels, and it is not desirable to see the pixel structure in the digital image. This image has 1150 rows and 650 columns. A small 16 × 16–pixel area from this image, indicated by the square white space, has been enlarged in Figure 3.4(b) so that the individual pixel structure is more obvious.

Each pixel has an average "grayness," or intensity associated with it. Each gray value can be associated with a number, as shown in the reference strip of Figure 3.4(e), which shows 16 discrete gray values between black and white and associated the numbers between 0 and 15, inclusive. This range of gray intensity values and associated numeric values is called a **gray scale**.

Gray Scale: Levels of gray tones covering the range from all white to all black.

If the gray level in the image is not one of these 16 selected values, it is assigned the number representing the gray level closest to it. By performing a visual comparison, we can associate a number with each pixel. The 16 × 16–pixel image segment in Figure 3.4(d) is taken from the crown of the statue. The most common background color for the sky behind the statue is a dark gray with a gray-scale decimal value of 4. An even darker pixel with a value of 3 is shown in Figure 3.4(c). Part of the crown of the statue is in direct light and has light-gray values such as 13. Other parts are shadowed and have a midlevel-gray value such as 9. The image can be re-created from the list of numbers by starting with the

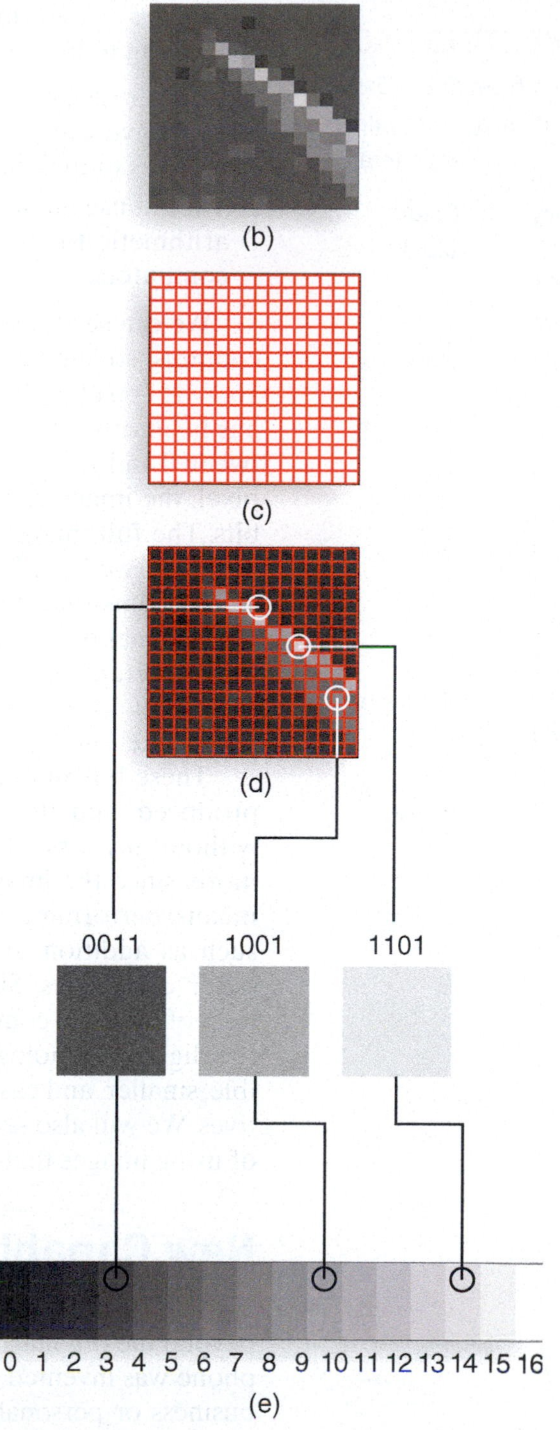

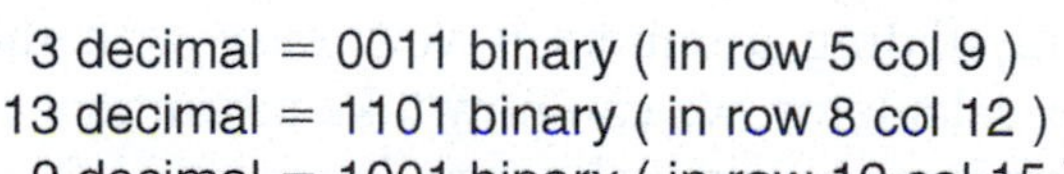

Figure 3.4 The digital image in (a) has 1150 rows and 650 columns. The enlarged area in (b) is 16 × 16 pixels. Each pixel represents the gray level, or intensity value, of one square of the regular grid shown in (c), and (d) shows this grid placed over the image in (b). In (e), the numeric value of three pixels from (b) is indicated and associated with the band showing all 16 possible gray values, from black to white.

empty grid of Figure 3.4(c) and painting each square with the gray value from the gray-scale reference band in Figure 3.4(e) that corresponds to the numeric pixel value. This approach is just like using a child's paint-by-number set, except that each area to be painted is just a small square.

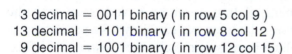

Here, we can also use the same binary-number representation that we introduced in Chapter 1, where a series of 1's and 0's is used to represent a numerical quantity. The margin insert on the next page lists the binary values for the decimal values from 0 to 15. We can see that if we have numbers that vary from 0 to 15, we need four bits to represent those numbers. Using the binary representation, we now convert each

INTERESTING FACT:

Binary values from 0 to 15 can represent 16 different gray level values in an image.

Decimal	Binary	Color
0	0000	black
1	0001	
2	0010	
3	0011	dark gray
4	0100	
5	0101	
6	0110	
7	0111	
8	1000	gray
9	1001	
10	1010	
11	1011	
12	1100	light gray
13	1101	
14	1110	
15	1111	white

pixel's gray value into a four-bit binary number. The motivation for converting to binary representation is twofold:

1. It is very easy to unambiguously represent and reproduce symbols that have only two options, 0 and 1. So with binary numbers, the chances of making an error are far lower.
2. We can use all of the existing digital technology that employs base-2 arithmetic to store and manipulate information in calculators and computers.

We can see that the grid of Figure 3.4(d) consists of 16 rows and 16 columns, so the total number of pixels is $16 \times 16 = 256$. The small 16×16–pixel partial image from the complete Statue of Liberty image has been converted into a table of 256 numbers that take on values between 0 and 15. Since the image has 256 pixels and 4 bits represent each pixel, the image can be represented by a sequence of $256 \times 4 = 1024$ bits. The full image is $1150 \times 650 = 747{,}500$ pixels, so, using four bits per pixel, we have 747,500 pixels $\times$ 4 bits/pixel $=$ 2,990,000 bits.

It seems that our approach using digital imaging satisfies most of our original design goals as long as we use enough pixels to represent the important details of the image and enough bits to have a gray-scale ramp that appears smooth. Later in this chapter, we will establish how many pixels and gray levels are "enough" with digital imaging.

These bits in our digital image can now be stored, transmitted, reproduced, and then converted back into the black-and-white image without any loss of quality if there are no errors in the bits. Furthermore, since the image now consists of numbers, processing the image means performing various arithmetic operations on these numbers, such as addition, subtraction, multiplication, division, and numerous other operations. Since manipulating numbers is very easy with the help of a digital computer or calculator, we can now take advantage of the digital technologies, which have become cheaper, faster, more flexible, smaller, and easier to use over the past 50 years, to meet our objectives. We will also see that digital imaging allows us to design new ways of using images that were not possible with photographic methods.

New Capabilities with Digital Imaging

Any good invention or engineering design always has implications well beyond the original intent of the inventor or the engineer. When the telephone was invented, it was viewed as a novelty toy, not an instrument of business or personal communication. Electronic computers were originally invented with funding from the U.S. Army during World War II to assist human computers who were calibrating the trajectories of artillery shells. Now, that same invention is not just powering the latest video games, but is also embedded in our entertainment systems, our cars, our household appliances, and our workplace instruments. The same statement is true for digital imaging. Here, we will explore some of the new capabilities and applications that digital imaging makes possible.

Image Enhancement: Processing an image to improve its appearance to a human viewer.

Image Enhancement Once images are represented by numbers, we can manipulate them in order to improve their appearance for a human viewer. This procedure is called **image enhancement**. For example, if we find an image's brightness or contrast not to our liking, we can lighten it, darken it, or increase the contrast by changing the values of the numbers that represent the image. We can also soften it by smoothing out sharp edges or sharpen it if the image seems too blurry. These techniques also are available with photographic methods, but you have to guess in advance how much you want to change the image and then wait for a chemical process to be completed before seeing the actual result. With digital imaging, you can see the results immediately and interactively make adjustments until you have exactly what you want.

Several examples of these techniques are applied to an image of two kayakers shown in Figures 3.5 and 3.6. The original image in Figure 3.5(a) is modified to increase the contrast in Figure 3.5(b). Increasing the contrast makes the dark pixels darker and the light pixels lighter so there are more extreme values of white and black than in the original image. In Figure 3.5(c), the original image has been darkened so that all pixels are a little darker, and in Figure 3.5(d) the original image has been brightened so that all pixels are a little lighter. Note that the bright parts of Figure 3.5(b) are similar to the bright parts of Figure 3.5(d), while the dark parts of Figure 3.5(b) are similar to the dark parts of Figure 3.5(c). Figure 3.6 shows the effect of sharpening and softening the same image of the kayakers. In the sharpened image in Figure 3.6(a), the edges are accentuated, but the smooth areas of the image are unchanged. Smoothing,

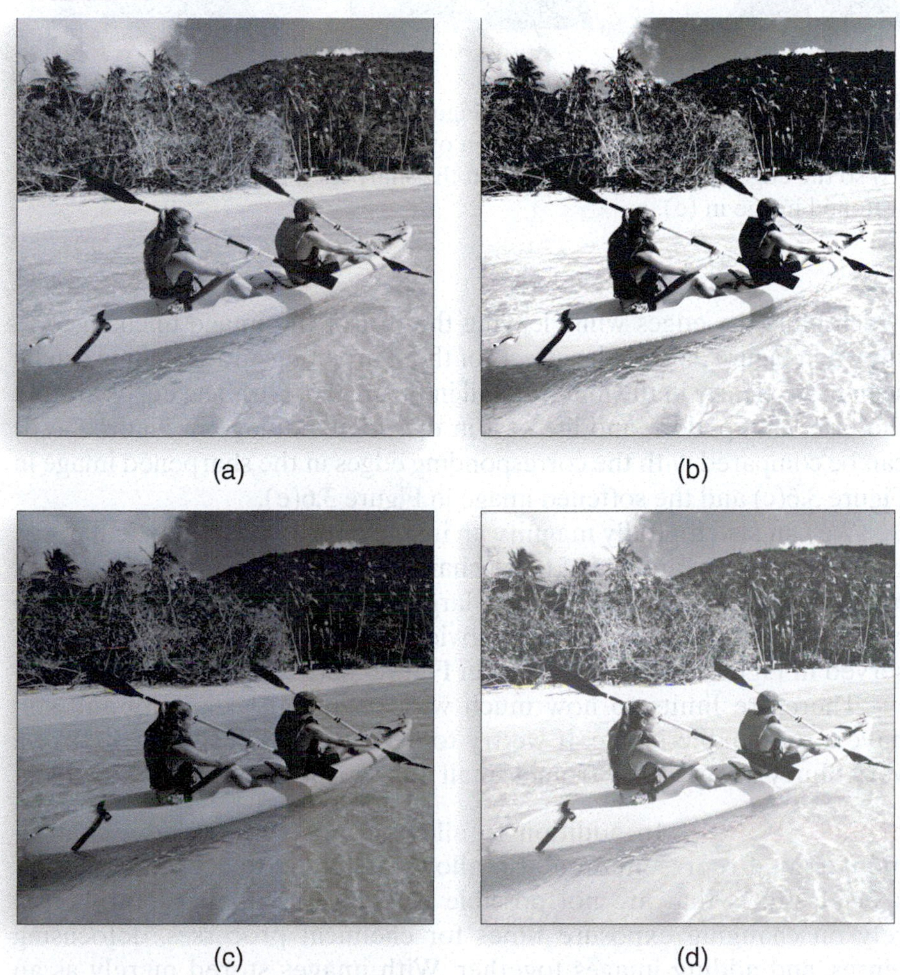

Figure 3.5 The original image in (a) is processed to increase contrast in (b). In (c), the image is darkened, and in (d) the image is brightened.

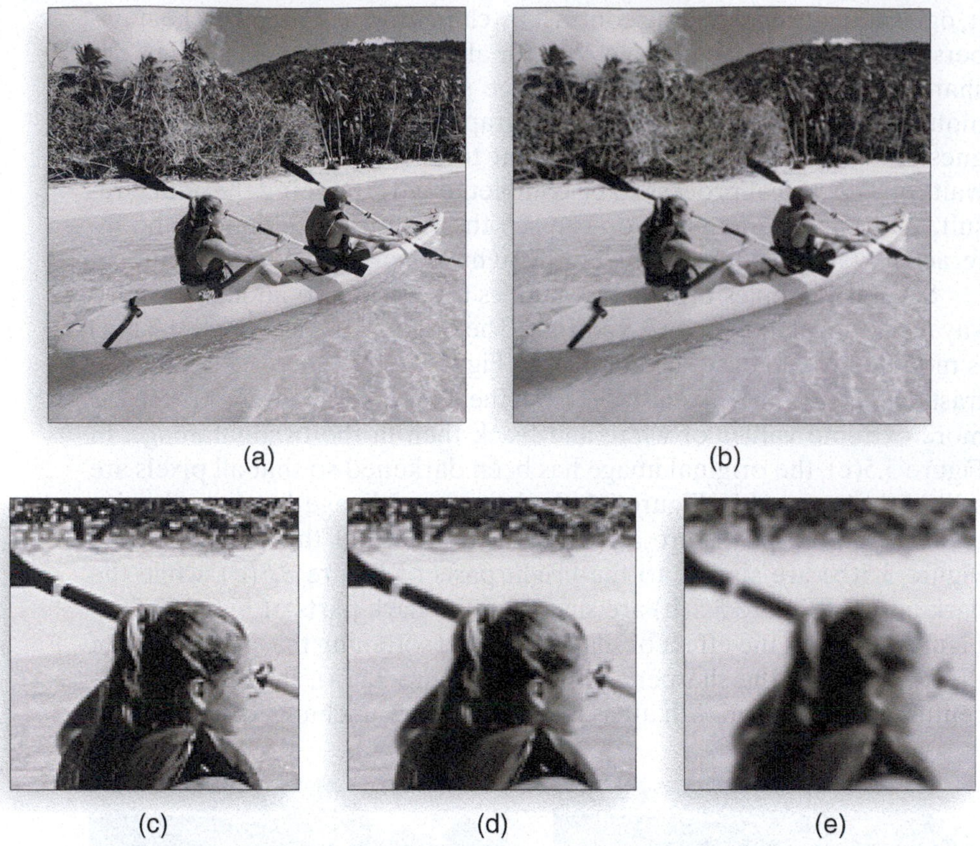

Figure 3.6 The original image from Figure 3.5(a) has been sharpened in (a) and softened in (b). A small section of the original has been enlarged in (d) so the edges can be compared with the sharpened image in (c) and the softened image in (e).

which blurs the edges while leaving the rest of the image unchanged, is shown in Figure 3.6(b). The effects of the sharpening and softening can be seen more clearly in the image detail in Figure 3.6(c–e). The edges around the face, hair, paddle, and life vest in the original image in Figure 3.6(d) can be compared with the corresponding edges in the sharpened image in Figure 3.6(c) and the softened image in Figure 3.6(e).

We can also digitally magnify an image or crop it to a particular size or shape. Some examples of these changes are shown in Figure 3.7. If an image is digitally magnified by a large factor, it is often necessary to smooth it in order to avoid the obvious pixel boundaries that we observed in Figure 3.4(b) and again in Figure 3.6(c–e).

There are limits to how much we can magnify digitally and still have a reasonable image. If we try to exceed this limit, we will have a very blurry image made from a small number of pixels.

Special Effects In addition to allowing real-time image enhancement, digital representation also allows us to create a wide variety of special effects that are not possible with photographic methods that rely on changing exposure times for chemical processes, defocusing lenses, and adding images together. With images stored merely as an array of numbers, there is virtually no limit to the kind of manipulations that can be performed with an image. Figure 3.8 shows a few

Figure 3.7 The original digital image of the Statue of Liberty in (a) is digitally magnified by a factor of 1.8 and cropped to a rectangular window in (b) to show more detail from the top half of the image. A digital zoom by a factor of five and an oval crop display detail of the torch in (c).

examples of digital special effects that simulate original art that are included with many basic image-processing applications. The kayak image from Figure 3.5(a) has been converted to a simulated pen-and-ink drawing in Figure 3.8(a) and a chalk-and-charcoal drawing in Figure 3.8(c). In Figure 3.8(b), the image has an embossed effect, and Figure 3.8(d) it has a canvas texture applied.

There is no reason to be limited by prepackaged special effects. We will see that, with a few simple mathematical tools, we can create our own customized special effects. In Figure 3.9, the special effects do not simulate traditional artistic techniques. In part (a), we see the effect of blur that would be caused by spinning the image around its center point. In part (b), we see the image as if viewed through a simulated spherical lens that magnifies the image content at the center and compresses the content at the edges. Note that the kayakers are larger, but the kayak appears rounded and the trees on the shore are no longer straight.

New Applications with Digital Imaging

Animation One of the most spectacular results of digitization of imaging systems has been the advent of computer **animation**. Animation, the creation of moving images, was demonstrated as far back as the early 1900s. In the early approaches, an artist carefully drew each frame in a movie. Walt Disney took this procedure to a masterful level in the 1930s with a series of animated feature-length films. Of course, animation was a highly skilled profession, and the whole operation was a very time- and money-intensive process. This era can be called the "analog age" of animation, because each image in the animation sequence was drawn or painted by a human animator.

Animation: *Creation of moving images by creating a sequence of individual frames that produces the illusion of smooth motion when shown in quick succession.*

Computers were first used to make animated sequences for training pilots and teaching special industrial skills. As computational power grew and costs decreased, computer-generated animation sequences

Figure 3.8 Four special effects simulating art techniques are applied to Figure 3.5(a). We see pen and ink in (a), embossing in (b), chalk and charcoal in (c), and canvas texture in (d).

Figure 3.9 Two custom special effects simulate the effects of (a) blur due to rotation and (b) use of a spherical lens.

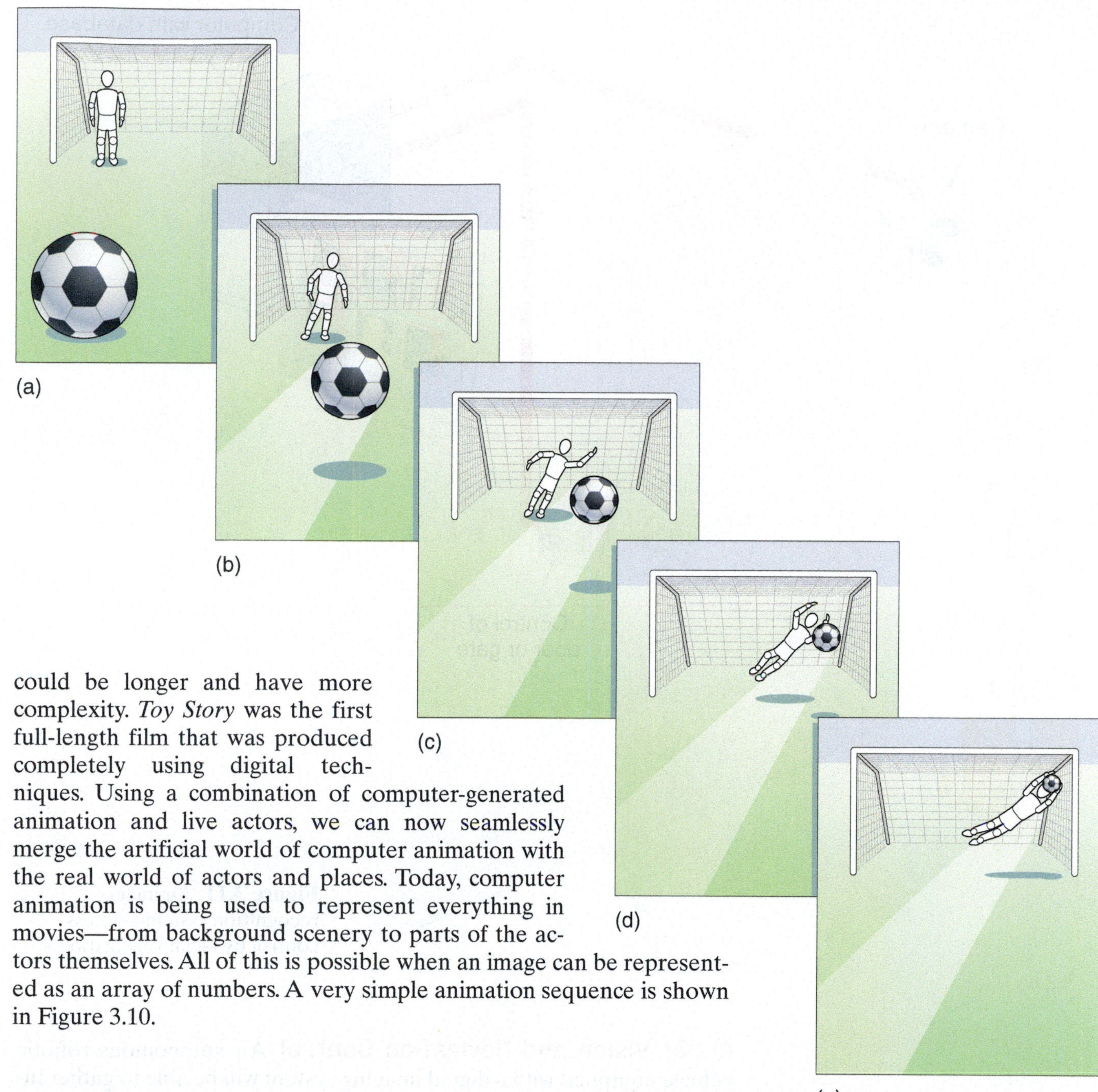

could be longer and have more complexity. *Toy Story* was the first full-length film that was produced completely using digital techniques. Using a combination of computer-generated animation and live actors, we can now seamlessly merge the artificial world of computer animation with the real world of actors and places. Today, computer animation is being used to represent everything in movies—from background scenery to parts of the actors themselves. All of this is possible when an image can be represented as an array of numbers. A very simple animation sequence is shown in Figure 3.10.

Figure 3.10 A simple animation sequence.

Access Control

A digital imaging system can be used for controlling access to places, as shown in Figure 3.11. A secure facility will have an image sensor to capture a digital image of the person desiring entry, as shown in Figure 3.11. The imaging device may look at the person's face, the retina of the eye, or a fingerprint. Then an image-processing system compares that image with images for the list of people with access privileges. If the person is included in the database of authorized users, the door is opened, allowing entry. If the person is not in the database, he or she is denied entry, and some form of alarm may be generated.

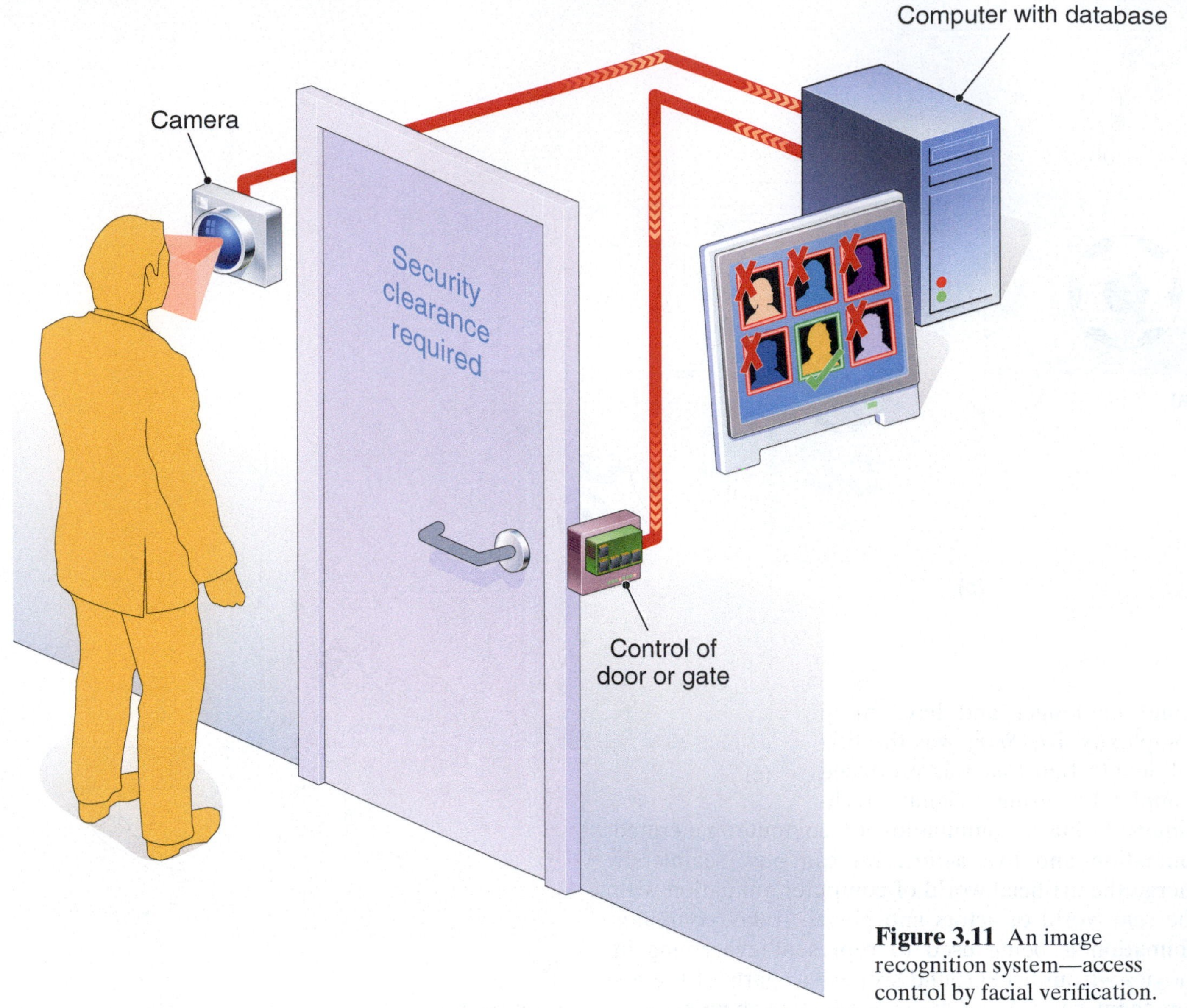

Figure 3.11 An image recognition system—access control by facial verification.

Robot Vision and Navigation Control An autonomous robotic vehicle equipped with a digital imaging system will be able to gather information about the surrounding area and use it to help make navigation decisions. This process attempts to imitate the capabilities of a human operator of a vehicle, who can look out through the windows and use the rearview mirror. Based on the image input and the goals that are programmed in the controller of the robotic vehicle, it can move around in order to accomplish the goals. Many examples of this type of system are in use or under development for industrial manufacturing, automatically controlled personal highway vehicles, and underwater and planetary exploration.

A spectacular example of digital imaging systems used in navigating unknown terrain can be found in the Mars *Pathfinder* rover shown in Figure 3.12. It is equipped with two cameras mounted on its side,

(a)

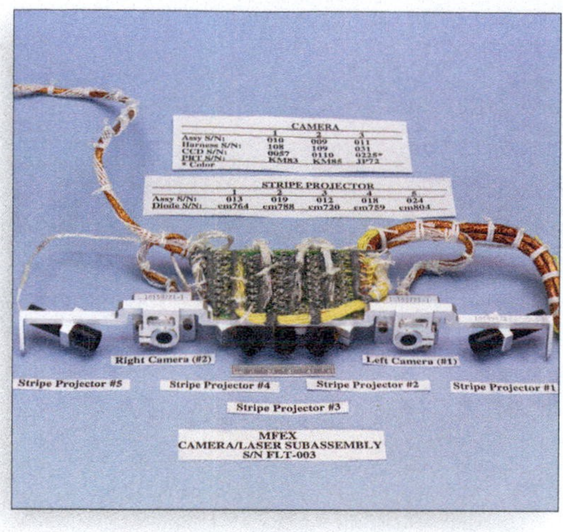

(b)

Figure 3.12 (a) The Mars *Pathfinder* rover. (b) The sensor assembly for the rover. Two cameras are contained in the aluminum housing adjacent to the main circuit board.The three black tubes in the center of the circuit board and the two mounted on the outside are laser stripe projectors.

pretty much like our two eyes placed side-by-side on our face. When more than one camera is aimed at the same object, it is possible to estimate the distance of the object just as humans do based on the two different images seen by the left and right eyes. In addition, the rover carries lasers that illuminate on the scene. These lasers project stripes of light. By analyzing how the stripes are distorted, the rover is able to determine the three-dimensional shapes of distant objects. An example of the striped illumination on objects is shown in Figure 3.13. This image analysis system allows the rover to accomplish simple tasks like detecting the presence of obstacles in its way, measuring the shape and the distance of the obstacles, and then navigating around the obstacles in order to reach the desired destination.

Exploration of Space Telescopes have been used to observe the stars and planets for hundreds of years. Digital imaging systems now allow us to further enhance the captured images in order to see the desired features. Figure 3.14(a) is an image of Saturn obtained by the Hubble telescope [shown in Figure 3.14(b)].

Medical Imaging Systems So far, we have discussed digital imaging systems that acquire images in the conventional manner of recording light reflected from objects. There is also a way of acquiring image information for which a significant amount of computation is needed before a recognizable image is available. Two examples of this kind of digital imaging system are computed tomography (CT) and magnetic resonance imaging (MRI).

The CT imaging system can create an image of an internal "slice" of a human body by taking external measurements of several X-ray shadows

INTERESTING FACT:

The main imaging mirror of the Hubble space telescope, as launched in 1990, had a microscopic flaw in its construction that gave the telescope blurry vision. A space shuttle mission was launched in 1993 to add corrective optics to the telescope in order to produce sharp images. However, until the shuttle launch, the blurry images were corrected using a digital computer.

INTERESTING FACT:

When stripes are projected onto an object, the distance to points on the object can be computed from a two dimensional photograph.

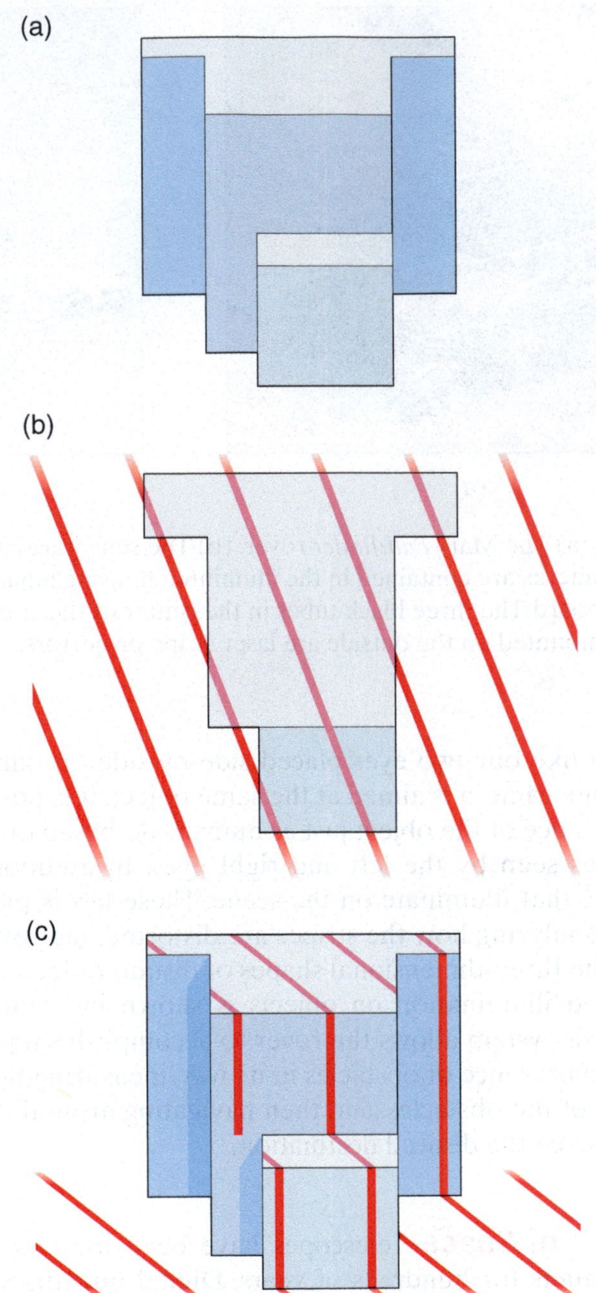

Figure 3.13 The simple block object in (a) is viewed from the front at a slight elevation. Because the image is only two dimensional, we do not know with certainty which parts are horizontal and which parts are vertical. The three-dimensional shape can be computed from an image if stripes are projected onto the object as shown in the view from above in (b). From the pattern of the stripes in (c) we can compute the distance to points on the object.

from different angles. These shadows are combined together in a computer to generate a full, three-dimensional view of the human body. Figure 3.15(b) shows images of slices through the top of a patient's head from two different types of medical imaging systems. In the lower images, the white ring is the bone of the skull surrounding the soft tissue of the brain.

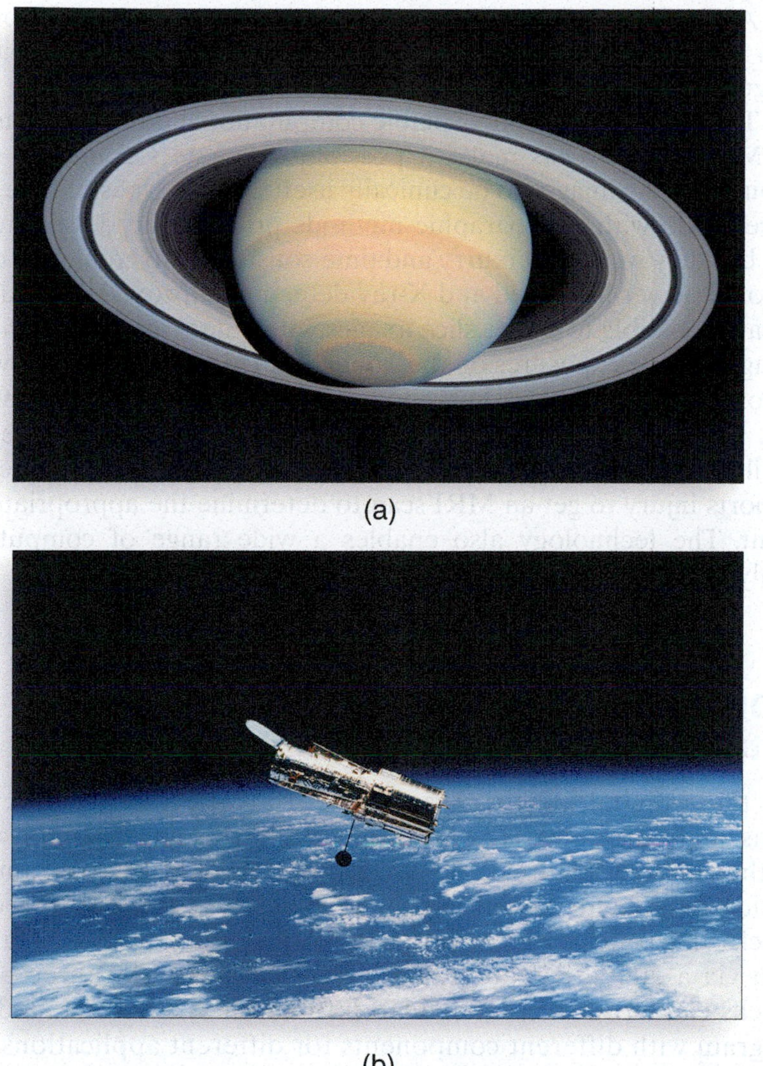

(a)

(b)

Figure 3.14 Imaging in astronomy: (a) a view of Saturn through the Hubble Space Telescope and (b) the Hubble Space Telescope.

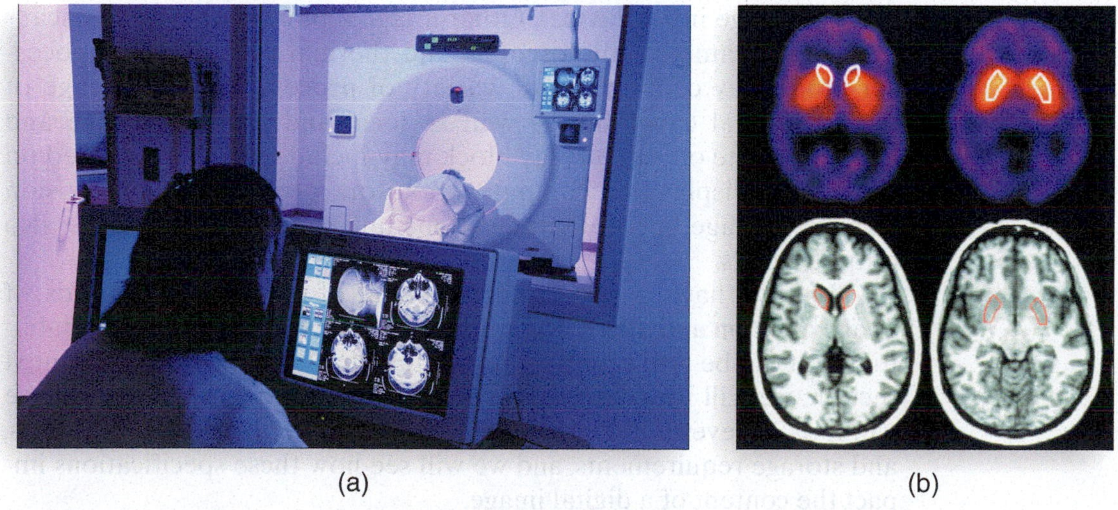

(a) (b)

Figure 3.15 A medical application of image processing: (a) CT imaging system and (b) four images of horizontal "slices" through the top of a patient's head.

An MRI system creates images by using the same type of computation as that used by a CT system, but the measurements are based on magnetic-field measurements rather than X rays.

The rapid increase in available computational power—as indicated by Moore's law—has make it possible to create CT images quickly enough for the images to be clinically useful. These types of slice images were made with photographic methods prior to the development of CT, but they were very blurry and time consuming to create. As computational power increased and X-ray detectors improved in the 1970s, it became possible to create slice images with digital methods rather than using multiple exposures on the same film. This technology allowed additional processing that sharpened the slice images, revealing more detail. CT and MRI systems provide information that was previously available only by surgical methods. Now it is routine for an athlete with a sports injury to get an MRI scan to determine the appropriate treatment. The technology also enables a wide range of computational analysis to be applied to the images.

Components of a Digital Imaging System

All these digital imaging systems have a variety of designs, operations, and objectives. However, they do have some common aspects to their design. Figure 3.16 shows a high-level block diagram of the basic components of all digital imaging systems. The three main blocks illustrate image acquisition from sensors, number manipulation via a digital computer, and an output stage. All three of these blocks can be further subdivided into another, more detailed, block diagram with different components for different applications. In the previous section, we explored three different sensors for inputs. A digital camera, the most obvious input device, is often connected directly to a computer, but it may store images internally for later use or transmit images via a communications link. The Hubble telescope and the CT scanner, as shown in Figures 3.14 and 3.15, respectively are also image input devices. Other image input devices include scanners, video cameras, microscopes, and thermal imagers. The processor block may contain a wide variety of processors with a range of computational capability as well as local and remote storage and databases. The output display block may include an image printed on paper; live displays on computer monitors, projectors, or television screens. Storage or a communications link to other systems are also output blocks.

Now we have a good understanding of traditional methods of image creation and distribution, and we have explored the new capabilities that can be achieved by using digital images and video. In the next section, we will investigate what it takes to create a digital image. We will look at several design specifications concerning image size, colors, and storage requirements, and we will see how these specifications impact the content of a digital image.

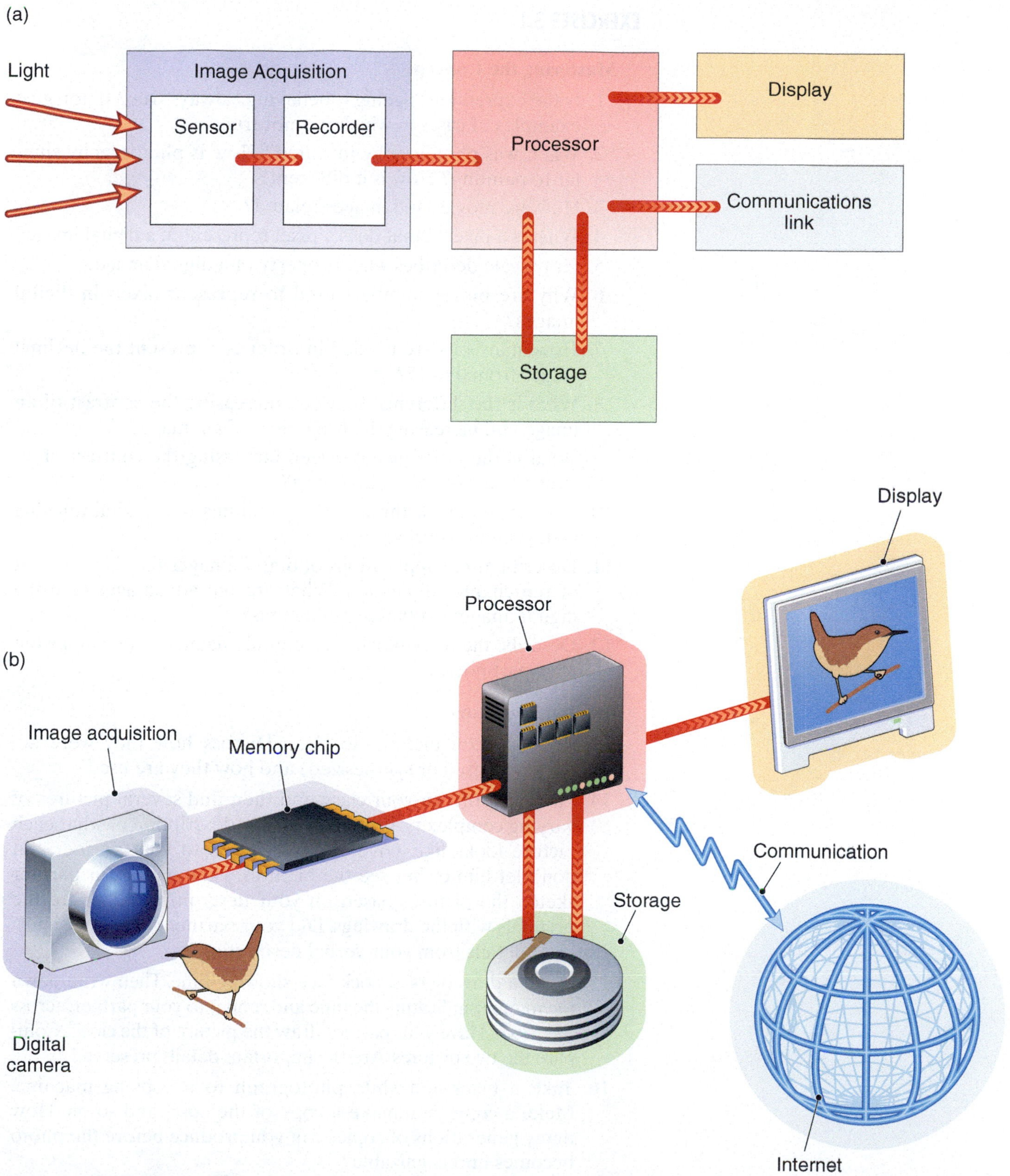

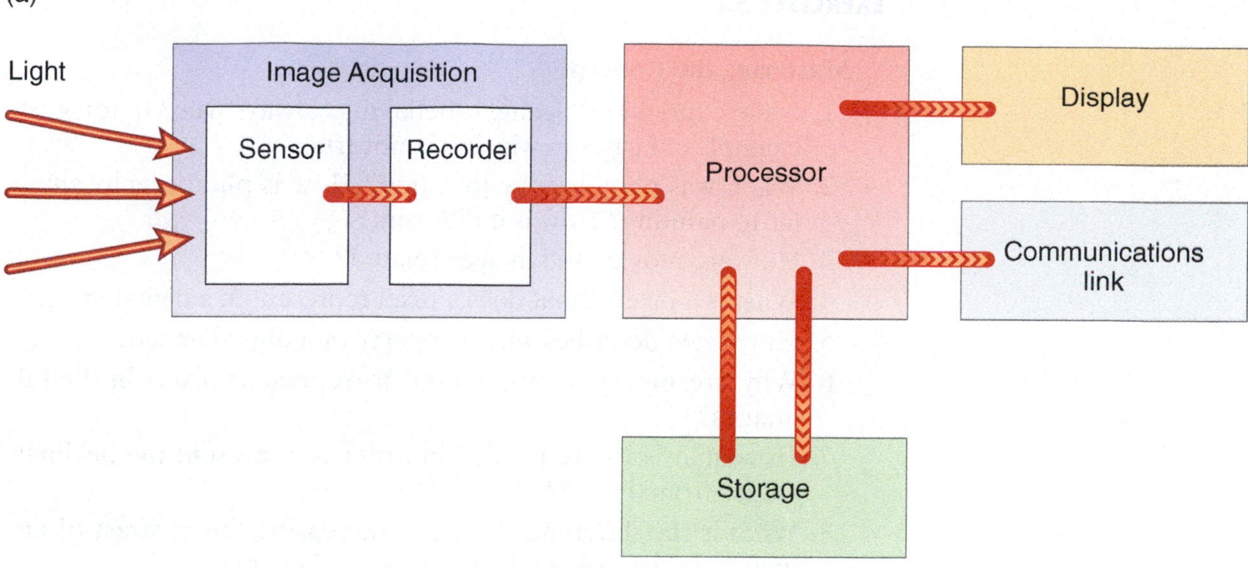

Figure 3.16 A high-level block diagram for all digital imaging systems with (a) showing the basic components. The sensor may respond to many types of input such as visible light, thermal energy, or X-rays. The storage might be film, tape, or digital memory. A block diagram for a specific system (b) uses a digital still camera for image input, a disk for storage, and a monitor for display. Image input could also come from a movie camera or a page scanner.

EXERCISES 3.1

Mastering the Concepts

1. Is the expression "Seeing is believing" always true? If not, give examples of cases in which it is not true.
2. When was photography invented? How is photography similar to painting? How is it different?
3. How are movies and images related?
4. What is a pixel? What does a pixel represent in a digital image?
5. Gray scale describes what property of a digital image?
6. Why are binary numbers used to represent pixels in digital images?
7. How many bits are needed in order to represent the decimal values from 0 to 15?
8. What is the difference between increasing the contrast of an image and increasing the brightness of an image?
9. What is the difference between increasing the contrast of an image and sharpening an image?
10. Give examples of three new capabilities that digital imaging systems have spawned.
11. Describe three applications of digital images that are not part of the chapter discussion. What are the advantages of using digital imaging in these applications?
12. Describe the three blocks of a digital imaging system and what each part does.

In the Laboratory

13. Bring several pictures to class. Discuss how they were acquired (sensed or synthesized) and how they are used.
14. Find a partner in your class, and then find several pictures of varying complexity. Write a one-page description of what each picture looks like. Give the description to your partner, but don't let him or her see the pictures. Then have your partner sketch the pictures based on your description. Compare the originals with the drawings. Did your partner get all the significant details from your verbal description?
15. Make a drawing of a clock face showing time. Then write down the number indicating the time and send it to your partner, across the room. Have your partner draw the picture of the clock. Compare the two pictures. Are the important details preserved?
16. Take a black-and-white photograph to a copying machine. Make a copy, then make a copy of the copy, and so on. How many generations of copies can you produce before the photo becomes unrecognizable?
17. Make a list of 32 random numbers between zero and nine. Make a copy, then make a copy of the copy, and so on. Make the same number of generations as it took in the previous exercise for the image to become unrecognizable. Can you still recover the original list of 32 random numbers? What charac-

teristics of the original list of numbers make the list more readable than the photograph after many copies are made?

18. Take a simple black-and-white picture of a single printed letter. Using tissue paper, create an 8 × 8 grid on top of it. Judge what average gray values are to be associated to each pixel, and assign a number between 0 and 15 to each one. Give the numbers to a friend or lab partner. Ask him or her to reproduce the picture according to the numbers in a kind of "color by numbers" fashion. Compare the two pictures.
19. Take an 8 × 8 grid of numbers between 0 and 15 provided by the teacher. Fill in the grid provided according to the numbers (black for 0, white for 15, and intermediate levels of gray for in-between values). Collect 16 grids filled in by 16 different students. Arrange them in an array of four rows and four columns to create a larger picture. How many pixels are in a row of the large picture? How many pixels are in each column? What is the total number of pixels in the large image?
20. Find an unusual example of a digital imaging system other than the ones discussed in this chapter. Write a short description of it, and draw its block diagram in a manner similar to that used in Figure 3.16.

Back of the Envelope

21. In an access control system, we would like the system to work, allowing recognized individuals access, even if an authorized person gets a haircut, new glasses, or changes his or her appearance in some other way. Discuss which facial features should be used when "recognizing" a person. Are there other ways of recognizing a person for access besides comparison with pictures of the person's face?

3.2 Digitizing Images

Making digital images involves lots of choices. In our simple example in Figure 3.4(c–d), we constructed a grid that had 16 rows and 16 columns, and the pixel intensity values were limited to 1 of 16 gray values (0 to 15). We also represented each number corresponding to a particular gray value in binary form, giving us four bits per gray value. Putting all these numbers together, we came up with 1024 bits representing the 16 × 16–pixel partial image. But why did we select 256 pixels for that small image segment? Why are there are 16 rows and 16 columns of pixels to cover that specific area within the full picture? And why did we select 16 shades of gray to create all the intensity levels we need? Obviously, there is nothing magical about these numbers. We could have chosen other values for the number of pixels or the number of bits per pixel.

In this section, we will study the impact of choosing different values for the grid size and for the number of gray values in a digital image. We will explore the impact on the picture quality if we choose a larger size grid and make more subtle distinctions among gray values. We will also examine how the factors increase the size of the picture in terms of the number of bits needed to represent the picture. Increasing the number of bits in an image increases the storage required as well as the time it takes to send the image in an e-mail message or view the image on a webpage.

We also will examine movies, which are simply sequences of pictures shown in order at a constant frame rate per second. We will study the impact of changing that rate on both the quality of the movies and the number of bits needed to represent the movies. Finally, we will take a look at the representation of color.

The specific questions we will address are as follows:

- How many different intensity levels, also called gray levels, are needed to accurately represent the image's content? These gray levels are also called gray values or gray scale.
- What should the size of the grid be? This size will determine the size of the image in pixels, because it will specify the number of pixels per row and the number of rows.
- What should the size of a pixel be? Or, more specifically, what is the size of the area in the real world that is represented by a single pixel? For example, if a page or photograph is scanned to form a digital image, a single pixel will represent a small area on the page. This size will be the fundamental limit on how much we can digitally zoom in, because we can never look inside a pixel.
- How can we handle color? How many different colors are needed to represent the image's content, and how will we specify what the colors are?
- What frame rate do we need for a movie? How fast must the sequence of still images be displayed so that we will experience visually continuous motion?
- What is the storage requirement for digital images? It is important that the previous questions be answered with values that are adequate for the image quality desired rather than generously above that level. Increasing the number of pixels or the number of colors beyond what is needed will not noticeably improve quality, but will certainly increase both the amount of storage required and the time needed to transmit the images from a website or a remote location. This would increase cost without increasing benefit.

How Many Bits per Pixel?

A pixel is the smallest element of a digital image. In a black-and-white image, we can measure the level of brightness, or "grayness," within each pixel. Then we can assign a numerical value to the pixel and represent that numerical value with a binary bit pattern. In an analog world, the gray values will be continuous, and we would simply make a qualitative statement in words as to where the gray value of a pixel stood in relation to the continuum. For example, we might use terms like "darkest," "somewhat dark," or "on the lighter side." However, since we want to use precise numbers to represent an image, we must break up the gray values into a finite number of discrete steps. This decision process is called **quantization**. Once we have quantized each pixel in a digital image, we have represented it using a number that can be remembered, stored, or transmitted.

Quantization: *The process of taking a set of continuous values and mapping them into a finite number of discrete steps represented by integers.*

Figure 3.17 shows a simple strip image that contains equally spaced gray values between complete black, at a value of 0, and complete white, at a value given by the highest integer we can represent with the

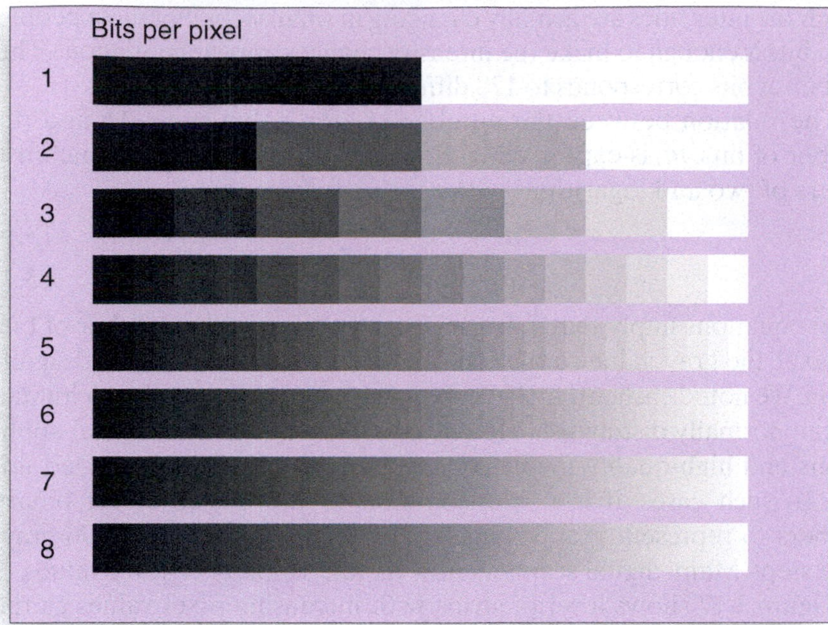

Figure 3.17 Bands of gray values that can be specified by a fixed number of bits in the binary representation of the pixels. With one bit, the top band can have only 2 different gray values. With eight bits, the bottom band can have 256 different gray values.

number of bits specified. The eight strips each use a different number of bits to represent the intensity values, so in each strip the range of gray levels is quantized to a different number of distinct levels, which appear as vertical stripes in each band. The fourth strip, with four bits per pixel, is exactly the same as the strip shown in Figure 3.4(e).

If the number of bits increases by one, the number of gray values will double. Increasing the number of bits from two, as shown in the second row of the figure, to three, as shown in the third row, increases the number of possible gray values from four to eight. These values are listed in Table 3.1, with names given to each gray level.

Figure 3.17 also shows us that, as the number of different gray levels increases, it becomes harder for us to visually distinguish the boundaries between the different levels. If the number of bits keeps increasing, eventually we will perceive a continuous smooth change from black to white, even

Table 3.1 Gray Levels and Corresponding Bit Patterns for Two Bits per Pixel (on the Left) and Three Bits per Pixel (on the Right)

Pixel Values for Two Bits per Pixel			Pixel Values for Three Bits per Pixel		
Decimal	**Binary**	**Color**	**Decimal**	**Binary**	**Color**
0	00	Black	0	000	Black
1	01	Dark gray	1	001	Very dark gray
2	10	Light gray	2	010	Darker gray
3	11	White	3	011	Dark gray
			4	100	Light gray
			5	101	Lighter gray
			6	110	Very light gray
			7	111	White

though the intensities are actually changing in small steps. For most people, seven bits is enough to make the intensity changes appear continuous. This number of bits corresponds to 128 different gray levels.

The relation between the number of gray value steps, M, and the number of bits, m, is expressed in Equations (3.1) and (3.2), which use powers of two and logarithms with respect to base two:

$$M = 2^m \tag{3.1}$$

$$m = \log_2(M) \tag{3.2}$$

These equations imply that, for a small extra price in the number of bits per pixel, the gray-value resolution of each pixel can be increased significantly. We noted earlier that 256 gray value steps are more than a human eye can normally distinguish. However, in scientific and industrial applications and high-quality media, a computer processes the digitized images. In such cases, it is not uncommon to use 10- to 12-bit binary numbers to represent gray values corresponding to 1024 to 4096 gray-value steps. Many digital scanners now handle 16-bit gray-level values.

Figure 3.17 shows a set of strips with increasing pixel values as the number of bits per pixel is increased. In these strips, the intensity always increases as we move from left to right. It is easy to see the abrupt changes in intensity when a small number of bits per pixel is used. In most images, there are much larger changes in intensity between adjacent pixels. To demonstrate the effect of changing the number of bits per pixel for a specific image, Figure 3.18 shows two 16 × 16–pixel areas selected from the Statue of Liberty image in Figure 3.4. Figure 3.18(a) displays the same 16 × 16 area that was used in Figure 3.4(b) with four bits per pixel. Figure 3.8(b) shows a second 16 × 16–pixel area taken from the rounded top of the crown, also displayed with four bits per pixel. With 16 different gray levels, some variation is visible in the dark background, which is the sky, and even more variation can be seen in the lighter grays that compose the crown. These variations help us perceive surface shape and surface detail. The same two 16 × 16–pixel areas are displayed in Figure 3.18(c) and (d), respectively, with only two bits per pixel. With only four gray levels, almost all of the sky is dark gray. The crown is light gray and white, and it is hard to see surface details. The numeric values for the pixels for Figure 3.18(b) and Figure 3.18(d) are shown in Figure 3.19(a) and (b), respectively. Note that all values from 0 to 3 in Figure 3.19(a) have been converted to 0 in Figure 3.19(b), and all values from 8 to 11 have been converted to 2.

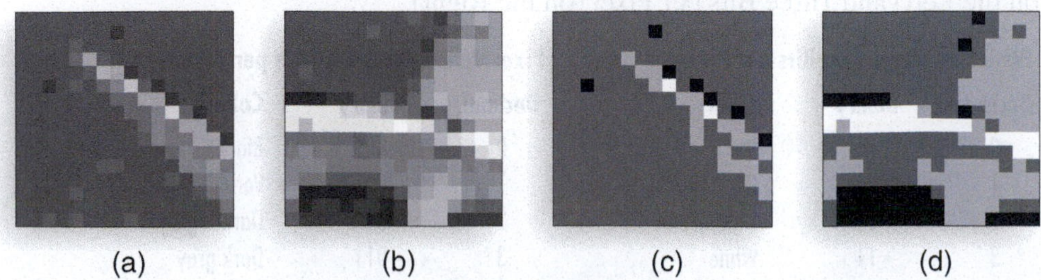

Figure 3.18 Two 16 × 16–pixel areas taken from the large image in Figure 3.4 are shown in (a) and (b) with four bits per pixel, which allows 16 different gray levels. The same two 16 × 16–pixel areas are shown in (c) and (d) using two bits per pixel, which allows 4 different gray levels.

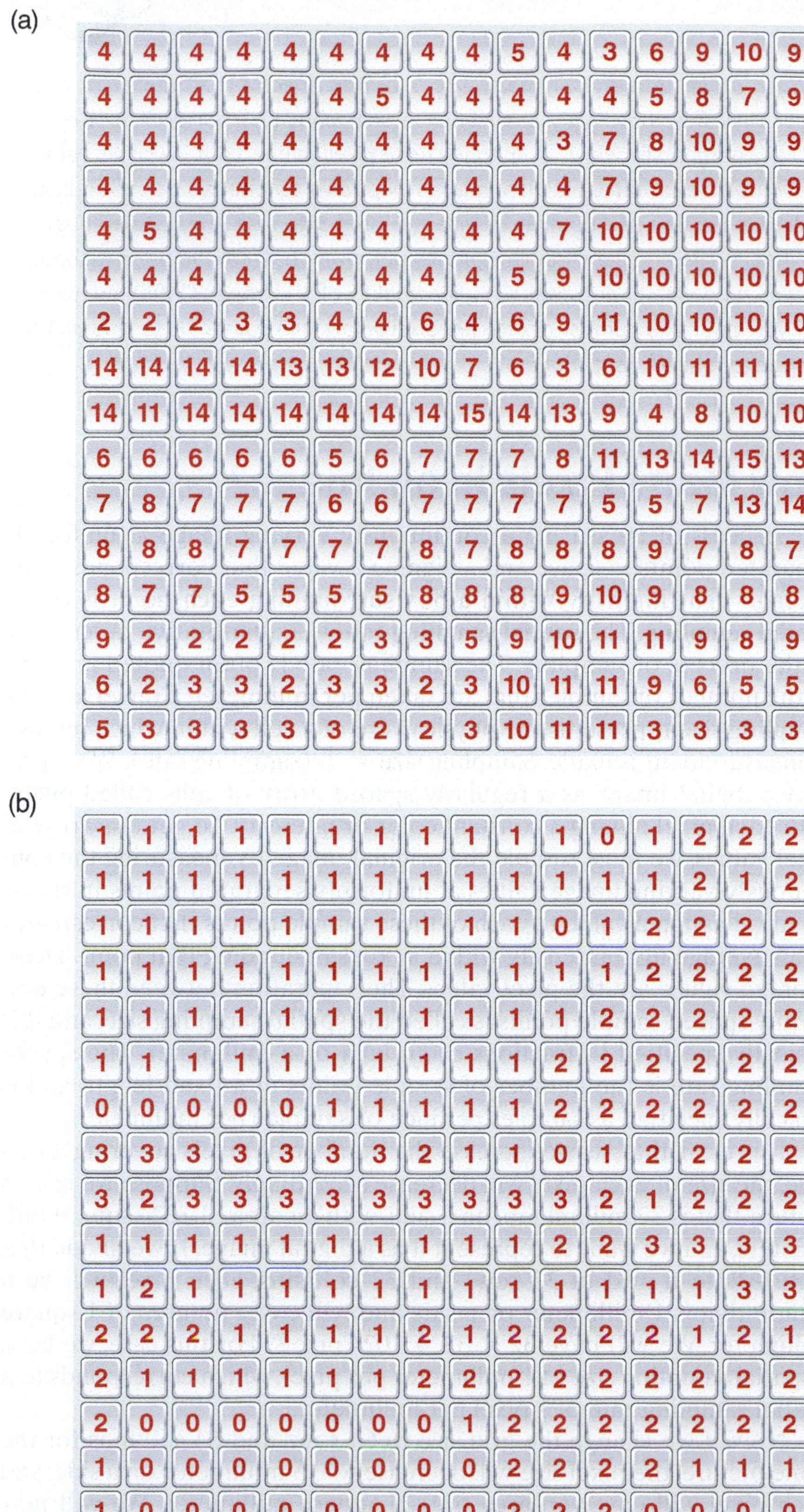

Figure 3.19 Corresponding numeric values for the images shown in (a) Figure 3.18(b) and (b) Figure 3.18(d). The range in (a) is from 0 to 15, and the range in (b) is from 0 to 3. Lower numbers correspond to darker values, and higher numbers correspond to lighter values.

Infinity Project Experiment: Image Quantization

When creating a digital image, the user can specify the number of bits per pixel. If we have gray-scale images with eight bits per pixel, it is relatively straightforward to change these pixel values to a lower bit value, such as seven bits, six bits, and so on, down to one bit per pixel. How does changing the number of bits per pixel affect a gray-scale image? What is the smallest number of bits per pixel that can be used without causing a significant change in the quality of the image? What happens if you lower the number of bits per pixel below this point? Does this point change for different images?

How Many Pixels in an Image?

Spatial Sampling: Measuring a continuously varying image at uniformly separated points in space and assigning a value that corresponds to an average light intensity within a small box surrounding each point.

Spatial Sampling Size: The width of one sample cell, or pixel. The spatial sample size in units of mm/sample is 1/(sampling rate) when the sampling rate is measured in samples/mm.

Spatial Sampling Rate: Number of samples of an image taken per unit of physical length of that image.

The process of measuring a continuously varying image at uniformly separated points in space and assigning a value that corresponds to an average light intensity within a box surrounding each point is called **spatial sampling**. The **spatial sampling rate** is a quantitative description of the process of sampling. The sampling rate is often quoted in terms of the number of samples that are taken per millimeter. Conversely, the **sampling size** represents the physical size of a cell within which intensity measurement is made. Sampling size $=$ 1/(sampling rate). We represent a digital image as a regularly spaced array of cells called pixels, with each pixel having a constant color or intensity. To get this array of pixel values, we must sample the original image by measuring the continuously varying image values at uniformly separated points in space. Then we assign a value to each of these sample points that corresponds to an average light intensity within a box surrounding that point. These average values are the pixel values. The separation between these uniformly spaced sample points is called the spatial sampling size, and it is typically specified in fractions of an inch or a millimeter. The spatial sampling rate is the number of sample points in a fixed length and is typically specified as samples per inch or samples per millimeter.

It is clear that, as we increase the number of pixels covering a picture, the size of each pixel will decrease accordingly. Suppose we have a picture that is 32 millimeters on a side, with an area of 1024 square millimeters, and we want to represent it as a digital image. If we divide that picture into a grid of 32 rows and 32 columns, each pixel will be a square that is 1 millimeter on a side and will cover an area of 1 square millimeter. We will have $32 \times 32 = 1024$ pixels. For this case, we have a spatial sampling size of 1 millimeter per pixel, which corresponds to a spatial sampling rate of 1 pixel per millimeter.

Now, if we change the grid size to 64 rows and 64 columns for the same picture, the pixel size will decrease to 0.5 millimeter on a side, and each pixel will cover an area of 0.25 square millimeter. We will now have 4096 pixels, with a spatial sampling size of 0.5 millimeter per pixel and a spatial sampling rate of 2 pixels per millimeter. By increasing the sampling rate by a factor of two in both the horizontal and vertical direction, we have increased the number of pixels by a factor of four, and each new pixel represents one-fourth of the area of the original pixel.

If instead we reduce the grid size to 16 rows and 16 columns, each pixel area will increase to two millimeters on a side, with an area of four square millimeters, and the spatial sampling rate will be reduced to 0.5 pixel per millimeter. We will now have only 256 pixels.

Often, the sampling rate is the same in the horizontal and vertical direction, but the two sampling rates can also be different. If X is the pixel width and Y is the pixel height, then the horizontal and vertical sampling rates are computed as shown in Equation (3.3). If the picture represented by the digital image has a width of W and a height of H, the number of pixels in a row, J, and the number of pixels in a column, I, are computed according to Equation (3.4). S_h indicates the horizontal sampling rate, and S_v indicates the vertical sampling rate. The equations are as follows:

$$\begin{aligned} S_h \text{ pixels/mm} &= 1/(X \text{ mm/pixel}) \\ S_v \text{ pixels/mm} &= 1/(Y \text{ mm/pixel}) \end{aligned} \tag{3.3}$$

$$J \text{ pixels/row} = S_h \text{ pixels/mm} \times W \text{ mm/row} = \frac{W \text{ mm/row}}{X \text{ mm/pixel}}$$

$$I \text{ pixels/column} = S_v \text{ pixels/mm} \times H \text{ mm/column} = \frac{H \text{ mm/column}}{Y \text{ mm/pixel}} \tag{3.4}$$

EXAMPLE 3.1 Determining Sampling Rate

A photograph four inches wide and six inches high is divided into an array of square cells that are 0.01 inch on a side. What is the sampling rate in the horizontal and vertical directions? What is the width and height of the pixel array?

Solution

The horizontal sampling rate S_h in pixels per inch will equal 1/(pixel width in inches). Here, the width is $X = 0.01$ inch/pixel, so

$$S_h \text{ pixels/inch} = 1/(0.01 \text{ inch/pixel}) = 100 \text{ pixels/inch}$$

Since the pixel height and width are the same, $S_h = S_v$.
The photograph itself is not square, so the number of pixels in a row and in a column will not be the same.

$$\begin{aligned} J \text{ pixels/row} &= S_h \text{ pixels/inch} \times W \text{ inches/row} = \\ &\qquad 100 \text{ pixels/inch} \times 4 \text{ inches/row} = 400 \text{ pixels/row} \end{aligned}$$

$$\begin{aligned} I \text{ pixels/column} &= S_v \text{ pixels/inch} \times H \text{ inches/column} = \\ &\qquad 100 \text{ pixels/inch} \times 6 \text{ inches/column} = 600 \text{ pixels/column} \end{aligned}$$

It is important to remember that each pixel has a single intensity value that represents the whole area of the pixel. It is necessary then to have enough pixels so that the picture intensity is not changing much over the area of a pixel in the digital image of the picture. In determining the gray value for each pixel, we ignore any variations within the pixel and assign a number that corresponds to the average gray value in the pixel area. If, for example, half of a pixel area is black, corresponding to the value 0, and half is completely white, corresponding to the value 15, we will assign an average value of 8 to the pixel. Although the average value of 15 and 0 is actually 7.5, we are using only four bits per pixel to represent values from

0 to 15, so we have to round off our average value to the nearest integer. As a result, some detail in the image is lost. The spatial sampling rate is the number of samples of an image taken per unit of physical length of that image. If we want to preserve fine details in an image, we need to make the individual pixels small enough in size so that values with large differences are not averaged together. As we discussed previously, making the pixels smaller is equivalent to making the sampling rate higher, which increases the total number of pixels required in order to represent an image.

For a simple demonstration of the problem of having pixels that are too large, let's consider an image of vertical stripes that is first sampled using a grid with eight rows and eight columns as shown in Figure 3.20. The pixel values are shown in Figure 3.20(a), and the intensity pattern is shown in Figure 3.20(b).

Now, if we decide to represent this image by a grid with four rows and four columns instead of eight, we will need to find a single intensity value to represent 2 × 2–pixel squares from the original image. If we take the average intensity over the area represented by each new larger pixel, then the resulting image is shown in Figure 3.21(b), with corresponding pixel values shown in Figure 3.21(a). All detail of the stripes is lost, because the spatial resolution has been lowered too much, and we can no longer represent the detail of the stripes.

This new image contains 16 pixels, all having the same intermediate-gray value of eight. The vertical bars in the original image now have disappeared completely. All four of the 8 × 8–pixel images in Figure 3.21(c–f) will have the same 4 × 4–pixel result shown in Figure 3.21(b) when the sampling size is increased. This kind of a loss of detail is due to inadequate sampling of an image.

A simple way of describing what has happened in this example is as follows: If the gray value in an image varies significantly over a small region of the original image, we need to ensure that the small region contains many pixels. In order for this to happen, the pixels themselves must be small. Since, in our digital image representation, all pixels are of the same

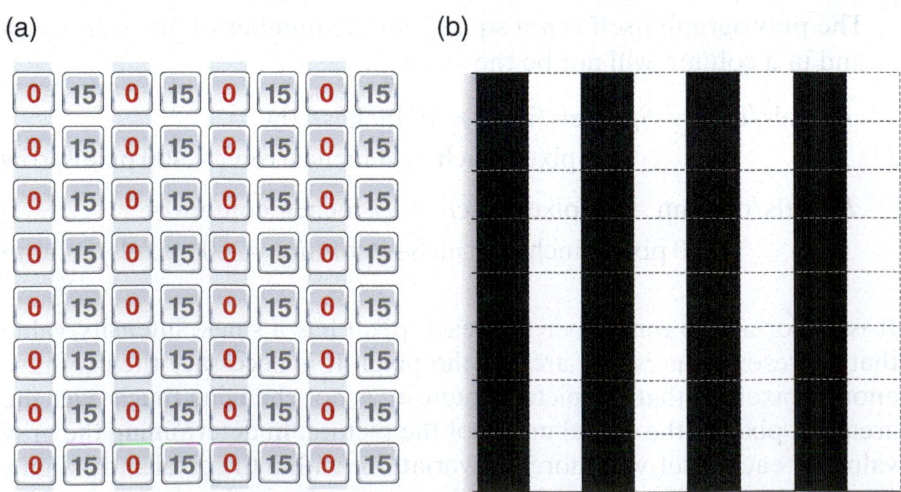

Figure 3.20 The intensity values for an 8 × 8–pixel image of vertical stripes with four bits per pixel are shown in (a). Each stripe has a width of one pixel. The corresponding gray-level image is shown in (b), with the actual pixel boundaries highlighted.

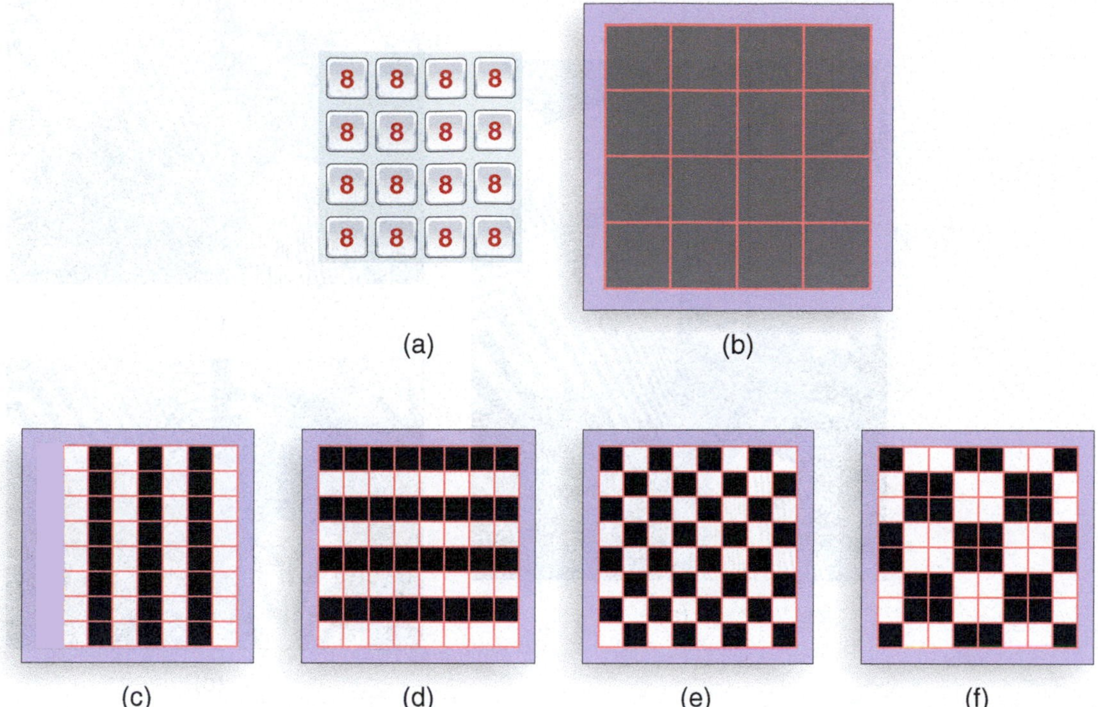

Figure 3.21 The resolution of the image shown in Figure 3.20(b) is reduced from an 8 × 8–pixel array to a 4 × 4–pixel array. In (a), the values for the new 4 × 4–pixel image are shown, and in (b) the image is shown with pixel boundaries highlighted. The same 4 × 4 image shown in (b) could have been created from any of the 8 × 8 images shown in (c–f), because each has two black pixels and two white pixels in the 2 × 2–pixel squares that correspond to one pixel of (b).

size, we must ensure that the whole image is covered with pixels that are sufficiently small to detect gray-value variations in all regions of the image. This rule of sampling will be formalized for both audio and video as the "sampling theorem" in Chapter 5.

When we choose pixel sizes that are not small enough, several types of image distortion may appear. The image may look blocky, and smooth object edges may appear staircased. In addition, the large pixels may trick our eyes into seeing connections and edges in the image that are not really there. This effect is called **spatial aliasing** because the sampled image's appearance is misleading. These distortions are also called **sampling artifacts**. Figures 3.22 and 3.23 demonstrate these effects.

Spatial Aliasing: Literally means "also appearing under a different name." For imaging, it occurs when the content in a sampled image takes on a misleading appearance compared with the original because the pixel size is too large.

Sampling Artifacts: Distortions arising in a sampled image when the sampling rate is too small to capture the finest detail in the input image.

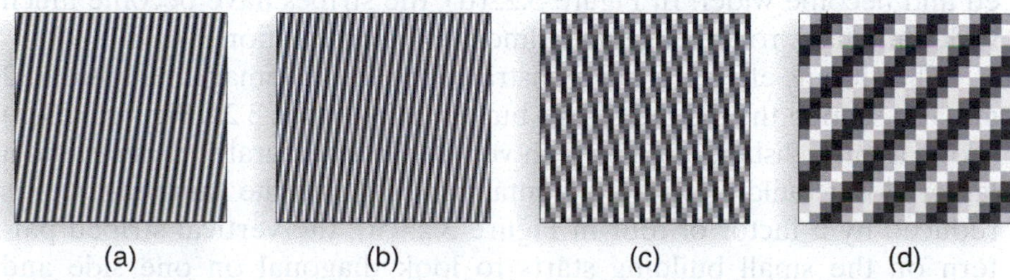

Figure 3.22 The test image in (a) has 160 rows and 160 columns of pixels. The diagonal pattern changes very little when the pixel size is doubled in (b). However, when the pixel size is doubled again in (c) and again in (d), the striped pattern changes stripe width and orientation. This demonstrates spatial aliasing due to reducing the spatial sampling rate.

Figure 3.23 The original image of two buildings in (a) has 812 rows and 650 columns. When the linear sampling rate is reduced by a factor of four in (b), the sampling rate is too low to accurately represent the vertical structures in the more distant building in the lower left, and diagonal bands begin to appear. In (c), the sampling rate is reduced by another factor of two, and these aliasing effects become more obvious.

Figure 3.22(a) shows a test image with a brightness pattern that has a slow periodic variation in the vertical direction and a faster periodic variation in the horizontal direction. The result is a striped pattern that is close to vertical. The horizontal and vertical sampling rates are reduced by a factor of two in Figure 3.22(b). The spatial sampling rate is still acceptable for this image, but some light gray diagonal stripes may be seen now going from the upper left to the lower right. Further reductions in the sampling rate by factors of two in Figure 3.22(c) and (d) result in pixel sizes that are too large to represent the fine detail of the horizontal gray-level variations. In Figure 3.22(c), the stripes have apparently rotated and become wider. In Figure 3.22(d), the stripes have become much wider and have rotated again to almost a 45° orientation.

KEY CONCEPT

Downsampling by a factor of two is the process of taking a sampled image and reducing the number of samples along rows and columns of a block of four pixels (two rows and two columns) to a single pixel in the new image.

The aliasing effects just demonstrated for the test image in Figure 3.22 are also seen in the picture of two buildings in Figure 3.23. Figure 3.23(a) shows a small distant building with vertical architectural structures and a larger closer building with horizontal bands. When the sampling rate is reduced by a factor of four in Figure 3.23(b), the vertical striped pattern on the small building starts to look diagonal on one side and curved on the other. A further reduction in the sampling rate by another factor of two in Figure 3.23(c) shows obvious curved patterns in the small building and the beginning of a curved pattern in the large building. The reason that these aliased patterns appear curved rather than like the straight patterns of the test image is that the tops of the build-

Infinity Project Experiment: Image Sampling

The choice of a sampling grid determines the size of the image and the way it looks to the eye. High-resolution digital images can be downsampled to produce low-resolution images with fewer numbers of rows and columns, using a process that is identical to that depicted in Figures 3.20 and 3.21. Try changing the sampling grid for various natural and artificial gray-scale images. How coarse can you make the sampling grid and still understand the content of the image? How does the choice of a sampling grid depend on the image's content?

ings are much further away than the bases. The perspective effect makes the spacing between the actual stripes decrease as the distance increases in that same way that railroad tracks seem to get closer together in the distance. The aliased spacing varies in a different way with distance, so apparent curved patterns appear. Often, these sampling artifacts can be seen when a person appearing on a TV show wears a jacket or a tie with a fine design such as stripes or a checkered pattern. As the person moves around, the stripes may appear to change direction, because of an inadequate sampling rate.

Sampling in Time for Movies

We noted earlier that a changing scene consists of images that are varying in time. When we record a changing scene in a movie, we end up recording discrete images, called frames, at fixed time intervals. If we record an image every second, we get a sampling rate of one frame per second. Recording images every tenth of a second, we get a sampling rate of 10 frames per second. This sampling rate in time is called a **temporal sampling rate**, to distinguish it from the sampling rate in space (spatial sampling rate) that we discussed earlier in this section. For spatial sampling within a frame, if the pixel size is small enough, our eyes do not perceive individual square pixels, but rather a continuous intensity pattern. Similarly, if the image frames are changed rapidly enough, our eye fails to perceive the individual discrete frames, and instead we have the sensation of viewing the smooth motion of objects in the scene. For a video rate of 60 frames per second, a new frame is acquired and displayed every one-sixtieth of a second.

Temporal Sampling Rate: Number of image frames taken per unit of time.

We discussed earlier the loss of fine spatial details in an image if the pixel size is not small enough. This effect was demonstrated in Figures 3.21–3.23. An equivalent phenomenon occurs in time if the objects in a scene are moving or changing substantially in a time much less than one-sixtieth of a second. We will then fail to perceive continuous motion and instead will see an image sequence with irregular or jerky motion. A characteristic of very old movies is that the movements of the actors looked jerky, because the frame rate was too low to represent continuous motion as the characters moved around the set. However, since motion pictures were such a novelty, this defect was accepted. In modern times, when news networks send reporters to remote locations for live interviews, sometimes the communication capability is limited, and the frame rate is too slow to capture continuous motion. This defect is minimized by having the reporters not move their bodies while reporting. That way, only the facial expressions and lip movements seem jerky.

Temporal Aliasing: For movies, this occurs when the sequence of image frames produces a misleading appearance because the frame rate is too slow and the time between frames is too long.

This problem of a slow frame rate can produce a **temporal aliasing** effect that is similar to the spatial aliasing effects in Figures 3.22 and 3.23. An example familiar to many is when the wagon wheels in a Western appear to rotate backward while the wagon is actually moving forward. This phenomenon can be explained with the help of Figure 3.24.

Figure 3.24 shows four frames of a movie, with the first frame at the top of the figure and the fourth frame at the bottom. Three wagons are traveling in the same direction at different speeds, and in the first frame, all the wheel markers for the wagons are in the vertical position. Wagon 2 is traveling twice as fast as wagon 1, and the speed of wagon 3 is seven times the speed of wagon 1. This condition can be observed easily in the second frame, because the wheels of wagon 1 have rotated 45°, or one-eighth of a full rotation, while the wheels of wagon 2 have rotated 90°, or two-eighths of a full rotation and the wheels of wagon 3 have rotated 315°, or seven-eighths of a full rotation. Since the speed of each wagon is constant, the wheel positions of the three wagons in the third frame will be 90°, 180°, and 630°, and in the fourth frame they will be 135°, 270°, and 945°, respectively. When the wagons are observed, it is clear that wagon 3 is traveling in the same direction as wagon 1, but at a much faster rate. But if only the wheels are observed, the wheel pattern of wagon 3 is a mirror image of the wheel pattern of wagon 1, and based on the wheels, it looks like wagon 3 is going the same speed as wagon 1 but in the *opposite* direction.

Figure 3.24 Four frames of a movie showing three wagons moving in the same direction at different speeds. Wagon 2 is traveling twice as fast as wagon 1, and wagon 3 is traveling seven times as fast as wagon 1. Temporal aliasing causes the wheels of wagon 3 to appear to be rotating at the same rate as the wheels of wagon 1, but in the opposite direction.

Infinity Project Experiment: Aliasing in Movies

The "wagon wheel" aliasing effect can make rotating objects seem to spin slowly forward and even backward in movies. The number of spokes and the frame rate determine the apparent wheel motion. The effect is particularly odd when the spinning wheel contains some painted spokes or reflectors. Take a look at a movie of a rotating bicycle wheel with reflectors as it spins slowly to a stop. How do the spokes move? How do the reflectors move? Is it possible for the spokes and reflectors to spin in opposite directions? What is going on?

Now the term aliasing can be understood, and the conditions under which it will occur can be predicted. The wheels of wagon 1 are rotating at a rate of one-eighth of a rotation per frame, so it will take eight frames to complete a full rotation of 360°. This rotation rate is slow enough relative to the frame rate to make the motion of the wheels appear fairly smooth. The wheels of wagon 2 are rotating at a rate of two-eighths of a rotation per frame, so it will take $8/2 = 4$ frames to complete a full rotation of 360°. This rotation rate is a little faster and might appear somewhat jerky, but the movie would not be misleading. Since wagon 3 has wheels rotating at a rate of seven-eighths of a rotation per frame, it will take only 8/7 frames to complete a full rotation of 360°. However, we will not see the image that would show the first full rotation, because we are sampling by taking images only at regular intervals in time, and the first full rotation occurs between frames one and two. In the four frames shown, we *see* the position of the wheels of wagon 3 at 0°, −45°, −90°, and −135°, not at 0°, 315°, 630°, and 945°. We have no way to know what actually happened between the frames, but we assume that if the frames are taken often enough to make the motion look smooth, then the slow rotation in the reverse direction is more reasonable that the fast rotation in the forward direction. From this example, we can see that if the wheels rotate more than half of a rotation between frames, we will form an incorrect opinion about the wheel motion, due to aliasing.

For some applications, an extremely high temporal sampling rate is needed, while, for others, a very low rate is acceptable. If the shattering properties of a new type of safety glass are to be evaluated from image sequences, then many image frames must be acquired in the small fraction of a second between the initial impact on the test glass and the completion of the shattering. In contrast, a video of a blooming flower could probably use frames acquired every 15 minutes.

The total number of frames contained in a video clip is determined by the frame rate and the time duration of the clip. A short, 10-second video will contain 600 frames if recorded at the full sampling rate of 60 frames per second. A full-length feature film typically is 120 minutes (or 7200 seconds) long and contains 432,000 frames. Now you can appreciate the tremendous advances in storage technology represented by such modern optical media as digital versatile discs (DVDs), which can store these many high-resolution frames and more on a single $13\frac{1}{8}$-cm-diameter piece of plastic.

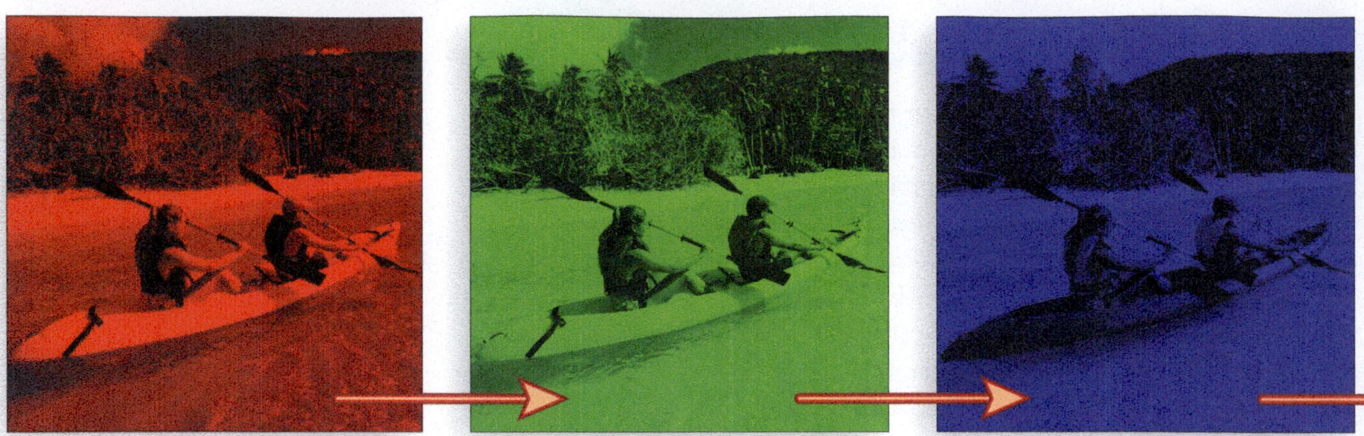

How to Add Color to Our Images

All the examples so far have involved black-and-white images that contained only intensity numbers for each pixel. Black-and-white pictures are now relatively rare, although they can still be found in art museums and hospitals (X rays). Black-and-white TV sets may still be found in security systems. Full color is what is required for all other applications of digital imaging systems.

KEY CONCEPT

Why only R, G, and B? The internal structure of the human eye and visual system allows us to represent all colors that we can see as mixtures of three colors: red, green, and blue. The visual systems of other animals are not the same as ours, and the three-color representation we use would not work well for many animals. For example, some birds can see infrared light, which is invisible to the human eye.

How do we handle color? In converting the gray-level information into a number, we used a strip like those shown Figure 3.17 for comparison. It might seem that, for color, we would have to take the same approach and have a reference strip for every color we can see. Then we would have tens of thousands of different colors, each at several levels of brightness. This approach is so much more complicated than black-and-white imaging that it is clearly unrealistic. Fortunately for us, the properties of the human visual system suggest an easier way to represent the colors that we can see.

What we call "color" is in reality a complex interplay between the physics of light waves, the design and construction of our eye, and the further processing that takes place in our brain. The net result of this interplay is that we can create a very large number of color shades by simply adding the light of three different colors: red, green, and blue (RGB). By varying the brightness of each of these colors, we create the perception of an almost infinite variety of colors. With only three colors, we can easily extend the representation we have used for black-and-white images to color images.

Color Planes: Every color can be created using three single-color images: one red, one green, and one blue. These three images are placed on top of each other and added together. These single-color components form the three color planes.

To create a color image, we create three separate images, one encoding the brightness of the red component of a scene, another the green component, and the third the blue component. These three colored images, called **color planes**, are added together to make a full-color image. Compared with black-and-white representation, color representation of an image is only three times as complex.

Each color image can be spatially sampled and quantized to discrete brightness levels as was done for black-and-white images. The number of brightness levels in each color image will determine how many different color combinations can be created. For example, if each color image is quantized to 16 levels, using 4 bits, for a total of 12 bits for all three color planes, then we will have $16 \times 16 \times 16 = 2^{12} = 4096$

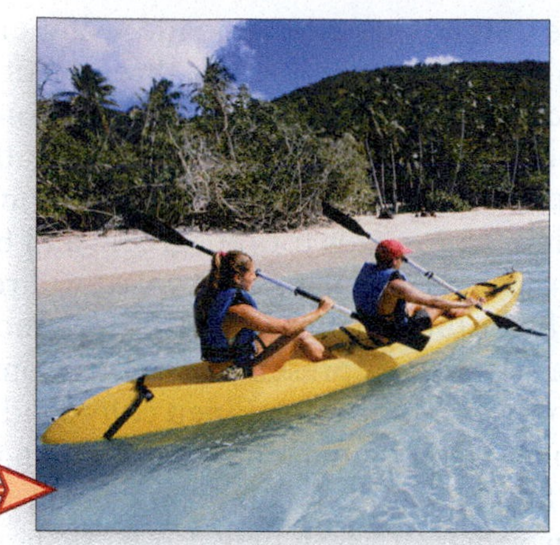

Figure 3.25 On the left, three separate red, green, and blue color planes are shown for the image of the kayakers used in Figure 3.5. On the right, we see the sum of the three color planes, which is a full-color image.

different possible color combinations. If, instead, each color in the image is quantized to an 8-bit binary representation, then a full-color image will have a much larger number of colors, 2^{24}, which is approximately 16 million color combinations. Figure 3.25 shows the three color planes and the superimposed full-color image of the black-and-white image of the kayakers from Figure 3.5. When maximum amounts of red, green, and blue light are combined, the sum is white light, so all three color-plane images are bright for the clouds and the river shore. When only red and green light are combined the sum is yellow, so the yellow back of the kayak is bright only in the red and green images. Note that the blue life vest is dark in the red image and the red cap is dark in the blue image.

In Figure 3.17, we illustrated the effect that increasing the number of bits per pixel had on the range of gray values. This concept was represented mathematically in Table 3.1. An analogous figure for three-component color representations would require a three-dimensional display with red, green, and blue intensity values changing along lines in three perpendicular directions. A one-bit-per-color-component RGB image, for example, would require three bits per pixel. In a one-bit-per-pixel black-and-white image, each pixel is either black or white. In a one-bit-per-color-component RGB image, there are three bits per pixel or eight possible colors. We can mathematically represent the eight possible colors as listed in Table 3.2. This was the color set used in the earliest color PCs.

INTERESTING FACT:

When we add the red, green, and blue color planes for an image display, we are adding light. When we add the maximum amount of each color, we get white light (see Figure 3.26, next page). When we combine ink colors, on the other hand, we are absorbing more light, not creating more light. Color printers also use three colors (yellow, magenta, and cyan) to create all the colors. But when we combine the maximum amount of each, in theory we will get black, not white. In practice, to get a crisp black, it is better just to have a fourth ink source, that is black.

Table 3.2 Colors and Corresponding RGB Bit Patterns

Pixel Values for Three Bits per Pixel, or Eight Color Images				
Decimal	**R**	**G**	**B**	**Color**
0	0	0	0	Black
1	0	0	1	Blue
2	0	1	0	Green
3	0	1	1	Cyan
4	1	0	0	Red
5	1	0	1	Magenta
6	1	1	0	Yellow
7	1	1	1	White

Figure 3.26 Sir Isaac Newton disperses sunlight through a prism. The white light separates into its underlying components—the colors of the rainbow.

Table 3.2 shows all the colors that can be made using either the maximum amount or zero amount of each of the three color components. We can use those colors to define a three-dimensional color cube that represents all possible colors within the cube in the same way that Figure 3.17 represents all possible gray values along a line. Figure 3.27(a) shows the three color axes for red, green, and blue. The coordinates of any point in the three-dimensional space will represent a particular color by specifying the amount of red, green, and blue to be added. If only two bits are used to represent each color component, then six bits will be used to represent the combined color, and $2^6 = 64$ distinct colors will be possible. These combinations are shown in the color cube in Figure 3.27(b). The four values of red only can be seen along the horizontal axis, and the four values of green only can be seen along the vertical axis. The maximum combination of red and green is yellow, which is visible in the upper right corner of the front face of the cube. The blue contribution increases along a direction into the page, but in this view of the cube it can't seen, as only 37 of the 64 colors are visible. In Figure 3.27(c), the cube has been cut into four slices to show all 64 colors. The front face of the cube is the slice on the left, and the rear face is the slice on the right.

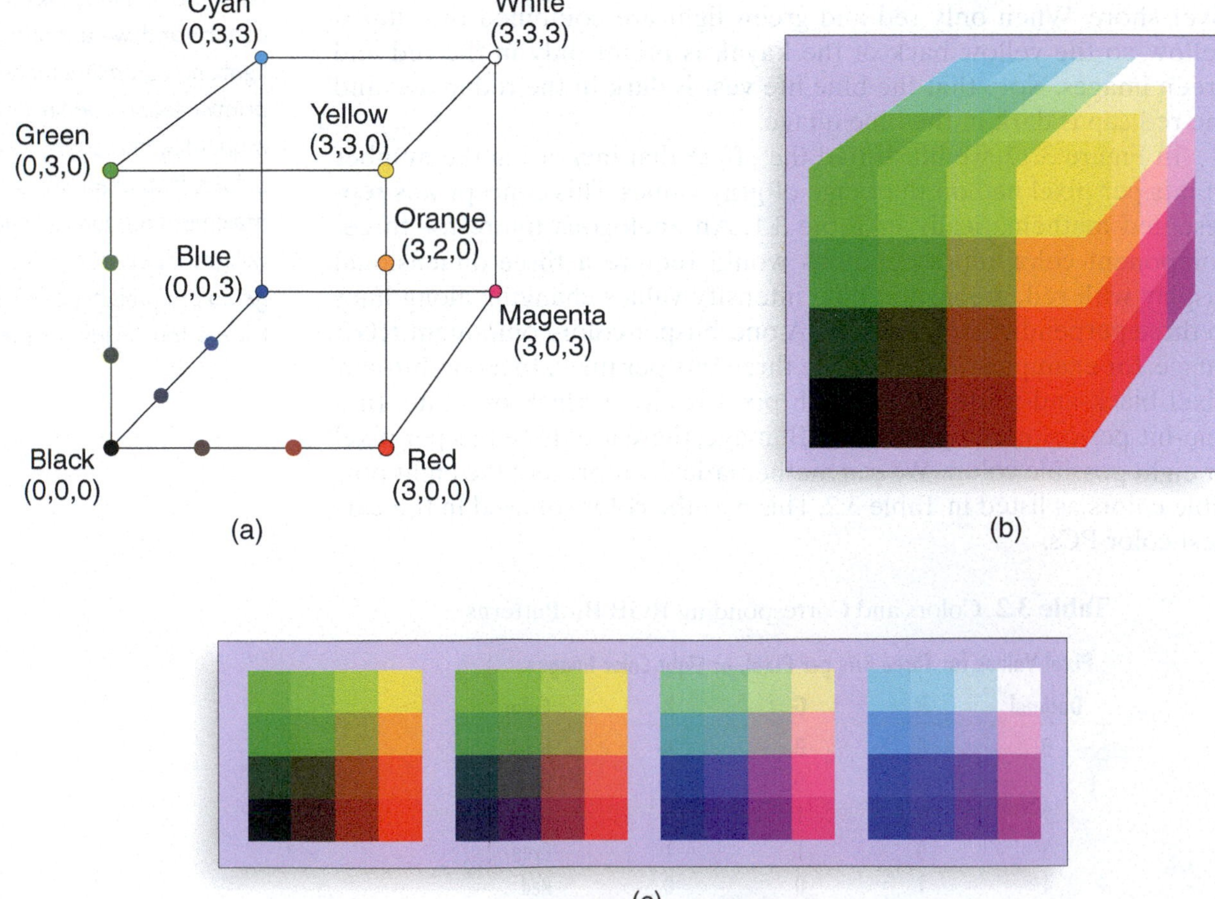

Figure 3.27 The three color axes are shown in (a), and a color cube for 64 colors is shown in (b). In (c), the color cube from (b) is sliced to show all 64 colors.

Infinity Project Experiment: Color Representation

Change the number of bits assigned to each color plane in a set of color images provided by your instructor or in a image from a digital camera or scanner. Remove one color plane completely, and observe the effect on the color. Change the numeric values associated with the red, green, and blue components of a region to produce the desired color effect.

EXAMPLE 3.2 Determining Colors and Required Bits

A color image uses four bits per color for each pixel. How many different colors can be represented? If the image is 200 pixels wide and 100 pixels high, how many bits are needed in order to store the image?

Solution

Each color can have $2^4 = 16$ different values, so the total number of colors is $(2^4)^3 = 2^{12} = 4096$ different colors. Each pixel needs 12 bits to represent the color, since there are 4 bits for red, 4 for green, and 4 for blue. The total number of bits needed is

$$(200 \text{ pixels/row}) \times (100 \text{ rows}) \times (12 \text{ bits/pixel}) = 240{,}000 \text{ bits}$$

EXAMPLE 3.3 Determining Pixel Bit Size

A color image containing 400 × 600 pixels requires 5,760,000 bits of memory in order to store all three RGB color planes. If each color plane uses the same number of bits per pixel, how many bits are used to represent a single blue pixel?

Solution

The 5,760,000-bit color image contains three different images, corresponding to the red, green, and blue color planes, respectively. If each color plane uses the same number of bits per pixel, then the amount of memory needed for a single color plane is one-third of the total:

$$(5{,}760{,}000 \text{ bits})/(3 \text{ color planes}) = 1{,}920{,}000 \text{ bits/color plane}$$

In each color plane, there are 400 rows × 600 pixels/row = 240,000 pixels. Therefore, the number of bits per pixel in each color plane is

$$(1{,}920{,}000 \text{ bits/color plane})/(240{,}000 \text{ pixels/color plane}) = 8 \text{ bits/pixel}$$

Each pixel color is represented by 8 bits of red, 8 bits of green, and 8 bits of blue for a total of 24 bits, no matter what the color is. The color is changed by changing the bit pattern, not the number of bits.

EXERCISES 3.2

Mastering the Concepts

1. What's the relationship between the number of bits per pixel and the number of gray levels for a pixel? Which number is larger?
2. How is pixel size related to spatial sampling rate?
3. What is spatial sampling? Given a black-and-white photo from a film camera, how do you use spatial sampling to create a digital image?
4. How are sampling artifacts created in a digital image? Are these artifacts caused by too fine of a sampling grid or too coarse of a sampling grid?
5. What is the effect of not sampling often enough in movies?
6. Why do we use a mixture of three different colors to represent one color in a digital color image?

Try This

7. A Pentium microprocessor uses 32 bits to represent a number. If we use 32 bits to represent the color of a pixel, how many colors will we have?
8. If a color image is represented by 16 bits per pixel, what is the total number of colors that can be obtained?
9. How many bits are needed to represent a pixel with 32 possible gray levels? How many bits will be needed for 16,384 gray levels?
10. An 8×8–pixel image is defined as follows: The four corner pixels all have a value of 15. The two central columns and the two central rows have a value of 15. All other pixels have a value of 0.
    a. Write this image as a grid of numbers and draw it. What is the total number of pixels in the image? How many bits are needed to represent the gray values of the pixels? What is the total number of bits needed to represent the image?
    b. If we reduce the sampling rate for the image by a factor of two along rows and columns, what will be the size of the new image? Write this image as a grid of numbers rounded to the nearest integer. Draw the new smaller image.
    c. Many different 8×8 images will look the same as the result that you found in (b) when the sampling rate is reduced by a factor of two. (*Note*: Refer to Figure 3.21.) Specify one of these 8×8 images that is as different from the image specified in (a) as you can make it. Write your image as a grid of numbers and draw it.
11. Draw four successive frames of a movie of a moving wheel containing one spoke. Select a frame sampling rate of two frames per second. Draw the diagrams for the following rotation rates of the wheel:
    a. One rotation every four seconds
    b. One rotation every second
    c. Two rotations per second
    d. Two and a half rotations per second
    e. Three rotations per second
    f. Four rotations per second

    Explain the effects observed.

12. A wheel is rotating at 15 revolutions per second. At what frame rates greater than 60 frames per second, but less than 100 frames per second, will the wheel appear to be stationary?

In the Laboratory

13. Every display screen—from the computer screen that you use in the computer lab to the small screen of a portable video game—is made up of pixels. Measure the dimensions of any computer screen that you use regularly. Then go into the "Display" settings and find the resolution at which the screen is operating. You should see this information in (number of columns) × (number of rows). What are the pixel dimensions of your screen?

Back of the Envelope

14. Determine the pixel size of a standard-definition and a high-definition television:
    a. All standard-definition television screens are measured by their diagonal measurement, which is the distance from any upper corner to the opposite lower corner. The ratio of the width to height of a standard television screen is 4:3 (so that a 20-inch screen has a width of 16 inches and a height of 12 inches). The number of horizontal lines of a standard-definition television is 480. Calculate the pixel size of a standard-definition television whose diagonal measurement is 20 inches. Repeat this calculation for a television with a 36-inch diagonal measurement.
    b. Repeat part (a) for a high-definition wide-screen television in which the number of horizontal lines is 1080 and the ratio of the width to height of the screen is 16:9.
15. You are designing an imaging system for a page reader that will image a page of text and convert the printed text into computer-readable text strings. Using this page as an example, estimate how many pixels per inch will be needed to create character images that you could read visually. Characters should not be broken or merged in the digital image. If you used this estimate, how many pixels would be needed for a full 8.5 × 11-inch page?

3.3 Putting It Together

So far in this chapter, we have developed methods for representing full-color, full-motion images as binary digits, or bits. Now, we will try to put everything together to get a good understanding of what resources we need for the type and quality of images and video we want to create. How many pixels are needed for different kinds of images? How many distinct intensity levels do we need in our quantizer? What kind of frame rate is adequate? All this translates into the total number of bits needed to store a digitized image or image sequence. If we want to transmit a digitized movie for real-time viewing, we need to know what data rates are required in order to accommodate the number of bits generated per second.

The finite resources available to us are

1. the storage capacity of our system, given in the number of bits; and
2. the data rate of our system, given in the number of bits per second that we can transmit.

Any engineering design problem involves making optimum use of available resources. After all, a good engineering design for a car, for example, leads to a product that delivers maximum performance for the lowest cost, weight, size, or amount of gas consumption. We will take a similar approach in analyzing our potential designs for a digital imaging system in order to decide how many bits we are willing to "buy" for the quality we desire.

Sampling and Quantization

First, let's consider the simplest case of representing a single black-and-white image. Two elements we need to decide on are the number of pixels in the rows and in the columns of the image. If I represents the number of rows and J represents the number of columns, the total number of pixels, N, is computed using Equation (3.5):

$$\begin{aligned} N \text{ pixels} &= J \text{ pixels/row} \times I \text{ rows} \\ &= I \text{ pixels/column} \times J \text{ columns} \end{aligned} \tag{3.5}$$

A very low-quality image generated by a Web camera may have 240 rows and 320 columns. Such an image will have $320 \times 240 = 76{,}800$ pixels. A higher quality digital image might have twice the number of rows and columns. The total number of pixels in this 640×480 image will be 307,200. A graphing calculator could have a small display containing an array of 96×64 pixels (6144 pixels total), while a high-resolution display might be as large as 2000×2000 (four million total) pixels or more.

In Equation (3.1), we used M to indicate the number of discrete steps for quantizing the gray values of each pixel and m to represent the number of bits needed to represent M gray values. If we want a very crude image, we could use only 16 gray-value steps per pixel, which requires 4 bits per pixel. On the other hand, a medium-quality image will use 256 gray-value steps per pixel, corresponding to 8 bits per pixel. With the increasing performance of digital electronics, high-quality professional equipment can quantize an image to as many as 12 bits (4096 gray values) to 16 bits (65,536 gray values). A high-contrast pure binary image will allow only 2 gray values per pixel, either black or white, and will require 1 bit per pixel to specify the pixel intensity.

Putting Equations (3.1), (3.2), and (3.5) together, we get the following equation for b, the total number of bits required to represent a black-and-white digitized image with I rows, J columns, and M gray-level values:

$$\begin{aligned} b \text{ bits} &= N \text{ pixels} \times m \text{ bits/pixel} = \\ &\quad I \text{ pixels/column} \times J \text{ columns} \times \log_2(M) \text{ bits/pixel} \end{aligned} \tag{3.6}$$

Byte: Eight bits equal one byte.

Note that in computer language, eight bits are called a **byte**, and bytes are often used to measure storage amounts. Bytes are convenient to use because data communications systems often transfer multiples of eight bits at a time. For large amounts of data storage, standard prefixes are used: MB represents a megabyte, which is a million bytes, and GB represents a gigabyte, which is a billion bytes. Equation (3.6) can be modified to compute bytes B by simply dividing the number of bits by eight, as shown in Equation (3.7):

$$\begin{aligned} B \text{ bytes} &= \frac{(N\text{pixels} \times m \text{ bits/pixel})}{(8 \text{ bits/byte})} \\ &= \frac{(I \text{ pixels/column} \times J \text{ columns} \times \log_2(M) \text{ bits/pixel})}{(8 \text{ bits/byte})} \end{aligned} \tag{3.7}$$

Table 3.3 Number of Bits and Bytes Needed to Store Images of Varying Quality

Columns = Pixels/Row	Rows = Pixels/Column	Number of Pixels	Gray Levels	Bits/Pixel	Total Bits	Total Bytes
J	I	N	M	m	b	B
320	240	76,800	2	1	76,800	9600
320	240	76,800	16	4	307,200	38,400
320	240	76,800	256	8	614,400	76,800
640	480	307,200	2	1	307,200	38,400
640	480	307,200	16	4	1,228,800	153,600
640	480	307,200	256	8	2,457,600	307,200

Now we will calculate the number of bits and the number of bytes required in order to represent digital images of varying quality. We consider two different sizes of pixel arrays, 320 × 240 and 640 × 480, and three different values for the number of steps of gray values, 2, 16, and 256. Table 3.3 gives the values for these choices.

Which Is a Better Picture?

As we examine Table 3.3, we find that the range of total number of bytes per image is quite striking—from 9600 bytes to 307,200 bytes. We also notice that different image sizes may require the same number of bits of storage. For example, a 320 × 240–pixel image with 16 gray values per pixel requires the same number of bytes as a 640 × 480–pixel image with 2 gray levels per pixel. The following question now arises:

> *If we have 38,400 bytes of storage at our disposal, should we use it to store a more detailed, higher resolution 640 × 480–pixel image, or should we choose the lower resolution 320 × 240–pixel image, with more gray values per pixel?*

The answer to this question is, "It depends on what the image contains!" If the original object has a naturally high contrast with only two gray levels, such as a computer-generated line drawing that was created on a 640 × 480 grid or a printed page with text in large and plain fonts, then obviously we should choose the more detailed image with more pixels and fewer gray levels. Reducing the number of gray levels will make little difference in a high-contrast image, since most of its pixels will have one of two extreme gray-level values. If we made the opposite choice and kept more gray levels at the price of having fewer pixels, we would have less detail about the lines in the drawing, and some of the fine features may be completely lost. On the other hand, if the original object were the photograph of a face, which has less fine structural detail, but many different intensity levels that help us perceive the shape of the face, then we would benefit by having more gray values per pixel rather than by having more pixels.

The effects on image content of reducing the number of gray levels or reducing the number of pixels are illustrated with the three images used in Figures 3.28–3.33. The image of the kayakers has fine detail structure in the trees and foliage that is represented by sharp changes in neighboring pixel gray-level values. Slowly varying gray-level values are found in the clouds and on the kayak. In the picture of the young woman's face with the ocean in the background, slowly changing pixel intensity levels define the facial structure and water surface, while the

Figure 3.28 The original 650 × 650–pixel image of the kayakers in (a) is shown in (b) with the number of pixels reduced by a factor of 4 in both the horizontal and vertical directions. In (c), the number of pixels is reduced by a factor of 8 in both directions, and in (d) the number of pixels is reduced by a factor of 16.

Figure 3.29 The original eight-bit-per-pixel image from Figure 3.28(a) is shown with four bits per pixel (a), three bits per pixel (b), two bits per pixel (c), and one bit per pixel (d).

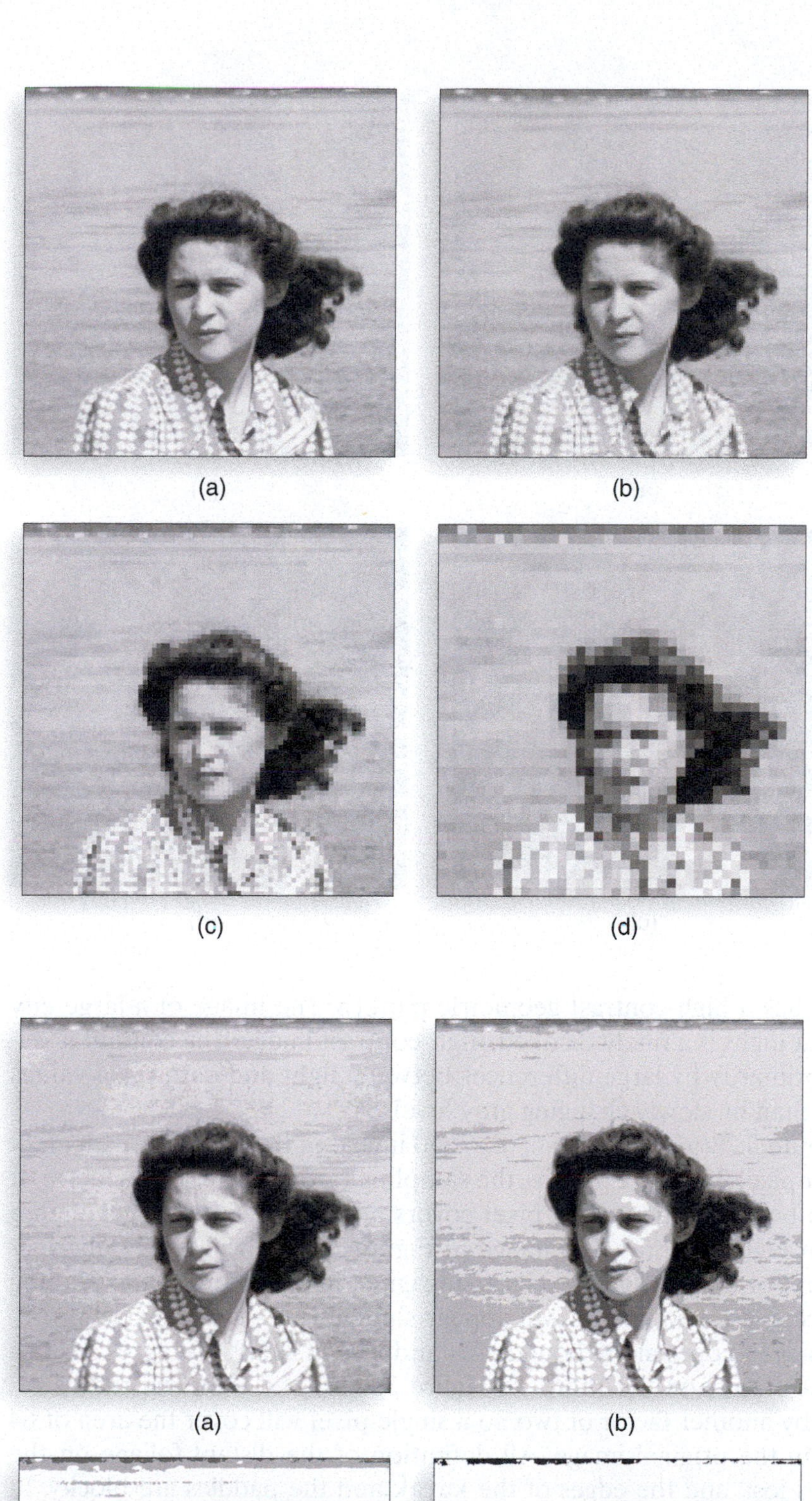

Figure 3.30 The original 650 × 650–pixel image of a young woman with the ocean in the background in (a) is shown in (b) with the number of pixels reduced by a factor of 4, in (c) with the number of pixels reduced by a factor of 8, and in (d) with the number of pixels reduced by a factor of 16.

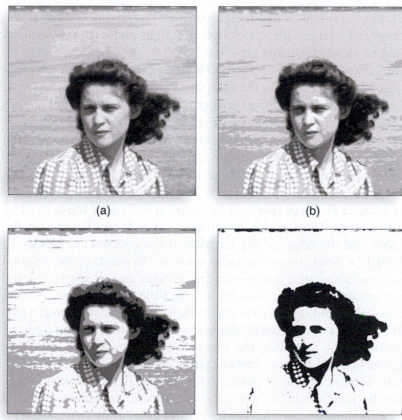

Figure 3.31 The original eight-bit-per-pixel image from Figure 3.30(a) is shown with four bits per pixel (a), three bits per pixel (b), two bits per pixel (c), and one bit per pixel (d).

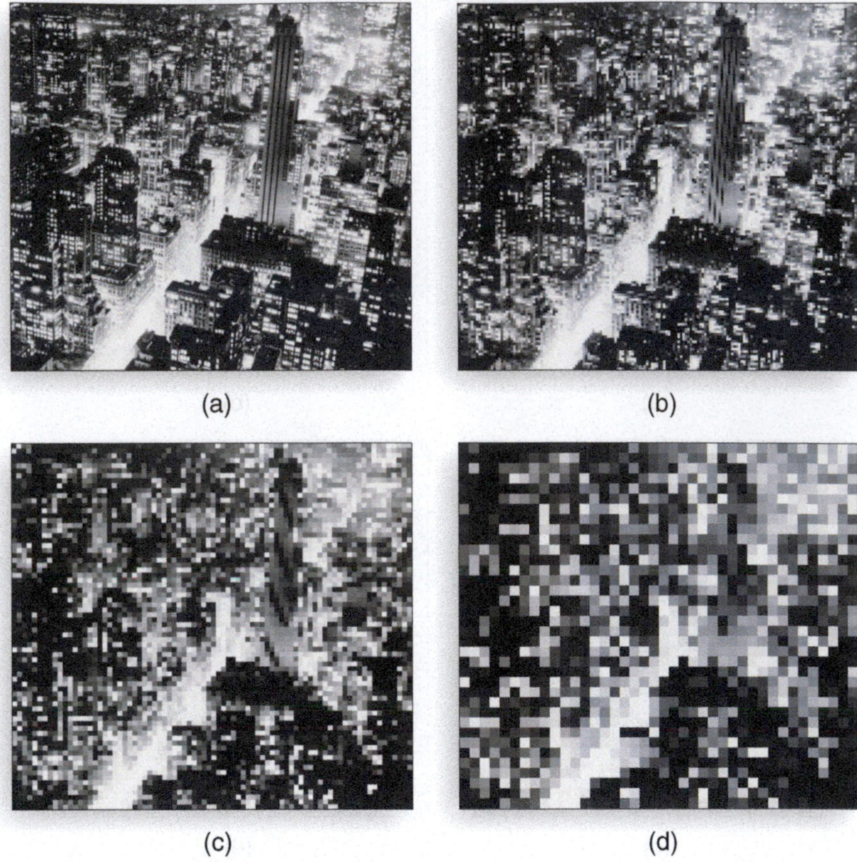

Figure 3.32 The original 650 × 650–pixel image of a large city street at night in (a) is shown in (b) with the number of pixels reduced by a factor of 4, in (c) with the number of pixels reduced by a factor of 8, and in (d) with the number of pixels reduced by a factor of 16.

blouse has a high-contrast geometric pattern. The image of a large city street at night is a finely detailed, high-contrast image. The content is defined primarily by large differences between light and dark pixel values rather than by slowly changing gray levels.

Figure 3.28(a) shows the 650 × 650 image of the kayakers with eight bits per pixel. In Figure 3.28(b), the sampling rate is reduced by a factor of four in both directions, so a pixel covers an area 16 times larger than a pixel in Figure 3.28(a). The image in Figure 3.28(b) will need only one-sixteenth of the storage used by the original image. Viewed at a reading distance, these two images will appear similar, but a close inspection reveals jagged edges along boundaries and loss of small structures such as the tree trunks on the shore. In Figure 3.28(c), the sampling rate is reduced by another factor of two, so a single pixel will cover the area of 64 pixels in the original image. All definition of the distant foliage on the shore is lost, and the edges of the kayak and the paddles are blocky. In Figure 3.28(d), a pixel covers the same area as 256 pixels in the original image. The individual pixel boundaries are obvious, and most image detail is lost. For example, we can no longer see that the kayakers are wearing life vests, and the heads of the kayakers, which are defined by about 6 pixels, could not be used to recognize their identity. However, if the figure is viewed from several feet away, the low-resolution images will look very much like the original, because from a distance the limits of our human visual system do not allow us to perceive fine details in the original image.

When the number of bits per pixel is reduced, as demonstrated in Figure 3.29, the image resolution for fine detail in areas of high contrast is not lost. Even though the number of different intensity levels is reduced,

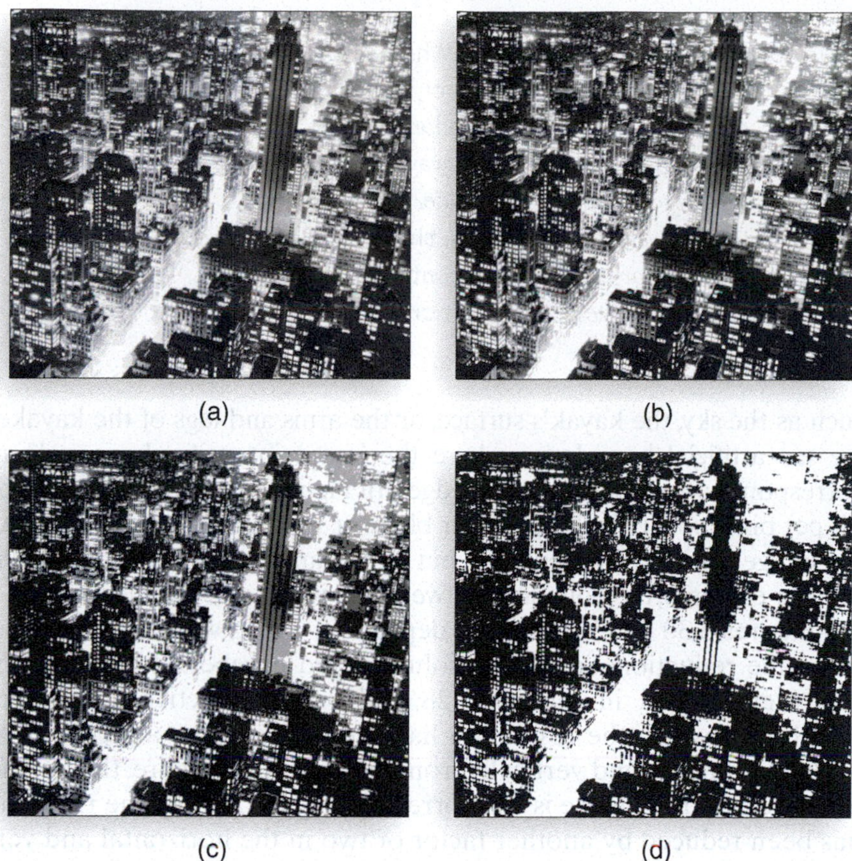

Figure 3.33 The original eight-bit-per-pixel image from Figure 3.32(a) is shown with four bits per pixel (a), three bits per pixel (b), two bits per pixel (c), and one bit per pixel (d).

the differences between pixels are still visible, because in high-contrast areas the differences between the original neighboring pixel values are quite large. However, in areas of the image where the intensity varies slowly, the effect of reducing the number of gray levels is much more obvious. In Figure 3.17, we saw that when there are more than 64 intensity levels, we perceive a continuous variation rather than individual step changes, but as the number of gray levels is reduced, the step changes become noticeable. In areas of the image that are slowly varying, the reduced number of gray levels results in either big intensity changes or no intensity changes. The noticeable steps due to big intensity changes are perceived as outlines of areas that do not have any structural meaning. This effect is often called false contouring or posterization. The contour lines are similar to a child's paint-by-number picture, in which regions of the painting are filled by a limited number of colors, and the boundary of a region to be filled by a single color is often not the same as a structural boundary.

In Figure 3.29, 650 × 650–pixel images of the kayakers are shown with four, three, two, and one bit per pixel, respectively. It is difficult to notice the effect of reducing the number of bits per pixel from eight to seven, six, or five, so these examples are not shown. In Figure 3.29(b), which uses eight gray levels, and Figure 3.29(c), which uses four gray levels, the structures of the edges of objects are still smooth. However, with such a small number of gray levels, all changes in pixel values are noticeable, and a noticeable change between gray levels always causes us to see edges of objects or patterns. This is not much of a problem for the finely detailed area of the trees and bushes on the shore or for the edges of the kayak and the shoreline. But for large areas with slowly varying intensity,

INTERESTING FACT: The term "contour" is used to mean a boundary line. When we see a boundary line in a photograph, we assume that it is the edge of an object or part of a pattern. When we see an area of slowly changing gray levels without a boundary, we assume that the area is part of a single object. When only a small number of gray levels is used, all changes from one gray value to another are large enough to be noticed, so they are all seen as boundaries. In image areas that originally were slowly varying, the noticeable changes appear to form connected boundaries, which are called false contours, because they are not edges of any objects. This effect is also called posterization.

such as the sky, the kayak's surface, or the arms and legs of the kayakers, we see artificial boundaries where the image intensity changes do not correspond to any real structural edges. In Figure 3.29(d), there is only one bit per pixel, so all pixels are either black or white. The general content of the image is still easily identified, but significant details are no longer visible. For example, the boundary between the water and the shore is lost.

Figures 3.30 and 3.31, which depict the young woman's face, show the same reductions in pixel resolution and number of bits per pixel that were shown in Figures 3.28 and 3.29, respectively. In Figure 3.30(b), for which the resolution has been reduced by a factor of two both horizontally and vertically from part (a) of the figure, the geometric pattern of the blouse is still correct. In Figure 3.30(c), the resolution has been reduced by another factor of two in the horizontal and vertical directions, and this reduced resolution is too low to represent the pattern accurately. Even at the lowest resolution, in Figure 3.30(d), the content of the image is clear, although when viewed at reading distance it is very blocky. Viewing from a greater distance makes these images look more alike. In Figure 3.31(a), the false contouring already can be seen in the water and on the face at four bits per pixel. In part (b), at three bits per pixel, the artificial lines in the face and the water are very obvious, but the high-contrast pattern of the blouse does not seem to change. In Figure 3.31(d), at one bit per pixel, the subject of the image is still clear, but all detail identifying the background as water is lost.

The content of the image shown in Figures 3.32 and 3.33 is very different from the image of the young woman's face. It has high contrast, which means that most of the pixels are either very dark or very light, and it is finely detailed, which means that the structures are outlined by just a few pixels. The two kayakers' heads use a 45×45–pixel square and a 55×55–pixel square, respectively, and the head of the young woman uses a square area that is about 250 pixels along a side. In the nighttime view of the city street, the windows of the buildings are covered by areas that are only 6×6 pixels or less. The same reductions in number of pixels and number of bits per pixel that were used in Figures 3.28 and 3.29 are used in Figures 3.32 and 3.33, respectively, but the effects are different. In Figure 3.32(c), which has a sampling-rate reduction of a factor of eight in both directions, the definition of individual buildings is lost and the diagonal stripes on the tall building show an aliasing effect. In Figure 3.32(d), the sampling rate is too low to show any of the image's content.

Although this image of the city lights suffers more than the previous two images when the sampling rate is reduced, it has fewer problems than the other two images when the number of bits per pixel is

Infinity Project Experiment: Resolution Trade-Offs

The visual quality of a digital image depends on many factors, including the number of bits per pixel, the number of rows and columns in the image, and the content of the image. Using a digital camera, create a digital version of a page from this textbook. How does sampling the image affects its quality? Can you reduce the number of bits per pixel and still read the page? Repeat this experiment using a digital image of your face or that of a friend. How much can you reduce the number of bits per pixel before the image starts to degrade?

reduced, as shown in Figure 3.33. Because the original image has such high contrast, the image looks very much like the original even at three bits per pixel in Figure 3.33(b). In Figure 3.33(d), with only one bit per pixel, most of the image detail is still preserved, although there is some loss of detail in the larger buildings and in the most distant part of the city street in the upper right corner of the image.

From these three examples, we can draw several conclusions about how to allocate bits for pixel resolution and gray levels. If we reduce the number of bits per pixel below six and there are large areas of slow variation, we will see the problem of false contouring. The noticeable changes in gray levels may be caused by a limited selection of intensity values instead of by the existence of structures in the image. If we reduce the number of pixels, we will not be able to represent the finest details, because the boundary of an object in an image needs at least two pixels in order to show the sharp change in gray levels.

Sampling and Quantization of Color Images

In this section, we consider the use of color in images. We learned earlier that adding color to images is easy, since we can create the sensation of a full range of colors by simply superimposing three primary-color images in red, green, and blue. Each of the three separate single-colored images is quantized to a specific number of intensity levels, so the complexity and storage requirement increases by a factor of three compared with that for a simple black-and-white image of the same size.

Previously in this section, we studied the trade-off between increasing the number of pixels and increasing the number of gray levels per pixel in order to achieve the best image for the same number of total bits. The same analysis can be applied to the use of color. In Equations (3.1) and (3.2), we showed that a black-and-white image with m bits per pixel would have M possible gray levels, where $M = 2^m$. If the number of bits per pixel can be increased from m to $3m$, is it better to use three times as many bits with a black-and-white image, which increases the number of gray levels from M to M^3, or is it better to add color to the image, with m bits per color for each pixel? And again, the answer is, "It depends on what the image is and what you want to do with the image."

It also depends on how many bits are available. If 24 bits can be used, it would not be reasonable to devote them all to increasing the number of gray levels. As we saw in Figure 3.17, visually we cannot distinguish the difference between 7 and 8 bits, so we would never see the visual effect of more than 8 bits. For computational purposes, we might want to use 12 or 16 bits of gray-level information, because some processing may effectively reduce the number of meaningful bits. If we used the 24 bits for color, with 8 bits each for red, green, and blue, we could make a much more visually interesting image using over 16 million colors instead of over 16 million shades of gray. If we are limited to a small number of bits, say 6 or fewer, then we cannot create enough different colors for most images, and it would make more sense to use the bits to define gray levels. Although a typical color image would look cartoonish if 64 or fewer colors were used, for presentation charts and graphics this small number of colors could still be a better choice than a gray-level display. When the number of bits is between 6 and 16, the image's content and application can make a big difference in whether color or gray scale is the best choice.

The two color images from Figure 3.2 will demonstrate the effects of using bits for color or gray shades. In Figure 3.34, a close-up view of the

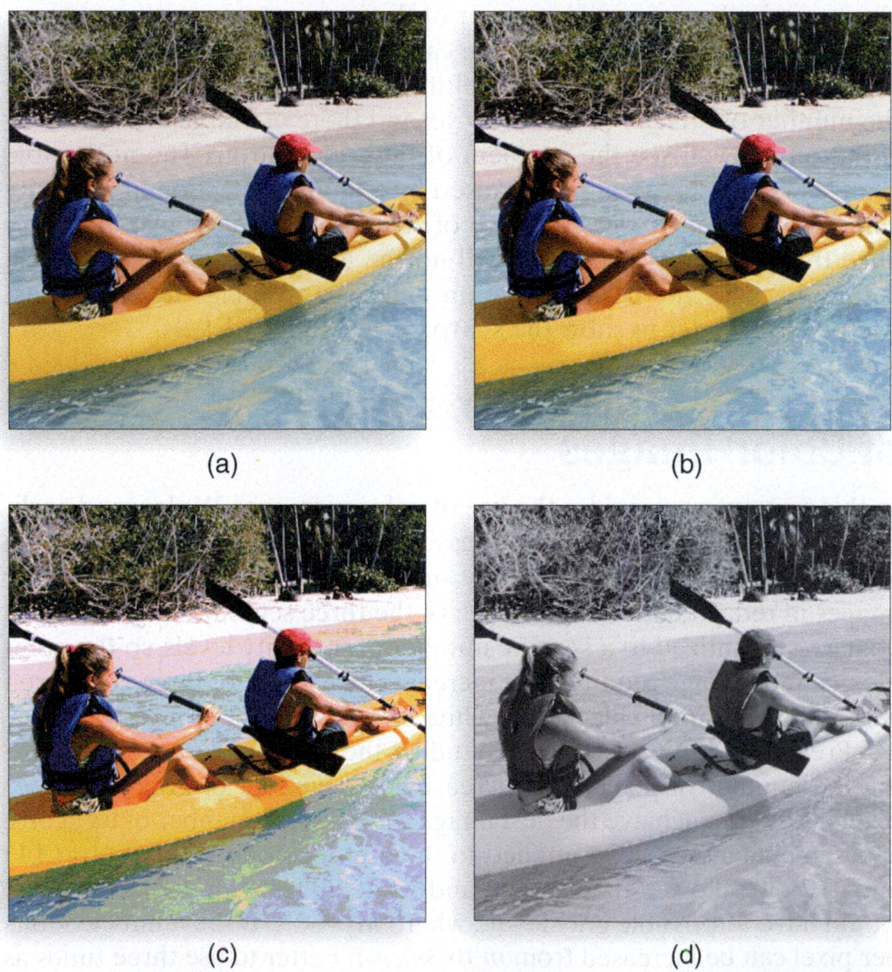

Figure 3.34 A close-up view of the kayakers from Figure 3.2 is shown with 24, 12, and 8 bits per pixel. In (a), the color image has 24 bits per pixel, with 8 bits for each of the red, green, and blue color components. In (b), only 12 bits per pixel are used, and each color has 4 bits. In (c) and (d), each pixel has 8 bits. In (c), 3 bits are used for red and green and 2 bits are used for blue. In (d), the 8 bits are used for gray levels.

kayakers can be seen. In Figure 3.34(a), the original image is shown, with the full 24 bits per pixel. When the number of bits is reduced to 12 in Figure 3.34(b), four bits are used for each of the red, green, and blue color components. At first glance, this image seems very similar to the original 24-bit-per-pixel image, but a closer examination will show the same problems with false contouring on the yellow kayak that were observed in Figure 3.29(a). When the number of bits is reduced to 8, the bits may be used for 256 gray levels, as shown in Figure 3.34(d), or for 256 colors with three bits each for red and green and two bits for blue, as shown in Figure 3.34(c). The black-and-white image has no distracting artificial contours due to a too-small number of bits, but from the image we cannot learn anything about the colors of the kayakers'clothing or of the kayak. In contrast, we do know something about the colors of objects in the image when we use 256 colors, but in this case we have both obvious false contouring and an unnatural pink color on the shoreline. The contouring is particularly noticeable on the kayak and on the arms of the kayakers.

In Figure 3.35, the orange fish from Figure 3.2 is displayed in the same four ways that are used in Figure 3.34. When eight bits are used

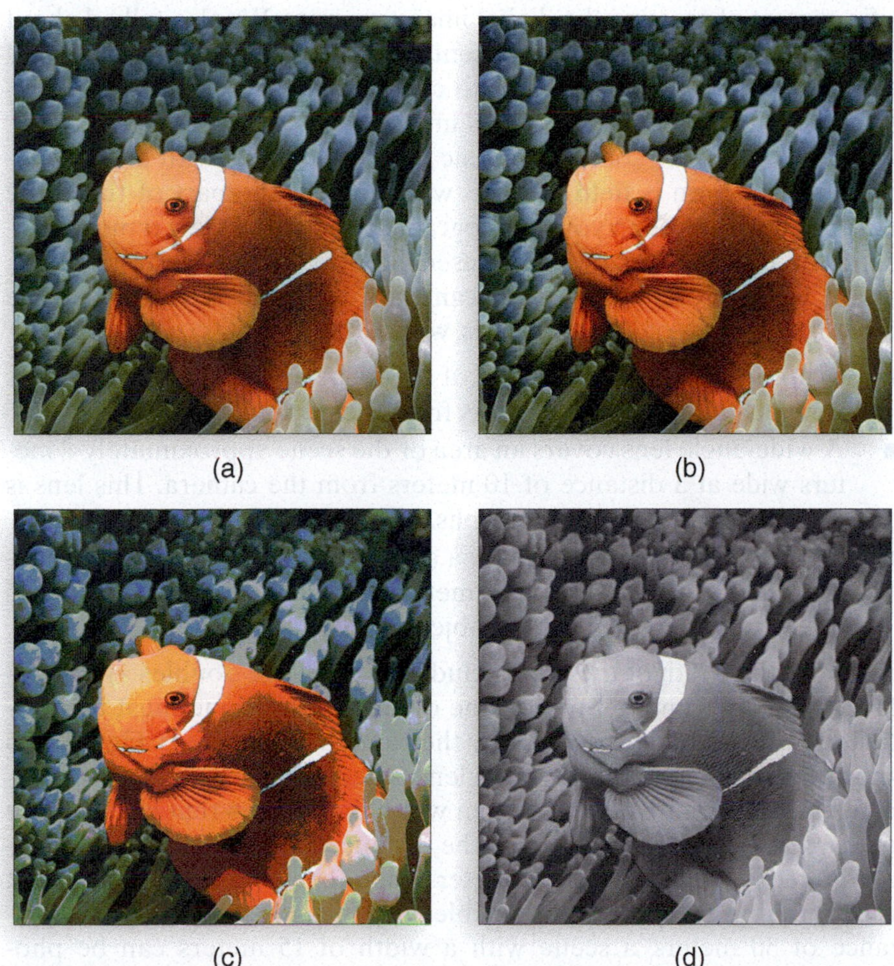

Figure 3.35 The orange fish from Figure 3.2 is shown with 24, 12, and 8 bits per pixel. In (a), the color image has 8 bits for each of the red, green, and blue color components, and in (b) each color has 4 bits. In (c) and (d), each pixel has 8 bits. In (c), 3 bits are used for red and green and 2 bits are used for blue. In (d), the 8 bits are used for gray levels.

for color in Figure 3.35(c), the colors do not look unnatural, but the abrupt changes in color cause some false contouring in the underwater plants as well as on the fish. When the eight bits are used instead for gray levels in Figure 3.35(d), the gray levels change smoothly, but there is no way to know the color of the fish.

These two examples show that it is difficult to draw a general conclusion about whether it is best to use bits for color or gray levels when the number of bits per pixel is between 6 and 12. The color and gray-level displays will be limited in different ways, and the effects will depend on the specific content of the image. In the next section, we will explore a more effective method of reducing the number of bits used for a color image while maintaining image quality.

Taking Digital Pictures

Let's return now to our original problem of proving to your friends that you were indeed close enough to your rock idol to see the whites of his or her eyes. In previous sections, we used technical terms like "sampling," "pixels," and "gray scale," but we still need to put these terms in the specific context of creating the digital image we want. We also talked about the design issues related to representing an existing picture as a digital image. Now, we will explore how to capture a view of the three-dimensional world in the form of a digital image.

Field of View: The area of the scene that will be captured by the camera.

Let's begin with some very basic terms associated with taking photos. The first term is **field of view**, which means the part of the scene that will be captured in your photograph. The field of view typically depends on the type of camera lens used, how far you are from the scene, and the format of the film or the camera's electronic sensor. There are three broad types of camera lenses, with different fields of view:

- A normal lens covers an area of the scene approximately 5 meters wide at a distance of 10 meters from the camera.
- A wide-angle lens covers an area of the scene approximately 8 meters wide at a distance of 10 meters from the camera. This lens is used for panoramic photographs.
- A telephoto lens covers an area of the scene approximately $1\frac{1}{2}$ meters wide at a distance of 10 meters from the camera. This lens is used to zoom in on a distant objects.

Since the width of the area included in an image doubles as the distance from the camera doubles, the objects in the scene must be positioned inside a triangle defined by the camera and the lens. Figure 3.36 shows these triangles for the three lenses described previously. The normal lens can capture a scene with a width of 5 meters that is a distance of 10 meters from the camera. If the distance of the objects to be photographed is doubled to be 20 meters, then the width of the scene that is captured in the photo also doubles to 10 meters. Similarly, at a distance of 30 meters, a scene with a width of 15 meters can be photographed. The height of the scene included in a photo will have the same relationship to distance as the width.

When we take a photo, the objects within the field of view are imaged on the sensor in the camera. The sensor may be either photographic film or an electronic sensor that contains an array of discrete pixels. A digital camera is usually specified in terms of how many pixels

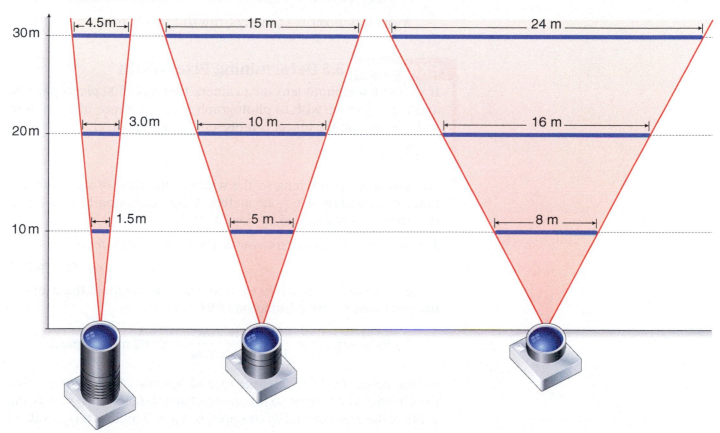

Figure 3.36 The width of the scene included in a photo increases as the distance of the scene from the camera increases. Distance is shown on the vertical axes for each of the three lens types: telephoto, normal, and wide angle. The widths of the scene included in a photo are shown for distances of 10, 20, and 30 meters. The triangle shown for each lens defines its field of view.

the electronic sensor contains. For example, a medium-quality digital camera will contain an image sensor that has 1500 columns and 1200 rows of pixels.

If we want to create a digital image with the necessary pixel resolution for the objects to be photographed, we have to relate the size of the area represented by a pixel to a physical object dimension out in the three-dimensional world. We can compute the width of the field of view at any distance by using the foregoing definitions and Figure 3.36. If d is the distance of the object from the camera and s is the width of the field of view for the particular lens at a distance of 10 meters, the field of view, f, is computed using Equation (3.8). Then the pixel width X_P is computed for J pixels per row as shown in Equation (3.9). This computation is very similar to Equations (3.3) and (3.4), except that, in the case of digitizing an existing photo, the width of the area covered by a pixel is constant, because the photograph is flat and the distance does not change. When we talk about making a picture from the three-dimensional world instead, the width of the area covered by a pixel must increase as the distance of the object from the camera increases. A similar computation is used for the height of the pixel area. The two equations are as follows:

$$f \text{ meters/row} = s \text{ meters/row} \times (d \text{ meters}/10 \text{ meters}) \quad (3.8)$$

$$X_P \text{ meters/pixel} = (f \text{ meters/row})/(J \text{ pixels/row}) \quad (3.9)$$

EXAMPLE 3.3 Determining Pixel Width

If we use a telephoto lens on a camera that has 1500 pixels per row and the object we wish to photograph is at a distance of 20 m, how wide an area on the object is represented by each pixel?

Solution

We have a telephoto lens, so the width of the field of view at a distance of 10 meters is $s = 1.5$ meters. Using Equation (3.8), we find that the width of the field of view at 20 meters is

$$f \text{ meters/row} = 1.5 \text{ meters/row} \times (20 \text{ meters}/10 \text{ meters}) = 3 \text{ meters/row}$$

Since we have 1500 pixels in each row, we can compute the width of the pixel area by using Equation (3.9):

$$X_P \text{ meters/pixel} = \frac{3 \text{ meters/row}}{1500 \text{ pixels/row}} = 0.002 \text{ meters/pixel}$$

For this range of values, it is easier to think in terms of millimeters rather than fractions of a meter, so for an object at a distance of 20 meters, the width of the area covered by one pixel is $X_P = 2$ millimeters/pixel.

If we want to get a good close-up image of the rock star, or any other person, we must decide on the proper resolution required for a clear image of the face based on the quality of the image of the eyes. The average size of the "white" of a human eye is approximately 3.5 centimeters = 35 millimeters wide and 1.5 centimeters =15 millimeters high. This means that if you take a digital picture of the performer under the conditions in Example 3.3, each eye of the star will get approximately 17 pixels horizontally and 7 pixels vertically. This might seem like enough, but the resolution is too low for the high-quality image we are hoping for. Figure 3.37 shows enlarged images of an eye where the width of the eye in pixels decreases by a factor of 2 moving from left to right across the figure. The eye in Figure 3.37(c) is about 25 pixels wide, which is a little better than the quality we might expect for our image of the rock star based on our calculations of an eye that is 17 pixels wide. Although the image can easily be identified as an eye, the eyelashes cannot be resolved

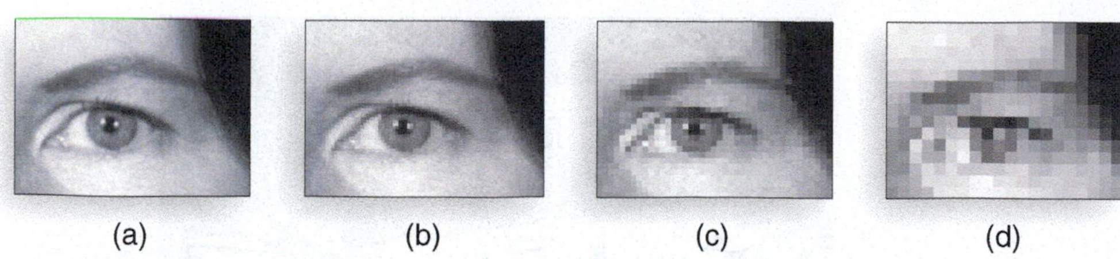

Figure 3.37 Enlarged images of an eye show the effects of pixel resolution. In (a) the eye is about 100 pixels wide. Progressing to the right, the width of the eye in pixels is reduced by a factor of 2 for each image. At a width of 25 pixels in (c) eyelashes cannot be distinguished and the edges of the eye are blocky.

and the edges of the eye are jagged. The lower resolution image to the right is unacceptably grainy or pixelated. In Figure 3.37(b), which has about 50 pixels across the eye, more details are visible and the curvature is more natural. The image on the left with 100 pixels across the eye shows the difference in shading that can convey expression. If we want to capture the expression and detail in the rock singer's eye, our image should have at least 50 pixels wide eyes.

We could double the resolution by getting twice as close, so the distance would now be 10 meters. Our field of view would decrease to 1.5 meters, and the width of the area covered by a pixel would decrease to 1 millimeter. The eye would now be imaged in an area that is 35 pixels wide and 15 pixels high. This would be similar to the images in Figure 3.37(b) and (c). So if the picture is taken from a distance of 10 meters or less, the resolution of the picture will be high enough to convince even the most skeptical friend of how close you actually were to the performer.

Capturing Events That Are Very Fast and Very Slow

The previous discussion on sampling in time was limited to the application of capturing real-time motion for human viewing. The frame rate of 60 frames/s was determined strictly by the limits on the human visual system's ability to notice changes in time. However, in some cases, we would like to record and study phenomena that occur too quickly for us to capture at this rate or too slowly for us to perceive the changes in a continuous manner. If we always use the same frame rate that is used for recording movies, either we will miss the details of motion completely for fast events or we will waste a significant number of bits for slow events by storing many images that are exactly the same and hence do not convey any additional information.

The capabilities of very high speed and very slow speed recording are not unique to digital imaging and are available in older, film-based technology as well. Use of digital technology gives us the same advantages in these special-purpose applications of real-time operation, flexibility in fixing any problems, easy sharing of information, and an almost infinite number of possible ways to process acquired data. The challenge of capturing a fast-moving object on film is quite familiar to everybody who has photographed sports events. The exposure time for a camera, that is, the time during which an

Figure 3.38 One of a series of photographs of a car as it crashed into a barrier in a test to determine the behavior of the structure and the safety of the occupant.

image is recorded, is determined by the light conditions and the sensitivity of our sensor, whether it is film or an electronic sensor. On old mechanical film cameras, the exposure time can be as short as a thousandth of a second. Special-purpose electronic cameras exist that have exposure times one-hundredth of that! The exposure time needed depends on the speed of the object to be photographed and the details on the object that we wish to preserve.

The mathematical relation is given by Equation (3.10), where T is the exposure time, V is the speed at which the object is moving past the camera, and D is the distance that the object can move while the image is being recorded:

$$T \text{ s} = \frac{D \text{ meters}}{V \text{ meters/s}} \tag{3.10}$$

EXAMPLE 3.4 Calculating Exposure Time

We wish to record the crash test of a car traveling at 36 km/hr (Figure 3.38). We wish to study the details on how the bumper crushes to within 1 cm of variation. What exposure time is needed so that the image will not be blurred and we can see the patterns as the bumper is crushed?

Solution

The speed of the car is $V = 36$ km/hr. We will first convert the speed into m/s:

$$V = \frac{36 \text{ km/hr} \times 1000 \text{ m/km}}{60 \text{ min/hr} \times 60 \text{ s/min}}$$

$$= \frac{36{,}000 \text{ m/h}}{(3600 \text{ s/hr})} = 10 \text{ m/s}$$

We can now compute an exposure time T such that the car moves less than $D = 1$ cm within that time, using Equation (3.10):

$$T \text{ s} = 1 \text{ cm} \times 0.01 \text{ m/cm}/(10 \text{ m/s}) = 0.001 \text{ s}$$

This exposure time of 0.001 s or 1 ms is much shorter than $\frac{1}{60}$ s = 16.7 ms

EXERCISES 3.3

Mastering the Concepts

1. How are the number of bits per pixel, the number of pixels per image, and the storage requirements for an image related?
2. What is false contouring? What causes false contouring?
3. What makes an image high contrast? Why is a high-contrast image less likely to show false contouring than a low-contrast image?
4. Does one have to choose the same number of bits per blue pixel as the number of bits per green pixel or per red pixel?
5. What is the field of view? When taking digital photographs, how does the field of view relate to the pixel resolution of the final picture?

Try This

6. Calculate the storage needed in number of bytes for 100 images with the following properties:
   a. 320 × 240 black-and-white images (one bit per pixel)
   b. 320 × 240 color images with 64 intensity levels for red, 32 for green, and 4 for blue.
   c. Medical X rays with 1500 × 2000 pixels and 1024 gray levels per pixel.
7. A color picture has a total of 2,097,152 distinct colors.
   a. How many bits are required for each pixel?
   b. If each of the red, green, and blue color planes has the same number of bits, how many bits per pixel are used for each color?
   c. If the red and green color planes each use eight bits per pixel, how many bits per pixel are left for the blue color plane?
8. You are taking a photograph of a butterfly that is five centimeters across. If your field of view exactly matches the size of the butterfly and the photograph has 2048 × 2048 pixels, what is the size of the smallest detail on the butterfly wing that you can resolve?
9. You are scanning 4 × 6–inch photos to make digital images. In this exercise, the field of view is specified relative to the photo size rather than to the objects in the photo.
   a. If you want to scan the whole photo with 24 bits per color pixel at 1200 pixels per inch, how big will the image file be?
   b. If you want to scan the photo for use on a webpage that limits the width and height of the photo to a maximum of 600 pixels each, what is the maximum number of pixels per inch you can use? How big will the image file be?
   c. You are scanning your photo according to the limits specified in part (b), but you want to use the full 1200 pixel/inch resolution on a small area of your photo. What is the width of the maximum area you can scan at 1200 pixels/inch and still meet the restrictions of (b)?

10. Show the trade-off between pixel resolution for an image and the number of steps in the gray-level values per pixel when the total number of bits is fixed. Assume that all images are square.
    a. If the number of bits in the image is limited to 131,072, compute the maximum number of pixels per row that can be used as the number of bits per pixel increases from 1 to 24. Plot these results with the pixels per row on the vertical axis and bits per pixel on the horizontal axis.
    b. Repeat part (a) for the case in which the number of bits in the image is limited to 2,097,152.
    c. Write a general equation for the number of pixels per row for a square image with b bits and m bits per pixel.
11. Compute the ratio of the size of the horizontal field of view to the distance to the camera for each of the three lens types.
12. In the text, we computed that, with the camera and lens from Example 3.3, an image taken from 20 m would be poor in quality and that, to get the minimum desirable resolution for an impressive photo, we would have to decrease the distance to 10 m. If we could not change the distance, how would the camera have to be improved in order to get the desired resolution?

In the Laboratory

13. For different images, we can explore the importance of color by converting a color image into a black-and-white image. One way to do this is by using a television set: Turn the "color" setting all the way down. How does the image differ from the color image? Another way to change a color image into a black-and-white image is to copy it on a copier; alternatively, you could average the red, green, and blue pixel values together before displaying the image. Make general observations on what kinds of images (such as medical, satellite, artistic, and advertisements) use color in a critical fashion.

Back of the Envelope

14. Give real-world examples of situations where the spatial resolution or number of pixels in the image is much more important than the number of gray-level values.
15. Give real-world examples of situations where the number of gray-level values is much more important than image resolution or number of pixels in the image.
16. For the following cases, state whether you would choose to use a color or black-and-white image, and justify your choice. For each case, specify the number of gray levels for a black-and-white image. This value will be the same as the number of distinct colors that is possible. If you choose a color image, indicate how you would assign the bits to the red, green, and blue color planes.
    a. You have 24 bits per pixel for an image of a sunset.
    b. You have 8 bits per pixel for an image of a car's license plate.

c. You have 3 bits per pixel for an image of bar charts of the number of students in each class.

d. You have 4 bits per pixel for an image of animals at the zoo.

e. You have 8 bits per pixel for an image of the finish line at a track meet as three runners approach it.

17. Look at ads from recent newspaper advertisements, and list 10 different digital cameras on the market. For each camera list the following information: vendor and model, number of pixels in the image sensor, optical zoom range, and cost. Try to find the greatest range of both number of pixels in the image sensor and cost. Make a plot of the cost on the vertical axis versus the number of pixels in the image sensor on the horizontal axis.

18. What kind of camera lens and what kind of image sensor are needed if you want to take the picture of a blue jay from a distance of 50 m while having at least a 5×5–pixel array covering the bird's eyes? Are the numbers reasonable? Find out which of the commercially available digital cameras can satisfy your requirements.

3.4 Better Design within the Bit Budget

In the previous section, we explored the design challenges of using a fixed number of bits to achieve the best image quality. From Equation (3.6), we found that the number of bits needed for an image was $b = I \times J \times m$, so if we decrease I and J, for example, we can increase m to keep b constant. We also explored the relative value of using the m bits per pixel to represent colors or gray levels. In several examples, we saw that decreasing the resolution or decreasing the number of bits per pixel resulted in a poor-quality image. Now that we understand the issues of image resolution and image quantization, we can ask from an engineering design perspective, if there is a better way to use a fixed number of bits than the ways we have already considered. We will look at three new approaches that can improve image quality in the context of image storage limitations.

Halftone Images

As the number of bits per pixel was reduced for a gray-scale image, distracting artificial boundaries, called false contours, often appeared. In the extreme case of one bit per pixel, all the light-gray pixels became white and all the dark-gray pixels became black, and significant information about the image's content was lost. Another approach to using one bit per pixel could set a pixel to white or black based not just on the gray value of the pixel, but also on the pattern of gray values around the pixel. This process, called dithering, could be accomplished so that the relative amount of black and white within a small area of the image is determined by the average gray value in that area of the original image.

Halftone Image: An image created with strategically sized areas of black and white to simulate gray levels to the human eye.

This type of image is called a **halftone image**. It creates the appearance of a gray-level image with many different shades of gray by using only pixels that are either black or white and strategically setting individual pixels to black or white to achieve the original average gray level. This pattern is interpreted as a gray level by the human eye when the pixels are too small to be noticed individually.

Halftone encoding was first developed for the newspaper printing industry and predates modern digital imaging. Halftone printing achieves varying amounts of gray values by controlling the size of the black dots printed in a region. Leaving the black dots out completely achieves a completely white area. On the other hand, making the black dots so big that they are overlapping each other, thus filling the entire area with black ink, results in a completely black area. Intermediate levels of blackness can be achieved by changing the size of the black spots and, therefore, the ratio of the dark space to the white space in a small square area. For example, if we want to create the perception of a pixel with a gray value that is halfway between black and white, we can use black dots with a diameter that is approximately three-fourths of the spacing between the dot centers. When we read the newspaper from a typical distance of 30 centimeters or more, our eyes are unable to resolve the individual black dots; the sensation created is of continuously varying gray values. A close examination of laser printer output of a gray-level image will show that a similar strategy is used to create the appearance of various shades of gray, using only black toner.

In Section 3.1, we learned how to sample continuous images by overlaying them with grids of regularly sized cells. If there was a variation of intensity within the cell, we simply calculated the average intensity within the cell and assigned that value to the corresponding pixel.

Now, we are reversing the process for displaying a digital image. We know that the human eye cannot distinguish very fine details in an image without magnification and that the human visual system simply responds to the average value in a region that it can resolve. Imagine that we have a grid containing 64×64 array of cells and that we allow only one-bit values for each cell. A cell can be either completely dark or completely bright.

Let's consider dividing this grid into a 32×32 grid of blocks where each block contains four cells arranged in a 2×2 square array of adjacent cells. Since each cell in this block can be either black or white, we can create five different average gray intensity values for each block. For example, we will get zero intensity when all four cells are black and an intermediate gray value when two are black and two are white. Calculating the average intensity of these four cells can be performed visually by simply holding the image a certain distance away so one cannot actually see the individual cells. It is the number of black cells in a block, rather than the exact details of the spatial arrangement of the black cells, that is important for determining intermediate intensity values. However, exercising care in designing these spatial patterns can minimize artifacts.

In Figure 3.39(g), a band of eight gray levels similar to one shown in Figure 3.17 has been converted to a halftoned image and then enlarged by a factor of 10 to explicitly show the patterns of black and white that appear as eight different gray levels when viewed from a distance. The percentage of white pixels increases steadily as we move through the

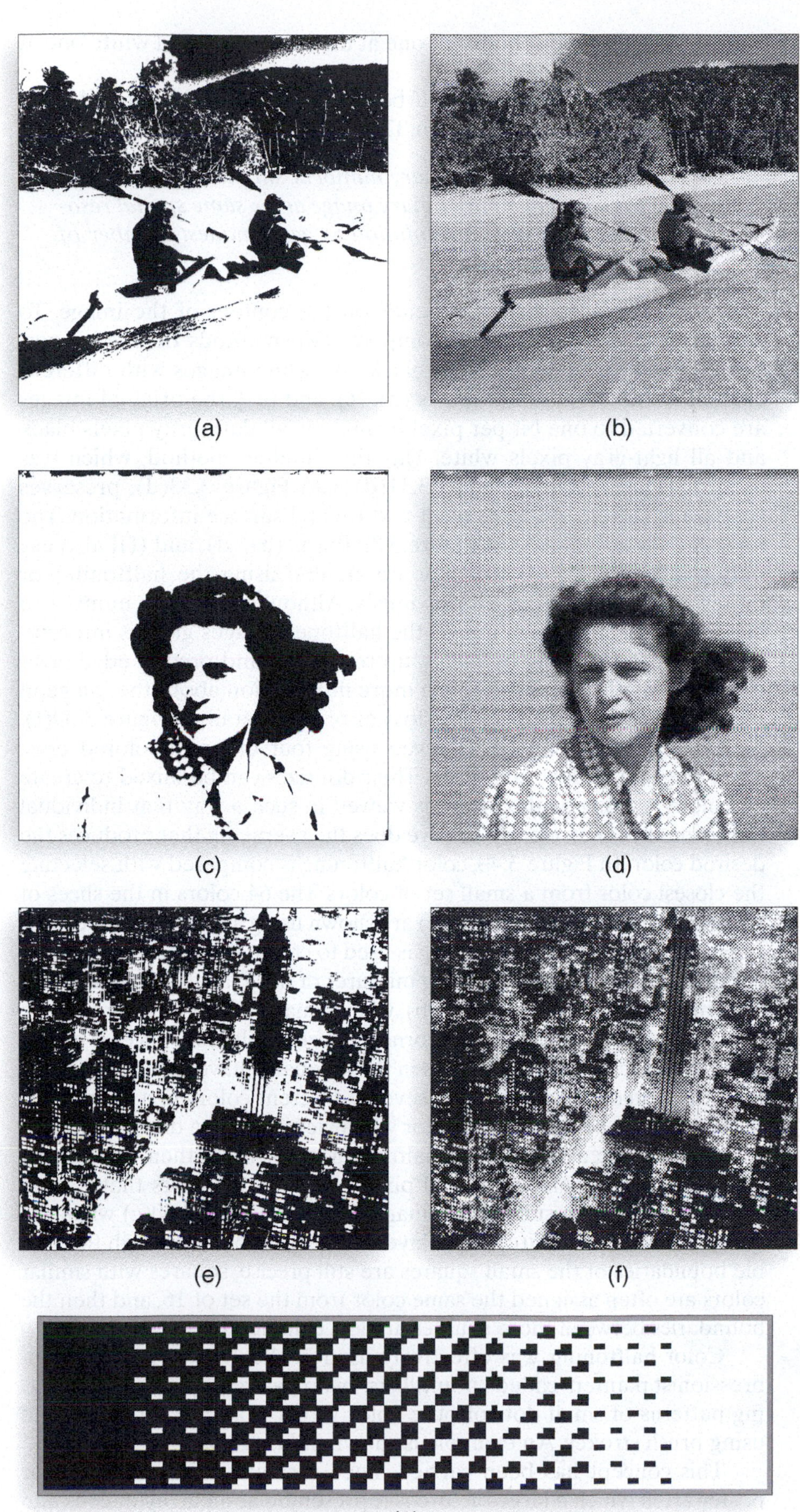

Figure 3.39 Two methods of creating one-bit-per-pixel images are compared. The images on the left [(a), (c), and (e)] were created by thresholding. These images were shown in Figure 3.29(d), Figure 3.31(d), and Figure 3.33(d), respectively. The images on the right [(b), (d), and (f)] were created by halftoning, or dithering. In (g), a band of eight different gray levels from Figure 3.17 has been halftoned and then magnified by a factor of 10 to show the pattern of black and white pixels for each gray level. A thresholded image of that band would look exactly like the top band in Figure 3.17.

eight bands from the solid black one at the left to the solid white one at the right.

Now, we can see the trade-off between spatial resolution and gray-scale resolution in a direct fashion. The question to ask now is as follows:

> *Given a 64 × 64 array of binary numbers, should we use our resources to represent a pure binary image at the same spatial resolution or sacrifice spatial resolution to get a modest number of gray levels?*

Again, the best answer depends on the content of the image. To demonstrate this, Figure 3.39 compares two methods of one-bit-per-pixel display for three different black-and-white images with different characteristics. On the left [parts (a), (c), and (e)], the original images are converted to one bit per pixel by making all dark-gray pixels black and all light-gray pixels white. This thresholding method, which was used in Figure 3.29(d), Figure 3.31(d), and Figure 3.33(d), preserves spatial resolution, but loses most texture and surface information. The images on the right side of Figure 3.39 [parts (b), (d), and (f)] also use only one bit per pixel, but they are created using the halftoning, or dithering, method discussed previously. Although the same number of bits is used to store each image, the halftoned images give us information about the shading by giving up resolution and precise edge positions. This method gives us much more information about the image in Figure 3.39(d), but causes some loss of precise detail in Figure 3.39(f).

Color printing can be achieved using four different-colored dots: cyan, magenta, yellow, and black. Their dot sizes can be mixed to create all colors. As long as the image is viewed in such a way that individual dots cannot be seen clearly, the eye does the averaging that produces the desired color. In Figure 3.40, color halftoning is compared with selecting the closest color from a small set of colors. The 64 colors in the slices of the color cube from Figure 3.27(c) are shown using a limited set of colors. In Figure 3.40(a), color halftoning is used to create the perception of colors that are not available by using mixtures of the available colors. This is seen more clearly in Figure 3.40(b), where small parts of the image have been enlarged. Each of the four corner squares of the four slices is one of the available colors, and all pixels in those squares have that color. All of the other squares have pixels of several different colors in a proportion that appears to be the desired color when viewed from a distance. If only a few colors are available and halftoning is not used, then the nearest available color is selected for all pixels. A set of 16 colors that was an early PC standard results in the image shown in Figure 3.40(c) when the image from Figure 3.27(c) is displayed in those colors. Although the visible boundaries of the small squares are still precise, squares with similar colors are often assigned the same color from the set of 16, and then the boundaries between those squares are lost completely.

Color halftoning was effectively used in the 1800s by French impressionist painters called "pointillists," who painted by carefully creating patterns of small dots of pure color rather than mixing paints and using brush strokes. An example is shown in Figure 3.41.

This concept has been carried much further in the very popular poster art in which a large picture is represented as an array of cells and each cell is, itself, a small image. When such a poster is viewed from far

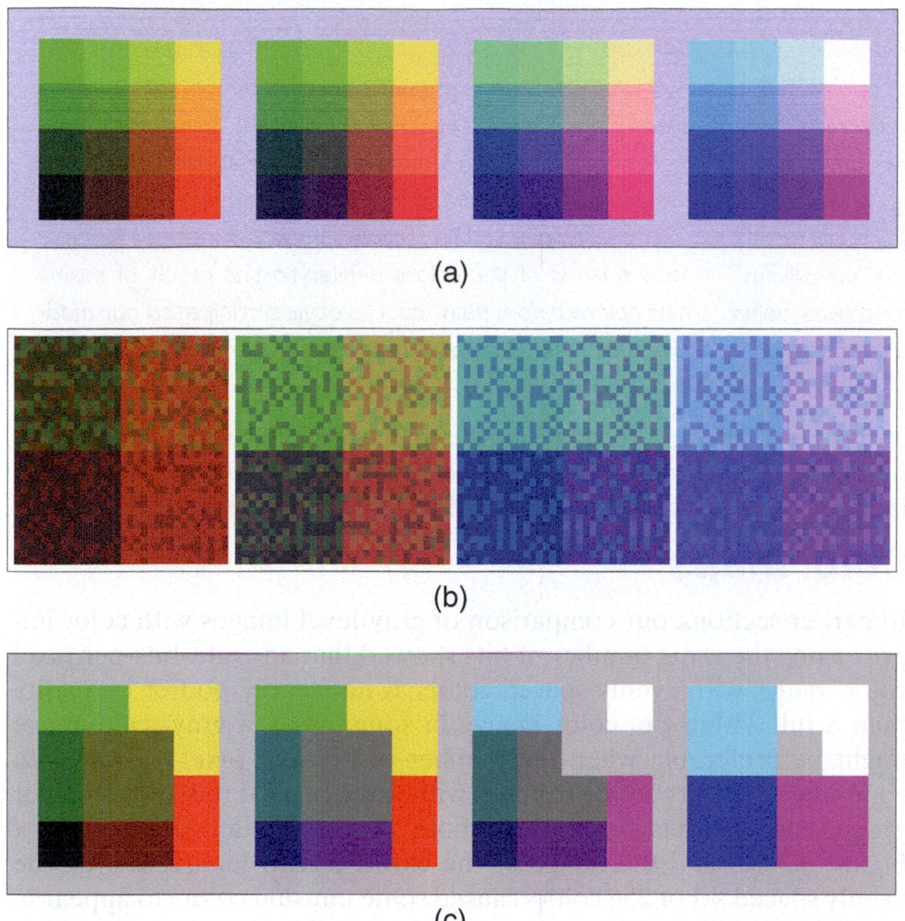

Figure 3.40 Color halftoning in (a) uses only a few colors to show the 64 colors in the slices of the color cube from Figure 3.27(c). In (b), an enlarged image of small sections from each of the four slides in (a) shows examples of the individual color mixtures. In (c), selecting the nearest color from a set of 16 colors that was an early PC standard causes many of the 64 small squares to have the same color when the image from Figure 3.27(c) is displayed.

Figure 3.41 Georges-Pierre Seurat's use of pointillism anticipated our modern methods of representing and printing images.

INTERESTING APPLICATION

Pointillism

We tend to think of dividing an image into discrete pixels of constant color as a technique derived from modern digital technology. However, over 100 years ago, a French impressionist painter, Georges-Pierre Seurat (1859–1891), developed a style of painting that he called "pointillism," Rather than paint with continuous brush strokes using mixed colors from a palette, he painted by covering a canvas with small dots of pure colors. Different-colored dots were placed close together to give the impression of a combined color. The dots were so small that a human viewer would not see the individual dots unless positioned extremely close to the canvas and would instead perceive only a blend of the colors similar to the result of mixing the colors before painting. His style anticipated our modern methods of representing and printing digital images.

away, the details in the small images melt away, leaving a general impression of a color and gray value in their place. A clever selection of individual images creates an aesthetically pleasing poster.

Colormaps

In earlier sections, our comparison of gray-level images with color images using the same number of bits showed that an eight-bits-per-pixel color image, with evenly spaced colors, is noticeably poorer in quality than a full 24-bits-per-color image. In some cases, a gray-scale image might be preferable when the number of bits per pixel is limited to eight and color bits are distributed with three bits for red, three bits for green, and two bits for blue. In both the kayak image example of Figure 3.34(c) and the image of the clown fish in Figure 3.35(c), the evenly spaced set of 256 colors caused some unusual colors to appear in some parts of the image; some artificial color boundaries were created. If the eight bits are divided into three sets used for red, green, and blue, respectively, for all images, then this loss of quality must be accepted.

An alternative method could limit an image to 256 colors, but allow the selection of the colors to be different for different images. With eight bits per color, 2^{24}, or over 16 million, possible colors are defined. For each image, the best set of 256 colors could be selected from the 16 million possible colors. This selection is called a **palette**. With this method, the value of a pixel specifies which color in a specific list of 256 should be used for the pixel rather than specifying the individual red, blue, and green components of the pixel.

Palette: The selection of colors available for a digital color image—similar to a painter's selection of colors on an artist's palette.

A separate list, called a **colormap**, of 256 colors would be made for each image. Typically, the colormap will specify the red, green, and blue components of its colors with eight bits each, regardless of the length of the colormap. Two simple color maps that allow eight colors for a three-bits-per-pixel image are shown in Table 3.4. Colormap 1 produces exactly the same colors that are listed in Table 3.2, where the three bits are used to represent the red, green, and blue components independently. A pixel with decimal value 1 will have a blue color if colormap 1 is used and a gray color if colormap 2 is used.

Colormap: A list of specific colors used in an image. Each color is defined by its red, blue, and green color content typically with eight bits of resolution for each color. But only a small number of different colors, typically 256, are allowed.

In Figure 3.42, the two methods of color display using only 256 colors are compared using three different color images. On the left [parts (a), (c), and (e)], the eight bits are divided among the three colors to produce 256 evenly spaced colors from the over 16 million possible colors using 24 bits per pixel. Three bits are used for red, three bits for green, and two bits for blue. A close-up view of Figure 3.42(a) is used in Figure 3.34(c), and

Table 3.4 Colors for Two Different Color Look-Up Tables for a Three-Bits-per Pixel Image (Colormap 1 Has the Same Result as Using One Bit Each for Red, Blue, and Green.)

Pixel Values for Three-Bits-Per Pixel, or Eight-Color, Image Using Colormaps								
	Colormap 1				Colormap 2			
Decimal Pixel Value	**R**	**G**	**B**	**Color**	**R**	**G**	**B**	**Color**
0	000	000	000	Black	000	000	000	Black
1	000	000	255	Blue	128	128	128	Gray
2	000	255	000	Green	000	000	128	Dark Blue
3	000	255	255	Cyan	000	128	000	Dark Green
4	255	000	000	Red	255	000	000	Red
5	255	000	255	Magenta	255	128	000	Orange
6	255	255	000	Yellow	255	255	000	Yellow
7	255	255	255	White	255	255	255	White

Figure 3.42(c) is also used in Figure 3.35(c). The 256 colors are widely separated, and they cannot represent the colors actually used in any of the three images. In (a), there is an unnatural pink in the sky, the shore, and the water. False contouring can be seen in large areas with slowly varying color such as the blue edges in the sky and yellow edges on the kayak. False contours can also be seen on the orange fish and the large peppers.

On the right [parts (b), (d), and (f)], a special palette of 256 colors was selected from the over 16 million colors for each image to best match the colors in that image. Because this method just uses a list of the best 256 colors without requiring evenly spaced colors, the large slowly varying areas will have more colors available for use, and infrequently used colors will not appear in the list. Note that in Figure 3.42(b), the colors look more natural and the false contouring in the sky has been eliminated. However, the red cap has been changed to an orange color, because there is such a small amount of red in the image that it was not selected as one of the best colors. In (f), many different shades of red and green were selected for the list of best colors, and the large peppers look much more natural.

In Figure 3.43, the individual color palettes of 256 selected colors for the three images of Figure 3.42 are shown. For reference, Figure 3.43(d) shows the uniformly spaced color palette. In order to display these colormapped images correctly, it is essential that the colormap be available, because the pixel value by itself no longer provides any information about the color. It only specifies the *position* in the colormap list of the correct color for the pixel. Figure 3.44 shows what can happen when the wrong colormap is used. In Figure 3.44(a) and Figure 3.44(b), the colormaps for the kayak and the clown fish have been switched. In the corresponding palettes from Figure 3.43, we can see that where the kayak palette has shades of yellows and whites, the orange-fish palette has orange shades. When the color maps are switched, the yellow kayak and white shore and clouds become orange, while parts of the claun fish turn yellow and white. This is what would happen in a paint-by-number set if the paints from one set were used with the numbers for another set. In Figure 3.44(c), the image of the peppers is displayed using the colormap for the kayak image. The colors in Figure 3.44(b) and Figure 3.44(c) are the same because the same colormap is used for both images.

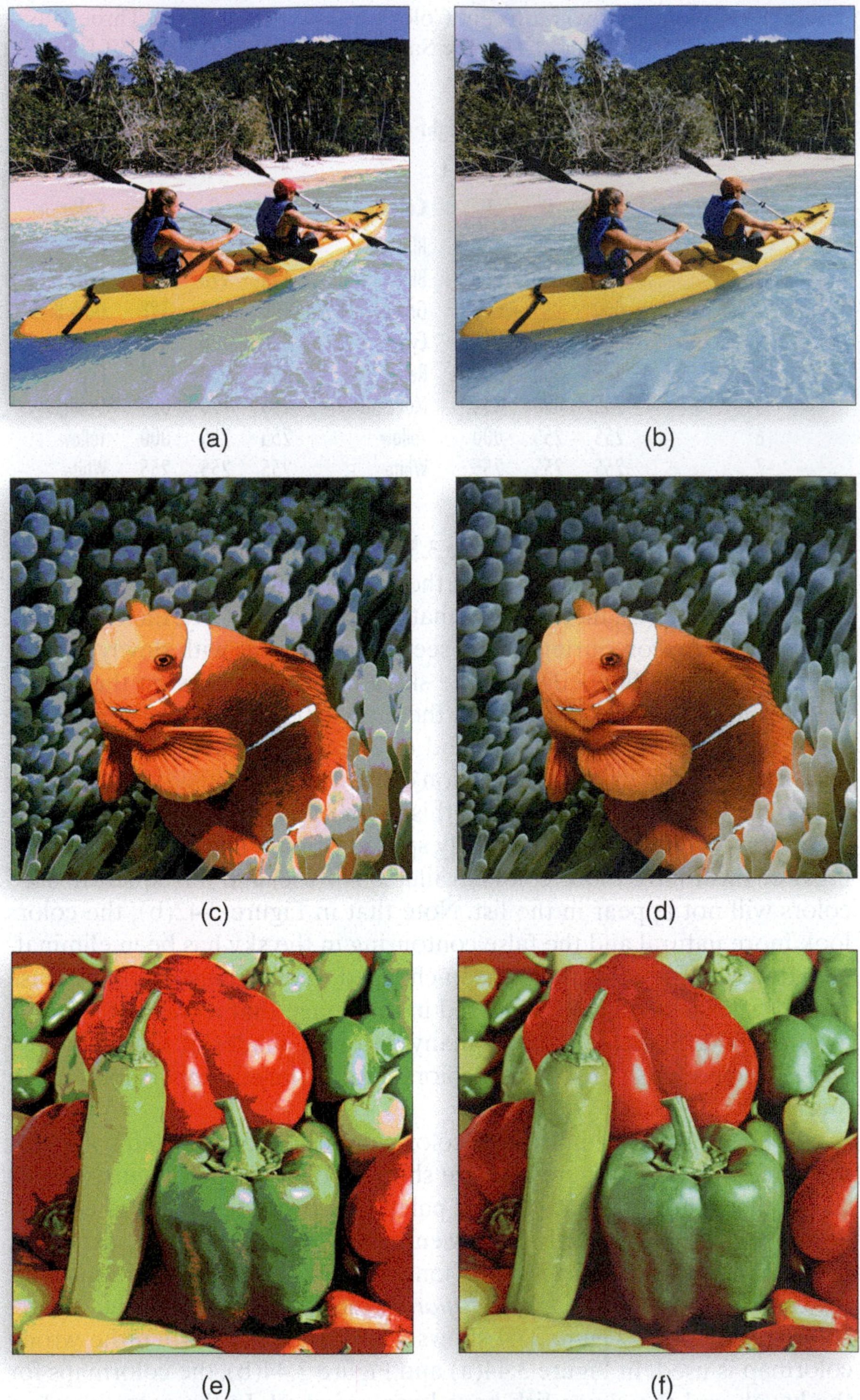

Figure 3.42 Comparison of two different eight-bits-per-pixel color display methods. On the left [(a), (c), and (e)], three bits each are used for red and green and two bits are used for blue. On the right [(b), (d), and (f)], a special palette of 256 colors was selected from the over 16 million colors for each image.

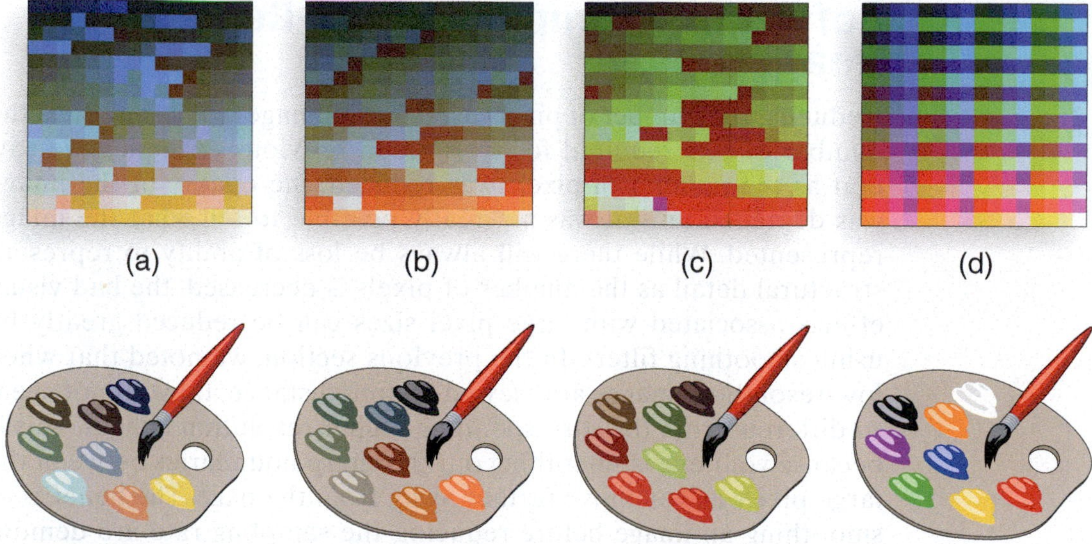

Figure 3.43 The palette with the 256 best colors for the kayak image is shown in the 16 × 16 square at the top of (a). The palette of the 256 best colors for the image of the orange fish, shown in the 16 × 16 square at the top of (b), is similar to the palette in (a), but has more shades of orange for the fish and fewer shades of the yellow needed for the kayak. In contrast, the palette for the red and green peppers has a very different set of colors that includes many shades of red and green. The evenly spaced palette using three bits for red, three bits for green, and two bits for blue is shown in (d). Below each square palette of 256 colors, a smaller artist's palette is shown with different color selections.

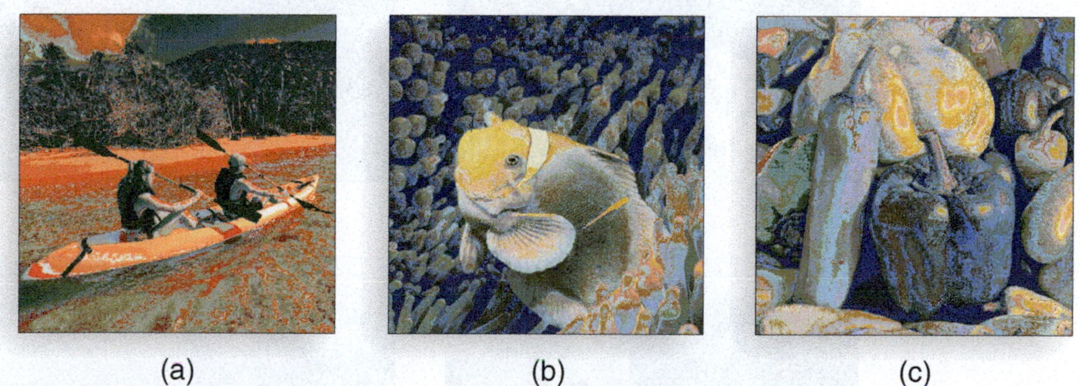

Figure 3.44 Three examples of displaying a colormapped image with an incorrect colormap. In (a), the kayak image is displayed using the palette for the image of the clown fish, and in (b) the image of the clown fish is displayed using the palette for the kayak image. In (c), the image of the peppers also uses the colormap for the kayak image, so the colors are the same as those used in (b).

Softening to Improve Low-Resolution Images

Reducing the number of pixels used for an image will also reduce the number of bits required for storage. In previous examples, we saw that as the number of pixels was reduced, the quality of the image was degraded until it was not really possible to tell what the image represented. While there will always be loss of ability to represent structural detail as the number of pixels is decreased, the bad visual effects associated with large pixel sizes can be reduced greatly by using smoothing filters. In the previous section, we noted that when low-resolution images are viewed from a distance, they do not seem so different from the corresponding higher resolution images. This is because your eye is smoothing out the sharp boundaries between the large pixels as you move further away from the page. The benefits of smoothing an image before reducing the sampling rate are demonstrated in Figure 3.45. When the image of the two buildings from Figure 3.23(a) is resampled at a rate reduced by a factor of eight, the image in Figure 3.23(c), with obvious aliasing, is created. That aliased image is shown again in Figure 3.45(a) in order to compare it with the image in Figure 3.45(b) which has the same reduced sampling rate. However, in Figure 3.45(b), the original image was smoothed before the resampling was done, and the effects of aliasing were reduced. The image in Figure 3.45(b) looks much less blocky and has better detail in the closer building and in the edges of the clouds. The curved-line artifacts in the smaller building are less intense as well.

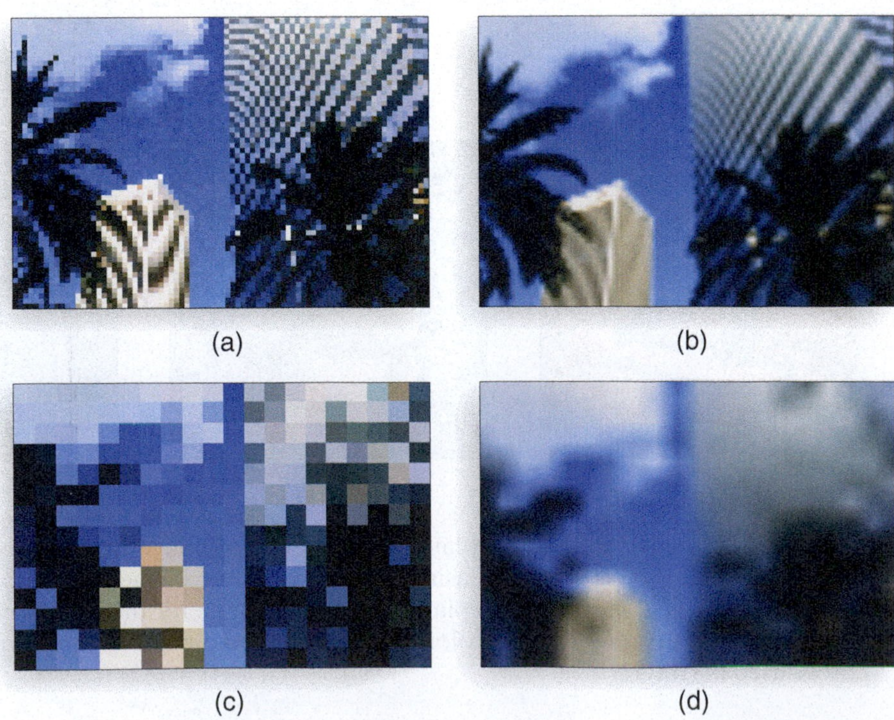

Figure 3.45 The original image from Figure 3.23 is shown with the horizontal and vertical sampling rates reduced by a factor of eight in (a) and by a factor of 32 in (c). Smoothing before the resampling operation results in the improved images in (b) and (d), which have the same sampling rates as in (a) and (c), respectively.

This image also looks blurred compared with Figure 3.45(a) when they are viewed at a reading distance.

If the sampling rate is reduced further by a factor of four, then, without smoothing, the image in Figure 3.45(c) is created, while smoothing before resampling results in the image shown in Figure 3.45(d). In both cases, reducing the sampling rate horizontally and vertically by a factor of 32 from that of the original image results in an enormous loss of image detail. That loss cannot be restored from the resampled image, because the pixel size determines our resolution, and we cannot look *inside* a pixel. The image in Figure 3.45(c) looks very sharp, but the content is not easily interpreted, because of the aliasing. In Figure 3.45(d), aliasing has been reduced by the smoothing that was done before resampling, and some parts of the image content can still be identified.

Smoothing prevents aliasing. It does not increase the image's resolution, but it greatly reduces the visual artifacts that come with a lower sampling rate. This process could allow a lower resolution image to be acceptable in some applications.

Summary

Your job is done! You now have an understanding of the technology that will allow you to take photos and correct mistakes or reduce the appearance of defects *after* they occur. You have also learned about the fundamental mathematical notions that are behind the current revolution in digital imaging, which really turns out to be a fancy name for "imaging by the numbers." You know how to figure out what kind of digital sensor you need and what kind of camera lens you need in order to take highly detailed pictures of your object. You also know how to spend your "bit budget" wisely.

Then you realize one thing: This business of converting pictures into numbers can have great potential. You already saw some examples of digital animation and digital special effects. You wonder how far you can push it. You start wondering if there is a way you can convince your friends that you actually *met* the rock star and even joined the group on the stage for one of the numbers. Well, you *can* do this by using color in a very special way and then using a digital computer to manipulate the images.

Ideally, what you would like to do is to take a photo of yourself, standing alone, and add it to a image of the rock star and his or her surroundings. You have probably seen some cheap, "analog" versions of this concept on street corners during your visit to tourist destinations such as Washington, DC. These versions used cardboard cutouts of the president and other celebrities. For a small amount of money, the roadside photographer would let you pose for a picture with the cardboard cutout. The quality of such photos is so poor that they look as phony as a three-dollar bill.

Another possible solution would be to take your photo, cut it out very precisely, and simply paste it next to the photo of the rock star. Then you could snap a photo of the composite picture you just made to create a passable forgery. This solution seems like a lot of work and requires too much skill. So you start looking for a digital approach to the problem of cutting and pasting.

In cutting out your image, you want to remove all the parts of the photo that are not *you*. There must be some factor that clearly distinguishes you

Figure 3.46 The blue screen effect. The composite in (d) takes pixel values from (b) only when they are not blue. When pixel values in (b) are blue, the corresponding pixel value in (a) is used for the composite in (d).

from the background. The best and the easiest way to separate the background from the main object is to use color. For example, you hang a drape in the background that has a color that is not to be found anywhere on you, and then you have a photograph of yourself taken standing in front of the drape. Typically, such a drape has a deep-blue color. So now, in the photo of you in front of the deep-blue background, wherever the computer finds the blue color, you can be sure that it has not found you. Next, select photo of a scene of exactly the same number of rows and columns as in the picture of you with the blue background. For example, you would like to add yourself to a group picture of a rock band. The computer can merge the two pictures according to the following recipe:

- The computer examines corresponding pixels, which occupy the same row and same column, from each of the two pictures.
- Wherever a pixel from the first photo is deep blue, exactly matching the color of the background that you had picked, the computer will select the pixel from the exact corresponding location in the second picture of the scene for placement in the composite picture.
- Wherever pixel from the first picture is *not* the deep-blue color of the background, the computer will select that pixel as the corresponding pixel of the new composite picture.

Figure 3.46 shows these operations with specific images. The photo of an actress dangling from a rope was taken in the studio with a blue-screen background. A photo of a sunflower field is used as the background of the composite. The composite photo in Figure 3.46(d) now shows the actress dangling over the sunflower field. So when the computer finishes examining all the pixels of the photo of you, the new composite photo will contain your photo seamlessly blended into the scene with the rock band. Through creative selection of the scene photo and your own photo, you can manipulate location, size, and pose. You could depict yourself shaking hands with the rock star or even giving him or her a high five!

We all know that computers are becoming more and more powerful and cheaper everyday. Even a relatively cheap computer can do this pixel sorting very fast. If the computer can do this operation within the time of one frame of a video, which we identified as one-sixtieth of a second, then instead of doing the operation on a single frame, the computer can take two separate videos, one with the rock band doing their number and the other with you doing your part of the number in front of a blue background, and merge them. This operation is routinely done on television weather segments of the news. The weather person is added to an enlarged image of a weather map from a small display screen. In the case of you and rock band, the result will be a composite video showing you joining the rock group on the stage in one of their numbers. Now *that* will definitely impress your friends . . . at least the ones who haven't read this book!

EXERCISES 3.4

Mastering the Concepts

1. What is a halftone image? How does halftoning create the illusion of gray values?
2. What is a pointillist painting?
3. How are colormaps used in representing digital color images?
4. Suppose you are viewing an image from further away than the intended distance, such as viewing a picture held with your arms outstretched. Will a smoothed version of the picture look the same? Why or why not?

Try This

5. In the halftoned image of the eight gray levels shown in Figure 3.39(g), determine the correct proportion of white for each level. Then determine the actual proportion of white by identifying the repeating pattern of white and black for each of the eight levels and then counting the small squares in the pattern that are white.

In the Laboratory

6. Halftone images are used in almost all newspaper pictures. Take a magnifying glass and examine a picture from a local newspaper. Using a ruler, determine the number of dots per inch in the halftone image. Compare this result with a normal pixel size of
   a. a computer screen
   b. a 600-dpi printed photo
7. Televisions have been designed with a specific viewing distance in mind. What image artifacts do you see if you move very close to a standard-definition television? How far away do you have to sit from the television and still not see the lines? Repeat this experiment for
   a. a high-definition television
   b. a computer monitor

Master Design Problem

Design an image and video system to record local athletic events and theater performances. You will need to consider the following basic questions for each specific application:

- What subjects do you want to photograph with still images and video?
- What is the maximum required field of view in the horizontal direction? What is the maximum required field of view in the vertical direction?
- What is the size of the smallest object or feature that you want to see clearly in your photographs?
- How many cameras are needed, and where should they be placed?
- What kind of lens should be used? Note that three general lens specifications were used in this chapter, but that other lenses with different ratios of field of view to distance can be used.
- What video frame rate is needed?

The answers to these questions are not independent of each other and will depend on the specifics of the application. For example, if multiple cameras are used, a lower resolution image might be acceptable, because the closest camera could be used, and not all cameras would have to capture the smallest important details. If multiple cameras are used, then the field-of-view requirement might be reduced as well. With suitable positioning of cameras, images from two different cameras could be joined together to provide a wider field of view than either camera could provide independently.

1. For the athletic events, consider football, soccer, and basketball games and track and gymnastics meets. You will want to be able to get good images of individual athletes, but you also will want to be able to see the whole field of play.
   - For each athletic event, what is the field of view of the whole game or meet? For a wide-angle lens, what is the *minimum* distance the camera must be from the field?
   - What is the size of the smallest feature that you should be able to see when viewing the whole field? For example, is it enough to be able to count the athletes? Or should you be able to see the markings on the uniforms or the facial features of the athletes? Consider a digital sensor that is 1000 pixels in width. What is the *maximum* distance the camera can be from the athletes? If the sensor is 2000 pixels wide, how will this distance change?
   - Based on the limits of the camera position for appropriate detail and a complete view of the meet or game, how many cameras would you use and how would you place them?

- Now consider images of individual athletes made with a telephoto lens. If the width of the field of view is the same as the height of the athlete, where should the camera be? What if the field of view is just the athlete's upper body? What is the smallest feature that should be seen clearly in the image? Should you be able to see the athlete's eyes or drops of perspiration? How large will these features be when measured in pixels in your camera system? How many pixels will you need in your image?
- Based on the analysis for images of both the full field and individual athletes, determine how many different cameras are needed and where they should be located. You can choose a lens that is adjustable from wide angle to telephoto if that reduces the number of cameras you will need.
- Consider what frame rate is needed for video. The standard real-time rate of 60 frames per second may be fine for viewing the action, but would you need a faster frame rate in order to determine which runner crossed the finish line first in a close race or whether an athlete's feet were or were not out of bounds at a crucial moment in a play? Determine how fast an athlete would be running or how long a goalie might be in the air in order to determine what frame rate you want to have. At 60 frames per second, how much blur would you see in the most rapid movements of the athletes?
- If you wanted to include swim meets, how would you handle images taken underwater?

2. For theater applications, you will want to be able to make photos of the entire stage and of individual performers. It will be important to have enough resolution to capture the expressions on the faces of the actors when individual performers or small groups of performers are photographed.
   - What is the field of view of the whole stage? For a wide-angle lens, what is the *minimum* distance the camera must be from the stage? Consider both horizontal and vertical field of view, since a performance might include balconies or beanstalks.
   - What is the size of the smallest feature that you should be able to see when viewing the whole stage? For example, if the entire cast is photographed at the end of a performance, should you be able to see the costume details or the facial expressions of all the performers? Consider a digital sensor that is 1000 pixels in width. If there are 25 performers in the front row at full-cast curtain call, how many pixels would you estimate would cover the width of each actor's head? What is the distance the camera can be from the stage for this resolution? If the sensor is 2000 pixels wide, how much more detail can be seen in each face at the same distance from the stage?
   - Based on the limits of the camera position for appropriate detail and a complete view of the stage, how many cameras would you use and how would you place them?

- Now consider images of individual actors or small groups made with a telephoto lens. What should the width of the field of view be? What if the field of view is just one actor's face? What is the smallest feature that should be seen clearly in the image? Should you be able to see the pupils of the actor's eyes or small curvature of the mouth? How large will these features be when measured in pixels in your camera system? How many pixels will you need in your image?
- Based on the analysis for images of both the full stage and individual performers, determine how many different cameras are needed and where they should be located. You can choose a lens that is adjustable from wide angle to telephoto if that reduces the number of cameras you will need.
- Consider what frame rate is needed for video. At 60 frames per second, how much blur would you see in the most rapid movements of a dancer?

Big Ideas

Math and Science Concepts Learned

In this chapter, we learned why we would want to use digital images, how to create digital images, and how to choose the correct digital image representation based on the image quality we want and the amount of storage we can afford.

We learned that digital images can be used to save photos and to share visual experiences, as well as for many other applications in industry, science, medicine, and space exploration. Digital images are desirable because we can easily produce exact copies of an image and they can be transmitted electronically for wide distribution.

Digital images can represent a view of the three-dimensional world as seen through a camera, a two-dimensional pattern, or an array of data measurements presented in a spatial array rather than as a list of numbers.

- When we create a digital image of a two-dimensional pattern, like a page of newspaper or a photograph, we must look at the level of detail we need in order to keep the important information. We choose a pixel size that is small enough so that the color does not change much within the area of an individual pixel. We talk about our horizontal and vertical sampling rate in terms of pixels per inch or pixels per millimeter.
- When we use a digital camera to create a digital image of a three-dimensional scene, we have to know the characteristics of the lens we are using in order to know how the pixel area will relate to the three-dimensional world. The width of the area on an object's surface represented by a single pixel will always increase as the distance of the object from the camera increases.

- When the image represents data measurements, such as a medical X-ray images or a CAT scan, the size of the area represented by a pixel is determined by the recorded measurements.

We also detailed our technique for converting still images or movies (black and white or color) into a sequence of 1's and 0's. This process, called digitization, involves two distinct steps:

1. sampling in space (creating an array of pixels) and sampling in time (creating a discrete sequence of frames);
2. quantization in amplitude (creating discrete amplitude values for each pixel). For a black-and-white image, a single-pixel array is used. Three-pixel arrays (one pixel for each of the three components red, green, and blue) are used for full-color images.

Once we have converted images into a binary representation using bits, we can use the full capabilities of the modern computing technologies of processing, storage, display, and communication to handle the images. Furthermore, binary numbers are easy to communicate, since they can tolerate a large amount of distortion and noise while retaining the essential information. The number of pixels in an image determines how much detail can be represented. Since a pixel will have a uniform color, representing the average color in the area covered by the pixel, it is important to choose a high-enough number of pixels to preserve detail.

The number of bits that represents the value of a pixel determines how many different colors or gray levels can be represented. If the number of bits per pixel is too low, there will be too large a change in colors in areas that should have slowly changing colors. We will perceive a line or contour when we switch to the next available color, but this line will not correspond to any object boundary or structural edge. It will be caused by having too few colors available for the image.

We can improve the quality of digital images to a point by increasing the number of pixels so as to increase the resolution of details or by increasing the number of bits per pixel so as to increase the number of colors. However, there is a limit to the amount of improvement we should make, due to the limitations of the human eye. At some point, we will no longer be able to notice the increased detail or the wider choice of colors. Adding more pixels or bits per pixel at this point increases the cost of storage and transmission without providing increased benefit for the human viewer. Such an increase is justified only if it is needed for more detailed image-processing steps.

If we do not have enough storage available for the quality of imaging we want, we can explore the possibility of reducing the resolution or the number of colors. If the number of pixels is reduced, the image should be smoothed first so that unusual patterns due to aliasing do not appear in the image. We can also use colormaps to get a limited set of colors that is the best set for a particular image. Black–and-white and color halftoning also allow us to make good use of a very limited number of colors.

Important Equations

The relationship between the number of gray-value steps and the number of bits is expressed as

$$M = 2^m$$

or

$$m = \log_2(M)$$

where m is the number of bits, M is the number of gray-value steps, and the logarithm is with respect to base 2.

There is a reciprocal relationship between the sampling rate S_h in the horizontal direction and the width of a sample, X. The same is true for the vertical sampling rate S_v and the height of a sample Y. These relationships are expressed as follows:

$$S_h \text{ pixels/mm} = 1/(X \text{ mm/pixel})$$
$$S_v \text{ pixels/mm} = 1/(Y \text{ mm/pixel})$$

Often, $X = Y$, but this is not always the case.

The first equation in Equation (3.4) shows the relationship between the spatial width of an image, W; the number of pixels in a row of the image, J; and the horizontal spatial sampling rate S_h. The second equation shows the same relationship in the vertical direction. The equations are as follows:

$$J \text{ pixels/row} = S_h \text{ pixels/mm} \times W \text{ mm/row}$$
$$I \text{ pixels/column} = S_v \text{ pixels/mm} \times H \text{ mm/column}$$

The total number of pixels in an image is given as

$$N \text{ pixels} = I \text{ pixels/column} \times J \text{ columns}$$
$$= J \text{ pixels/rows} \times I \text{ rows}$$

The number of bits in an image can be calculated as

$$b \text{ bits} = N \text{ pixels} \times m \text{ bits/pixel}$$
$$= I \text{ pixels/column} \times J \text{ columns} \times \log_2(M) \text{ bits/pixel}$$

The number of bytes in an image can be calculated as

$$B \text{ bytes} = (N \text{ pixels} \times m \text{ bits/pixel})/(8 \text{ bits/byte}) =$$
$$(I \text{ pixels/column} \times J \text{ columns} \times \log_2(M) \text{ bits/pixel})/(8 \text{ bits/byte})$$

When taking a digital photo with a lens that has a horizontal field view of s meters at a distance of 10 meters, Equation (3.8) can be used to compute the horizontal field of view f at a distance of d meters:

$$f \text{ meters/row} = s \text{ meters/row} \times (d \text{ meters}/10 \text{ meters})$$

At a distance of d meters, X_P is the width of the area covered by one pixel for a camera with J pixels/row.

$$X_P \text{ meters/pixel} = \frac{(f \text{ meters/row})}{(J \text{ pixels/row})}$$

The exposure time T that will show an object travelling at V meters/s through a distance of D meters is computed as

$$T\text{s} = \frac{D\text{ meters}}{V\text{ meters/s}}$$

Building Your Knowledge Library

Cornsweet, Tom N., *Visual Perception*, Academic Press, New York, 1970.

This is a very complete book describing the physiological basis of human visual perception of light, color, edges, and movement.

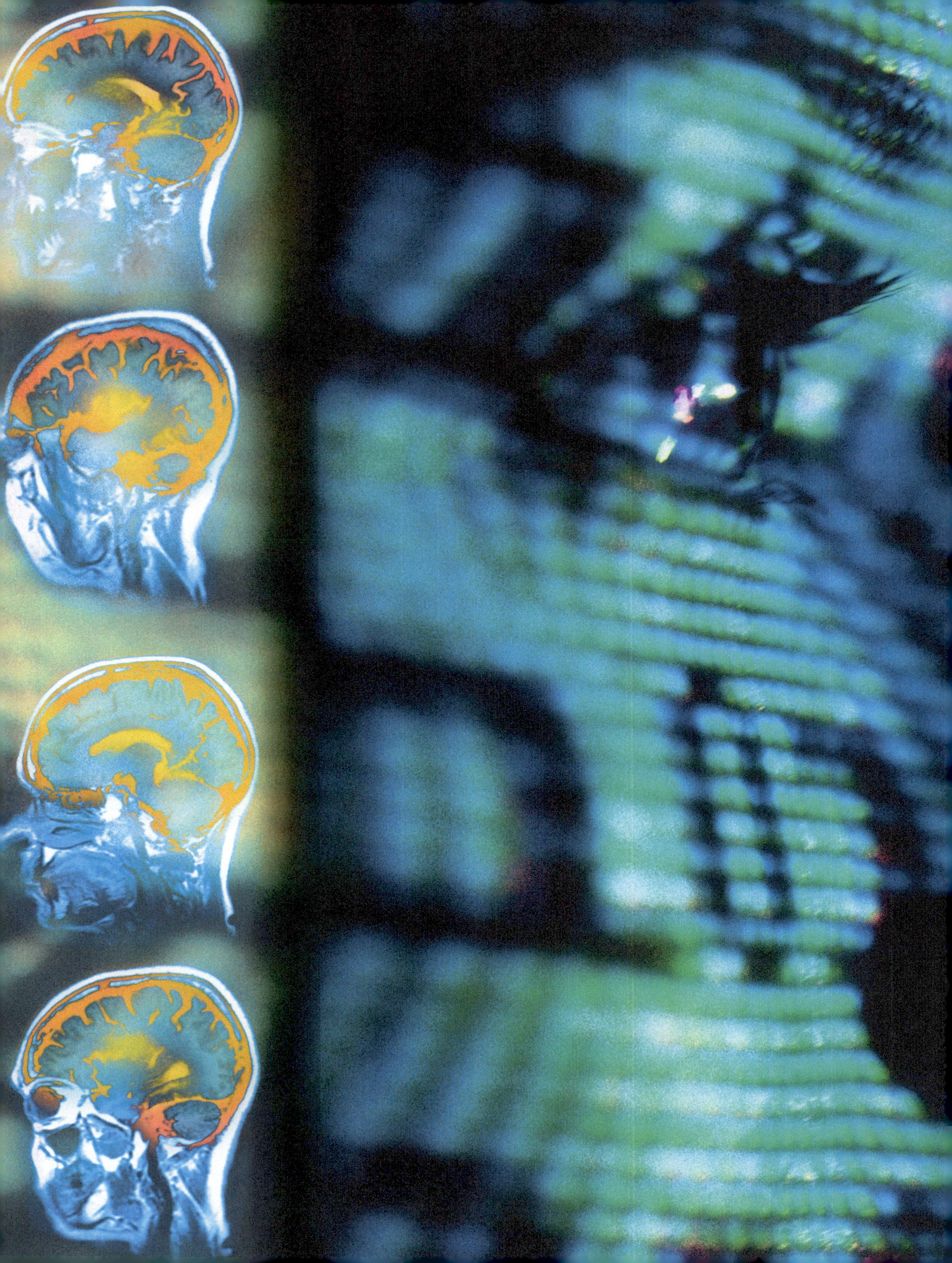

chapter 4

Math You Can See

OUTLINE

- How Can We Use Digital Images?
- A Digital Image Is a Matrix
- Digital Darkroom Tools
- Understanding Images from Robot Eyes
- Designing Simple Vision Systems

Suppose you would like to spend some time with your friends at the beach, a concert, or sporting event, but long-neglected routine household maintenance tasks like laundry and cleaning have to be done first. As you start your chores, you ask yourself, "Isn't there a better way to do these things?" After all, each individual task by itself seems rather simple. It is not very hard to find all the books and CDs on the floor and put them on a shelf, or stack all the papers and magazines, or pick up clothes and sort them by color for the laundry. Small game pieces and other junk lying on the carpet can easily be picked up and put in a drawer. Shouldn't we be able to buy a robot that can do boring work like this so that we can spend our time on more interesting activities?

From the previous chapters, we have learned to approach our needs for new capabilities from an engineering design perspective. We need to identify the problem to be solved and then define a basic block diagram that will accomplish our goals. Then we can look at each of the blocks,

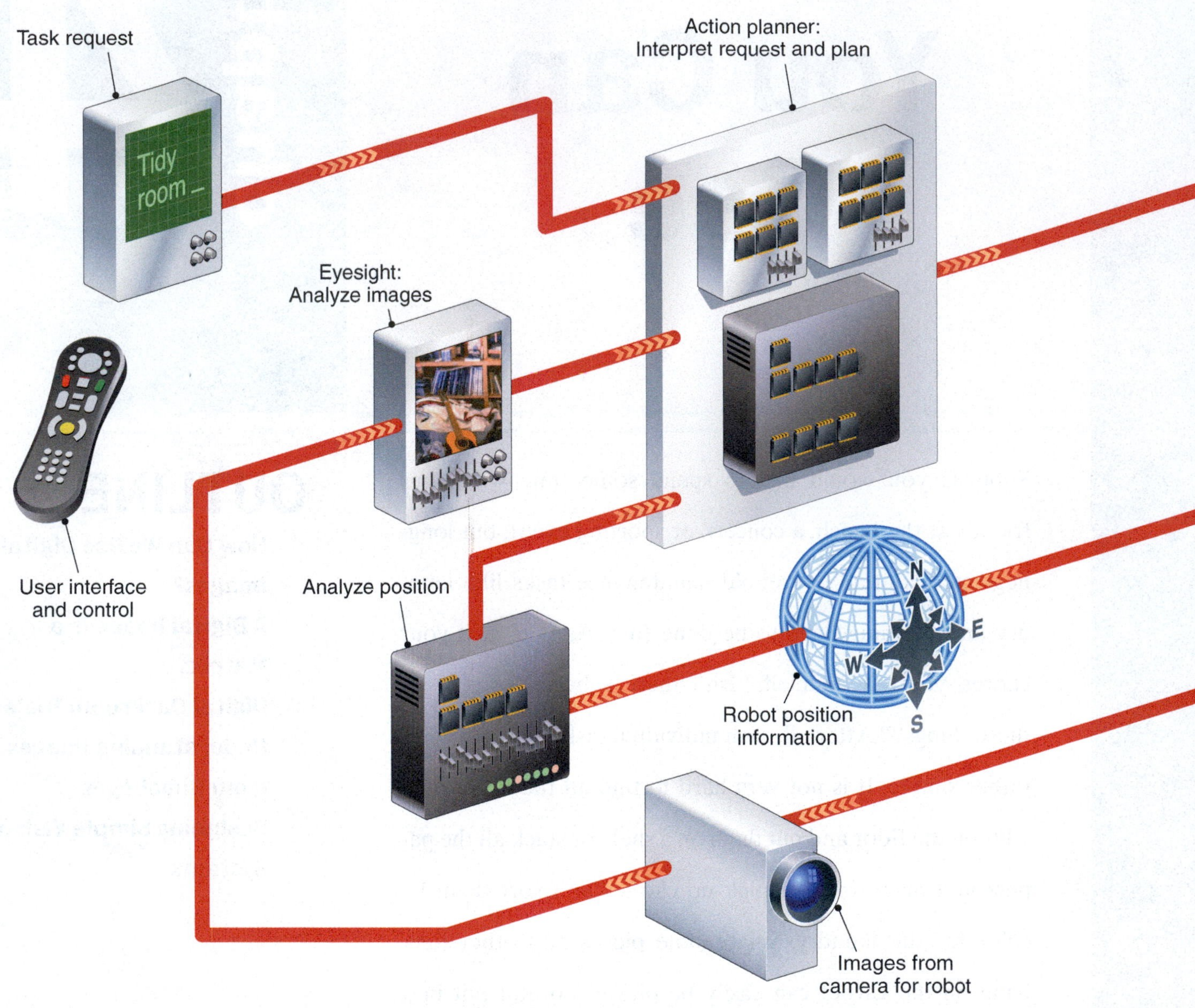

Figure 4.1 Block diagram for a "robo-helper," a robotic household assistant.

subdivide them into smaller, more manageable tasks, and explore different solutions for each one.

The high-level block diagram of a "robo-helper" is shown in Figure 4.1. The first block in the upper left corner is the user-interface block. The design of this block determines how you tell the robot what to do. This block is an important part of the design. It should be easy for you to give instructions to the robot, otherwise the robot will not be able to help you very much. The robot must be able to interpret your requests and then take action.

This brings us to the center block in Figure 4.1. This block interprets your instructions and then plans the individual mechanical actions to carry them out. The mechanical design of the physical part of the robot and its control system are also extremely important. It is a real challenge to make a robot with mechanical hands that have the sensitivity and flexibility to pick up small fragile objects like CDs and the strength to pick up heavy objects like loaded backpacks.

reate control signals

Robot with camera

The action planner needs more input than just our task request. Our robot needs a camera, as well as other devices, to recognize and manipulate its environment. Without some kind of eyesight, the robot cannot see, locate, or recognize objects, so it would not be able to move objects to straighten a room. The robot has to know where it is relative to these objects, where it needs to go, how to navigate a reasonable path around the objects, and how it will determine when it has arrived at specific destinations. This factor is the reason for the two other input blocks for the center block (the action planner) in Figure 4.1.

The block that analyzes images gets image data from the robot's camera system. The block that analyzes the position of the robot uses input from devices on the robot or sensors positioned in the room. The individual blocks in Figure 4.1 are all very complex, despite their simple functional descriptions.

In this chapter, we will focus on just one block, the "eyesight" block that analyzes and interprets images from a camera to help guide the robot. We can use what we learned in Chapter 3 about digital images to decide how many pixels and how many colors or gray levels our camera will need so that the objects of interest are clearly visible. We also may need more than one camera. When we are sure that our camera images can show the objects of interest to the robot accurately, we will have to design processes that actually perform the object identification and location from the images. Many other imaging applications also require interpretation of images in order to control objects or make decisions; several were described in Chapter 3. From these other applications, we can learn useful things to make our robohelper design better. Some specific applications related to our robotic-vision problem are as follows:

- finding a good landing site for a Mars landing pod;
- determining from satellite images if crops in a field need water or fertilizer;

- recognizing faces for access control;
- using motion detectors in video security systems;
- distinguishing between normal and abnormal X-ray images;
- using small cameras in internal medical examinations;
- controlling automated manufacturing processes;
- analyzing integrated circuits to find manufacturing defects;
- training robotic toys and robotic pets to perform specific tasks;
- controlling the autonomous highway vehicle of the future, which automatically will drive you where you want to go.

Some of these applications are more of a challenge than others. Some are currently in use today, and others are currently in the process of development. A few need more futuristic technologies to arise before they can be built. However, they all need a reliable way to interpret image data.

Once we convert images into numbers to create a digital image, we can use mathematics to describe image manipulations in terms of numeric operations developed by mathematicians. The image manipulations then can be implemented using electronic digital technology, which has seen explosive growth in computational capability over the past 50 years.

In this chapter, we will learn how some simple numeric operations affect digital images so that we can use them to interpret image content automatically. Then we will explore how we can combine these simple operations in various applications to design solutions to such problems.

Let's explore two specific applications briefly to see how automatic interpretation of images can be useful. Examples of one such application, security cameras, are found everywhere, from museums and convenience stores, to banks and parking garages. It would be difficult to have people watching all of the cameras' video outputs all of the time, because this arrangement would require employing a very large number of people. Just staring at a video monitor where nothing is changing most of the time would be a really boring job. It would be better to automatically interpret the video from all the cameras so that a human viewer would be called only when something unusual happened. A typical commercially available system might use a simple black-and-white camera with a wireless transmitter and a video cassette recorder. The recorder would be triggered only when the camera and associated processor detect motion in the sequence of images. A more sophisticated security system, with much more complex processing, was described in the previous chapter. That system examines images of people entering a facility and compares their faces with those in a database of people authorized to enter.

Our robo-helper vision system will be even more complicated than these examples, because the robot itself will be in motion and will be moving objects around. We will place cameras on the robot that model our own visual system, but we may also need to put small cameras in many permanent places in the room to transmit video information to the robot so that it knows exactly where it is.

A different kind of application is the growing selection of robotic "pets" that are designed to imitate the behavior of real pet dogs and cats. In addition to small motors that enable them to move around, these toys may have simple vision systems so that they can run around without bumping into things. Many have additional touch sensors to help them learn about their environment, and some can even learn to understand spoken commands. Figure 4.2 shows an example of a robotic pet dog.

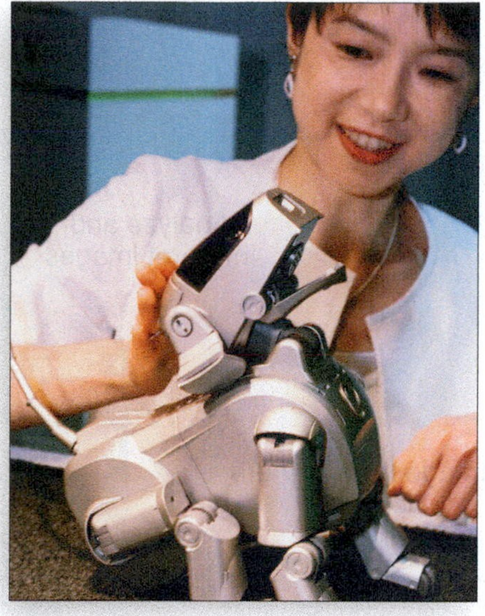

Figure 4.2 Aibo, a robotic pet dog with simple vision capability.

Design Objectives: The Automatic Vision System

The complete design of our robo-helper would require detailed attention to all of the blocks in Figure 4.1, which is more information than can be included in a single chapter. In this chapter, we will focus our attention on a single block, the image-interpretation block that gives the robot "eyesight." We will apply our design strategy from the previous chapters to this block alone. Since the block is just one of several, we will also have to be sure that the output from our image-interpretation block is in exactly the right form for input to the action-planner block. If all the blocks are designed separately so that the output of each block is exactly what the connecting blocks need, then all of the blocks can be connected together to make a functioning robo-helper.

- **What problem are we trying to solve?** We want to design an imaging system that will provide information about the objects around the robot for use by the robot's action planner. This design will include the number, type, and placement of cameras, as well as how to interpret the information coming from the cameras.
- **How do we formulate the underlying engineering design problem?** We will need to answer the following questions in order to determine specifications by which we will design the camera system and the image analysis process:
  1. What is the smallest object that the robot will need to see?
  2. What is the size of the area that the robot will have to see?
  3. How accurate must the information about the size and location of objects be?
  4. How much light will the area of operation have?
- **What will we achieve if our design goals are met?** We will have a vision system that can be part of a wide variety of robotic assisting devices used in personal daily living, the workplace, or the community. If we can design a robo-helper to clean our room, we could also design one to mow the lawn, sweep the sidewalk, or assist a disabled person.
- **How will we test our design?** We will have to test the individual components of our design first to make sure that it can find and locate objects. We will need to be sure that it does not interpret several different objects as one or one object as many. We will need to test it at the extremes of the lighting conditions. Finally, we will have to list situations that might cause confusion—for example, distinguishing a blue book on a blue carpet from a pattern on a blue carpet—and make sure that the vision system makes the most reasonable choice in those cases.

In Figure 4.3, we look inside the image-interpretation block of the complete block diagram of Figure 4.1 to see that three blocks are needed to locate and identify objects. The first step improves the digital

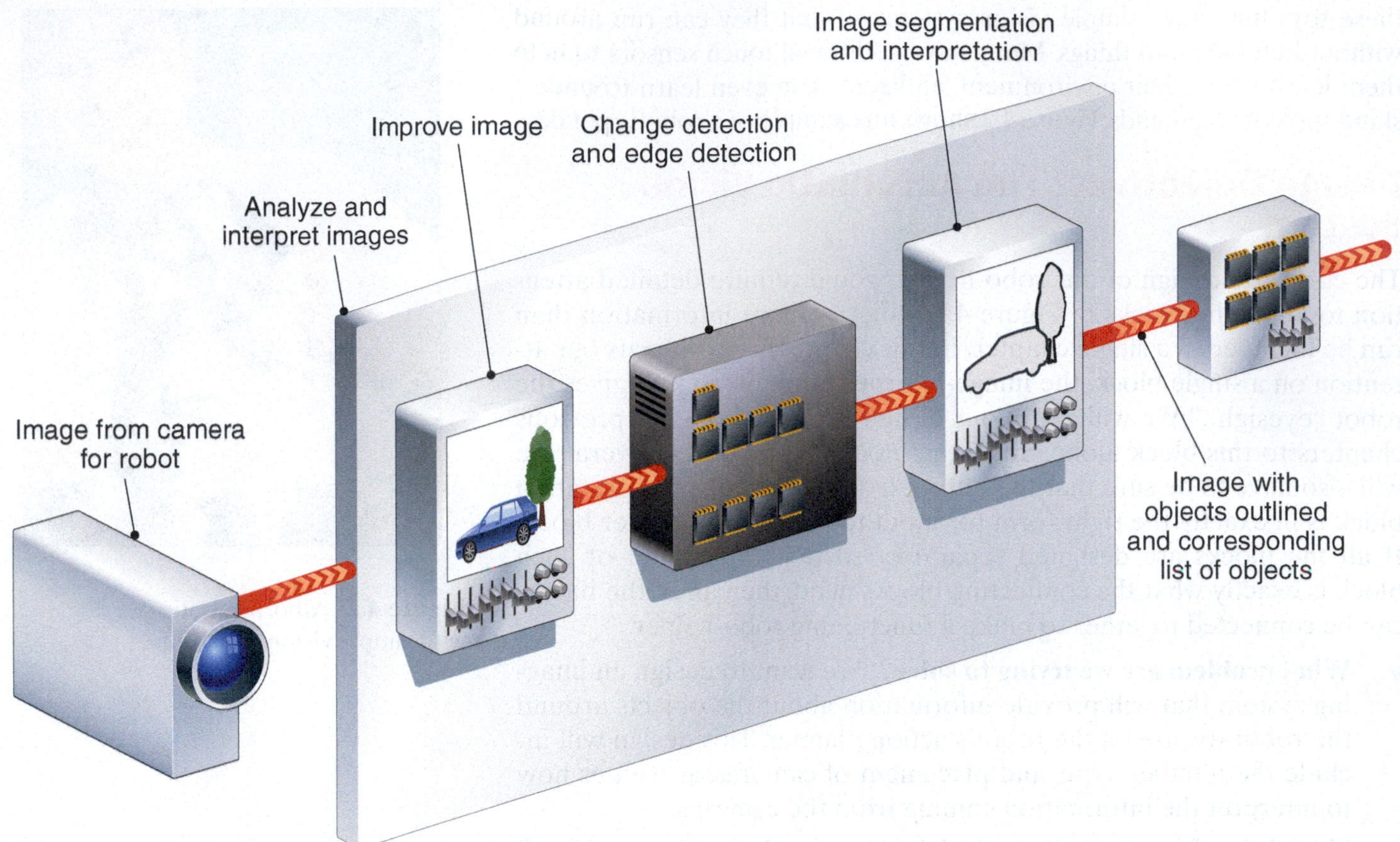

Figure 4.3 Block diagram of the robo-helper's vision system, which corresponds to the "Eyesight: Analyze images" block in Figure 4.1.

image as much as possible in order to make our other blocks more successful. For example, the improvement might require adjustments to brightness or contrast, or corrections for focusing problems.

The second step would use change detection to look for edges of objects, objects that are in the wrong place, or objects that are moving. For example, if we compare an image of a scene with an image of the same scene taken at a different time, the difference will show us the objects that have moved. If we look for abrupt changes of color or intensity within an image, we can find the edges that can help reveal the shapes of objects and their location.

The third step uses this image-change information to make a list of the objects and their locations. The list is sent to the robotic control system, which will use this information to accomplish whatever task was requested by the user.

Image Processing: Mathematical computation using pixel values to create a new image or compute information about objects within an image.

The blocks in Figure 4.3 represent **image-processing** functions, or operations that use mathematical computation on image pixel values to create a new image or compute information about objects within an image. We will discuss several simple image-processing operations that are used to build these blocks. Then, in the next section, we will learn the simple mathematics for each operation. We will see that relatively simple mathematical operations that change the values of pixels or create new images are the basis for powerful image-processing operations such as edge detection and noise reduction.

4.1 How Can We Use Digital Images?

Image Improvement Operations

Several image-processing operations can be used to improve an image so that the performance of later processing steps is improved. The amount of improvement will always be limited by the resolution and the number of gray levels in the input image. For example, no amount of image processing will tell us where all the football players are if we start with a 10-pixel square image of the whole field. However, in many cases, the content of an image can be used more easily if object edges are emphasized or if noise is reduced.

Noise Reduction The process of creating a digital image with a camera or a scanner may not produce a perfect result. Sometimes, an image has variations that are not related to the image's content. These variations, which are called **random noise**, are particularly noticeable when an image is captured on film or with a digital camera under low light conditions. While noise is a hard term to define precisely, most of us intuitively understand the difference between noise and what we are trying to observe. In digital imaging, we can often use mathematical operations to reduce the noise after the image is recorded.

Random Noise: Variation in a signal that is not related to the signal's information. Each pixel can have a different random variation added to its value.

Figure 4.4(b) and (c) show two different kinds of noise added to the image of the Statue of Liberty in Figure 4.4(a). In Figure 4.4(b), a small random change has been added to each pixel in Figure 4.4(a). In Figure 4.4(c), 3% of the pixels have been selected at random and their values have been changed to black. Another 3% of the pixels selected at random have been changed to white. This type of noise is called **impulsive noise**. In both cases, we can still see the statue, but the edges in the images are irregular and the grainy variations or spots are distracting. The enlarged images in Figure 4.4(d)–(f) show this effect more clearly. Both kinds of noise might be caused by a poor communications system or a problem when images are acquired, so we would like to have a noise reduction operation that could change the images in Figure 4.4(b) and (c) to look as much like the image in Figure 4.4(a) as possible.

Impulsive Noise: When a small percentage of pixels selected at random are set to the brightest or darkest values.

Image Softening It is often desirable to make images look softer. For example, a portrait image of a person's face might be softened intentionally by introducing a small amount of blur to reduce age lines or other imperfections while retaining the personal characteristic of the face. Image softening was introduced in the previous chapter as an image enhancement option. Later in this chapter, we will develop the mathematics to create the desired amount of softening. This operation is related to noise reduction in some cases, but since the objective of image softening is different from the objective of noise reduction, the mathematical processes used may also be different.

Figure 4.5 shows three different softened images created from Figure 4.4(a). In Figure 4.5(a), a small amount of softening is used. The effect is subtle, but noticeable, and all the significant details of the original image are still visible. The increased amount of softening in Figure 4.5(b) causes loss of fine detail in many parts of the image, and even more softening in the image in Figure 4.5(c) causes so much blur that important structures become ambiguous.

Figure 4.4 The visual effect of noise in images. The image in (b) is the image from (a) with one kind of added noise that causes small variations all over the image. The noise in the image in (c) causes 6% of the pixels to be changed to white or black. The enlargements in (d)–(f) show the effects of the noise more clearly. Many factors may cause noise to be added to an image when a photo is taken.

Image Sharpening Most interesting images contain many different light and dark areas. The changes between these areas can be very gradual or rather abrupt. Qualitatively, human observers prefer images where the transitions between light and dark areas are crisp. It is often said that images should look sharp. An image might lose sharpness for many reasons, such as lens defects or poor focus. Blurred images can occur if the camera moves or if the objects shown in the image are moving while the image is being taken. Inexpensive Web cameras may take so long to capture an image that motion blur is a serious problem. Images can also lose sharpness if an image-softening process has been used to reduce noise or make object surfaces look smoother. Although sharpening and softening have opposite effects on an image, doing one does not perfectly undo the other. If the two processes are done in sequence, the result is rarely the same as the original image.

With film-based analog imaging technology, many costly film-processing steps are required in order to improve the appearance of a blurred image. One technique, called **unsharp masking**, combines an image and a blurred negative of the image to highlight edges. For film-based methods, the extent and type of improvements that can be achieved are severely limited. With digital images, corrections for blurred images are easier to make and more correction is possible. If a fuzzy image cannot be reac-

Unsharp Masking: The process of adding a blurred negative to an image to make the edges more crisp and noticeable.

Figure 4.5 Three images created from Figure 4.4(a) by softening. The loss of sharpness and detail increases as the degree of softening increases.

quired with better camera settings, it can be processed digitally to make it sharper. As we discussed in Chapter 3, when the Hubble telescope first began sending images back to Earth, there was an obvious problem with the camera's ability to focus. It took several years to send a manned mission into space to improve the telescope's physical optics. Those early images had to be improved using digital image processing.

Figure 4.6 shows the sharpening operation applied to the same image that was used in Figure 4.4(a). The small amount of sharpening in Figure 4.6(a) makes the edges of the statue and the folds of the robe more distinct. As the amount of sharpening increases in Figure 4.6(b) and (c), the smooth surfaces of the statue that should have a slowly varying intensity become speckled and noisy, because small variations that we normally would not notice have been emphasized. Sharpening brings out the details on surfaces and emphasizes the edges; if too much sharpening is added, then small details that we do not want to see are emphasized too much. Sharpening also was discussed briefly in the previous chapter as an image enhancement option. Later in this chapter, we will learn how to use sharpening to achieve specific goals. This operation must be done carefully if the image has noise, because the sharpening operation may intensify the noise.

Finding Edges and Changes

Finding Edges of Objects Locating and identifying objects in an image starts with finding all the edges within that image. In Figure 4.6, we saw that the image-sharpening operation makes edges brighter and leaves

Figure 4.6 Three images created from Figure 4.4(a) by sharpening. As the degree of sharpening increases, details of the statue, such as the folds of the gown and the base of the torch, are more noticeable.

most of the gradually varying areas in the image unchanged. This operation is related to edge finding; however, if we want to determine the size, shape, and positions of objects in the image, we need only the bright edges. The object's surface color and intensity do not matter. The edge-finding process will retain the object's bright edges and render all other areas dark. This operation is called **edge detection**. Our eyes perform a similar edge-finding process. The signals that our eyes send to our brain have a strong emphasis on the edges where intensity levels change sharply.

Edge Detection: The process of looking for intensity differences between pixels that are near each other in an image. After edge detection, the resulting image shows the edges of objects and patterns, but not the areas where the color or intensity does not change.

Figure 4.7 illustrates the result of simple edge finding. To make the edges easier to see, the pixels at edges are made dark and the rest of the image pixels are made light. Because the resulting images, called edge images, have no information about the color or brightness of surfaces, they look very different from the images used to create them. However, when we look at edge images, we know from the outlines, what objects are in the images.

In Figure 4.7(a), we see a photo of a room with a variety of objects. In Figure 4.7(b) and (c), we see simple edges from two different areas of that image. In both cases, it is easy to see which part of the original image is represented by the edge images, even though there is no information about color. Notice that large objects like the guitar, shelves, and picture frames are clearly outlined. The books on the shelves are defined well enough for us to know whether there would be space for new books to be added to the shelves, and we would not confuse the irregular outlines on the trophy shelf with books and papers.

(a)

(b) (c)

Figure 4.7 Photo of a room and two enlarged sections of an edge image computed from the photo.

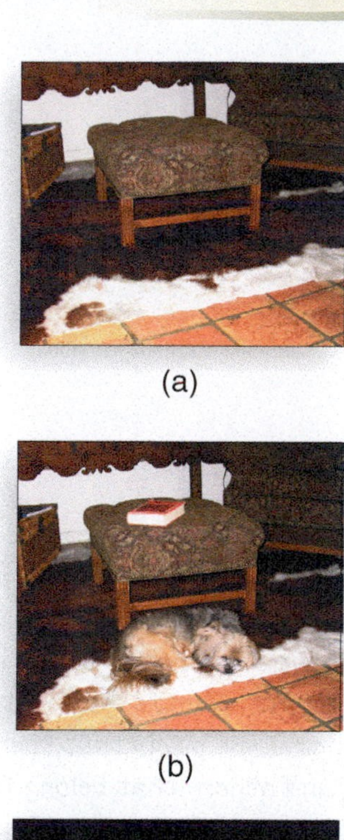

(a)

(b)

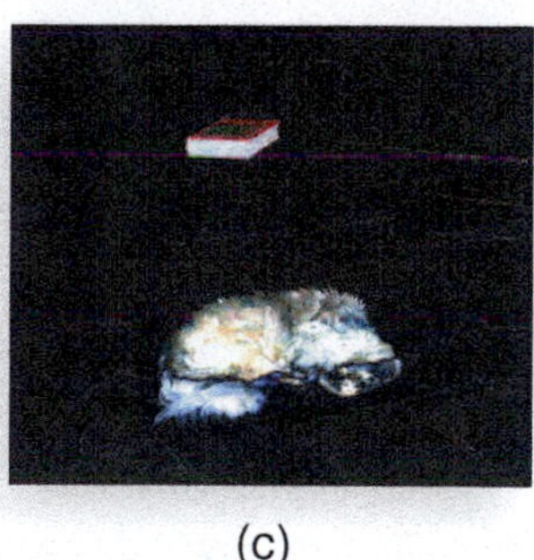

(c)

Figure 4.8 Two photos of the same scene are shown in (a) and (b). In the difference image in (c), content common to both images is dark. Only objects that are in one image but not the other appear in the difference image.

Change Detection Edge detection is a process that looks for changes in intensities of pixels that are near each other in a single image. In contrast, **change detection** looks for changes in intensities of pixels that are in exactly the same location in two different images. This process can show if an object in an image has grown, shrunk, moved, or changed color. For example, when analyzing images of the Earth from space, we are often more interested in determining what has changed in the scene than in the complete contents of the scene. This type of analysis can provide important information about the weather, water levels in rivers and lakes, or the extent and health of forests and crops. The same type of change detection can be used by security cameras to detect an intruder in a museum or a small child near a swimming pool. Again, the comparison with human vision can be noted. It has been demonstrated that our eyes are most sensitive to changing light intensity. If we hold our eyes perfectly still and look at a perfectly still scene, then the image we see fades and seems to gradually disappear.

Change Detection: A process that looks for changes in intensities of pixels found at the same location in two different images of the same scene taken at different times. After change detection, the resulting image shows the objects that have moved, but not the objects that have stayed in the same place.

In Figure 4.8 we see two images of the same scene taken at different times. The third image shows only the change. Some time after the first photo was taken, a book was left in the room and the pet dog came in for a nap. The robo-helper can locate these changes in the difference image because the parts that are common to both images are dark. The difference image shows the content of both images where there is a change, so the boundary between the dark and light part of the rug can be seen through the dog in the third image.

Identifying Objects: Image Segmentation

Image Segmentation: A process that divides an image into distinct areas that each corresponds to an object in the image or a specific region in the image. For example, segmentation might find some pixels that belong to a book and others that belong to the carpet-covered region of a floor.

A typical image seen by the camera on our robo-helper contains many objects. Consider all the objects shown in the photo of the room in Figure 4.7. If we want to design a system that automatically locates and identifies these objects, it is necessary to separate the objects from each other and from the background. This process, called **image segmentation**, is used by a wide variety of applications. For example, we might want to count the red blood cells in the image in Figure 3.3. The boundary of each cell must be found, and then the number of cells can be counted. If we want to design a system that automatically examines integrated circuits for manufacturing defects, then we would use image segmentation on an image showing the structure of individual circuit components, which are often less than one micron in width. In this case, the segmentation would find the circuit components and make sure that the size, shape, and location are correct.

Although segmentation based on object edges is used in a wide variety of applications, it always includes the same basic steps. It starts with edge detection to define small segments of the object boundaries that separate regions of different color or intensity in the image. Then these boundary pieces are connected to get complete outlines of the objects. Several small objects may be combined to form a larger object. For example, a bookcase could include the frame, the shelves, the books, the CDs, and any other objects on the shelves.

Figure 4.9 shows an example of a segmentation operation based on color applied to a photo of a stairway with several items on the stairs that might be of interest to our robo-helper. It should be noted that the separation between the objects and the background can be based on simple characteristics, such as intensity, color, or edge positions. In some cases, more complex characteristics, such as the texture pattern of carpet, are used to define objects. Segmentation of an image into meaningful objects is a very simple task for a person because, in addition to separating areas by color or texture, we use judgments based on our experience to determine what small objects are part of bigger objects. Segmentation can require many steps for digital image-processing systems, and the final result will depend on what rules are used to combine objects into larger objects.

Other Image-Processing Applications: Computer Graphics and Special Effects

Computer Graphics: The process of creating digital images from descriptions or models of the image content.

The three basic blocks in Figure 4.3 can be accomplished with combinations of simple image-processing operations described in the next section. These same simple operations are often applied to other applications that use digital imaging. In our robo-helper application, we want to interpret image content so that a picture is converted into a description. Then the description is used by the robot to decide where to go and what to do. Other applications start with a description and make an image. This type of processing is generally called **computer graphics**. Computer graphics is used to create images for games and flight simulators, as well as for adver-

(a)

(b)

Figure 4.9 The image of the cluttered stairs in (a) is segmented by color to find objects that are red and objects that are blue. The two results are combined in (b).

tising, movies, and creative art. Other applications start with images, but then change or use them to create other images for creative effects. One example of such applications is image morphing.

Image Morphing You have probably seen image **morphing** in movies, videos, and advertisements. An object or a person slowly changes into another object or person, and the shapes in all the intermediate frames are reasonable. The term morphing is a shortened version of metamorphosing, and it literally means "changing from one form to another." The first extensive use of morphing in a popular movie was in *Terminator 2* (Figure 4.10) released in 1991, although it had been used in many commercials and short features before that. Morphing slowly makes small changes that will gradually change one object into another. Any two successive frames differ only by a small amount, but the last image of the sequence is completely different from the original image. Each intermediate frame shows an object with a reasonable structure because morphing changes object boundaries and colors. In contrast, blended images, which look like double exposures,

Morphing: A slow change of one object to another in an image. Several important corresponding points on the two objects are matched throughout the change.

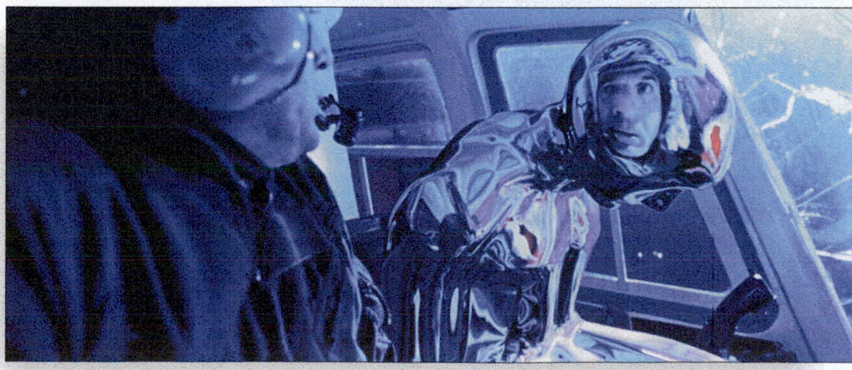

Figure 4.10 An example of morphing from *Terminator 2*.

simply add two images without considering the structure of objects in the image. Morphing can also be used to create an intermediate image that fits between two images in a sequence taken at different times. Similarly, if two different camera views of a scene are available, morphing can be used to create an image that would be seen by a third camera if it were placed between the two cameras.

Morphing is now widely used because digital images and the computational resources needed for morphing are readily available. Digital morphing is much more versatile, effective, and efficient than making physical models of all the intermediate stages of the morphing operation and then photographing each stage as a single frame in a movie.

Special Effects Digital special effects are now so common that they are expected in movies and video games. Techniques from computer graphics, which are used to create images from models, can also use actual digital images of people and scenes to create a wide range of effects for both passive viewing and interactive participation in a virtual reality. In addition to entertainment, virtual reality can be used for training. Practice for driving a car, flying an airplane, or using complex equipment can be much less costly and dangerous when the initial training is done on a simulator with realistic surrounding images created by computer graphics and image processing. Using surgical simulation with medical images from a specific patient, doctors could explore several different possibilities for a complex surgery before deciding which procedure to use. An example of a flight simulator is shown in Figure 4.11.

Designing Image-Processing Systems

The examples and applications discussed in this section are just a small selection from the rapidly growing number of digital-imaging applications made possible by the smaller, lower cost electronics used by digital cameras, the computational hardware for image processing, and the transmission capabilities available for image data. There are now a large number of systems for which the image processing is so integrated into the system that it is not apparent to a casual observer. It should always be remembered that very sophisticated applications are built up from some very simple and basic processing blocks. In the next section, we will learn how to make blocks for those elementary operations and then how to connect them to build interesting and useful applications.

Figure 4.11 An example of a flight simulator.

EXERCISES 4.1

Mastering Concepts

1. Describe the two different kinds of noise in Figure 4.4.
2. In what ways do image sharpening and image smoothing have opposite effects?
3. What is the difference between edge detection and change detection?
4. Name three properties of images used for segmentation.
5. List at least three specific actions that the robo-helper would take in response to a command to pick up all shoes and put them in the closet. For each action describe the input needed from the image-analysis part of the robot's control system.
6. What is the difference between image processing and computer graphics?

Try This

7. A camera is sending images of a checkerboard every 30 seconds. There are no checkers on the board, and no players are near the board.
   a. What would you expect to see if an edge-detection process were applied to the images?
   b. What would you expect to see if a change-detection process were applied to the images?
8. A camera takes a picture of a checkerboard from a game in progress after every move is complete and the players' hands are no longer near the board. What would you expect to see if a change detector processed these images?
9. What property of an image can we use to segment an image that shows green apples, bananas, and oranges in a brown wooden bowl on a light-blue countertop?

In the Laboratory

10. You are concerned about the erosion of a hillside near your home. To monitor the situation you set up a camera in a permanent place that will photograph the hillside every morning at 10 o'clock.
    a. Which of the basic operations discussed in this chapter would you apply to these images to detect and measure the erosion?
    b. Identify two problems with taking the photographs that might make your application a little more complicated than you originally expected.
11. Make a block diagram for a home alarm system that would detect any movement of a gate that prevents very young children from leaving their own yard. The alarm should sound as soon as the gate is moved rather than after a child has already gone through it.

12. Make a block diagram for a system that would count the cars traveling down a one-way residential street.
13. Make a block diagram for a system that would plot the paths taken by pool balls. How would you use color? How would you use change detection?

Back of the Envelope

14. Identify other applications for digital imaging systems from the entertainment, defense, and medical fields.
15. Give examples of types of noise commonly encountered in TV images.
16. List some specific structures that are clear in Figure 4.4(a) and (d), but are not distinguishable in Figure 4.5(c) and (f). Look at the other parts of Figure 4.5 and determine when the structures were lost.
17. List some artificial structures or surface patterns in Figure 4.6 that were added by the sharpening. Do they appear more noticeable as the amount of sharpening increases?
18. Give examples of the most common problems you have when you take photos. What kinds of image processing might improve them?

4.2 A Digital Image Is a Matrix

A black-and-white digitized image is just a two-dimensional array of numbers that tells us the gray-level intensity of each pixel in the image. This array can be represented mathematically as a **matrix**. Since the pixel values are just numbers, we can use mathematics to change the numbers and create a new array of numbers. Then we can view the new array of numbers as a new image. We can also use the pixel values in an image to compute information about the objects in the image. This type of computation using the pixel values of an image is what happens inside an image-processing block.

Matrix: Like pixels in an image, numbers in a matrix are arranged in rows and columns.

Chromakey: A method of extracting a foreground subject from one image and placing it on a background that comes from another image. The foreground subject is usually photographed against a solid blue background so that the subject and background are easily separated. For this reason, chromakey is also called a blue-screen effect.

Matrix representation is a convenient way to manipulate images, and many digital systems and calculators include standard functions for matrices. Since a digital image is a matrix, it is easy to use these functions to perform image-processing tasks. A simple sequence of computations will allow us to increase the contrast of an image, brighten the edges of objects, or create the blue-screen **chromakey** example described at the end of Chapter 3 in Figure 3.46.

Element: An element of a matrix is a single value of the matrix. Each element is identified by its row number and column number.

Like pixels in an image, numbers in a matrix are arranged in rows and columns. These numbers are called **elements** of the array or matrix. An example of a matrix that contains four rows and three columns is shown as matrix **A** in Equation (4.1). (We use capital letters in bold type to represent matrices.) On the left, the matrix elements are shown with symbolic names, such as $A(3,2)$, and an example of a matrix with specific values assigned to each element is shown on the right. Note that an individual element of the matrix is just a single number. The number corresponding to element $A(3,2)$ is 16 in the example shown.

In general, an element is identified by its location in the matrix and is written as $A(i, j)$, where i is the row number and j is the column number. In this example, $A(3, 2)$ is the element in the third row and second column of matrix **A**. For a color image, three separate matrices would be needed, one for each of the red, green, and blue color components.

$$\begin{array}{cccc} \text{col. 1} & \text{col. 2} & \text{col. 3} & \end{array}$$

$$\mathbf{A} = \begin{pmatrix} A(1,1) & A(1,2) & A(1,3) \\ A(2,1) & A(2,2) & A(2,3) \\ A(3,1) & A(3,2) & A(3,3) \\ A(4,1) & A(4,2) & A(4,3) \end{pmatrix} \begin{array}{l} \text{row 1} \\ \text{row 2} \\ \text{row 3} \\ \text{row 4} \end{array} \quad \mathbf{A} = \begin{pmatrix} 12 & 9 & 0 \\ 5 & 6 & 23 \\ 8 & 16 & 27 \\ 4 & 7 & 19 \end{pmatrix} \quad (4.1)$$

Numbers in a matrix can be represented either as decimal numbers or as binary numbers, and in general the numbers may be positive or negative. Since an image matrix is an array of numbers representing light-intensity values, its values can never be negative. All elements of an image matrix will be either positive or zero.

KEY CONCEPT

All elements of an image matrix are either positive or zero.

The size of a matrix is computed from the number of rows and number of columns the matrix contains. Matrix **A** in Equation (4.1) is called a 4×3 matrix, because it has four rows and three columns. This matrix has a total of 12 elements. In general, an I by J matrix will have $I \times J$ elements arranged in I rows and J columns. Each pixel is represented by a matrix element, so the number of elements in an image matrix is exactly the same as the number of pixels in the image. The rows and columns of a matrix can be associated with the two spatial directions in an image. Moving along the x axis in the horizontal direction of an image corresponds to moving along a row of an image matrix, and moving along the y axis in the vertical direction corresponds to moving along a column. A very simple matrix that contains only one row and one column is called a **scalar**. Since a scalar is an array of one number, it can be treated just like any other number when doing arithmetic.

Scalar: A matrix with only one row and one column, and thus only one element. A scalar is just a simple number.

EXAMPLE 4.1 Calculating Pixels and Matrix Elements

An image has 16 pixels in the horizontal direction and 24 pixels in the vertical direction. How many pixels are in this image? How many elements are in the matrix that represents this image, and how are they arranged (i.e., how many rows and columns are there?)

Solution

The image has 16 pixels per row and 24 rows, so we calculate that

$$16 \text{ pixels/row} \times 24 \text{ rows/image} = 384 \text{ pixels/image}$$

Since each pixel corresponds to one element of the image matrix, there are 384 elements of the image matrix arranged in 24 rows and 16 columns.

Simple Arithmetic Matrix Operations

We will define some basic and useful matrix operations for matrices of black-and-white image pixel values. These operations can be applied to any matrix, so they can be applied easily to each of the three separate matrices used by a color image for the red, green, and blue intensity values.

Simple arithmetic operations applied to matrices are the building blocks for our image-processing goals just as three simple arithmetic operations applied to signals in Chapter 2 allowed us to create complex musical sounds from simple sines and cosines. Three categories of basic operations for images will be described. Each operation will have a mathematical definition and will be represented by a characteristic block. By connecting blocks, we can show how to use the simple operations to do more complicated image-processing tasks without having to write complex equations for the whole process.

1. *Mapping*: Mapping operations create pixel values for a new image matrix by employing a computational rule that uses only the value of the corresponding pixel in the input image. The same rule is applied to all pixel values no matter where they are located in an image. Changing brightness or contrast are examples of mapping operations.
2. *Arithmetic combination of two image matrices*: These operations create pixel values for a new image by applying a simple arithmetic function to corresponding pixel values in two different images. Adding two images to form a new image is an example of arithmetic combination.
3. *Filtering*: Filtering an image produces pixel values for a new image matrix computed from groups of pixels in a single input image. These groups are usually located close to the position of the pixel being computed, so we think of them as forming a "neighborhood." For that reason, filtering is sometimes called a **neighborhood operation**. Often, filtering will include combinations of mapping and arithmetic functions. Edge finding, for example, is a filtering operation, because we have to look at at least two pixels next to each other to see if the change in the intensity is enough to call the area an object edge.

Neighborhood Operations: Operations that compute each output pixel value based on a group of input pixels with row and column numbers close to those of the output pixel. The neighborhood is defined by the location of these neighboring pixels used in the output pixel computation.

In all cases, we will need to consider how we will handle the result of a mathematical computation if the result is not a possible pixel value. For example, when we brighten an image, we cannot brighten the whitest white. An image that has 8 bits per pixel can have $2^8 = 256$ different pixel values, interpreted as the integers from 0 to 255. The result of a mathematical computation might be one of those 256 integers, but it might also be −23, 300, or 6.7, for example, which are not possible values for 8-bit pixels. The way these results are treated might be different for different applications. One simple thing to do is to change the result to the nearest value that is an acceptable pixel value, so −23 would become 0, 300 would become 255, and 6.7 would become 7. This process of assigning all possible numbers to the nearest number in a limited list is often called **quantization**. When numbers are larger than 255 or smaller than 0, it is called **clipping**. We also will find some situations where choosing the nearest number does not produce the image result we want. In those cases, other rules can be used.

Quantization: Replaces the computed value of each *n*-bit pixel with the closest value than can be represented by just *n* bits.

Clipping: Sets values larger than the maximum allowed value to the maximum value. Sets values smaller than the minimum allowed value to the minimum value.

Mapping

Mapping: An image mapping operation uses a computational rule to change each pixel value based only on its value, not its location. The rule is defined by a table, a graph, or a mathematical function.

Since mapping operations depend on the values of pixels, but not the pixels' locations in the image, a **mapping** can be described simply as the relationship between an input pixel value and an output pixel value. Each input pixel value has some corresponding output pixel value, so the relationship can be described by a mathematical function or a graph. The number of input values for an m-bit-per-pixel image is limited to 2^m, so the relationship can also be defined by a list of all 2^m possible input values and the corresponding output values. The block for an image-mapping operation will have a single image for the input and a single image for the output. What the block does depends on its function and some adjustable constant values in the block. The size of the output image in pixels will be the same as the size of the input image.

Changing Brightness: Adding or Subtracting a Constant Value to or from a Matrix One of the simplest ways to improve an image is to brighten an image that is too dark or to darken an overexposed image. Changing brightness is also used to adjust the values of pixels so that other processing operations will work better. The brightness can be changed by adding or subtracting a constant value to all pixel values. Figure 3.5(c) and (d) show an example of darkening and brightening, respectively a black-and-white photo of kayakers.

Table 4.1 shows two mappings for brightening and darkening a 3-bit-per-pixel image **A**. Values for the 3-bit-per-pixel brightened output image **B**, in the center of the table, are computed by adding 2 to each input pixel value. Values for the 3-bit-per-pixel darkened output image **B**, on the right side of the table, are computed by subtracting 2 from each input pixel value. The mapping from this table is applied to all pixels in image **A**, no matter how many pixels the image has.

When the sum or difference is not a possible pixel value, the nearest possible value is used. This change normally has very little effect on the appearance of the output image, because, for example, if we are brightening a dark image, we do not expect to find many pixel values in the dark image that are close to the maximum allowed value.

Equation (4.2) shows the arithmetic equation for adding a constant k to each element of the matrix **A**. The amount of change in the brightness can be adjusted by changing the constant that is added or subtracted. The constant k is positive to brighten an image and negative to darken an image.

Table 4.1 Maps for Brightening and Darkening a Three Bit per Pixel Image

Input Pixel Values $A(i, j)$		Brightened Output Pixel Values $B(i, j) = A(i, j) + 2$		Darkened Output Pixel Values $B(i, j) = A(i, j) - 2$	
Decimal	*Binary*	*Decimal*	*Binary*	*Decimal*	*Binary*
0	000	2	010	0	000
1	001	3	011	0	000
2	010	4	100	0	000
3	011	5	101	1	001
4	100	6	110	2	010
5	101	7	111	3	011
6	110	7	111	4	100
7	111	7	111	5	101

Because the scalar constant value k is just a single number, it is written in lowercase letters to distinguish it from a matrix of values.

Note that each individual element, such as $A(i, j)$ is just a single number, so the element addition in Equation (4.2) is just an ordinary arithmetic addition of two numbers resulting in the sum $B(i, j)$, which is one element of the image matrix **B**.

$$B(i, j) = A(i, j) + k \tag{4.2}$$

The mapping for changing the brightness of an image also can be represented by a graph. In Figure 4.12(a), the red plot shows the brightening of an 8-bit-per-pixel image by adding 50 to each pixel value, and the blue plot shows the map for darkening by subtracting 50. Figure 4.12(b) shows the general processing block for adding a constant k to input image **A** to create output image **B**.

The brightness of color images can be adjusted by adding a constant to each of the three matrices for the red, green, and blue color components. The 24-bit-per-pixel color image from Figure 4.13(a) is brightened in Figure 4.13(b) by adding 60 to all three color components. The darkened image in Figure 4.13(c) is obtained by subtracting the same constant. If different values are added to each of the red, green, and blue components, then the change in the relative contributions of the component colors will cause changes in the output image colors in addition to changing the overall brightness. In Figure 4.13(d), the green and blue components are brightened by adding 60, while the red component is darkened by subtracting 60. This combined effect brightens the trees

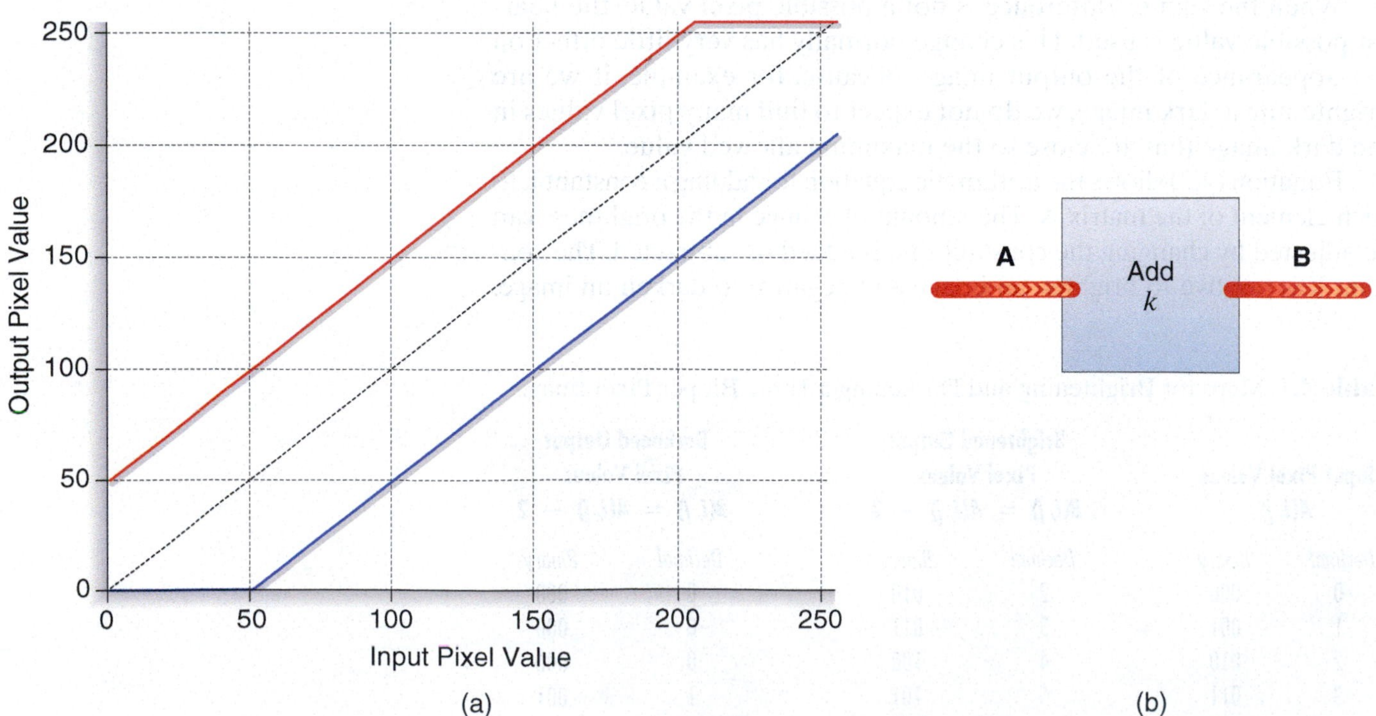

Figure 4.12 The graph in (a) shows the mapping in red for brightening by adding 50 to each pixel value, and the mapping in blue for darkening by adding −50 to each pixel value. Values outside the 0 to 255 range for 8-bit pixels are set to the nearest possible pixel value. The black dashed line added for reference shows the mapping for no change. The block symbol for adding a constant to all pixel values is shown in (b).

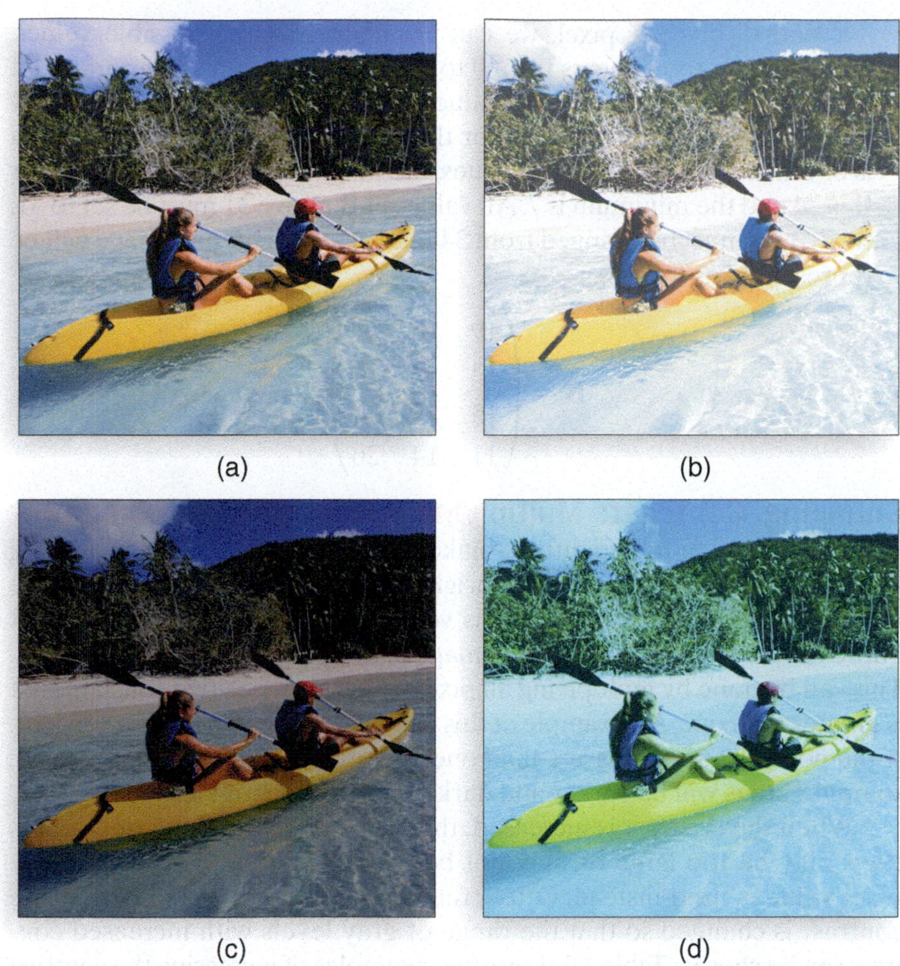

Figure 4.13 In (b), the image has been brightened by brightening each of the three color planes, and in (c) the image has been darkened by darkening all three color planes. In (d), green and blue have been brightened, but red has been darkened, so the image is brighter and the colors have changed.

and the sky, makes the water look more blue–green, makes the kayak look less orange, and reduces the suntan of the kayakers.

EXAMPLE 4.2 Brightening an Image

If matrix $\mathbf{A}$ in Equation (4.1) represents a 5-bit-per-pixel image, find matrix $\mathbf{B}$ after the image represented by $\mathbf{A}$ is brightened by adding the constant 7 to it. What are the maximum and minimum values possible for a 5-bit-per-pixel image? What are the maximum and minimum values in $\mathbf{A}$? What are the maximum and minimum values in $\mathbf{B}$? Which values had to be changed because they were not possible pixel values?

Solution

The matrix $\mathbf{B}$ computed by adding 7 to each pixel value in $\mathbf{A}$ is

$$\mathbf{B} = \begin{pmatrix} 19 & 16 & 7 \\ 12 & 13 & 30 \\ 15 & 23 & 34 \\ 11 & 14 & 26 \end{pmatrix}$$

With 5 bits per pixel, we can calculate $2^5 = 32$ possible values, which will be the integers from 0 to 31. The maximum value for a 5-bit pixel is 31, and the minimum value is 0. In the image represented by matrix **A**, the maximum value for the 12 pixels is 27 and the minimum is 0. When we add 7 to these values, the maximum value computed for **B** is 34 and the minimum is 7. All values in **B** above 31 must be set to 31, so $B(3,3)$ must be changed from 34 to 31. The brightened image will be represented as follows:

$$\mathbf{B} = \begin{pmatrix} 19 & 16 & 7 \\ 12 & 13 & 30 \\ 15 & 23 & 31 \\ 11 & 14 & 26 \end{pmatrix}$$

Changing Contrast: Multiplying a Matrix by a Constant Value Increasing the brightness makes all pixels brighter, but increasing the contrast makes bright pixels brighter and dark pixels darker. If the range of pixel values in an image is very narrow, increasing the contrast improves the visual quality of the image so that more details can be seen. This can be done by multiplying all pixel values by a positive constant that is greater than one. An example of this operation is shown in Figure 3.5(b). If the positive constant is less than one, then the contrast will be reduced, so bright areas will be darker and dark areas will be brighter.

When the contrast is increased, the result of some of the multiplications may be too large or may not be integers, so the nearest possible pixel value is used instead. Often, the brightness is adjusted before the contrast is changed so that the range of gray levels with increased contrast can be chosen. Table 4.2 shows two examples of increasing the contrast of a 3-bit-per-pixel input image. In the center of the table, the contrast is increased by multiplying each input pixel value by the constant value 2.7. This operation spreads the darkest four input pixel values over the full range of brightness, from 0 to 7, and makes the other four pixel values 7, which is white. On the right side of the table, the image is darkened first by subtracting 2, and then the darkened pixels are multiplied by the constant value 2.7. In this case, it is the middle four pixel values that are spread over the full range of brightness, from 0 to 7.

Table 4.2 Maps for Increasing the Contrast of a 3-Bit-per-Pixel Image

Input Pixel Values $A(i,j)$		Increased-Contrast Pixel Values $B(i,j) = 2.7 \times A(i,j)$			Pixel Values for Increased Contrast after Darkening $B(i,j) = 2.7 \times (A(i,j) - 2)$		
Decimal	Binary	Decimal result computed	Decimal pixel value	Binary pixel value	Decimal result computed	Decimal pixel value	Binary pixel value
0	000	0	0	000	−5.4	0	000
1	001	2.7	3	011	−2.7	0	000
2	010	5.4	5	101	0	0	000
3	011	8.1	7	111	2.7	3	011
4	100	10.8	7	110	5.4	5	101
5	101	13.5	7	111	8.1	7	111
6	110	16.2	7	111	10.8	7	110
7	111	18.9	7	111	13.5	7	111

Equation (4.3) shows the mathematical equation for multiplying a matrix by a constant scale factor s, using the style of Equation (4.2). The amount of change in the contrast can be adjusted by changing the scale factor s. The positive constant s is greater than one to increase the contrast and less than one to decrease the contrast.

$$B(i,j) = s \times A(i,j) \tag{4.3}$$

The mapping for changing the contrast of an image can also be represented by a graph. In Figure 4.14(a), the red plot shows the increase in contrast for an 8-bit-per-pixel image obtained by multiplying each pixel value by 1.8. The blue plot shows the map for decreasing the contrast by dividing by 1.8. Usually, it is not desirable to reduce the contrast of an image in this way unless the output image pixels are represented by fewer bits than the input image.

Figure 4.14(b) shows the general processing block for multiplying an input image **A** by a constant s to create an output image **B**. In Figure 4.14(c), the contrast is changed for the central range of values by first changing the brightness. This operation requires two blocks, as shown in Figure 4.14(d), and results in the middle gray value of the input becoming the middle gray value of the output.

Figure 4.15 illustrates the effect of using the two different methods shown in Figure 4.14. In Figure 4.15(a), only the multiply block is used, with $s = 1.3$. The contrast is increased for the darkest areas of the trees and bushes, but details in light areas such as the clouds and the shoreline are lost. In Figure 4.15(b), a constant of $k = -29.5$ is added before the multiplication by $s = 1.3$. This operation moves the high-contrast region to the middle-intensity colors so that more details are seen in the water and on the kayakers.

EXAMPLE 4.3 Changing Image Contrast

If matrix **A** in Equation (4.1) represents a 5-bit-per-pixel image, find matrix **B** after the image contrast is increased by scaling by a factor of two. What are the maximum and minimum values in **B**? Which values had to be changed because they were not possible pixel values?

Solution

The value of **B** computed by multiplying each pixel value in **A** by two is

$$\mathbf{B} = \begin{pmatrix} 24 & 18 & 0 \\ 10 & 12 & 46 \\ 16 & 32 & 54 \\ 8 & 14 & 38 \end{pmatrix}$$

Since all values above 31 must be set to 31 so that they will be the maximum value for a 5-bit pixel, $\mathbf{B}(2,3) = 46$, $\mathbf{B}(3,2) = 32$, $\mathbf{B}(3,3) = 54$, and $\mathbf{B}(4,3) = 38$ must all be set to 31.

Thus, the matrix with increased contrast will be

$$\mathbf{B} = \begin{pmatrix} 24 & 18 & 0 \\ 10 & 12 & 31 \\ 16 & 31 & 31 \\ 8 & 14 & 31 \end{pmatrix}$$

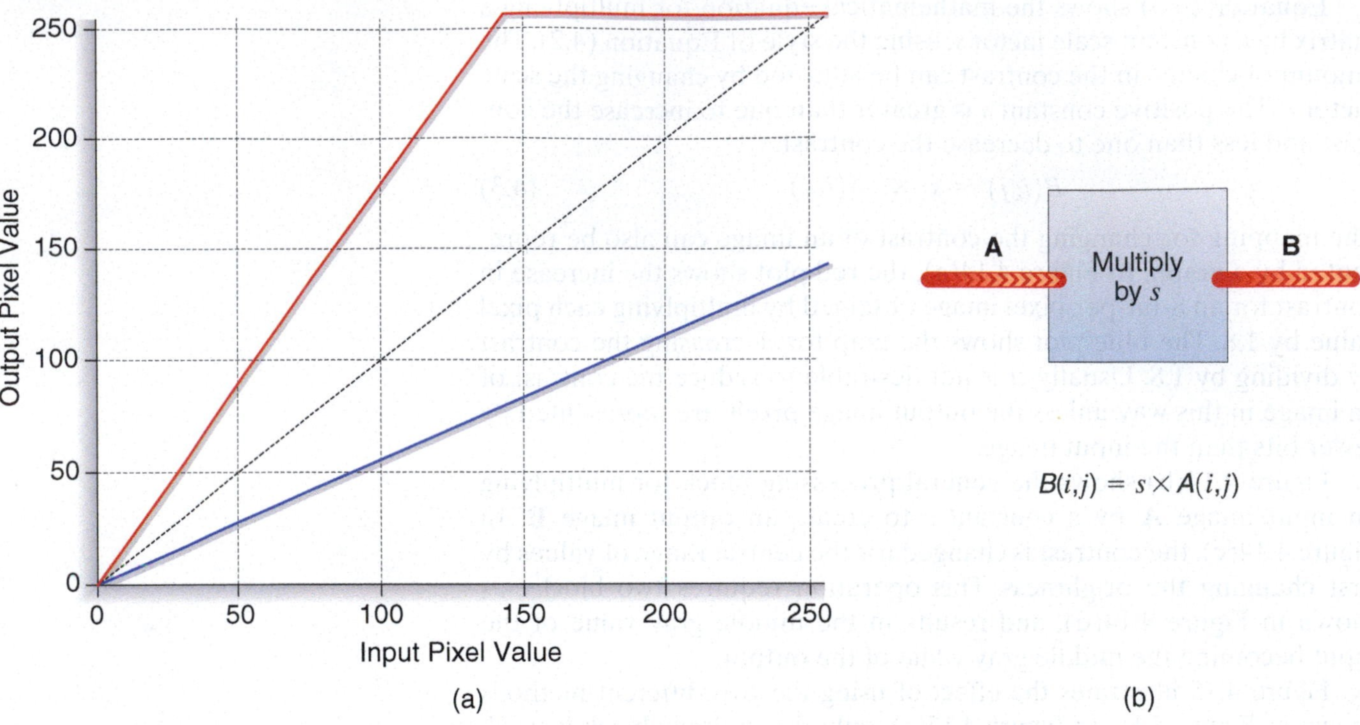

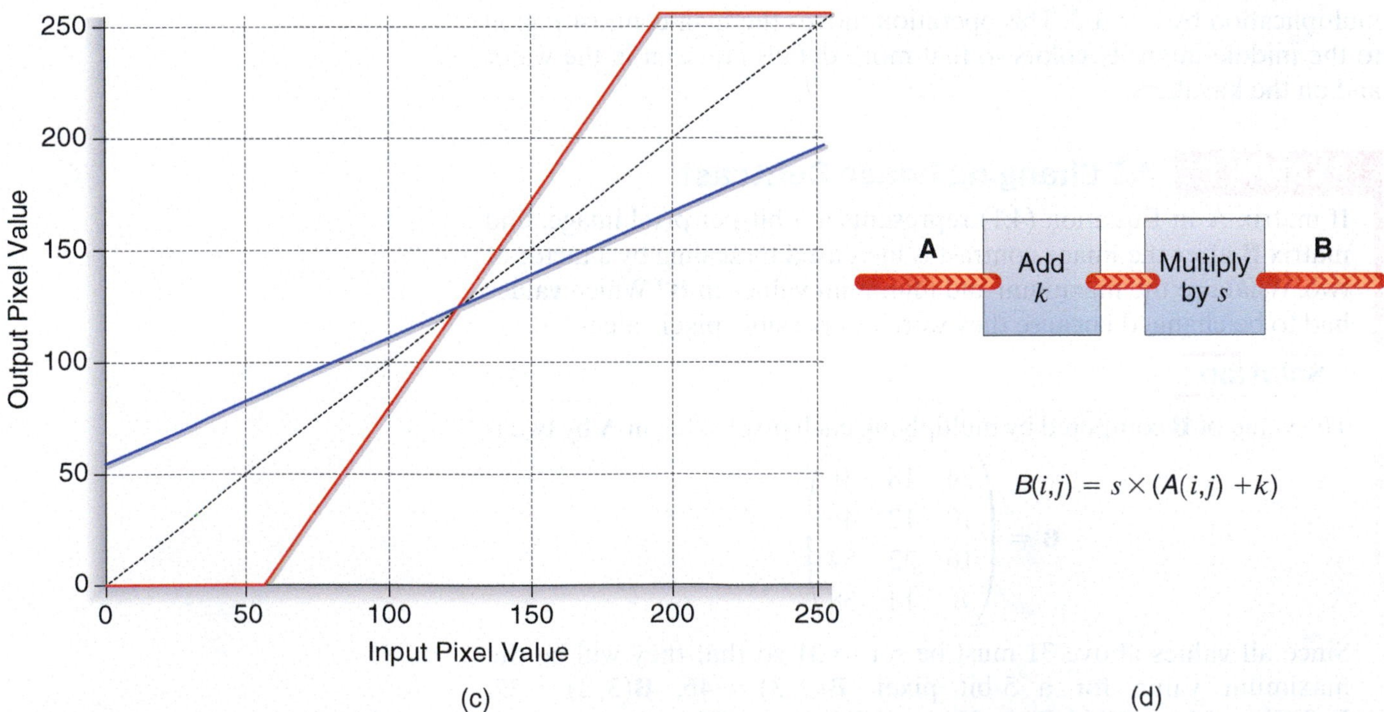

Figure 4.14 The mapping in red in the graph in (a) shows an increase in contrast from multiplying each pixel value by 1.8 and selecting the closest integer between 0 and 255. The mapping in blue shows a decrease in contrast from multiplying each pixel value by 1/1.8. The black dashed line added for reference shows the mapping for no change. The block symbol for multiplying all pixel values by a constant is shown in (b). In (c), the mappings for increasing and decreasing the contrast use a combination of adding a constant and then multiplying by a constant to position the contrast change at the center of the intensity values. The red mapping uses $k = -57$ and $s = 1.8$, and the blue mapping uses $k = 102$ and $s = 1/1.8$. The two blocks needed for this are operation shown in (d).

Infinity Project Experiment: Brightness and Contrast

Using input from a camera or image files, adjust the brightness and contrast of an image by adding a constant or multiplying by a constant. Note the difference between multiplying first and adding first.

(a)

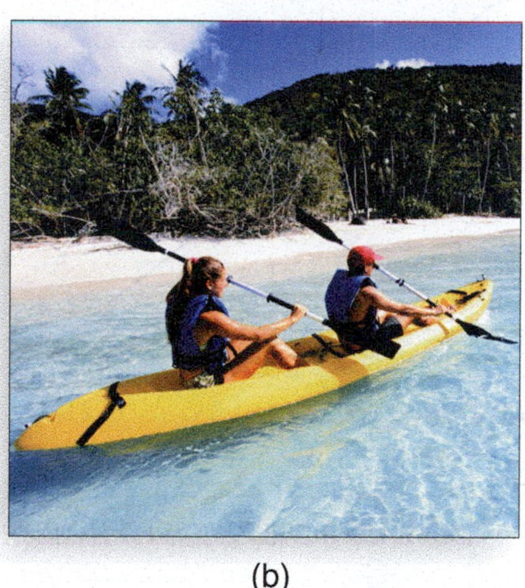

(b)

Figure 4.15 Two examples of increased contrast. In (a), the contrast is increased for the darkest regions, and the lighter regions lose detail. In (b), the contrast is increased for colors at middle intensities.

Negative Images With film-based cameras, the film is the negative of the photograph we look at, which is sometimes called a positive. Where we want white on the photographic print, the negative must be black, and where we want black on the print, the negative must be transparent. With digital images, we still often use the negative of an image in processing steps even though a film negative, itself, never exists. For an 8-bit-per-pixel image, we can create the negative by subtracting each element of the input image **A** from the maximum possible value for an 8-bit pixel, which is 255. This operation is shown in Equation (4.4):

$$B(i,j) = 255 - A(i,j) \tag{4.4}$$

The negative function is so common that we will use the block in Figure 4.16(b) to represent it. It could also be created by using a multiply block with $s = -1$ followed by an add block with $k = 255$. The graph of the negative mapping in Figure 4.16(a) shows that, for the negative function, all output pixels will always be acceptable pixel values. If we had 4-bit pixels instead of 8-bit pixels, the negative's values would be computed by subtracting from 15 instead of 255, and the graph in Figure 4.16(a) would go from $(0, 15)$ to $(15, 0)$.

Figure 4.17 shows the negative image for both the black-and-white and color images of the kayakers. The black-and-white negative is easy to understand, but the color negative is not immediately so intuitive. For example, the yellow in the kayak is a combination of red and green,

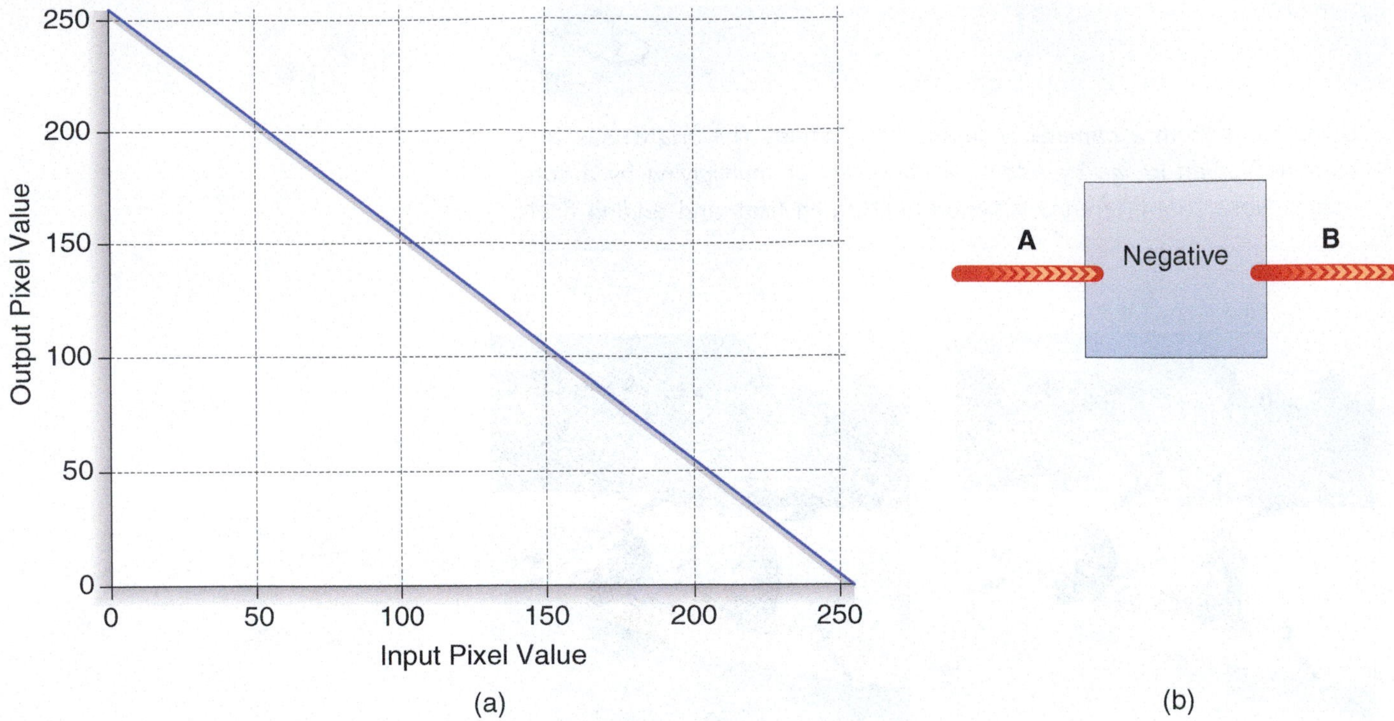

Figure 4.16 The graph for the mapping that produces a negative for 8-bit pixels is shown in (a), and the block symbol for the negative operation is shown in (b).

but not blue, so the negative will have blue, but not red and green. Similarly, the red hat has red, but not green or blue, so the negative will have blue and green, but not red.

Threshold: A value used to separate pixels in an image into two groups. All pixels with values above the threshold value are set to one value, and all pixels with values at or below the threshold value are set to a second value. These two values are often 0 and 1, or 0 and the maximum value.

Thresholded Images A very simple mapping technique that is used in many applications is called **thresholding**. This mapping creates an output image that has only two intensity values and typically is used to detect the presence of some important part of the image rather than to display all of the image's content. For example, if edges have been accentuated by a processing block, then a threshold might be used to keep only the bright edges and leave the rest of the image black. The

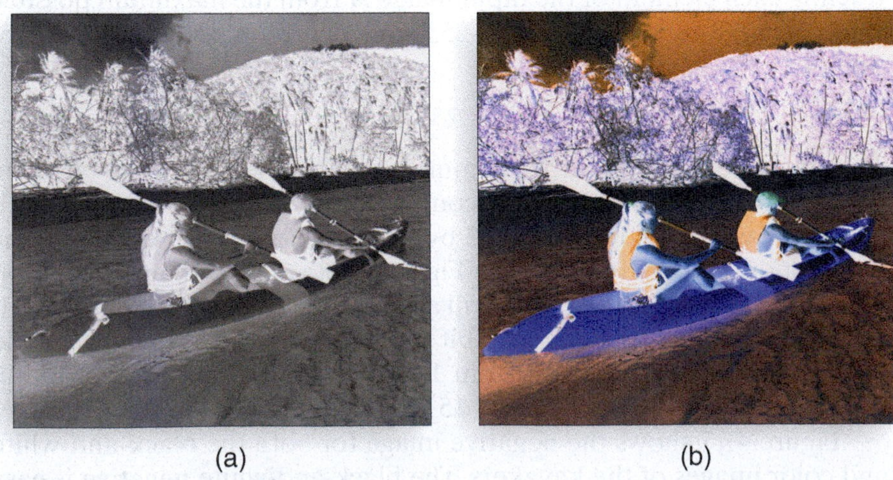

(a) (b)

Figure 4.17 Negative images of (a) a black-and-white photo and (b) a color photo.

edge image in Figure 4.7 was created this way, and then a negative mapping was used to make the edges appear black on a white background instead of white on a black background.

Equation (4.5) shows the simplest threshold operation, which sets all pixels with values at or below a selected threshold value v to 0 and all pixels with values above v to the maximum value, which would be 255 for 8-bit pixels:

$$B(i, j) = \begin{cases} 255 & \text{if} \quad A(i, j) > v \\ 0 & \text{if} \quad A(i, j) \leq v \end{cases} \tag{4.5}$$

For 1-bit pixels, the maximum value would be 1.

Several variations of the simple thresholding of Equation (4.5) may be better matched to some specific applications. A softer threshold has a high threshold value v_h for pixel values set to 255 and a low threshold value v_l for pixels set to 0. In between the two threshold values, the pixel values are spread out between 0 and 255. This variation is the same as increasing the contrast with the method in Figure 4.14(d), using a very large value for s. Figure 4.18 shows three different threshold mapping graphs and the corresponding thresholded images.

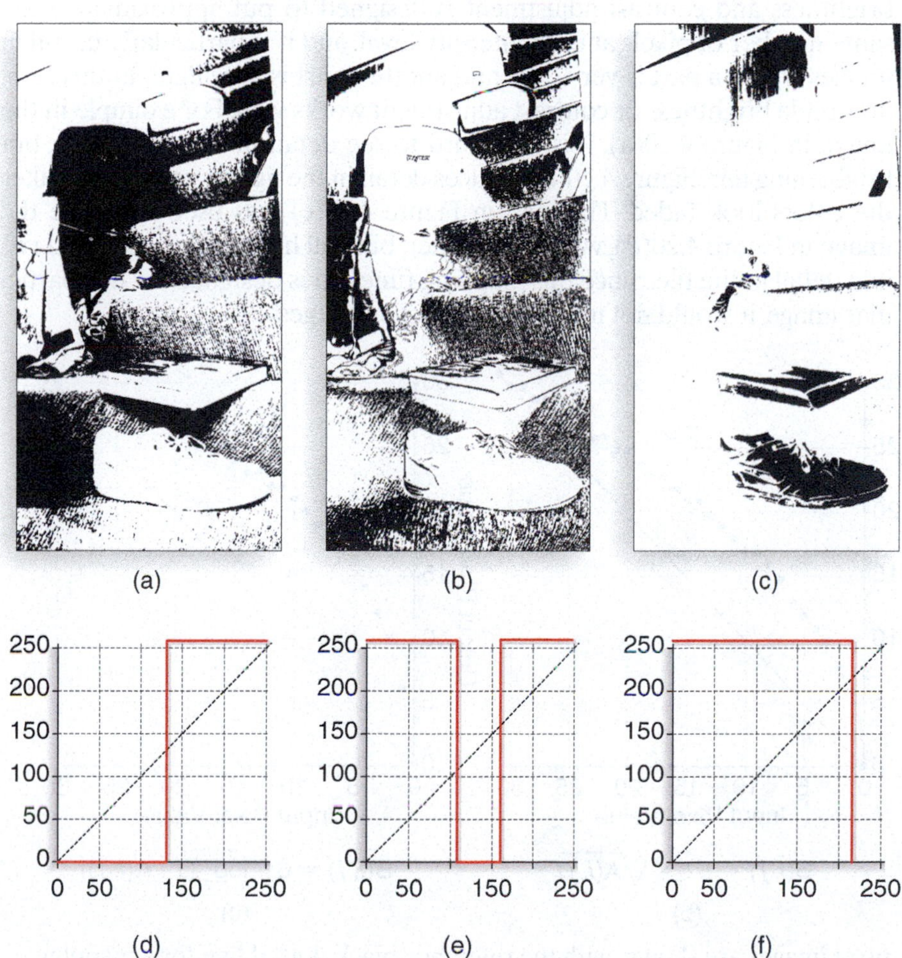

Figure 4.18 Graphs of three thresholding maps and the corresponding black-and-white images.

Other Mappings We have already discussed the most common mappings, but there is no limit to the variety of custom mappings that can be defined to solve a particular problem. Any mathematical function can be used, and mappings that do not have a simple arithmetic or logical expression can be described by a graph or a table that lists all possible input pixel values. For example, nonlinear brightness and contrast adjustments might be needed to adjust for lighting conditions, electronic-sensor characteristics, or changes in a photographic print due to aging.

Figure 4.19 shows the mapping graphs for three functions using 5-bit pixels. The graph for the mapping is 32 discrete points, which are restricted to pixel values between 0 and 31, but the continuous function defining the mapping is also shown for reference as a dotted line. The effect of using these three different functions is either to increase the contrast of the bright pixels and compress the range of the dark pixels, using a function like the square function, or to increase the contrast of the dark pixels with a function like the square-root function or logarithm function.

A special map can also be made for one particular image instead of using a general mathematical function such as those in Figure 4.19. A special map allows gray-level adjustments for the best possible improvement in brightness and contrast for a particular image. Creating such a map is similar in some ways to defining a colormap for one particular image to make the best use of a limited palette of 256 colors. The map for customized brightness and contrast adjustment is designed to put approximately the same number of pixels at each intensity level, and it is particularly useful in images where a part is very dark and another part is very light. In this case, no simple brightness or contrast adjustment works well. For example in the image in Figure 4.20(a), it is very hard to see detail in the dark areas, but brightening it in Figure 4.20(b) reduces detail in the lighter areas and makes the colors look faded. The map in Figure 4.20(c) was used to create the image in Figure 4.20(d), which is brighter, but still has strong colors and visible detail in the tile. Since this mapping function is designed for this particular image, it would not improve most other images.

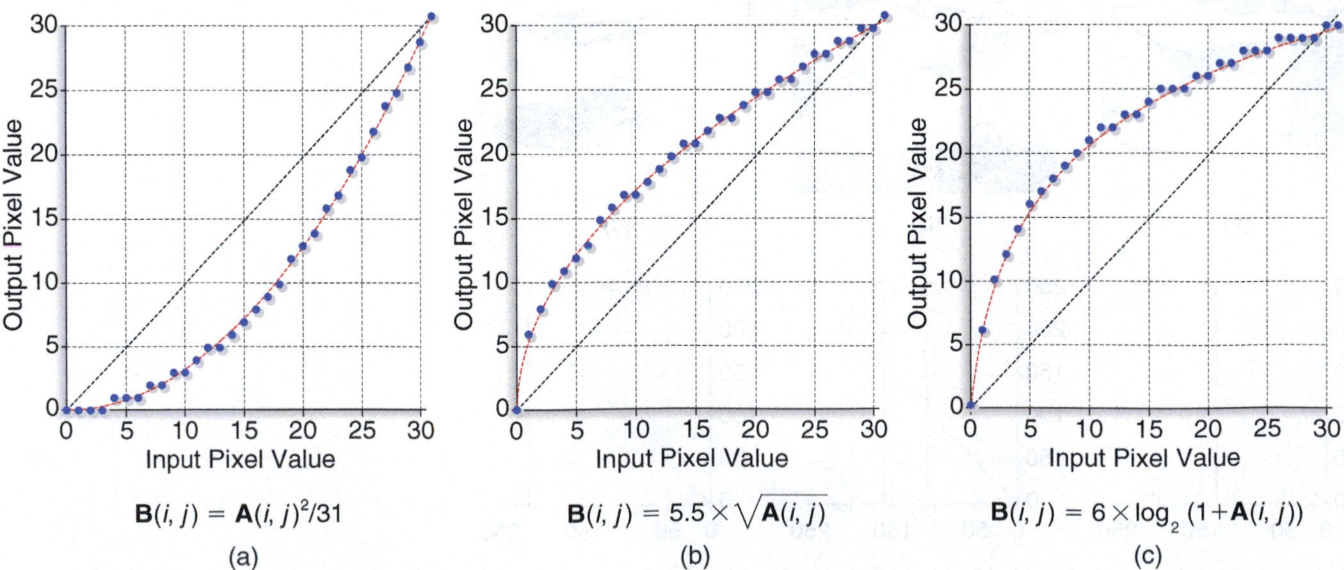

Figure 4.19 Three mappings for 5-bit-per-pixel images are shown, with the reference black dotted line for a mapping that causes no change. The continuous functions are shown in red, and the actual correspondence between the 32 input pixel values and their output values are shown with blue circles. In several cases, many input pixel values are given the same output pixel value.

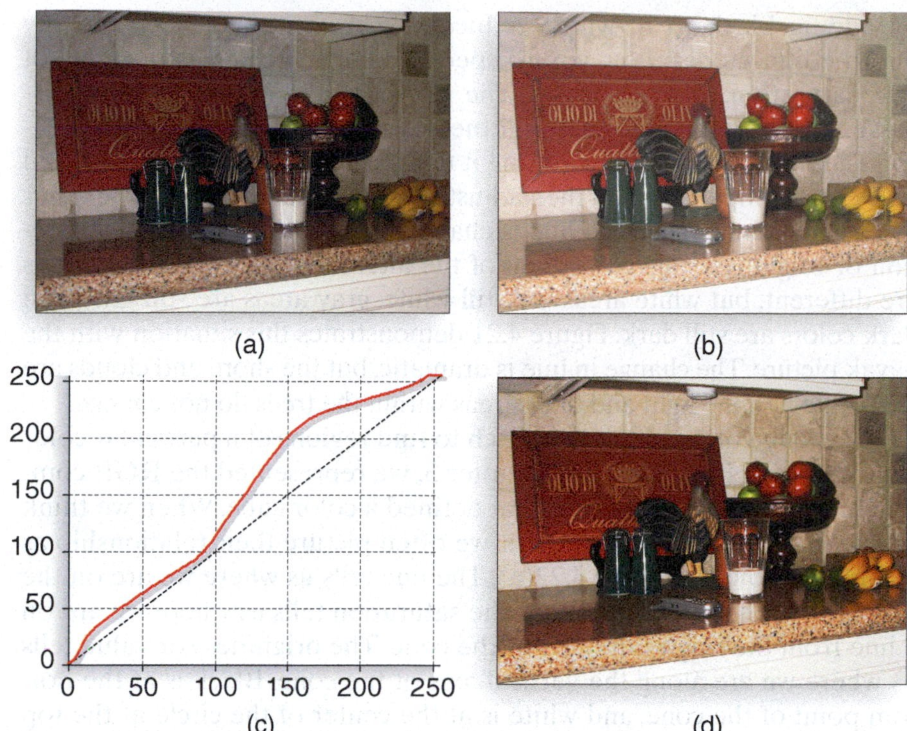

Figure 4.20 The image of the kitchen counter in (a) taken with a digital camera is a little dark, but when it is brightened by adding 60 to each pixel, the colors look faded. When the custom map in (c), which is designed for this image, is used instead to make the image in (d), the image is brighter, but the colors are still strong.

Infinity Project Experiment: Threshold and Negation

Using input from a camera or image files, experiment with thresholding an image and taking the negative. What happens to the output image as the threshold is increased from zero to the maximum value? What is the effect of using the same threshold value on both an image and its negative? What is the effect of forming the negative of a thresholded image?

Other Ways to Specify Color

When different mappings are applied to the three independent color components—red, green, and blue, (RGB)—the colors we see in the resulting image may be changed. This happened in Figure 4. 13 when green and blue were brightened, while red was darkened. We can also get unexpected results when we threshold a single color component. For example, if we try to find bright blue in an image by thresholding the image to keep pixels with large values for the blue component, we will select many pixels that are not blue as well. As we saw in Chapter 3, the value of the blue component is also large in bright cyan, bright magenta, and bright white.

An alternative way of specifying color is similar to the way artists specify color. Even people who are not artists will identify a color by its hue, such as red, yellow, or blueish green. This is the pure color of the pigment an artist starts with. The color can be lightened by adding white to it, which reduces its saturation, and the color can be darkened by adding black to it, which reduces its brightness or value. Many photo-editing tools allow you to specify colors or changes in terms of hue, saturation, and brightness. The mapping operations we have discussed can be applied to the brightness and saturation.

The hue is computed by setting it to the largest of the red, green, and blue color components and then adjusting it by the difference of the other two components. When the hue is changed without changing the saturation or brightness, our perception of the altered image is that the colors are different, but white areas are still white, gray areas are still gray, and dark colors are still dark. Figure 4.21 demonstrates this situation with the kayak picture. The change in hue is dramatic, but the shore and clouds are still white and the light and dark areas within the trees do not change.

The value of using this approach to find regions of a particular color can be seen in Figure 4.22. In Chapter 3, we represented the RGB components of color as three axes that defined a color cube. When we think of hue, saturation, and brightness, we often picture their relationship as a cone, as shown in Figure 4.22(a). The hue tells us where we are on the ring around the top of the cone. The saturation tells us where we are on a line from the ring to the axis of the cone. The brightness or value tells us where we are along the vertical axis of the cone. Black is at the bottom point of the cone, and white is at the center of the circle at the top of the cone. In Figure 4.22(b), the hue of each pixel is not changed, but the brightness is set to the maximum and the saturation is thresholded at 30% of the full value. It would be much easier to use hue to find objects in this image than it would in the original image of the kayakers. Figure 4.22(c) shows a version of the image of the cluttered stairs from Figure 4.9 in which both saturation and value have been set to the maximum. Without some threshold for saturation, every pixel that is not a shade of gray has some bright hue. Parts of the white shoes and the walls are blue, because they are slightly more blue than red or green. When the saturation is thresholded at 50%, as shown in Figure 4.22(d), the blue backpack, blue clothes, and the blue book remain. The segmented images in Figure 4.9 were obtained using this approach and some smoothing operations that will be described in a later section.

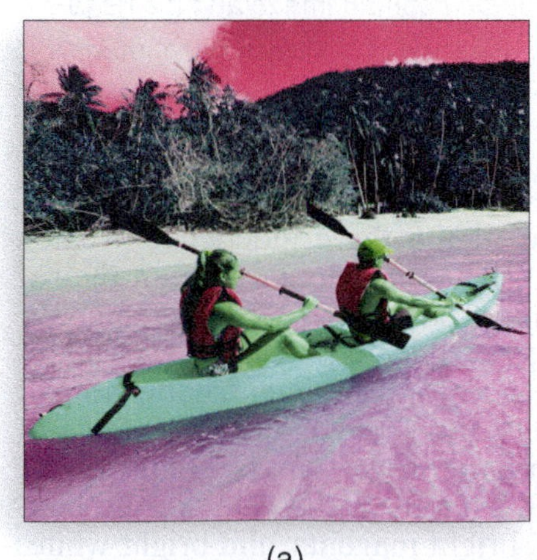

(a)

(b)

Figure 4.21 In these images of the kayakers, the RGB components have been switched so that the saturation and brightness for each pixel stay the same, but the hue is changed. In (a), red is displayed as green, green in displayed as blue, and blue is displayed as red. In (b), the colors are switched in the opposite way.

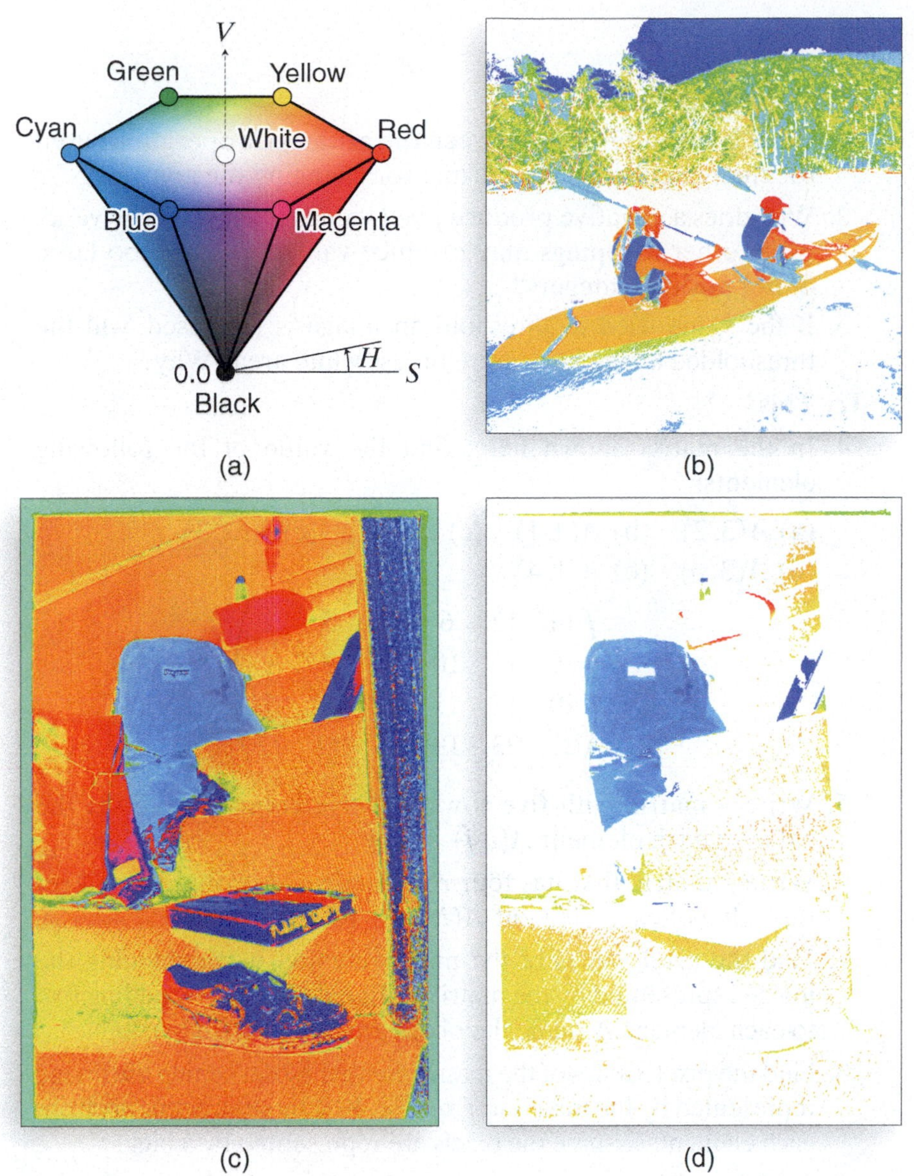

Figure 4.22 Instead of the RGB color cube, the HSV color cone in (a) can be used to specify color. Hue (H) is measured around the circle at the top, value (V) is the vertical distance, along the cone's axis, and saturation (S) is the horizontal distance from the axis. When S and V are changed to the maximum value in the images in (b–d), all colors are "pure" hue. In (b), the saturation is thresholded at 30% and in (d) the saturation is thresholded at 50%.

Summary

The mapping operations we have discussed give us powerful, but mathematically simple, ways to improve images and separate parts of images, based on color and brightness. An image is just a matrix of pixel values, and the mappings create new pixel values as a function of the original pixel values. We will see in the next section that, when we can also use simple mathematical operations to combine two different images or to combine several pixels in the same image, we will be able to create even more tools for our digital darkroom.

EXERCISES 4.2

Mastering Concepts

1. What is the difference between the arithmetic operations used for changing the brightness and for changing the contrast?
2. Why does a negative produce pixel values that are alway valid, while other mappings may produce values that are too large or not positive integers?
3. If the value used to threshold an image is increased, will the thresholded image have more or less white area? Why?

Try This

4. In the matrix shown here, find the value of the following elements:

   (a) $A(3,2)$ (b) $A(4,1)$ (c) $A(1,5)$ (c) $A(2,3)$
   (d) $A(3,6)$ (e) $A(1,4)$

$$\mathbf{A} = \begin{pmatrix} 14 & 12 & 6 & 7 & 5 & 2 \\ 16 & 13 & 10 & 4 & 3 & 1 \\ 20 & 15 & 11 & 9 & 8 & 5 \\ 27 & 23 & 19 & 18 & 17 & 12 \end{pmatrix}$$

5. Write a matrix with five rows and five columns, and set the value of each element $A(i, j) = 7 + i - j$.
6. Write a matrix that has four rows and seven columns, and set the value of each element $A(i, j) = 10 \times i + j$.
7. Find the pixel values of the matrix created by brightening the image represented by the matrix in Equation (4.1) by adding five to each element. Assume the pixels are represented by 5 bits.
8. Find the pixel values of the matrix created by darkening the image represented by the matrix in Exercise 4.2.4 by subtracting six from each element. Assume the pixels are represented by 5 bits.
9. Find the pixel values of the matrix created by increasing the contrast of the image represented by the matrix in Equation (4.1) by multiplying each element by 1.5. Assume the pixels are represented by 5 bits.
10. Find the pixel values of the matrix created by decreasing the contrast of the image represented by the matrix in Exercise 4.2.4 by multiplying each element by 0.4. Assume the pixels are represented by 5 bits.
11. Write a table for a map that changes a 3-bit-per pixel image into a 5-bit-per-pixel image. Remember that the whitest white value will be the maximum value possible for each image, and the blackest black value will always be 0.
12. If the contrast of an image is increased by using $s = 85$, how many of the input pixel values will be set to values larger than 0 and smaller than 255?
13. Compare the map for increasing the contrast with $s = 255$ and $v = k$ in Figure 4.14(d) to the threshold in Equation (4.5).

14. Find the pixel values of the matrix created by taking the negative of the image represented by the matrix in Exercise 4.2.4. Assume the pixels are represented by 5 bits.
15. If a negative scale value is allowed, find the values of k and s to create a negative image by scaling and then adding a constant. How do the values of k and s change if a constant is added first and then the scaling is done?
16. For a threshold value of $v = 9$, find the threshold image from the image represented by the matrix in Exercise 4.2.4.

Bach of the Envelope

17. If you want to rotate an image 90° to the right, what would you have to do to the image matrix? Write a rule or a mathematical expression that tells you how to change the row numbers and column numbers for each pixel in order to perform this operation.
18. If you want to make a mirror reflection of an image, what would you have to do to the image matrix? Write a rule or a mathematical expression that tells you how to change the column numbers for each pixel in order to perform this operation.

4.3 Digital Darkroom Tools

If we want to smooth or sharpen images, reduce noise, find edges, or create a new image from more than one image, our digital darkroom will need more than the mapping operations from the previous section. Mapping computes new pixel values for a single image, based only on the original pixel values. However, edge finding and smoothing are not possible unless we look at the values of several nearby pixels at the same time. Also, change detection and other image comparisons require computation that uses more than one image.

In this section, we will add these new types of operations to our set of digital tools for processing images. We will see that we still use very simple mathematics on our matrices of pixel values and that we can create powerful tools with combinations of just a few of these simple operations.

Arithmetic Combination of Two Image Matrices

Two image matrices of the same size can be combined to make a new image matrix by applying a simple arithmetic operation to pairs of elements from the two matrices that have the same row and column number. Since the arithmetic operations are applied to individual elements, we use ordinary arithmetic to create each element of the new image matrix. When we view the new matrix, we will see the new image we have created. Although these operations are so simple that they may seem uninteresting by themselves, we can combine them to do much more complex and interesting operations.

The new matrix will be the same size as the two matrices used to compute it. Depending on the type of arithmetic operation and the number of bits available for the pixels of the new image, it may be necessary to change some computed values in order to stay within the range of possible pixel values of the new image.

Addition and Subtraction of Image Matrices When two image matrices are added, elements that are in the same row and column positions of the two input matrices are added together. The sum will be the value of the element of the output image with that same row and column position. Equation (4.6) shows the equations for image-matrix addition, both in the whole-matrix form and in individual element form. Both of these equations mean that each element of the image matrix **A** is added to its corresponding element in the image matrix **B** to make an element in the output image matrix **C**. The matrix form is more compact, but the element form shows us exactly how to do the computation. The element form also reminds us that matrices **A** and **B** must be the same size.

$$\mathbf{C} = \mathbf{A} + \mathbf{B}$$
$$C(i,j) = A(i,j) + B(i,j) \tag{4.6}$$

Images may need to be added in a variety of applications. If the brightness adjustment needed is not the same for all parts of the image, then instead of adding the same constant to all pixels, a matrix can be created with an individual brightness adjustment for each pixel. When the brightness-adjustment matrix is added to the image, we will have a new image with different corrections applied to different parts of the image.

Image additions can also be used to create a transparency effect of being able to see one image through another or to overlay two images or layers from different sources. We might also want to add components from two different images, as we did in the blue-screen Chromakey example in Figure 3.46. Sometimes, images taken at different times from the same camera viewing the same scene are added to reduce the effects of random variations in the digital image. This procedure might be needed for images acquired with very low light levels.

We can subtract two images in the same way by replacing the addition sign in Equation (4.6) with a subtraction sign, as shown in Equation (4.7). Image subtraction could be used to design a change detector if **B** is an image matrix of a scene and **A** is an image matrix of the same scene taken at a later time. The output matrix **D** would have large values where the current image matrix **A** and the earlier image matrix **B** are most different because some change has occurred, and it would have values close to 0 where image matrix **A** and image matrix **B** are most alike.

> **KEY CONCEPT**
>
> The sum of two images can be multiplied by 0.5 to keep the pixel values from getting too large for the number of allowed bits per pixel.

$$\mathbf{D} = \mathbf{A} - \mathbf{B}$$
$$D(i,j) = A(i,j) - B(i,j) \tag{4.7}$$

Since both **A** and **B** are image matrices with integer values, we can be confident that the result of addition or subtraction of the image-matrix elements will be an integer. However, it might not be possible to represent the integer result with the number of bits available for **C** or **D**. In the case of image addition, there are many combinations of values for elements of **A** and **B** that will produce a result that is too large. If we take the nearest acceptable value, all of these combinations that produce a sum above the highest value possible would be set to that value (for example, 255 for 8-bit pixels), and they would not be distinguishable in the result. Another way to handle large values of the sum would be to divide the sum by two so that no result would be too large and each of the two input images would contribute equally.

In Figure 4.23, two 24-bit per pixel color images are added using Equation (4.6) for each of the three RGB color planes. In (c), all values that exceed 255 in any of the three color planes are simply set to 255. Since many

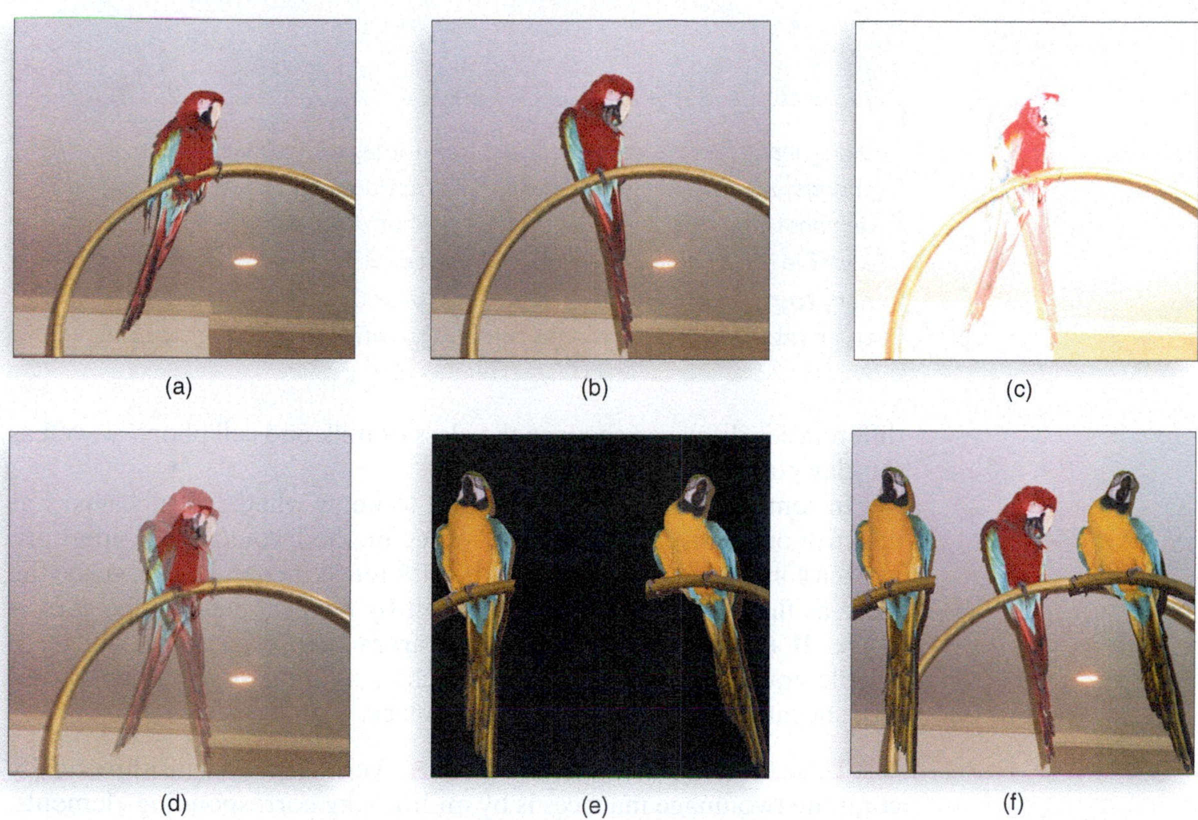

Figure 4.23 Two 24-bit color image matrices **A** (part (a)) and **B** (part (b)) are added to create an image matrix **C**. In (c), the image represented by **C** is displayed after setting all values above 255 to 255. In (d), the image represented by **C** is displayed after dividing all the values by two. The image in (f) is the sum of the image in (e) and the image in (b) after some pixels are set to zero.

of the values of the sums are too large, most of the image is white. In part (d), the sum is divided by 2 and set to the closest integer. The common background of two images is unchanged. The two birds look lighter where each is added to the light background. Figure 4.23(f) shows a sum of the two birds shown in part (e) added to a version of part (b) that has all pixels set to 0 if the corresponding pixel in part (e) is not zero. In this case, no pixel in the sum is too large because at least one pixel in each sum is 0. This sum looks more like our intuitive idea of adding two images.

When subtracting one image from another, the result will never be too large, but it might be negative. If all negative values are set to 0, which is the closest value, we will only see half of the changes. Areas of great difference with a negative sign would look just like areas with no difference. A better solution for a change-detector design is to ignore the sign of the elements in the output matrix and keep only the magnitude of the result of subtraction, as in the example in Figure 4.8. Ignoring the sign is called taking the absolute value. Then areas of great difference will always be bright, and areas of small difference will always be dark.

Figure 4.24 shows two images of a kitchen counter. The robot should put away the cell phone and put the glass of milk in the sink so that the counter will look like the image in part (a). The two images are subtracted to delete any change. In (c), all negative values are set to 0. The change due to the milk is not visible because when part (b) is subtracted from part (a) the high pixel values for the white milk make the difference negative. We only see the clay pot behind the glass. In (d), the absolute value of the

KEY CONCEPT

When subtracting two images, the absolute value can be used to make negative pixel values positive.

Infinity Project Experiment: Adding and Subtracting Images

Using input from a camera or a stored image sequence, implement the change detectors using image subtraction described previously and demonstrated in Figures 4.8 and 4.23. What do you see in the different images? Is it necessary to take the absolute value of the output? Is it necessary to increase the contrast of the output? Change the subtraction block to an addition block. How does this change affect the output image?

difference is displayed showing the glass of milk and cell phone as well as the clay pot behind the glass.

In some cases, it may be important to know whether the change was negative or positive, so the absolute-value method would not be appropriate. If a constant of 255 is added to the difference, then the range will be the same as that for the sum, and dividing it by two will put it in the correct range. If this approach is used, the areas with no change will be a middle-gray value, with large positive changes being shown as white and large negative changes being shown as black.

Multiplication of Image Matrices Another way to arithmetically combine two image matrices is by multiplying corresponding elements. This operation is useful when brightness adjustments are not the same for all parts of an image. Multiplication also allows us to select part of

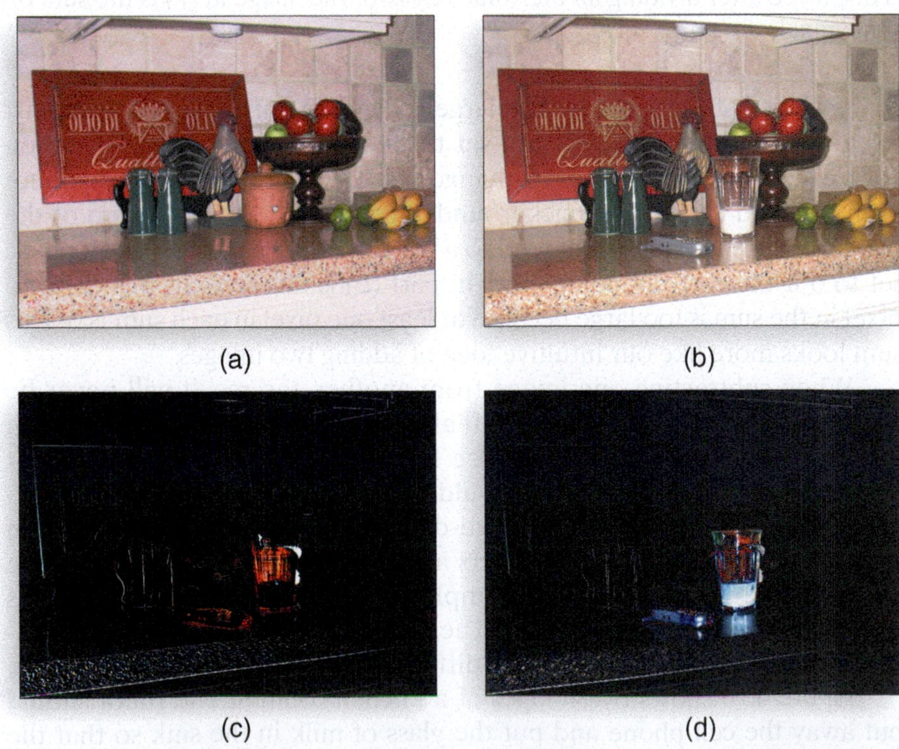

Figure 4.24 The 24-bit color image matrix **B** (part (b)) is subtracted from **A** (part (a)) to create a color image matrix **D**. In (c), the image **C** is displayed after setting negative values to 0. In (d), the image matrix **D** is displayed after negative values are made positive by ignoring the sign.

an image and remove the rest by multiplying the part to be removed by zero and the selected part by one. This was used to change Figure 4.23(b) before adding it to Figure 4.23(e).

The image multiplication operation defined in Equation (4.8) has a form similar to that of the addition and subtraction operations, but the arithmetic operation applied to the individual elements here is multiplication. While there is only one way to multiply two scalar values, many different multiplication methods may be defined for matrices. The specific matrix multiplication method shown in Equation (4.8) uses element-by-element multiplication. We will use the symbol $\otimes$ to represent this operation when we use the matrix form. In the second form of the equation, where we are multiplying individual elements, we use the ordinary multiplication symbol, since we are just multiplying two numbers. The image-matrix product obtained from this multiplication will be of the same size as the two input matrices.

$$\mathbf{G} = \mathbf{A} \otimes \mathbf{B}$$
$$G(i,j) = A(i,j) \times B(i,j) \tag{4.8}$$

When one of the input matrices has only 1 bit per pixel, it is often referred to as a **mask**. Where the mask has a value of 1, the output looks exactly like the other input image. Where the mask has a value of 0, the output image is also 0.

Mask: An image with 1 bit per pixel that is used to select part of another image and remove the rest, using multiplication.

These three arithmetic combinations of matrices can be represented by blocks in the same way as mapping operations, as shown in Figure 4.25. For the arithmetic operations, a block has two input images and one output image. All the images must be the same size, although, in some cases, they might have different numbers of bits per pixel. The name of the block explains what function it performs, and there are no adjustable constants. The blocks in Figure 4.25 can be combined with mapping blocks as demonstrated in the following example.

EXAMPLE 4.4 Drawing a Block Diagram

Suppose that objects of interest in a particular 8-bit-per-pixel image **A** have pixel values between 100 and 180. Pixels with values above 180 or below 100 should be set to 0 to make the objects easier to locate. Using blocks for multiplication, thresholding, and taking the negative, draw the block diagram that will do this task.

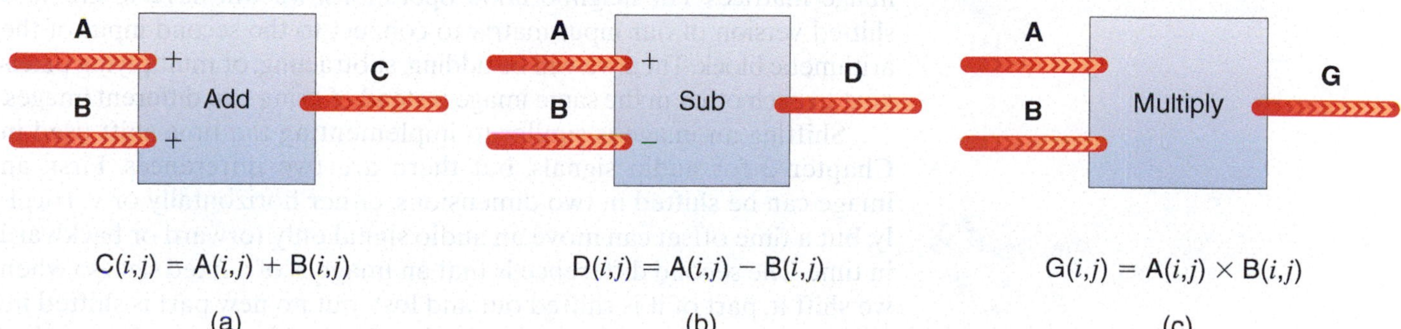

Figure 4.25 Image-processing blocks for arithmetic combinations of image matrices.

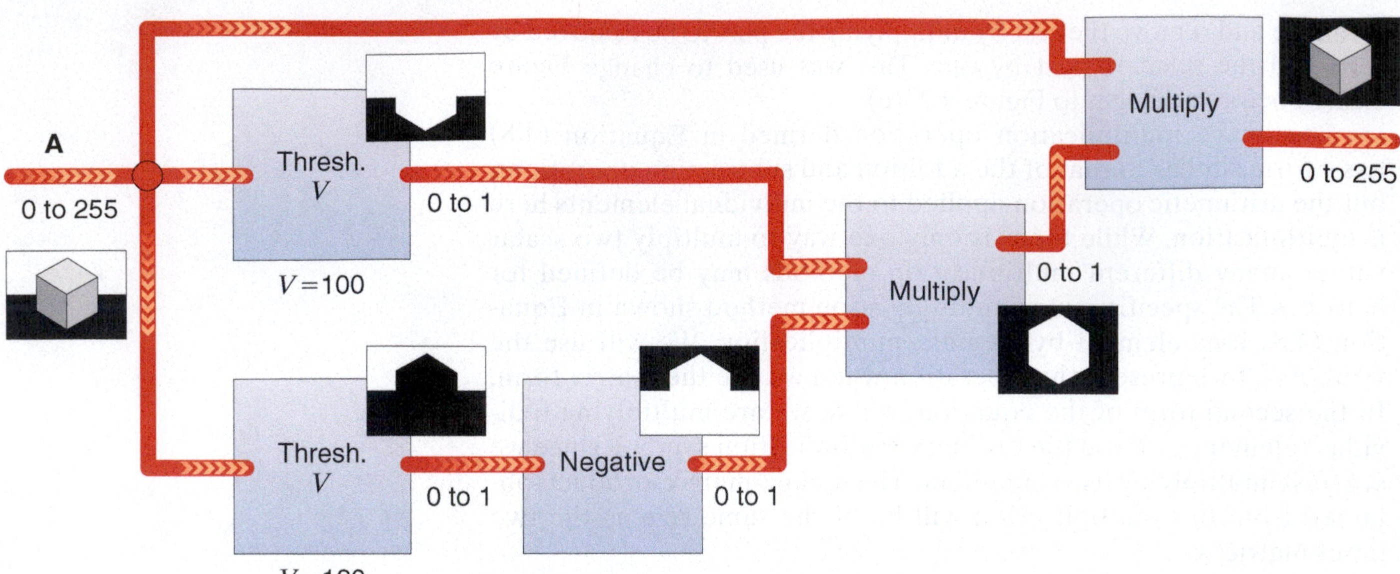

Figure 4.26 Block diagram to display only pixels with values greater than 100 but less than or equal to 180. The range of pixel values for the output image is shown for each block.

Solution

First, we will use the input image matrix **A** to make a mask of 1 (white) for input pixel values greater than 100 and 0 (black) for all other input pixel values by thresholding the image at $v = 100$. Next we will make a second mask that is 1 for input pixel values less than or equal to 180 and 0 for all other input pixel values by thresholding the input image at 180 and then taking the negative. Now we need to multiply the two masks to get a mask that is 1 where pixels in **A** are greater than 100, but less than or equal to 180. This third mask is multiplied by our input image. This procedure is illustrated in the block diagram in Figure 4.26.

Filtering: Neighborhood Operations

The third type of operation uses a group of elements in the input matrix to compute the value of a single element in the output matrix. The elements in the group of input-matrix pixels usually are adjacent to each other, so they are thought of as a neighborhood. Neighborhood operations on a single image can use the same arithmetic operations of addition, subtraction, and multiplication that we have already used to combine two different image matrices. For neighborhood operations, we will have to create a shifted version of our input matrix to connect to the second input of the arithmetic block. Then we will be adding, subtracting, or multiplying pixels next to each other in the same image instead of using two different images.

Shifting an image is similar to implementing the time shift used in Chapter 2 for audio signals, but there are two differences. First, an image can be shifted in two dimensions, either horizontally or vertically, but a time offset can move an audio signal only forward or backward in time. The second difference is that an image is of a fixed size, so when we shift it, part of it is shifted out and lost, but no new part is shifted in. However, we assume that an audio signal exists for a long time, and so when we have a time offset, it changes which part of the signal we see.

Shifting of Images Shifting an image horizontally or vertically simply moves all of the pixels one pixel position over along a row or one pixel position up or down along a column. If **H** is an image matrix created by shifting image matrix **A** horizontally by one column to the right, then $H(i,j) = A(i,j-1)$. By shifting the 4×3 example that was used in Equation (4.1) one column to the right, we obtain the result in Equation (4.9):

$$\mathbf{H} = \begin{pmatrix} H(1,1) & H(1,2) & H(1,3) \\ H(2,1) & H(2,2) & H(2,3) \\ H(3,1) & H(3,2) & H(3,3) \\ H(4,1) & H(4,2) & H(4,3) \end{pmatrix} = \begin{pmatrix} ??? & A(1,1) & A(1,2) \\ ??? & A(2,1) & A(2,2) \\ ??? & A(3,1) & A(3,2) \\ ??? & A(4,1) & A(4,2) \end{pmatrix} \tag{4.9}$$

Since the elements of **H** are the same values that were found in **A**, all pixel values in **H** will be acceptable pixel values. However, we do have to consider what to do about the first column of **H** since it does not correspond to any column of **A**. Unknown values are shown as question marks. Often, the first column of **H**, where $j = 1$, either is set to a value of 0 or is made the same as the second column, which is the first column of **A**. If we shifted one column to the left instead, we would not have values to shift in for the rightmost column of the shifted image.

The result of a horizontal shift by only one column in a large image is hard to notice visually, although we will see later that combinations of images with such small shifts can have very noticeable effects.

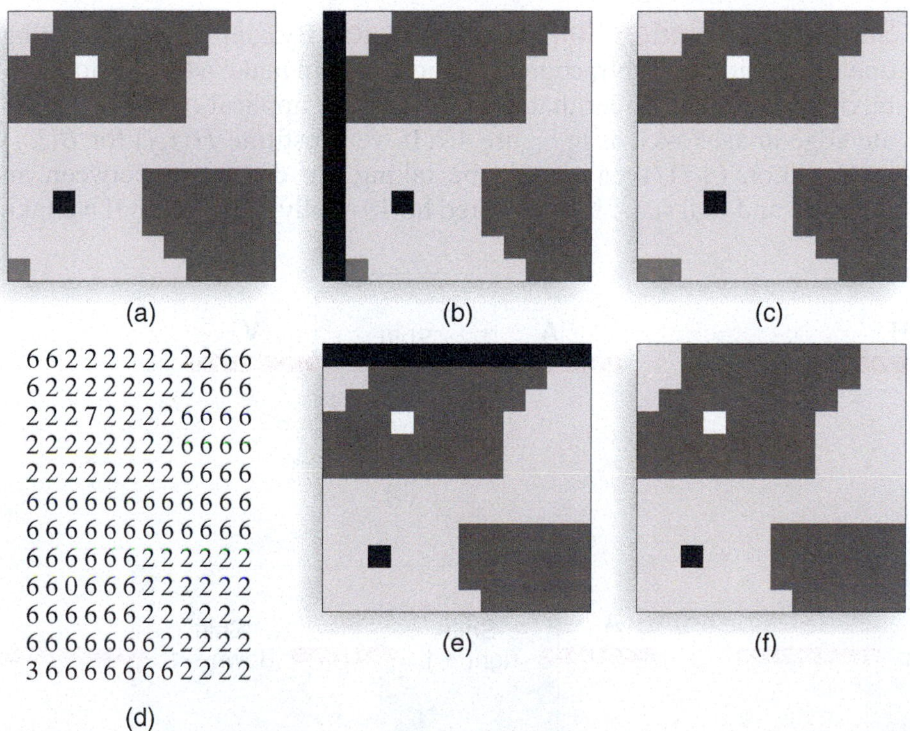

Figure 4.27 The 12×12 pixel image with 3 bits per pixel in (a) is shifted to the right by one column in (b) and (c). In (b), the unknown values in the left column are set to 0, and in (c) the unknown values in the left column are set to be the same as the values in the second column. The image is shifted down by one row in (e) and (f), with the unknown top row being set to 0 in (e) and to the same values as in the second row in (f). The individual pixel values of the original image are shown in (d).

Figure 4.27 demonstrates shifting images for a very simple 12×12–pixel image with 3 bits per pixel. The effect of both settings of the first column is shown. Often, neither choice is good, and, instead of using either, the shifted image just does not use that first column. This approach makes the shifted image one column narrower than the original image. It must also be remembered that when we shift an image to the right in this way, we will lose the rightmost column of the original image matrix **A**, because the last column of the shifted matrix **H** will be the next-to-the-last column of **A**. An image that is vertically shifted down by one row can be created in a similar way by using $V(i, j) = A(i-1, j)$. Equation (4.10) shows the mathematical expressions for a shift of one pixel position to the right, left, down, and up:

$$
\begin{aligned}
H(i,j) &= A(i,j-1) && \text{Shift } \mathbf{A} \text{ to the right}\\
H(i,j) &= A(i,j+1) && \text{Shift } \mathbf{A} \text{ to the left}\\
V(i,j) &= A(i-1,j) && \text{Shift } \mathbf{A} \text{ down}\\
V(i,j) &= A(i+1,j) && \text{Shift } \mathbf{A} \text{ up}
\end{aligned}
\tag{4.10}
$$

The block diagrams for shifting an image one column to the right and one column down are shown in Figure 4.28(a) and (b). The number inside the block shows how many pixel positions the image is shifted, so to shift left or up the number would be -1 instead of $+1$. A diagonal shift can be created by using a horizontal shift followed by a vertical shift, as shown in Figure 4.28(c) and (d).

Simple Edge Finder: Shift and Subtract By combining the shifting operation with the subtraction operation, we can make a simple horizontal or vertical edge finder that is useful in many applications and can create edge images such as in Figure 4.7. If we substitute $H(i, j)$ for $B(i, j)$ in Equation (4.7), then we will be taking the difference between an image **A** and that same image shifted horizontally to the right. If we take

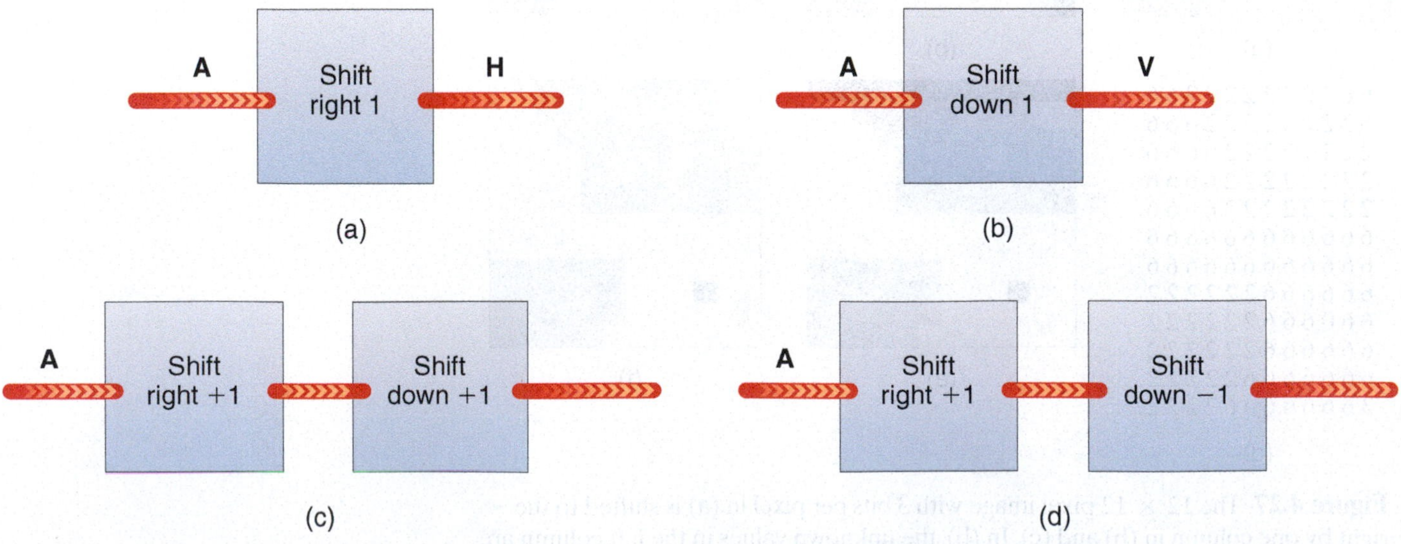

Figure 4.28 Block diagram for (a) a shift of one pixel position to the right and (b) one pixel position down. Two of the four possible diagonal shifts, shown in (c) and (d) are created with a horizontal shift followed by a vertical shift. The other two diagonal shifts are created by starting with a horizontal shift of -1 pixel instead of $+1$ pixel.

the absolute value of the difference, then the output will be bright where the original image has vertically oriented edges, because of the large change in the intensity values over a small horizontal distance. The output will be dark where there are no edges or where there are only horizontal edges, because the horizontal movement will not cause much of a change in intensity.

Such a vertical-edge image is shown in Figure 4.29(c) for the same simple 12 × 12 test image used in Figure 4.27. To find the horizontal edges, we would substitute $V(i, j)$ for $B(i, j)$ in Equation (4.7) so that a vertically shifted image is subtracted from the unshifted image. The vertical edge image is shown in Figure 4.29(d). Note that the diagonal edges and the single pixels of different gray levels create both horizontal- and vertical-edge pixels. To find all the edges of an object, we would need to have both horizontal and vertical edges. In both examples, the image displays are a column narrower or shorter than the original image, because we have lost a column or row due to the shifting.

Edge finding, as shown in Figure 4.29, is very similar to change detection, as shown in Figure 4.8. For the change detector, two different images are the inputs for the subtractor. The images come from the same camera,

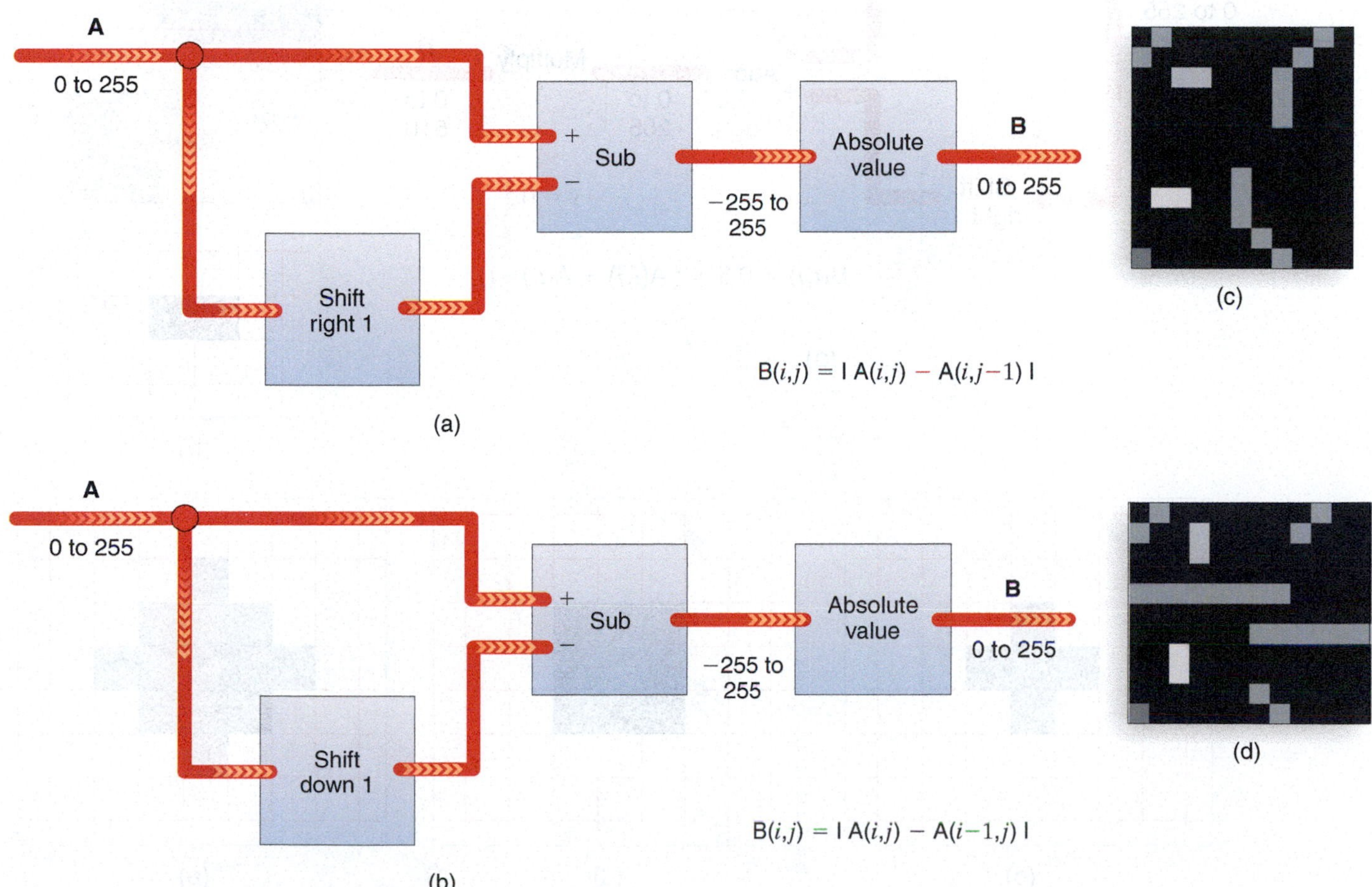

Figure 4.29 Block diagrams for vertical- and horizontal-edge finders for 8-bit-per-pixel images are shown in (a) and (b) with their respective mathematical equations. The images in (c) and (d) are created by the edge finder when **A** is the 12 × 12 image from Figure 4.27 with a pixel range of 0 to 7. In (c), the left column, where $i = 1$, is not displayed, and in (d) the top row, where $j = 1$, is not displayed.

but are taken at different times. For the edge finder, only one input image is used. The second input image is created from the first by shifting it.

Simple Image Softener: Shift and Add A simple image smoother can be made with an addition block and a shift block in a manner very similar to that used to make the simple edge finder. Here, we are adding instead of subtracting, but we use the same horizontal shift to provide the second image input to the addition block. Since the range of outputs is different from that produced by the simple edge finder, the absolute-value block of the simple edge finder will be replaced by a block that multiples the sum by 0.5. Figure 4.30 shows the block diagram for this simple image smoother and the output from the simple test image in Figure 4.27(a). The 2-pixel neighborhood of **A** used for the computation of each output pixel, $B(i, j)$ is shown below the output image. Adding a vertically shifted image instead of a horizontally shifted image would cause the horizontal edges to be blurred instead, and the 2-pixel neighborhood for the computation would be two pixels aligned vertically with $A(i, j)$ in the lower position. This is the neighborhood that was used in Figure 4.29(b).

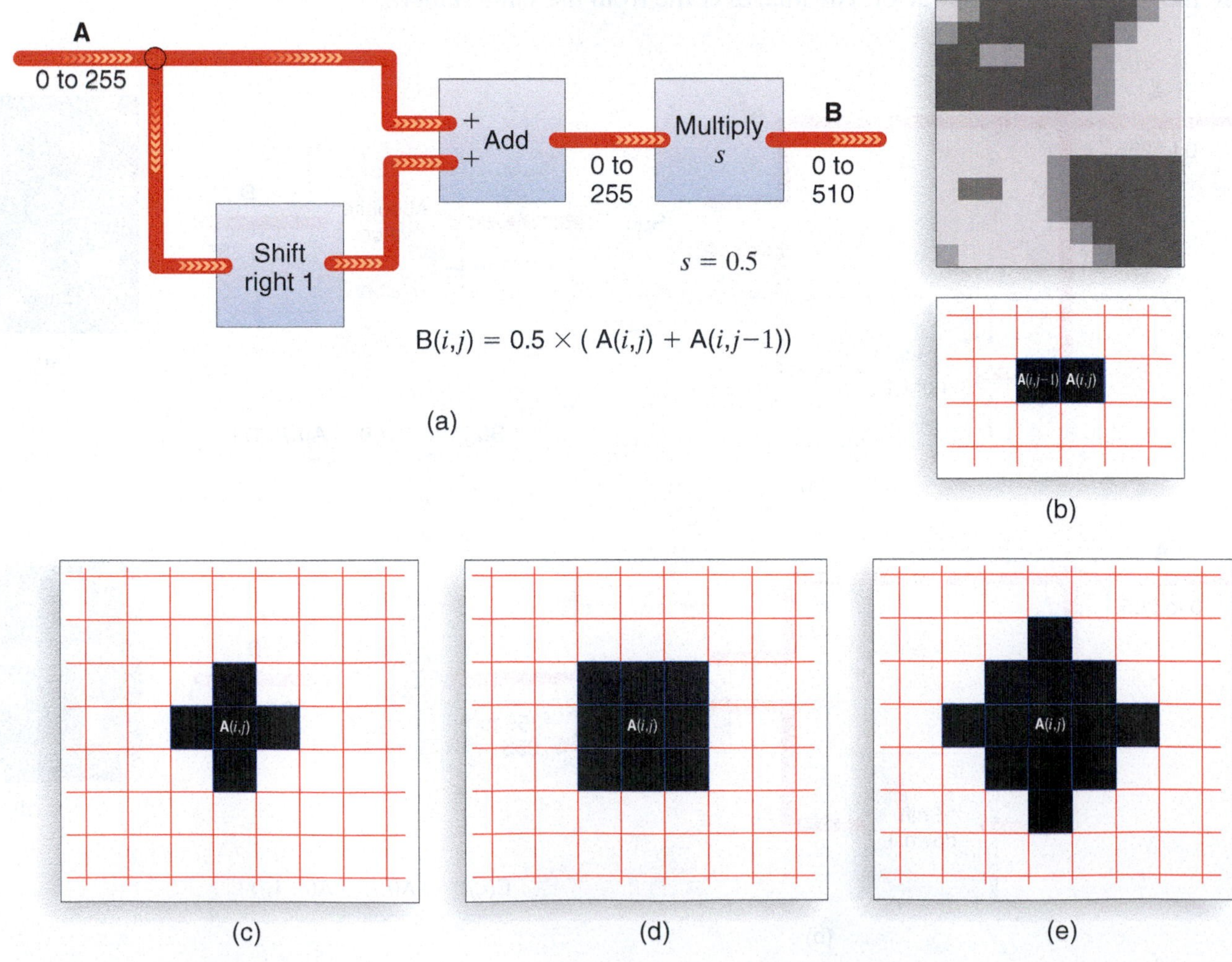

Figure 4.30 The block diagram for a simple smoother that adds an image to a shifted version of the same image is shown in (a) with its mathematical equation. The image in (b) is created by the smoother when **A** is the 12 × 12 image from Figure 4.27(a). The left column, where $j = 1$, is not displayed because it will depend on what value was chosen for the left column of the shifted image. In (c–e), three large neighborhoods are shown that might be used for image softening.

Infinity Project Experiment: Adding and Subtracting Shifted Images

Using input from a camera or a stored image, experiment with shifting the image and adding the result to the unshifted image. What happens as the amount of the shift increases?

The Neighborhood Defined by Shifting By using sequences of horizontal and vertical shifts, a shift by any amount in any direction is possible. The different shifted images that are added or subtracted define the neighborhood that is used for filtering operations. In the two examples in Figures 4.29 and 4.30, the neighborhood used to compute output pixels was just 2 adjacent pixels. Figure 4.30(b) shows the horizontal 2-pixel neighborhood used in both Figure 4.29(a) and Figure 4.30(a). Larger neighborhoods that might be used are shown in Figure 4.30(c)–(e). If the neighborhood covers 5 pixels instead of 2, then four shifted images will be added to or subtracted from the unshifted input image. The position of the unshifted image relative to the shifted images is shown by the pixel labeled $A(i, j)$ in the neighborhood diagram. Usually, the shifted images are multiplied by different scale factors before the addition in order to create different kinds of output images. In the next section, we will find that, to reduce noise or sharpen images for our robo-helper's vision system, we will use neighborhoods such as the one shown in Figure 4.30(d).

EXERCISES 4.3

Mastering Concepts

1. What is the difference between the arithmetic operations used in edge detection and in change detection?
2. When arithmetic operations are used to combine two image matrices, why do the matrices have to be the same size? What is the size of the output image?
3. What is a neighborhood? How is it defined?
4. How is shifting an image different from shifting an audio signal?

Try This

5. Add the following image matrix **B** to the image matrix **A** from Exercise 4.2.4:

$$\mathbf{B} = \begin{pmatrix} 1 & 2 & 3 & 3 & 2 & 1 \\ 2 & 3 & 4 & 4 & 3 & 2 \\ 2 & 3 & 4 & 4 & 3 & 2 \\ 1 & 2 & 3 & 3 & 2 & 1 \end{pmatrix}$$

6. Multiply the mask **M1** by the image matrix **A** from Exercise 4.2.4:

$$\mathbf{M1} = \begin{pmatrix} 1 & 1 & 1 & 0 & 0 & 0 \\ 0 & 1 & 1 & 1 & 0 & 0 \\ 0 & 0 & 1 & 1 & 1 & 0 \\ 0 & 0 & 0 & 1 & 1 & 1 \end{pmatrix}$$

7. Multiply the mask **M1** from Exercise 4.3.6 by the image matrix **B** from Exercise 4.3.5.
8. Shift the image matrix **A** from Exercise 4.2.4 one column to the right, and put a value of 0 in any element that is unknown.
   a. Which values from the original matrix are lost?
   b. What happens if the shifted matrix is shifted again one column to the left, with a value of 0 used for any elements that are unknown?
9. Find the matrix obtained by shifting the matrix **A** from Exercise 4.2.4 to the left by one column and then down by one column. Use a 0 for any element that is unknown.
10. Compute the value of the matrices obtained for the horizontal and vertical shifts as shown in Figure 4.28 for the matrix **A** given as follows:

$$\mathbf{A} = \begin{pmatrix} 4 & 4 & 4 & 7 & 7 & 7 \\ 4 & 4 & 6 & 7 & 7 & 7 \\ 0 & 0 & 0 & 0 & 7 & 7 \\ 0 & 0 & 0 & 0 & 0 & 0 \\ 7 & 7 & 7 & 7 & 2 & 2 \\ 7 & 7 & 6 & 5 & 4 & 3 \end{pmatrix}$$

11. Arithmetic operations are used to combine an image and two shifted versions of the image, where one is shifted one pixel to the left and the other is shifted one pixel to the right. Draw the 3-pixel neighborhood in the style used in Figure 4.30.
12. Arithmetic operations are used to combine an image and four shifted versions of the image, where the shifted versions are shifted one pixel diagonally in each of the four diagonal directions. Draw the 5-pixel neighborhood in the style used in Figure 4.30.
13. If the neighborhood in Figure 4.30(c) is used for arithmetic combination of images, how many shifted images will be combined with the unshifted input image?
14. If the neighborhood in Figure 4.30(d) is used for arithmetic combination of images, how many shifted images will be combined with the unshifted input image?

4.4 Understanding Images from Robot Eyes

The robo-helper has to be able to find objects before it can pick them up. It also has to avoid running into things or running over things. To do this, it must be able to interpret the images provided by its camera, as shown in Figure 4.3. Although locating edges in an image and then segmenting the image into areas that correspond to individual objects may seem simple, it is a very challenging task for automatic image processing. There is so much variability in how an object will look in terms of image pixels that we have to consider the following issues when we design image-processing systems:

- Object edges may be irregular in an image because it is extremely rare that the boundary of an object will be aligned perfectly with the edge of a pixel. An object has to be identified in spite of these irregularities.

- Some edges are more important than others. For example, object edges like the outline of a guitar or the corner of a wall are useful for navigation and planning the actions of the robot. Other edges (such as those due to patterns in the weave of a carpet, folds in curtains, or shadows) should be ignored, since they are not useful for controlling the robot.
- An edge might be spread out over several pixels because the image is blurred by the motion of the camera, the motion of the object, or poor focus. When an edge is blurred, the difference between nearby pixels is reduced and may not be large enough for the edge to be considered a strong edge. A blurred edge is also spread out over several pixels, which makes it hard to distinguish the exact edge.
- The color of an object may vary due to the curvature of its surface or lighting variations, and we would not want these differences to be interpreted as a false edge.
- The image may have noise that could create many small false edges.

Figure 4.31 shows an example of irregular edges that occur because the object edge is slanted and of false edges due to the uneven surface of the carpet on a staircase.

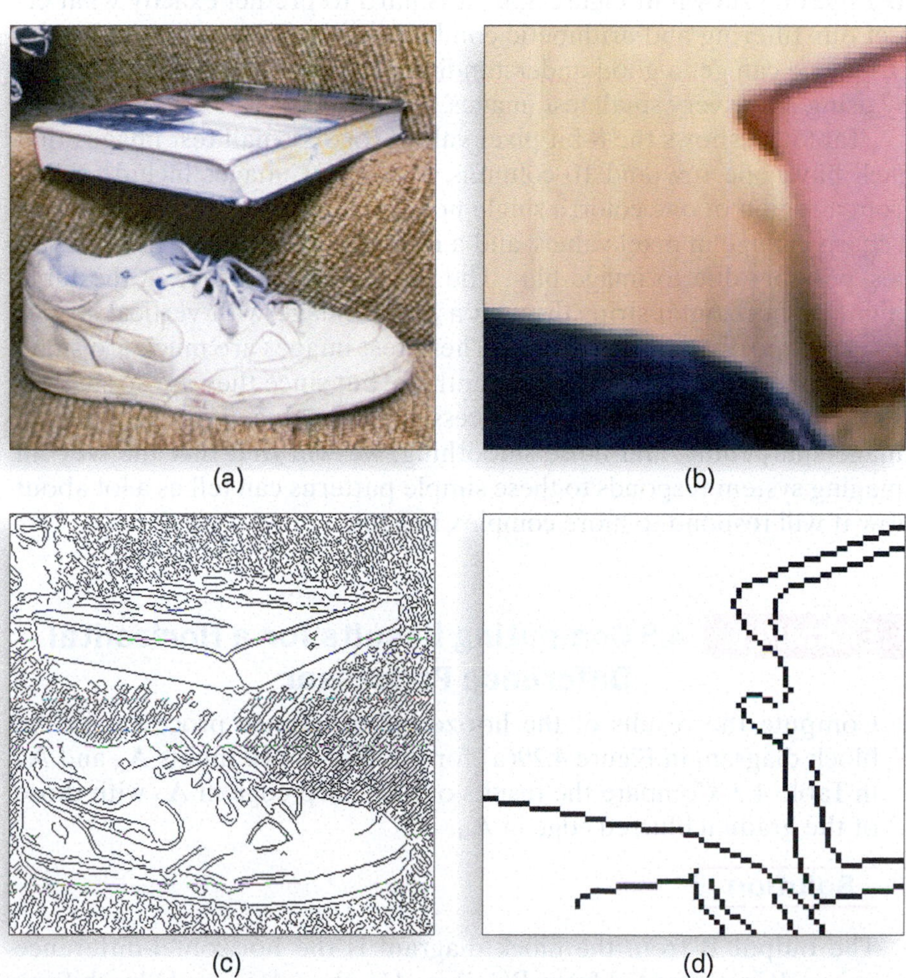

Figure 4.31 Two enlarged sections of the image in Figure 4.9 are shown in (a) and (b). The edge images in (c) and (d) show true object edges as well as false edges. In (c), the variations due to the texture of the carpet cause numerous false edges around the book and the shoe. In (d), the diagonal edges are irregular and a false edge is created on the wall from a shadow.

KEY CONCEPT

Smoothing increases blur, reduces the sharpness of edges, and reduces noise.

KEY CONCEPT

Sharpening reduces blur, increases the sharpness of edges, and increases the noise.

In order to improve the ability of the processing system to locate and identify objects within as image, the image may need to be sharpened in order to accentuate the edges before the edge-location process is started. The image may also need to be smoothed in order to reduce false edges due to noise. We will see that smoothing to reduce noise also reduces the sharpness of edges, and sharpening an image to make edges easier to see also increases the noise. For each application, these processes must be balanced to work best with the specific types of images that are expected.

The trade-offs between smoothing and sharpening are illustrated in Figure 4.32. The edges in the sharpened image in (b) are more noticeable than in the original image in (a). However, when the image has noise, as in (c), the sharpened image in (d) is far more noisy. In this image, it is very difficult to separate true object edges from small false edges due to the noise. Smoothing the noisy image, as in (e), before sharpening, as in (f), reduces the noise effects in the sharpened image, but they are not eliminated completely. In addition to reducing noise, the smoothing also softenes the object edges, making them less distinct.

For images that have many different objects and background areas, like the one shown in Figure 4.32, it is hard to predict exactly what effect our filtering and arithmetic combinations will have on the image. Often, we can get a good understanding of what our processing will do by using a few very small test images.

Table 4.3 shows the 8-bit pixel values for six small test images that each have one row and 10 columns. These test images include a line from a region of one color; a single noise pixel in a region of one color; a sharp change in pixel values; and a more gradual change in pixel values, possibly due to image blur. The last two test images in the table might be horizontal strips through a larger image with vertical stripes that are 1 pixel and 2 pixels wide. These test images are much too small to be perceived as an image of something, but since they are so small, it is easy to compute and list the processed pixel values. For edge finding, image sharpening, and noise smoothing, we will find that the way an imaging system responds to these simple patterns can tell us a lot about how it will respond to more complex images.

EXAMPLE 4.5 Computing Results for a Horizontal Difference Processor

Compute the results of the horizontal-difference processor in the block diagram in Figure 4.29(a) for the test-image inputs $\mathbf{A}_3$ and $\mathbf{A}_4$ in Table 4.3. Compare the results of the sharp edge in $\mathbf{A}_3$ with those of the gradual blurred edge in $\mathbf{A}_4$.

Solution

The output **B** from the block diagram is the horizontal-difference image. It is computed from $B(i,j) = A(i,j) - A(i,j-1)$ by shifting the test image one column to the right and then subtracting it from the unshifted image.

When $\mathbf{A}_3$ is the input image, we compute the shifted image of pixels $A(i,j-1)$ by moving each pixel one position to the right.

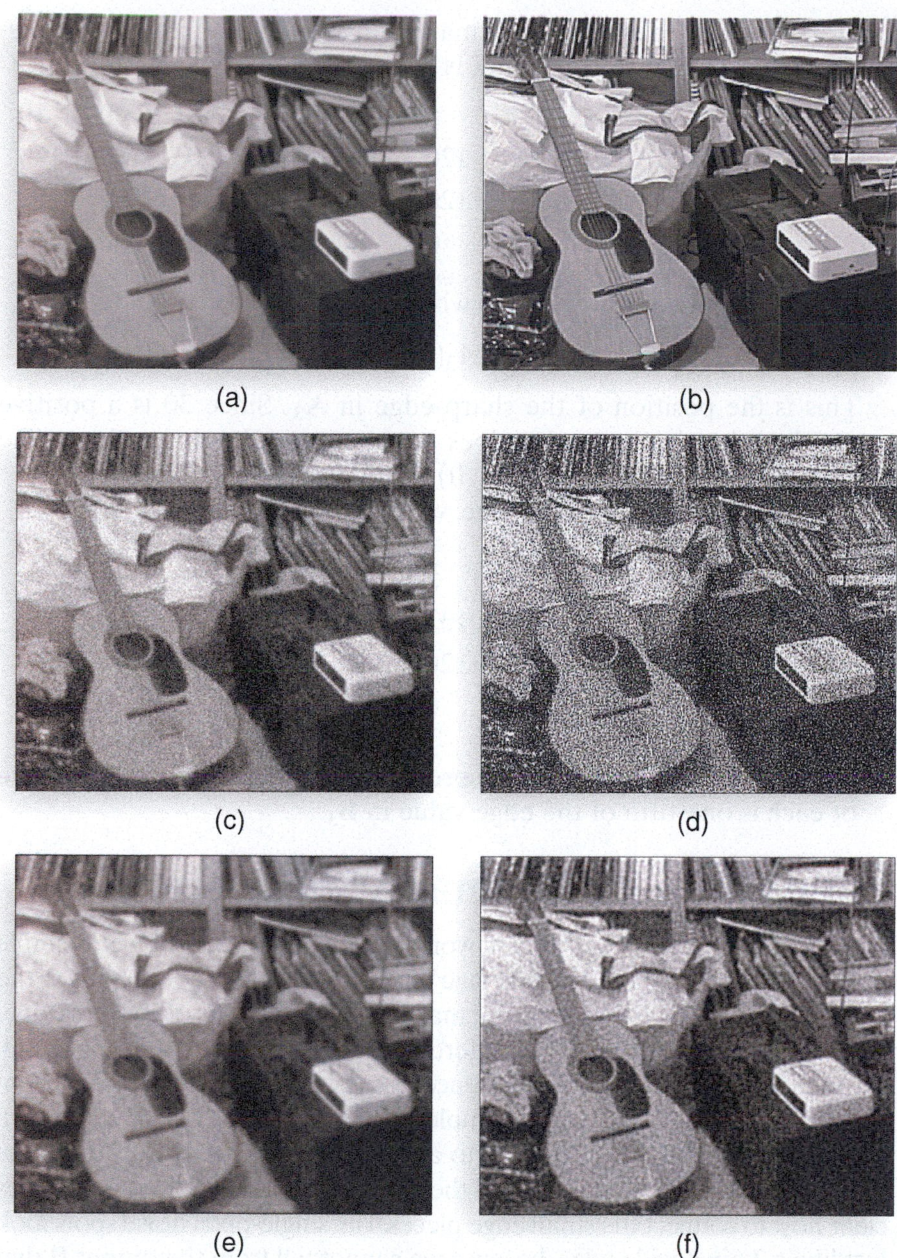

Figure 4.32 When the image in (a) is sharpened in (b), the edges are more noticeable. Sharpening the noisy image in (c) results in the much noisier image in (d). When the image in (c) is smoothed in (e) before sharpening in (f), the noise is reduced, but not removed completely.

Table 4.3 Six Simple Test Images of One Row and 10 Columns

$\mathbf{A}_1$ = [120 120 120 120 120 120 120 120 120 120] Constant value	$\mathbf{A}_2$ = [120 120 120 120 120 150 120 120 120 120] Noise pixel on a constant value background
$\mathbf{A}_3$ = [120 120 120 120 120 150 150 150 150 150] A sharp edge	$\mathbf{A}_4$ = [120 120 120 126 132 138 144 150 150 150] A gradual edge
$\mathbf{A}_5$ = [120 150 120 150 120 150 120 150 120] Narrow vertical lines	$\mathbf{A}_6$ = [120 120 120 150 150 120 120 150 150 150] Wider vertical lines

Here $i = 1$ because the test images have only one row. The question marks show that nothing is shifted into the vacant column 1:

Column number: $j =$	1	2	3	4	5	6	7	8	9	10
$A_3(1,j) =$	120	120	120	120	120	150	150	150	150	150
$-A_3(1,j-1) =$	??	−120	−120	−120	−120	−120	−150	−150	−150	−150
$= B_3(1,j) =$	??	0	0	0	0	30	0	0	0	0

The output image $\mathbf{B_3}$ is 0 everywhere, except at

$$B(1,6) = A_3(1,6) - A_3(1,5) = 150 - 120 = 30$$

This is the position of the sharp edge in $\mathbf{A_3}$. Since 30 is a positive number, the absolute value does not change it. We do not know the value of $B(1,1)$, because $A_3(1,0)$ is not part of the test image.

When $\mathbf{A_4}$ is the input image, we have

Column number: $j =$	1	2	3	4	5	6	7	8	9	10
$A_4(1,j) =$	120	120	120	126	132	138	144	150	150	150
$-A_4(1,j-1) =$	??	−120	−120	−120	−126	−132	−138	−144	−150	−150
$= B_4(1,j) =$	??	0	0	6	6	6	6	6	0	0

In B_4, the gradual edge is spread out over 5 pixels, and the value of each is one-fifth of the edge value in $\mathbf{B_3}$.

Edge Finding and Image Sharpening

How would an ideal edge finder work? At first, it seems to be a simple problem of finding parts of the image where there are large changes in the value of nearby pixels. It should not matter what the values of the individual pixels are, because it is only the difference between them that is important. The horizontal and vertical difference processors from Figure 4.29 seem well suited to this task, and Example 4.5 has shown that the difference processor gives a strong response to a sharp change. But if we reexamine Figure 4.29, we notice that since all the background is black, it is not always clear how to connect the small edge pieces. The single-pixel noise spots look just like a 2-pixel-wide edge, because we cannot tell from the output **B** that the input image had the same value on both sides of the noise pixel.

Another way to find edges is to sharpen the image so that edge differences are larger, but in a way that causes no change in areas where all the pixels have the same color. Keeping information about the pixel values can help with decisions about how to connect the edge pieces. Equation (4.11) will sharpen vertical edges in this way. The block diagram in Figure 4.33 shows how to build this sharpening filter, using two shifted versions of the input image matrix **A**. A typical row of an image is shown above the block diagram to indicate exactly how pixels from a 1×3 neighborhood are combined to create each output pixel.

$$B(i,j) = -A(i,j-1) + 3 \times A(i,j) - A(i,j+1) \quad (4.11)$$

The plots in Figure 4.32(b)–(g) show the outputs of the sharpening filter in Equation (4.11) for each of the six test images in Table 4.3. The output images are shown for only 8 pixels because the two shifted images each lose one column. The output for a stronger sharpening filter also is

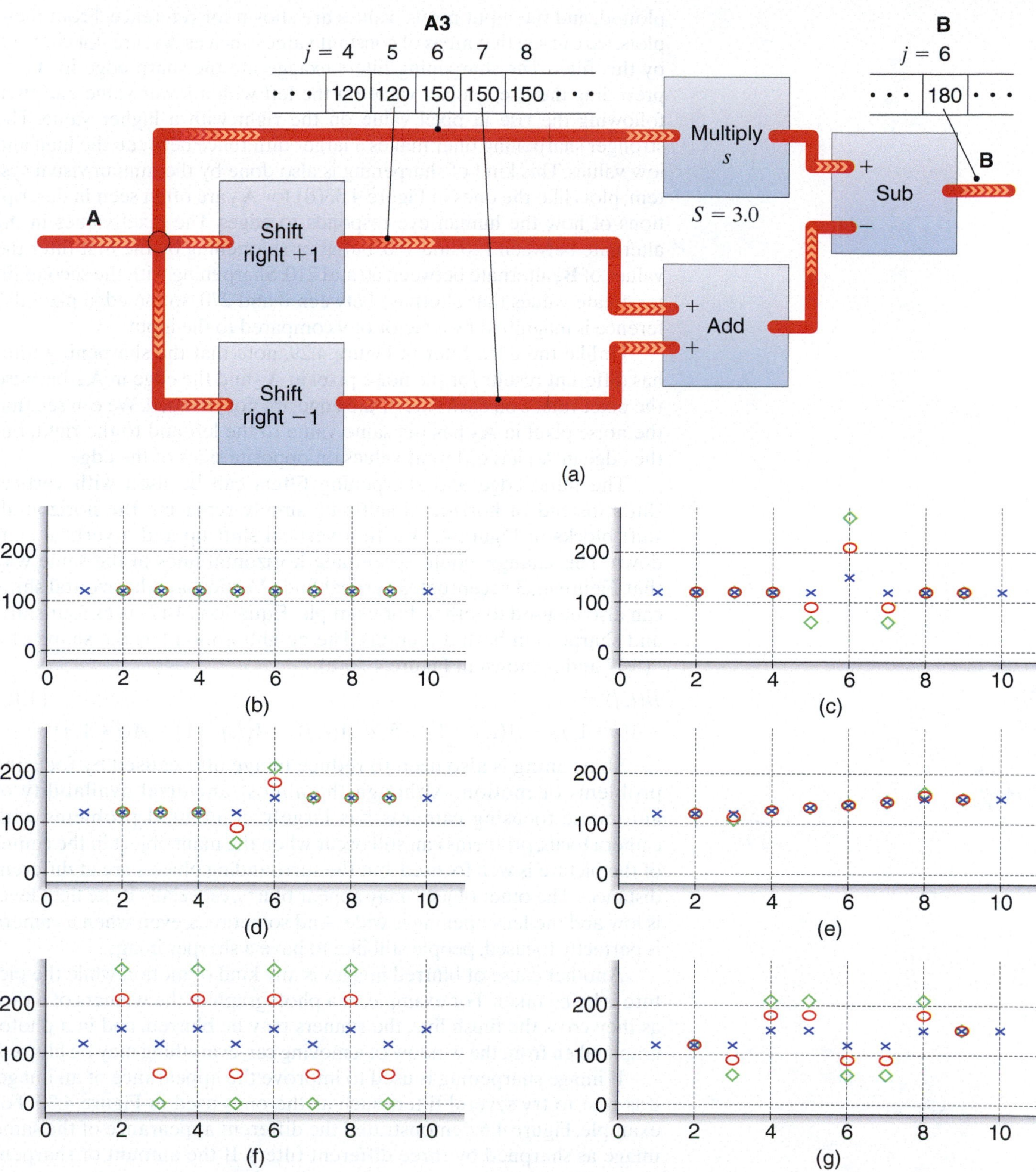

Figure 4.33 The processing blocks for image sharpening using $B(i, j) = -A(i, j - 1) + 3 \times A(i, j) - A(i, j + 1)$ are shown in (a). A specific example using input test image matrix $\mathbf{A_3}$ at column $j = 6$ is also shown. In (b–g), outputs from two different sharpening filters are shown for all six text images. Output pixel values for $B(i, j) = -A(i, j - 1) + 3 \times A(i, j) - A(i, j + 1)$ are shown as red circles, and output pixel values for $B(i, j) = -2A(i, j - 1) + 5 \times A(i, j) - 2A(i, j + 1)$ are shown as green diamonds. The input pixel values are shown as blue x's, for reference. Note that the second sharpening filter, shown in green, has a stronger sharpening effect than the first filter, shown in red. In (b), where there is no variation in the input, the sharpened images are exactly the same as the input image.

plotted, and the input pixels' values are shown for reference. From these plots, we can see that areas of constant values, such as $\mathbf{A_1}$, are not changed by this filter. The sharpening filters exaggerate the sharp edge in $\mathbf{A_3}$ by preceding the rise in pixel value on the left with a lower value and then following the rise in pixel value on the right with a higher value. The stronger sharpening filter makes a larger difference between the high and low values. This kind of sharpening is also done by the human visual system; plots like the ones in Figure 4.33(d) for $\mathbf{A_3}$ are often seen in descriptions of how the human eye responds to edges. The pixel values in $\mathbf{A_5}$ alternate between 120 and 150, but after sharpening by the first filter the values of $\mathbf{B_5}$ alternate between 60 and 210. Sharpening with the second filter create values that alternate between 0 and 270, so the edge pixel difference is magnified by a factor of 9 compared to the input.

Unlike the edge filter in Figure 4.29, note that the sharpening filter has different results for the noise pixel in $\mathbf{A_2}$ and the edge in $\mathbf{A_3}$, because the pixel values on each side of the edge are not changed. We can see that the noise pixel in $\mathbf{A_2}$ has the same value to the left and to the right, but the edge in $\mathbf{A_3}$ has different values on opposite sides of the edge.

The same edge and sharpening filters can be used with vertical shifts instead of horizontal shifts by simply replacing the horizontal-shift blocks in Figure 4.33 with a vertical shift up and a vertical shift down. This change would accentuate horizontal lines in the same way that Figure 4.33 accentuates vertical lines. Vertical and horizontal shifts can also be used together. For example, Equation (4.12) uses four shifts and sharpens in both directions. The neighborhood has the shape of a "(+)" and is shown in Figure 4.30(c).

$$B(i,j) = \tag{4.12}$$
$$-A(i-1,j) - A(i,j-1) + 5 \times A(i,j) - A(i,j+1) - A(i+1,j)$$

Sharpening is also used to reduce image blur caused by focusing problems or motion. Although the almost universal availability of automatic-focusing cameras has largely eliminated problems with camera focus, problems can still occur when the main object in the center of the picture is well focused, but the surrounding objects are at different distances. The other objects may appear blurry, especially if the light level is low and the lens opening is wide. And sometimes, even when a camera is perfectly focused, people still like to have a sharper image.

Another cause of blurred images is any kind of motion while the picture is being taken. For example, in a photograph of the winners of a race as they cross the finish line, the runners may be blurred, and in a photograph taken from the window of a moving car, everything may be blurred.

If image sharpening is used to improve the appearance of an image, it is best to try several filters such as the ones used in Figure 4.33. For example, Figure 4.6 demonstrated the different appearance of the same image as sharpned by three different filters. If the amount of sharpening is adjustable, then the result that has the best balance between increased contrast and increased noise can be selected.

Noise Reduction

The images shown in Figure 4.32 caution us to avoid sharpening noisy images. We need to have some method to reduce the noise before edge sharpening can be used. But how can we reduce noise? In most situa-

Infinity Project Experiment: Sharpening Images

Using input from a camera or image files, try applying the different sharpening filters described in this section and observe the effects they have on various parts of the image. What happens in very smooth areas of the image? What happens to patterns like woodgrain on a tabletop or printed patterns on clothes? What happens when there is a lot of small detail like in grass or gravel?

tions, we have only a general idea about the nature of the objects in an image. After all, if we knew exactly what the image was going to look like, why would we need to record it? So, for any particular pattern of pixel values in an image, we can never know with absolute certainty whether that pattern is part of the desired image content or is an undesirable artificial structure that should be rejected as noise.

In a diagnostic medical image, a few unusual pixel values might be due to noise, or they might indicate a small tumor or a torn ligament. The judgments that will be made about the pixels in a medical image will not necessarily also apply to other applications such as the Mars Rover images or the robo-helper images. For example, if the robo-helper sees lots of specks on the floor, it might be seeing image noise, but it might also be viewing dirt, indicating that it is time to get out the vacuum cleaner.

The problem of reducing noise may seem impossible if we do not know how the image should look or what the noise looks like. We often refer to image noise as random noise because its location and intensity values are not predictable. But we do know that there must be some difference between the desired image content and the noise, because if there were no differences, the desired image and the noise would be so similar in appearance that we would not notice the noise.

If we look again at the images in Figures 4.4 and 4.32, the noise is obvious. Usually, objects that we want to identify in an image are large enough to be covered by many pixels, but the variations caused by noise would create false objects that would only be a few pixels in size. Smoothing a noisy image could reduce the variations caused by noise while leaving most of the pixels in larger objects unchanged, because the objects are so much larger than the noise spots. We just have to choose a neighborhood that is bigger than the noise spots and smaller than the objects we want to find.

Although very complex filters can be designed to smooth noise and keep objects of interest, there are two simple ways of making filters that work well in many applications. One is an averaging filter, and the other is a median filter. Both will be described and demonstrated next.

Averaging Filters The average of a set of values tells us something general about the set, but it does not tell us any specific values. For example, the average daily rainfall for a month does not tell us how much it will rain on a particular day, but it can help us decide what kind of plants we should put in our garden. Similarly, the average height of players on a basketball team does not tell us the height of any individual player, but the average

Infinity Project Experiment: Edge Detectors

Using input from a camera or image files, try several edge detectors. What happens in very smooth areas of the image? What happens to patterns like woodgrain on a tabletop or printed patterns on clothes? What happens when there is a lot of small detail like in grass or gravel? How is the edge detector different from the sharpening filter?

heights of teams do tell us whether one team is likely to have a performance advantage over another team.

Averaging pixel values over neighborhoods smaller than the size of objects we want to locate, but larger than noise spots, can be used as a smoothing operation for images. Simple averaging often reduces noise, but does not eliminate it. To form the average, the values of all the pixels in a neighborhood are added together, and then the sum is divided by the number of pixels that were added.

Average: The average of a set of values is the sum of all the values divided by the number of values. It gives us a general idea about the values in the set, but it does not tell us any specific value.

To make a new image where each pixel is computed as the **average** of three pixels in a row in the input image, we would have to add our image to a version shifted one pixel to the left and another version shifted one pixel to the right. Then we would divide the sum by three to get the average, as shown in Equation (4.13). This averaging filter uses exactly the same neighborhood as the sharpening filter in Equation (4.11). The different effects of the averaging and sharpening filters on an image are due to the different constant values that each of the shifted images are multiplied by. The averaging filter multiplies all three images by $\frac{1}{3}$, while the sharpening filter multiplies the shifted images by -1 and the unshifted image by 3.

$$B(i, j) = \tfrac{1}{3} \times (A(i, j - 1) + A(i, j) + A(i, j + 1)) \quad (4.13)$$

EXAMPLE 4.6 Finding Average Values

Find the average value of the following list of five numbers:

5 7 3 9 6

If the second value in the list, which is now 7, changes to 7007, how does the average change?

Solution

To find the average of five things, we add them up and divide by five. The average value is $(5 + 7 + 3 + 9 + 6)/5 = 30/5 = 6$. If 7 is replaced by 7007 in the list of five numbers, the average value is $(5 + 7007 + 3 + 9 + 6)/5 = 7030/5 = 1406$. Because 7007 is so different from all the other values in the list, the average value is not near any of the five actual values. In the same way, when a pixel has a very large value due to noise, an averaging filter reduces the noise, but does not eliminate it.

If we wanted to average over a neighborhood that is 3 pixels wide and 3 pixels high, we would need eight shift operations to make nine pixels values to be averaged for each output pixel. Equation (4.14) shows how each output pixel would be computed. For each pixel in **B**, the pixel values of a 3 × 3 pixel area of **A** are averaged.

$$B(i,j) = \tfrac{1}{9} \times (A(i-1,j-1) + A(i-1,j) + A(i-1,j+1) + A(i,j-1) + A(i,j) + A(i,j+1) + A(i+1,j-1) + A(i+1,j) + A(i+1,j+1)) \quad (4.14)$$

Median Filters Another common value that is used to indicate the general behavior of a group of elements is the **median**. This value is often used in economic reporting, such as when the median income or the median house price might be compared for several counties. The median of a group with an odd number of elements is found by arranging all the elements in order of increasing value and then identifying the one that is in the center of the list. This value is the median. If the number of elements is even, then the average value of the two elements on either side of the list center is the median value.

Median: The median value of a set of numbers is found by putting the numbers in a list and then sorting the numbers in ascending order (from the lowest value to the highest value). The median value is the value in the center of the list.

The median is different from the average in several ways, although both are used to calculate a single number that is generally representative of a group of numbers. The median does not use the same computation of addition and division that the average uses. However, it does require that the values in the list be sorted, and for long lists this operation can take more time than the computation of the average value.

Unlike the average, the median value for a list with an odd number of values always will be one of the numbers in the list. Probably the most important difference for noise reduction is that a single value in the list that is much larger than all the other values will cause the average value to be much larger than most of the elements in the list. On the other hand, a single very large value will not affect the median, because the large value will always be at the end of the list and the median is taken from the center of the list. There is no standard arithmetic symbol for the median, so we will write it as shown in Equation (4.15):

$$B(i,j) = \text{median}(A(i,j-1), A(i,j), A(i,j+1)) \quad (4.15)$$

EXAMPLE 4.7 Finding Median Values

Find the median value of the following list of five numbers:

5 7 3 9 6

If the second value in the list, which is now 7, changes to 7007, how does the median change?

Solution

To find the median of five numbers, we sort the five numbers in ascending order (from the smallest to the largest). Then we select the center value. The median of the ordered list (3, 5, 6, 7, 9) is 6, because 6 is the value in the middle of the list. If 7 is replaced by 7007 and the list is sorted into ascending order again, then our new list is (3, 5, 6, 9, 7007), and the new median is still 6. When one value is very different from all the others, it usually does not affect the median value. The very different value will always be at one end of the list or the other, so it will not have much effect on which value is at the center. In a similar manner, the median filter is good at eliminating impulsive noise where only a few pixels have noise.

Figure 4.34 illustrates the difference between the median filter from Equation (4.15) and the average filter from Equation (4.13) on the six

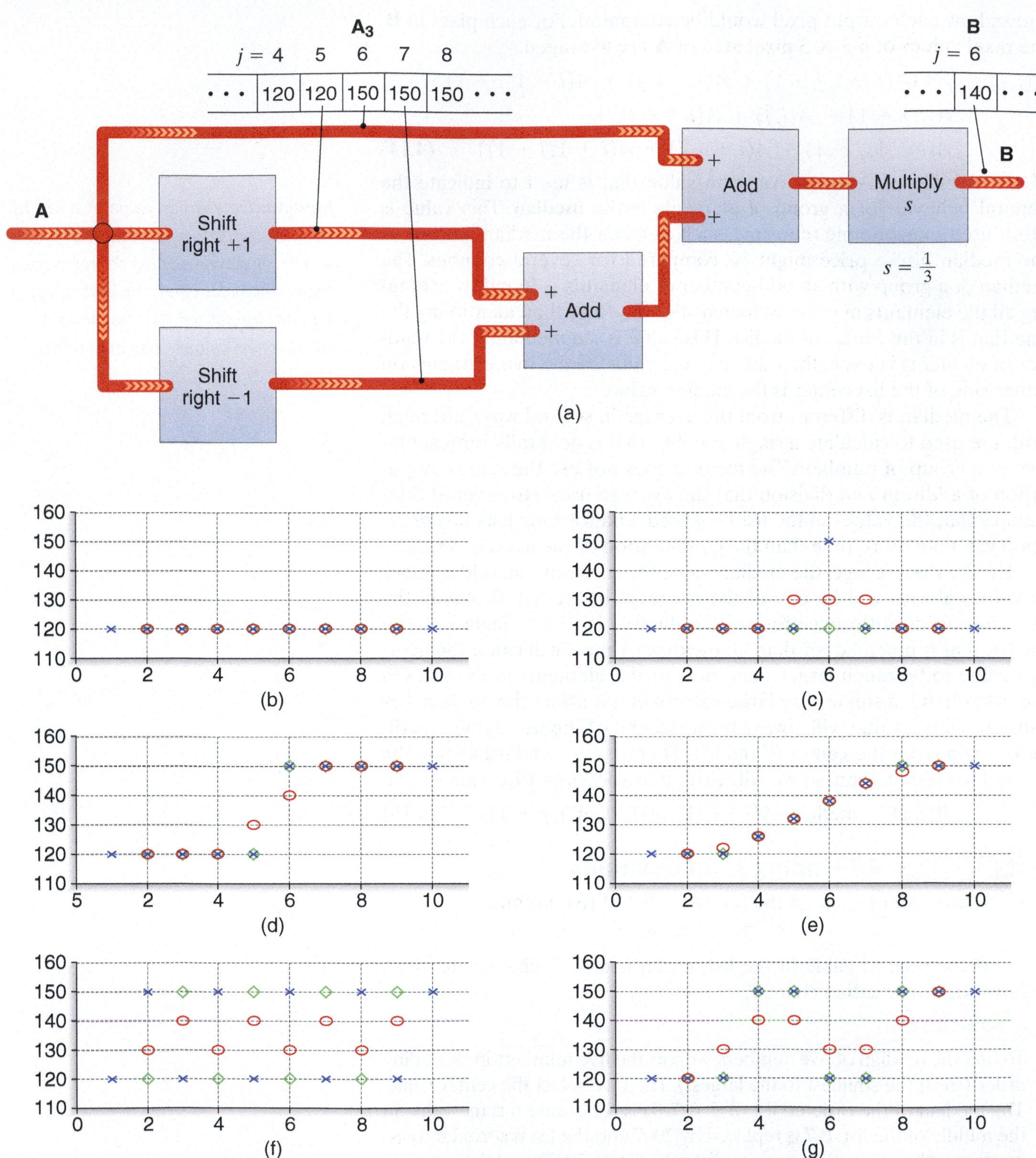

Figure 4.34 The processing blocks for image averaging using $\mathbf{B}(i, j) = \frac{1}{3}(\mathbf{A}(i, j-1) + \mathbf{A}(i, j) + \mathbf{A}(i, j+1))$ are shown in (a). A specific example using input test image matrix **A3** at column $j = 6$ is also shown. In (b–g), outputs from the averaging filter and a median filter with the same neighborhood are shown for all six text images. Output pixel values for $\mathbf{B}(i, j) = \frac{1}{3}(\mathbf{A}(i, j-1) + \mathbf{A}(i, j) + \mathbf{A}(i, j+1))$ are shown as red circles, and output pixel values for $\mathbf{B}(i, j) = \text{median}(\mathbf{A}(i, j-1) + \mathbf{A}(i, j) + \mathbf{A}(i, j+1))$ are shown as green diamonds. The input pixel values are shown as blue *x*'s for reference. Note that the averaging filter softens edges by spreading them over three pixels instead of 1 in (d) and reduces the amount of variation in (f) and (g). The median filter completely removes the single different pixel in (c) and keeps the sharp edges in (d).

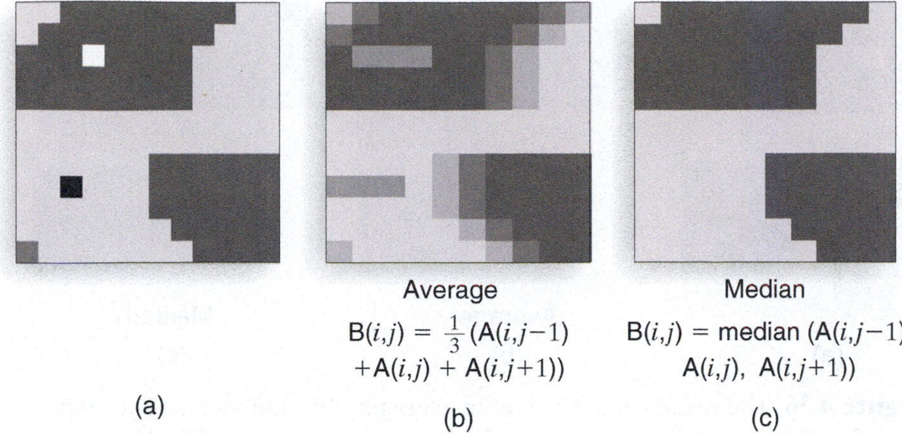

Figure 4.35 Horizontal averaging and median filtering using the image from Figure 4.27.

simple test images from Table 4.3. Note that the median completely removes the single noise pixel in $\mathbf{A}_2$, but the average only reduces the difference between the noise and the background by a factor of three and spreads it out over three pixels. In $\mathbf{A}_3$, the median keeps the sharp change in values, while the average makes the edge more gradual. In $\mathbf{A}_5$, where the alternating pixel values might be a single row from an image of vertical stripes, the median reverses the order of the pixels, but does not change the values. The averaging filter reduces the difference between the pixels by a factor of three.

Although the simple test images give us an idea of the performance of the median filter from Equation (4.15) and averaging filter from Equation (4.13) for a 1×3 neighborhood, the effect of a 3×3 neighborhood is demonstrated more clearly using the 12×12 pixel image from Figure 4.27. In Figure 4.35, the 1×3 neighborhood is used again. Both the median and the averaging image results are only 10 pixels wide because the leftmost and rightmost columns cannot be computed, due to unknown pixel values at the left and right sides of the shifted images. The effects of the two filters on the two noise pixels and the vertical edges are exactly as expected from the results of the same filters on test images $\mathbf{A}_2$ and $\mathbf{A}_3$. Since the neighborhood is horizontal, there is no averaging in the vertical direction, and horizontal edges are not blurred in either image.

In Figure 4.36, the 3×3 neighborhood is used, and the resulting images are 10×10 instead of 12×12. Note that the median filter completely removes the two noise pixels, but also removes the corner pixels from square boundaries. For the averaging filter in Figure 4.36, the smoothing has spread out the object boundaries.

In general, the median filter works well on noise with isolated values that are very different from the rest of the pixel values. An example of this kind of noise was shown in Figure 4.4(c). The averaging filter works best on noise that has small variations added all over the image, like the noise shown in Figure 4.4(b).

Image Segmentation

The images in Figures 4.7 and 4.8 are very rich in detail and include a complex background and many objects. We have explored the process of edge finding with sharpening and smoothing filters, but now we need

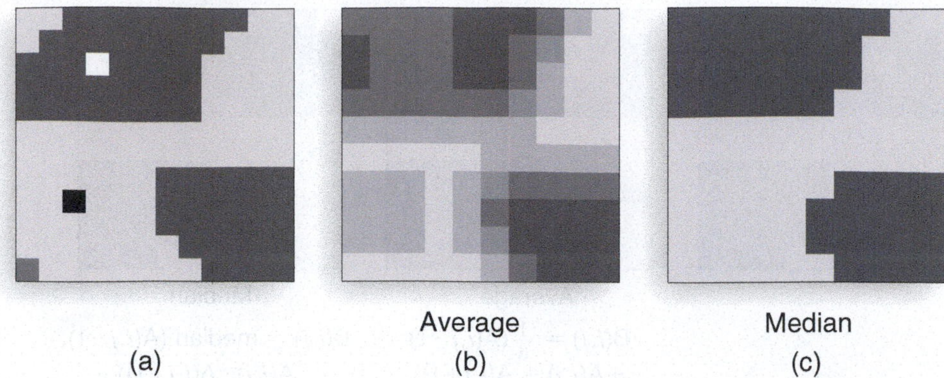

Figure 4.36 The results of a 3 × 3 area averaging (b) and median filtering over 3 × 3 neighborhood (c), using the image from Figure 4.27. The neighborhood for both is shown in Figure 4.30(d).

Infinity Project Experiment: Averaging and Median Smoothing Filters

Using input from a camera or image files, try applying the averaging and median filters described in this section, and observe the effects they have on various parts of the image as you adjust the size of the neighborhood. What happens in very smooth areas of the image? What happens to patterns like woodgrain on a tabletop or printed patterns on clothes? What happens when there is a lot of small detail like in grass or gravel? Then add random noise or impulsive noise to the image, and observe the smoothing. How is the size of the filter neighborhood related to the size of the noise for the best noise reduction?

to locate and identify the objects in the images. We need to separate the background from the objects by segmenting the image into distinct groups of pixels that represent a single object. This operation assumes that we have some idea about what distinguishes the background from the objects we are trying to identify. We can use several different attributes individually or in combination to separate objects. Intensity, color, edges, and texture can be the image parameters that will allow us to separate the background from the objects.

Color seems to be a simple parameter for segmentation, but color is actually three parameters. If an object to be segmented from an image has a particular color, then thresholds that select a specific range of red, green, and blue values can be used to isolate that color. At the end of the previous section, we learned that if we converted the red, green, and blue color components into an alternative description using hue, saturation, and value, then the hue was very similar to our perceptual sense of color and much easier to specify. For this reason, hue is often used in segmentation, as was demonstrated in Figure 4.9 and 4. 22. The hue distinguishes red from green, for example, but does not distinguish red from pink. Usually, for segmentation based on our perception of the color of an object, both a range of hues and a range of saturations are specified. Since the value is related to the brightness of a color, it is most affected by changes in lighting conditions. Since we normally want to segment objects in the same way under different lighting conditions, the brightness component is used less often for segmentation.

Texture is another parameter that can be used to separate objects in an image. We know what texture means when we touch a surface. It can be smooth, soft, or rough, for example. Although we associate texture with our sense of touch, different surface textures will also look different in images of the surfaces. In an image, texture refers to variations in the pixel values caused when light is reflected from surfaces that are not smooth such as woven fabrics, mowed grass, or sand on the beach. In Figure 4.9, we can see the texture pattern of the carpet on the stairs. The amount of variation and the pattern of the variation define the texture. We can define the hue of a single pixel, but since we have to look at many pixels near each other to comprehend the texture, using texture for segmentation will require neighborhood operations.

Summary

In this section, we have learned how to use the basic operations discussed the two previous sections to make image-processing systems that will sharpen the edges of images or smooth out variations cause by noise. These are the two most basic functions needed for the guidance and control of our robo-helper and for many other image processing applications. We also showed that the behavior of image-processing on complex images could be predicted based on what happened when that processing was used on relatively simple test images. In the next section, we will look at three specific design problems and use the methods developed in this section to design an image-processing solution to those problems.

EXERCISES 4.4

Mastering Concepts

1. How does a sharpening filter help with edge finding?
2. How can an averaging filter and a sharpening filter use the same neighborhood of pixels and still have such different effects on an image?
3. Name three ways in which the median is different from the average.

Try This

4. Find the output for the horizontal difference processor of Figure 4.29 when the input is test image $\mathbf{A}_5$. Would you be able to tell that the edges came from alternating lines?
5. Find the output for the horizontal difference processor of Figure 4.29 when the input is test image $\mathbf{A}_6$. Would you be able to tell that the edges came from alternating lines?

6. Apply the sharpening filter $B(i, j) = -0.5 \times A(i, j-1) + 2 \times A(i, j) - 0.5 \times A(i, j+1)$ to test images $\mathbf{A_3}$ and $\mathbf{A_5}$, and compare your results with those shown in Figure 4.33.
7. Apply the sharpening filter $B(i, j) = -3 \times A(i, j-1) + 7 \times A(i, j) - 3 \times A(i, j+1)$ to test images $\mathbf{A_3}$ and $\mathbf{A_5}$ and compare your results to those shown in Figure 4.32.
8. Apply the averaging filter from Equation (4.13) to the 1×20 image matrix

$$\mathbf{A} = (20\ 20\ 20\ 20\ 20\ 200\ 20\ 20\ 20\ 20\ 20\ 20\ 20\ 200\ 200\ 200\ 200\ 200\ 200\ 200).$$

   a. Compute the pixel values of the resulting 1×18 output image matrix **B**. Which pixels have the same value as the corresponding pixels in the input matrix **A**? Which have changed? How wide is the edge in the smoothed image? What happens to the isolated pixel with a value of 200 surrounded by pixels of value 20?
   b. Now take the average of an average. Use the output image matrix **B** from part (a) as the input to the averaging filter from Equation (4.13) and compute a new output. Which pixels change and which ones stay the same? How wide is the edge in the doubly smoothed image?
9. Repeat Exercise 4.4.8, but use a median filter instead of an averaging filter.
10. When a sharpening filter is used after a smoothing filter, it does not restore the original image perfectly.
   a. Use test image matrix $\mathbf{A_2}$ as an input. Use the smoothing filter from Equation (4.13) to compute a new image, and then use the sharpening filter from Equation (4.11) on the smoothed image to create a sharpened smoothed image. Plot the results in the style of Figures 4.33 and 4.34. How does the sharpened smoothed image compare with the original image matrix $\mathbf{A_2}$?
   b. Repeat (a), but reverse the order of smoothing and sharpening. How do these results compare with the results of (a)?

Back of the Envelope

11. If you could see only one pixel in an image, would you be able to tell if its value were correct or if it had been changed because of noise? How much of an image would you need to see to know if noise were a problem? How would you decide how to reduce the noise?

4.5 Designing Simple Vision Systems

We started this chapter with the specific objective of designing the vision system for a robot that can clean up the house or do the laundry so you can go see a movie with your friends. In previous sections, we have defined simple operations for image matrices, used the operations in combination for more complex image-processing tools, and created two basic preprocessing operations to reduce blur or reduce noise. Now we will put these basic operations and additional tools together to design some of the components needed for our robot vision system.

The robot we want to design must be capable of capturing a scene—say, the living room—and locating and identifying objects lying on the carpet that need to be picked up and put away. It should be "smart" enough not to try to "pick up" the design on the rug, but it should pick up the CDs and shoes left on the floor. It should have the capability of separating light clothes from dark clothes before putting them in the laundry. It should know better than to try to put your sleeping dog in the closet.

We now have the building blocks to begin the design of our robohelper's vision system, but it will be very complex, and the complete design is more than we can include in a single chapter. However, to get us started on our design, we will look briefly at three smaller scale problems that are similar to parts of the robot image-system design. In all three cases, we will start with a very simple approach and then try to figure out how to improve the performance.

Designing an Object Counter

An object counter can be used to count red blood cells on a microscope slide, cars on a highway, or shoes on the floor that should be put away. In order to build a system to do this, we must be able to list what characteristics distinguish the objects we want to count from other objects and background structures. Then we can process the image to keep only the pixels that have the characteristics of the objects we are counting. We could consider several characteristics that we might use individually or together:

- **Color or hue:** If the objects of interest were all similar in color, we could separate them from objects of a different color. This operation could be done using red, green, and blue color components, or it could be done with a combination of components that defines the hue, or perceptual color, of the objects.
- **Brightness:** What are the maximum, minimum, average, and median intensities expected for the objects we want to count?
- **Shape:** Is the region symmetrical from left to right or from top to bottom? What is the ratio of the area to the square of the perimeter? The area can be estimated as the number of pixels in the region, and the perimeter can be estimated as the number of edge pixels.
- **Size:** The size of an object in the image can be estimated by the width and height of a region in pixels or by the number of pixels. However, some knowledge about the distance of the object from the camera will be needed in order to separate objects, based on their actual physical size.
- **Texture:** How much variation is there in intensity and color? Is there a regular pattern?

Our first simple approach will be to define the objects we want, based on either color or intensity. Then we can threshold the image to get a 1-bit-per-pixel image that has a value of 1 where our objects are located and a value of 0 everywhere else. Then we can analyze the thresholded image to find connected pixels and determine how many objects were in the image and where each is located. This process is shown in the upper part of Figure 4.37.

As we think about how this preliminary design might actually work, many potential causes of failure come to mind. What will happen if the object edges are a little irregular and small pieces of objects break off and become separate objects? This factor would also cause problems for locating the objects and accepting or rejecting them based on size. We would probably want to smooth the image before we do any other processing, and we would specify our imaging system so that objects of interest would be at least 10 pixels in size so that the smoothing would not alter their shape or size. The characteristic used for separating the objects from the background might not be a simple threshold applied to intensity or color components. We might have to design a special mapping operation for our input image so that a simple threshold would work on the image after the pixel values are changed by the mapping function. An improved design is shown in the lower part of Figure 4.37.

Other complications that might require further design improvements include variations in lighting or the handling of objects that touch each other and appear to be a single object. This is where size and shape analysis might prove useful. Also, we may need to calibrate our system for camera position, because the distance of the camera and the direction it is pointing can change the size and shape of objects in an image.

Designing a Motion Detector

The numerous applications of motion detectors have been discussed in many places in this chapter and the preceding chapter. Our preliminary design of a motion detector could be based on the simple change detector of Figure 4.8. If we have an image sequence from a camera, then we only have to subtract the previous frame from the current frame to get a different image. Then an absolute-value operation will make all pixels positive, and a

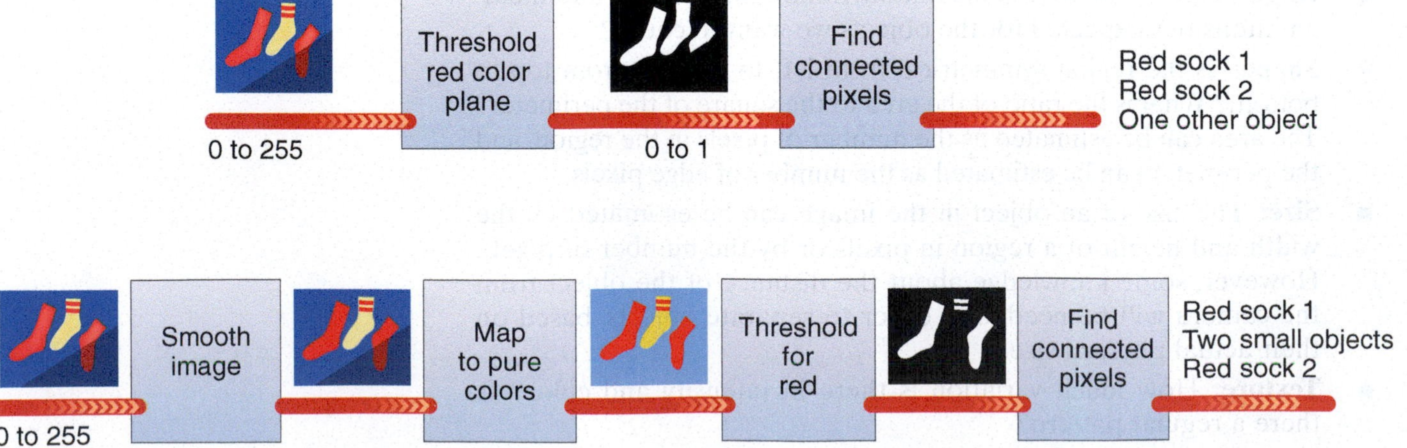

Figure 4.37 Block diagram of a simple object counter and an improved object counter with an example of counting red socks. The simple counter misses part of the rightmost red sock because a shadow makes part of it too dark. The yellow sock is classified as red because the color yellow has a strong red component.

Infinity Project Experiment: Design of an Object Counter

Using input from a camera or stored image files, define characteristics of some objects, such as marbles or coins, and count the objects.

threshold can be set for the amount of change that should be detected. This procedure is shown in the block diagram in the top half of Figure 4.38.

In some applications, the change detector may not be effective because changes happen too slowly and the amount of change never rises above the threshold for detection. A second subtractor could then be added that would take the difference of the current frame and an image from several frames before to detect slow changes. A bigger problem can arise if the camera itself is moving, because changes in the image could then be due entirely to repositioning of the camera. For our robot system, either we would have to find a way to adjust the images for the motion of the robot or we would have to rely on several fixed cameras that view the robot and the room it was cleaning.

Another refinement of the system might require that the motion be detected for some sustained time interval in order to avoid responses to noise. This could be done by comparing two or three of the thresholded outputs to see if the changes detected were close to each other. The system might also require that the size of the object in motion be above a particular limit or be limited to particular areas of the image.

There are two important parameters in this design that control the sensitivity to change. One is the time delay between frames that form the difference image, and the other is the threshold value that controls what part of the difference image is ignored and what part is accepted as a moving object. If the sensitivity is too low, some moving objects will escape detection, but if the sensitivity is raised too much,

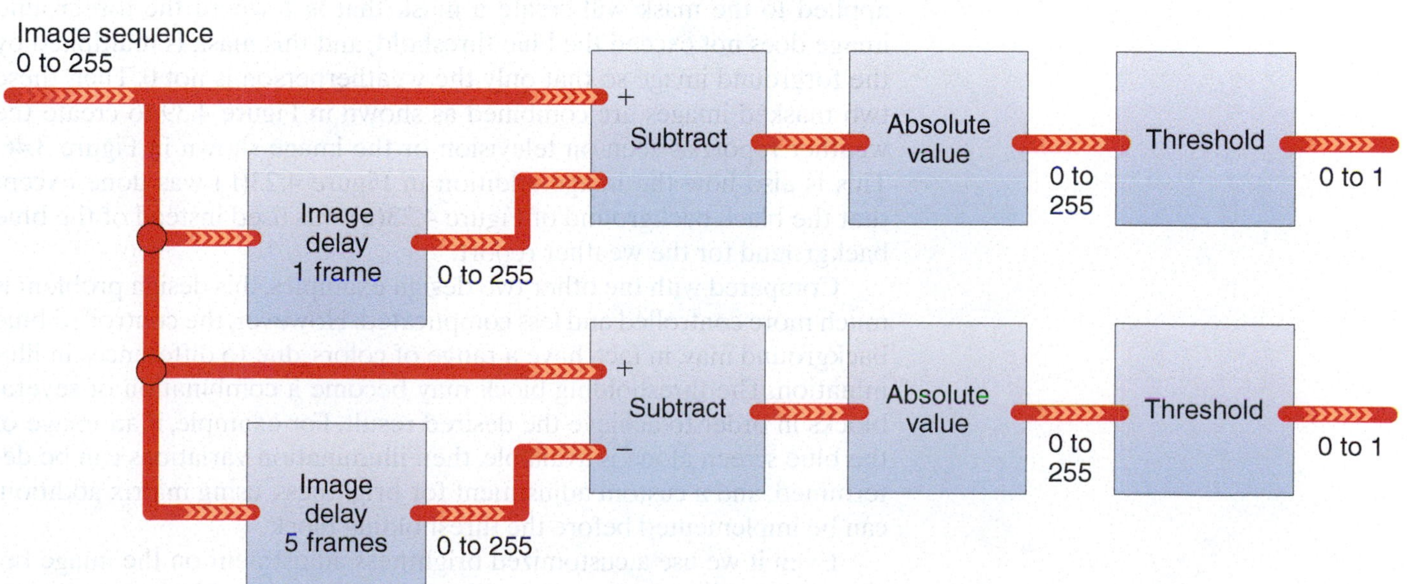

Figure 4.38 Block diagram for a motion detector.

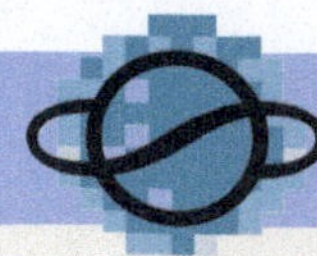

Infinity Project Experiment: Design of a Motion Detector

Using input from a camera or a stored image sequence, detect people entering and leaving a room.

false objects will be detected. Again, a design trade-off must be made for performance in a particular application. For example, in a security context, a system that generates a high number of false alarms will start to be ignored or at least will not be taken as seriously as it should be.

Designing a Blue-Screen Chromakey System

The blue-screen chromakey example from the end of Chapter 3 was created for personal entertainment, but the technology to segment and recombine images in this way has much broader applications. We have all seen the weather reports on television in which the person reporting the weather appears to be directly in front of a large sophisticated weather display map. In fact, the person stands in front of a solid-colored wall and then gets added into an image in which the size of the weatherperson and the smaller map display have been matched. Other examples of blue-screen Chromakey applications include virtual reality and training simulators.

With the simple operations developed in this chapter, the blue-screen problem seems very straightforward. First, we have to have some way to segment our foreground figure (e.g., the weatherperson) from the background, which we have chosen to be a solid bright-blue wall. We can do this with a simple thresholding operation. Then the image that is 1, where the foreground image is blue and 0 everywhere else is used as a mask that is multiplied by the desired background image (e.g., the weather map).

In the weatherperson example, the result of this product is a weather map with a weather person-shaped black hole in it. A negative operation applied to the mask will create a mask that is 1 where the foreground image does not exceed the blue threshold, and this mask is multiplied by the forground image so that only the weatherperson is not 0. Then these two masked images are combined as shown in Figure 4.39 to create the weather report as seen on television or the image shown in Figure 3.46. This is also how the image addition in Figure 4.23(f) was done except that the black background of Figure 4.23(e) was used instead of the blue background for the weather report.

Compared with the other two design examples, this design problem is much more controlled and less complicated. However, the controlled blue background may, in fact, have a range of colors, due to differences in illumination. The thresholding block may become a combination of several blocks in order to achieve the desired result. For example, if an image of the blue screen alone is available, then illumination variations can be determined, and a custom adjustment for brightness, using matrix addition, can be implemented before the thresholding block.

Even if we use a customized brightness adjustment on the image before thresholding so that all the blue background would be very close to the same color, just thresholding the blue component of a color image will

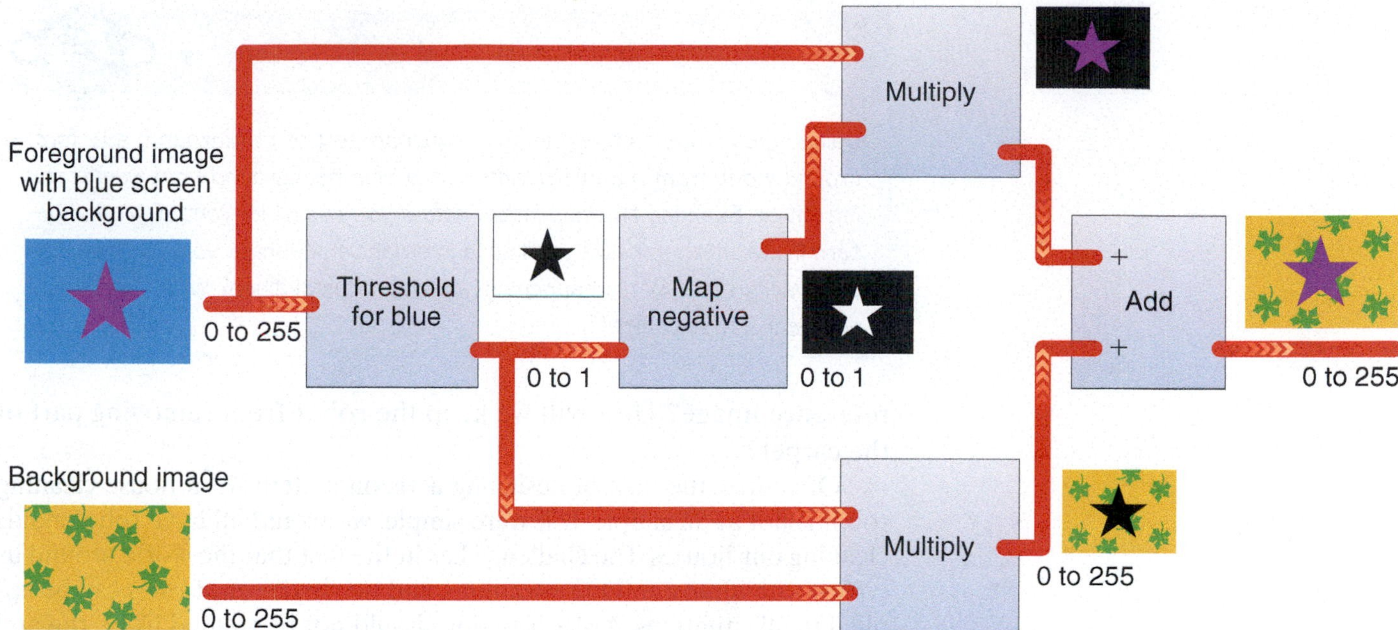

Figure 4.39 Block diagram of the blue-screen chromakey process.

not produce the mask image we need. As we noted when we discussed representing colors by hue, saturation, and brightness, the blue component is very high for bright blue, but it is also high for bright white, bright magenta, and bright cyan. If the weatherperson wore a white shirt, it would be treated just like the blue-screen background, and we would see only the head and hands of the weatherperson on the map. We could find three ranges of values for each of the three color components so that we create a mask only with bright blue. Alternatively, we could compute the hue and apply the threshold to it to select blue only. And finally, we note that it is important that the weatherperson not wear blue clothing.

Summary

We started this chapter thinking about the kind of robot vision needed to perform the simplest and most mundane household chores. This task initially seemed relatively simple. We have now designed most of the basic components of the system and explored other applications for them. But even in the smaller scale applications we just completed, there were many design choices that were application dependent, and there were many complications that could be anticipated.

Now we can think about how we would design the robot vision system, using these smaller applications as models. Although object identification seems manageable, how will all the possible objects that could be in a room be identified? It might be easier just to know what the room should look like after all the objects have been removed by keeping reference images of the cleaned room taken from established viewing positions. Then everything that is not in the reference image could be removed by the robot even if the objects were not precisely identified. That moves the problem of identifying all the objects to the later stage of sorting the removed objects. But what will happen if a drink has been spilled on the rug, making a stain that was not in the

Infinity Project Experiment: Design of a Blue-Screen Image System

Using input from a stored image sequence as the background, add foreground input from a camera aimed at a blue background that you stand infront of. Examine the blue image before you stand in front of it, and determine what thresholds you'll need in order to separate yourself from the blue background. What happens if you wear blue clothes? What happens if you wear a white shirt?

reference image? How will we keep the robot from removing part of the carpet?

Of course, the task of designing a vision system for a house-cleaning robot is not at all simple. If it were simple, we would all have robot maids cleaning our houses. The challenge lies in the fact that the world around us comes with almost infinite variations, and no single set of rules can be applied to all situations. A sleeping dog should not be treated like a rug and vacuumed, and a picture on the wall should not be mistaken for a distant object. These judgements are the things that we humans accomplish effortlessly and without receiving formal training based on years of daily living experience. However, we must teach the robot how to do these things.

Current image analysis systems can operate very effectively in environments in which the variability is somewhat limited or controlled, such as the Chromakey blue-screen example. Image analysis systems can make accurate medical diagnoses from X-ray or magnetic resonance images. Systems can guide robotic devices on a manufacturing floor to do tasks that are hazardous to humans. Systems to recognize faces in a crowd or control entry to a restricted area have become very effective. As processing capability continues to increase according to Moore's law, the complexity that image-processing systems can handle also increases. But the creation of a computer vision system as powerful and flexible as a human being's vision system is still in the future.

EXERCISES 4.5

Mastering Concepts

1. In a blue-screen chromakey application, what would happen if the subject in front of the blue screen were wearing bright-blue clothes?
2. In Figure 4.39, two images are added, but the result is not divided by two. In previous examples, we divided image sums by two to make sure the values of the sums were not too large to be represented by the number of bits available to each pixel. Why is there no division by two here? What would happen if we did divide by two?
3. Draw a block diagram for a system that could count people entering a room. In addition to blocks used in this chapter, you can use a block that counts every time it sees a new image frame with an average level above a threshold value. How often would you make a new image frame? How would you adjust the threshold value?

Master Design Problem

An image recording system for a wild-animal refuge must be designed to count the number of animals that enter and leave several important sites within the refuge. This system will allow the administrators to monitor and improve the conditions for the animals without placing human observers near the sites, which might result in changes in the animals' natural behavior. The sites to be monitored include trail crossings, water holes, areas of abundant food, and natural shelters. For each site, a log will be kept that can be examined locally or sent to a central site for the refuge.

You will need to begin by answering the following basic questions:

- How many different kinds of animals will be monitored? How big are they? How fast can they move? Will you include animals that fly? Will you include animals that burrow in the ground? Will you include animals that spend a lot of time in the water?
- How will you identify the animals? Typically, animals look a lot like their environment in order to survive. How different will animals look in daytime images taken with reflected light compared with nighttime images taken with an infrared heat-sensing camera?
- How will you detect changes? The animals might come to a site and stay there for some time, so although you can detect the change when they move, how will you know if they are entering the site, leaving the site, or simply moving around in the site? A single image of the site with no animals will not be very useful, because changes caused by the seasons, the weather, or the animals themselves would also be detected as changes. Average images of a site over long time intervals might be used as a reference.
- How will you determine what animals caused the image changes you detect?
- How will you determine the number of animals you are interested in?
- How will you distinguish movement of animals from movement of trees and brush caused by the wind?
- Could you use audio input from microphones in addition to the camera input to make a better system?

Start with one site and two or three kinds of animals for your first design. Then evaluate your design and find two ways to improve it.

Big Ideas

Math and Science Concepts Learned

In this chapter, we learned that images are arrays of pixel values that can be thought of as matrices. Since computation with matrices can be done easily with a calculator and computer libraries, we can do a wide variety of image-processing tasks by using easily available computational tools.

There are three types of simple image-processing operations:

1. A mapping operation changes pixel values based only on the value of the pixel and not on where the pixel is located in the image. These mapping operations are defined by lists, graphs, or functions and are used for tasks such as brightening images or creating a negative.

2. Arithmetic operations combine two images by adding, subtracting, or multiplying them. Masking an image is a special case of image multiplication.
3. Neighborhood operations involve a group of pixels in a single image that are used to compute a new pixel value. These are powerful operations that allow us to smooth noise or sharpen edges.

With our computations, we must be careful to ensure that our final results can be represented with the number of bits available for each pixel. If our computations produce some results that cannot be represented with the available number of bits, we can scale the result, take the absolute value of the result, or simply set all results to the nearest allowable pixel value.

Combinations of the three types of image-processing operations can be used for a wide variety of image smoothing and sharpening tasks. The performance of these tasks is based on the size of the neighborhood and the constant values used to multiply the shifted versions of images before they are added together.

Important Equations

A matrix **A** is an array of numbers arranged in rows and columns. An example of a general matrix with 4 rows and 3 columns and a matrix with specific element values is shown:

$$\begin{matrix} & \text{col. 1} & \text{col. 2} & \text{col. 3} & \\ \mathbf{A} = & \begin{pmatrix} A(1,1) & A(1,2) & A(1,3) \\ A(2,1) & A(2,2) & A(2,3) \\ A(3,1) & A(3,2) & A(3,3) \\ A(4,1) & A(4,2) & A(4,3) \end{pmatrix} & \begin{matrix} \text{row 1} \\ \text{row 2} \\ \text{row 3} \\ \text{row 4} \end{matrix} \end{matrix} \qquad \mathbf{A} = \begin{pmatrix} 12 & 9 & 0 \\ 5 & 6 & 23 \\ 8 & 16 & 27 \\ 4 & 7 & 19 \end{pmatrix}$$

Four basic examples of mapping are changing brightness, changing contrast, creating a negative, and thresholding. The matrix **B** is computed by adding the constant value k to each element of the matrix **A** to change brightness:

$$B(i,j) = A(i,j) + \mathrm{k}$$

The matrix **B** is computed by multiplying each element of the matrix **A** by the constant value s to change contrast:

$$B(i,j) = s \times A(i,j)$$

The matrix **B** is the negative of the 8 bit-per-pixel image matrix **A**:

$$B(i,j) = 255 - A(i,j)$$

The matrix **B** is computed by thresholding the matrix **A**:

$$B(i,j)\begin{cases} 255 & \text{if} \quad A(i,j) > v \\ 0 & \text{if} \quad A(i,j) \le v \end{cases}$$

The matrix **C** is computed by adding corresponding elements of the matrix **A** and the matrix **B**:

$$\mathbf{C} = \mathbf{A} + \mathbf{B}$$
$$C(i,j) = A(i,j) + B(i,j)$$

The matrix **D** is computed by subtracting elements of the matrix **B** from the corresponding elements of the matrix **A**:

$$\mathbf{D} = \mathbf{A} - \mathbf{B}$$
$$D(i,j) = A(i,j) - B(i,j)$$

The matrix **G** is computed by multiplying corresponding elements of the matrix **A** and the matrix **B**:

$$\mathbf{G} = \mathbf{A} \otimes \mathbf{B}$$
$$G(i,j) = A(i,j) \times B(i,j)$$

The matrix **H** is obtained by shifting the matrix **A** to the right by one column:

$$\mathbf{H} = \begin{pmatrix} H(1,1) & H(1,2) & H(1,3) \\ H(2,1) & H(2,2) & H(2,3) \\ H(3,1) & H(3,2) & H(3,3) \\ H(4,1) & H(4,2) & H(4,3) \end{pmatrix} = \begin{pmatrix} ??? & A(1,1) & A(1,2) \\ ??? & A(2,1) & A(2,2) \\ ??? & A(3,1) & A(3,2) \\ ??? & A(4,1) & A(4,2) \end{pmatrix}$$

Matrices can be shifted horizontally or vertically:

$$\begin{aligned} H(i,j) &= A(i,j-1) && \text{Shift } \mathbf{A} \text{ to the right} \\ H(i,j) &= A(i,j+1) && \text{Shift } \mathbf{A} \text{ to the left} \\ V(i,j) &= A(i-1,j) && \text{Shift } \mathbf{A} \text{ down} \\ V(i,j) &= A(i+1,j) && \text{Shift } \mathbf{A} \text{ up} \end{aligned}$$

The matrix **B** is obtained by a moderate amount of sharpening of vertical lines in the image matrix **A**:

$$B(i,j) = -A(i,j-1) + 3 \times A(i,j) - A(i,j+1)$$

Other sharpening filters using the same neighborhood are given as unnumbered equations in the chapter:

$$B(i,j) = -2 \times A(i,j-1) + 5 \times A(i,j) - 2 \times A(i,j+1)$$
$$B(i,j) = -0.5 \times A(i,j-1) + 2 \times A(i,j) - 0.5 \times A(i,j+1)$$

The matrix **B** is the image matrix **A** after sharpening both horizontally and vertically:

$$B(i,j) = -A(i-1,j) - A(i,j-1) + 5 \times A(i,j) - A(i,j+1) - A(i+1,j)$$

The matrix **B** is the average of one-row-by-three-column neighborhoods of the image matrix **A**:

$$B(i,j) = \tfrac{1}{3} \times (A(i,j-1) + A(i,j) + A(i,j+1))$$

The matrix **B** is the average of three-row-by-three-column neighborhoods of the image matrix **A**:

$$\begin{aligned} B(i,j) = \tfrac{1}{9} \times (&A(i-1,j-1) + A(i-1,j) + A(i-1,j+1) \\ &A(i,j-1) + A(i,j) + A(i,j+1) + \\ &A(i+1,j-1) + A(i+1,j) + A(i+1,J+1)) \end{aligned}$$

The median filter can also be used for smoothing:

$$B(i,j) = \text{median}(A(i,j-1), A(i,j), A(i,j+1))$$

chapter 5

Digitizing the World

OUTLINE

We hear it all the time: "Everything is going digital." What does this really mean? Is this good? What benefits do we get from digital data?

Well, we have already learned through the study of Moore's law that digital computing power and storage capacity are increasing at a remarkable rate, doubling every two years. This puts a tremendous amount of power in the hands of modern and future engineers to use in creating all sorts of new devices and solving all sorts of problems. However, for engineers to be able to use all of this power, we must be able to convert information and processes into a digital or binary form to run all these devices—we must "go digital."

But how do engineers do it? How do engineers take information such as sounds, pictures, and text and turn them into numbers that can be processed by our ever-increasing computing capabilities?

In this chapter, we will study precisely how engineers convert various types of information into a purely digital form. This critical process, used in devices such as digital

Digitization: Combined operations of sampling and quantization. Also called analog-to-digital conversion.

cameras, cell phones, and CD players, is called **digitization**: the process of representing various types of information in a form that can be stored and processed by a digital device such as a computer.

The benefits of digitization are tremendous and widespread. Once information is in a digital form, we can use our digital-processing power to improve it, such as by reducing unwanted sounds, combining sounds and images from different sources, or adding special effects. We can easily duplicate digital information so that we can share it with friends and move the information from place to place with great efficiency. Try doing that with music stored on a vinyl album or a movie recorded on film!

INTERESTING FACT:

Why does everyone want to "go digital"? With digital technologies, we easily can store, reproduce, modify, and send a wide variety of information, such as books, music, and movies.

You experience the benefits of digitization every time you listen to music from a CD or watch a movie from a DVD. For both CDs and DVDs, the player simply converts the numbers stored on them back into the original music or movie. Engineers have designed this process so well that your ears and eyes aren't able to tell the difference between the digital information and its original real-world source.

Digital Signal: A set of sampled values represented in binary form as bits.

Sample: A numerical value representing the height of a waveform at a particular time, or the brightness of an image at a specific point.

Digitization of information produces a **digital signal**, which is simply a list of numbers that can be turned back into the original form of the information at the time of your choice. Whenever the information in an analog, or continuous, signal is digitized, two important things must happen. First, we must convert the analog, or real-world, signal (such as music) into a list of numbers. Each number is called a **sample**. Next, we need to put these numbers into binary, or base-two format so that they can be stored and processed by our modern digital devices. We also need to be able to convert between the binary format, which is the natural format for digital storage, and the decimal (base-10) format, which is more natural to us. You already have seen these ideas introduced in Chapters 1 and 3. In this chapter, we will study the details of how these basic ideas can be applied to all types of information, including sound, moving images (such as video), and text.

Design Objective for Digitizing the World: The Digital Yearbook

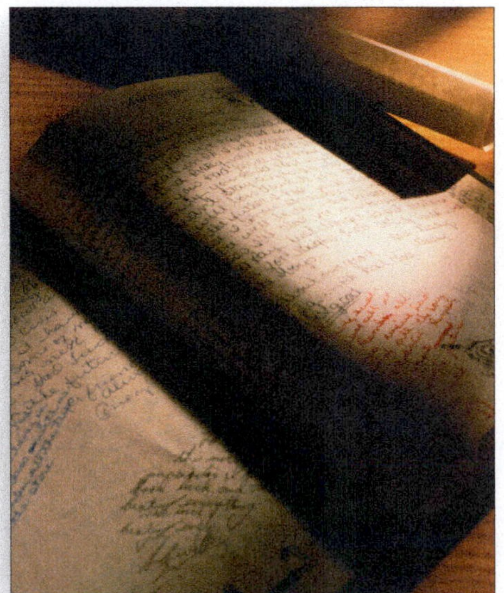

Figure 5.1 A typical yearbook.

Wouldn't it be great if your yearbook (Figure 5.1) could capture the complete story of your school and friends—from pictures and movies to your favorites songs, important world or local events, and stories and letters about sporting events and dances? Well, let's think about how engineering advances in new digital-processing and storage technologies can help you to make a yearbook that decades from now would create a rich sense of what you and your friends experienced as students.

So how do we do it? Simple enough—let's go digital. A digital yearbook could offer many important possible advantages. In designing a digital yearbook, we should ask ourselves the following questions:

- **What benefits or added value will the new technology give us?**
  1. We can include more types of memories of our year in a digital yearbook. A traditional yearbook has printed text and still images. A digital yearbook could include sounds and video.

2. A digital yearbook can be a "living document" that grows over time. For example, information from class reunions can be added easily.
3. A digital yearbook can be more versatile. The content display could be enlarged or reduced, and optional background materials can be included.
4. A digital yearbook can be more portable. We can access it over the Web or make copies available on CD or DVD.
5. The cost of creating copies of a digital yearbook will certainly be less than that of creating copies of a traditional yearbook, so replacing a lost yearbook will be easy.

INTERESTING FACT:

A typical 500-page book with color pictures on every page might require as many as 300 Mbits to store digitally. While this might seem like a big number, by modern standards this is a relatively small amount of information.

- **How will we use the new technology?**
  1. To design the digital yearbook, we will have to look at the component parts of the yearbook individually and see how we will acquire or create them and how we can store them. The traditional yearbook has "page budgets" for text and images. We will have "bit budgets" for text, still images, sounds, and videos. We will see that when each of these components is represented by bits, we can estimate the number of bits for the following items to determine the level of quality we want for our yearbook:
     a. a page of text,
     b. a small still image,
     c. a large still image
     d. one minute of speech,
     e. one minute of music, and
     f. one minute of video.
  2. Once the yearbook committee has agreed on a format and a quality level, we can estimate the number of bits we will need by adding up the numbers of bits for each component and then adding 10 to 15% extra for overhead and unforeseen needs. Then we can see if our digital yearbook will fit in a convenient package, such as a CD or DVD.

- **What questions or problems might the new technology raise, and how would we address those issues?**
  1. Will the quality of the images in the digital yearbook be the same as in a printed yearbook? We have explored the relationship between the number of pixels and the number of bits per pixel in Chapter 3. We can get digital images of any resolution and quality that we want, but we will need to budget enough bits for the images.
  2. Not everyone may be able to view the digital yearbook on a DVD player. Some may want to have the traditional bound-book version. It would be possible to create a printed version of the static parts of the yearbook at an additional cost.
  3. Many students get classmates to sign (and write a message in) their yearbooks. The ability to attach a personal message would be a new feature to be added to a digital yearbook that would require some engineering design. In addition to typing a message,

INTERESTING FACT:

Your friends could "sign" your yearbook digitally by putting written notes, spoken words, or pictures into your yearbook.

students could also append images or speech notes that would not be possible to include in the traditional yearbook.

4. As time passes, the technology used for the digital yearbook may become obsolete (like the $3\frac{1}{2}$" floppy disk did, for example). However, as technologies change, converting information for new digital storage technologies usually remains simple and inexpensive. For example, to a large extent, CDs have replaced vinyl records, and most recordings from vinyl records have been made available on CDs.
5. Can the new project be completed in time? Years of experience with the traditional yearbook have resulted in a predictable schedule, but when using new technology and processes, some things will be faster and some will be slower. The quality of all the yearbook components must be verified, and then the quality of the user interface must be tested to make sure students can go easily to the part of the yearbook they want to experience.

- **What investment would we have to make in order to use the new technology?**
  1. Buying the tools and equipment needed for the new technology is an obvious investment. You might need new digital cameras, new computers, and new digital audio- and video-editing software.
  2. In addition, the cost in time to learn to use the new technology should be considered. If the yearbook staff has to learn to use a lot of new equipment, then the time to produce the yearbook might not be enough.

5.1 Introduction

Where Do We Start?

If we want to explore the possibility of using digital technology to make a yearbook, where do we start? The first thing we have to explore is how to represent various types of information in digital form. From Chapters 3 and 4, we understand how to represent image information as numbers, but now we need to find out how to represent sound and text as numbers as well. As with images, we will find that the number of bits needed to store sounds will depend on what the sounds are and what quality we desire in our reproduction. We will find that there are some simple rules that will tell us how to convert sounds into bits and how to compute the number of bits we need to use. Once we know how to digitize information, it will be up to you and your yearbook committee to determine what information you want to digitize.

KEY CONCEPT

Digitization uses bits and the binary number system to store nearly all forms of information.

EXERCISES 5.1

Mastering the Concepts

1. A digital yearbook would contain what types of information?
2. Name five possible advantages of a digital yearbook over a hard-copy yearbook.
3. Name three possible difficulties in producing a digital yearbook.
4. Why might you need to iterate within the design process?

Back of the Envelope

5. Approximately how many pictures were in last year's yearbook for your school?
6. About how many letters (characters) of text were in last year's yearbook for your school? Come up with a strategy for estimating the number of characters in the yearbook, given some sample pages.
7. Given the number of students and the extent of extracurricular activities in your school, how many minutes of speech, music, and video do you think the digital yearbook should contain? Try to take into account all of the activities that go on at your school, such as sporting events, plays and musicals, concerts, speeches, and assemblies.
8. List some ways that you and your friends can personalize your yearbook, such as with a digital "signature" option.

5.2 From the Real World to the Digital World

From Information to Numbers

Our perceptions of the sights and sounds that we experience in the real world are analog streams of information. That means we see and hear continuous variations in colors, light, and sound. When there is a sudden change in any of these perceptions, such as a sound starting or stopping, we immediately pay attention to it.

The world of digitized information is much different. Digital devices can record and process only sequences of numbers, not the original analog sounds, pictures, and movies. For example, an audio CD stores sound as a list of numbers. If that list of numbers does not contain enough information, the reproduced sound will not sound the same as the original.

You are probably asking yourself, how is it possible that information such as music or movies can be turned into a list of numbers without losing any quality? We will see that there are two basic parts to the answer to this question, and engineers have to deal with both. The first is how to design and build the necessary new technologies. The second, often called "human factors," is how to determine what the limits are for human perception.

INTERESTING FACT:

One of the first applications of digitization was with radar during the 1950s. Digital information from radars allows military and civilian air-traffic controllers to use computers to help keep track of the location of flying airplanes.

INTERESTING FACT:

Nielsen Media Research monitors a small number of people for what TV shows they watch and then estimates the total number of viewers for TV shows. This is called "sampling" the TV audience.

To give you some sense of how engineers might get started, let's consider a simple experiment. Let's say we want to capture visual experiences through a sequence of pictures to be stored as matrices in a digital camera. The camera we'll use is set up to take and store a picture every 15 minutes.

First, we point the camera at a flower that is slowly blooming. This flower might take a day to completely open. In that time period, we might have taken as many as 100 pictures, which, when played back as a movie, would give us a very good fast-motion film of the blooming flower.

On the other hand, what would happen if we pointed this same camera at a professional football game? In this case, the game lasts only three hours. If we have the camera set up to take a picture every 15 minutes, the camera would take a total of only 12 pictures in those three hours. How well would these pictures tell the story of what happened during the football game? Clearly, we would get very little information about what happened during the game; more importantly, playing back this very short film would not provide a satisfactory football-watching experienced.

So, what can we learn from this experiment? Sequences of snapshots (or "samples") can give us a faithful re-creation of the original action scene if we take pictures often and fast enough. But how fast is "fast enough"? Should we just take pictures or measurements as fast as possible? Well, we wouldn't need to take pictures of the blooming flower 1000 times a second, because the pictures would all be nearly identical. More importantly, we would have to store all of those pictures, which would increase both the size and the cost of the digital device. We experienced the same issue in looking at the number of gray levels in an image in Figure 3.17: We could use more than 8 bits of gray level, but we would not be able to see any difference, and we would just use more storage space for the additional bits.

In this section, we'll discuss how often we should take measurements, such as samples of a waveform or pictures in a video, to ensure that the numbers allow us to faithfully re-create the original information.

Sampling: The process of recording values (samples) of a signal at distinct points in time or space. Waveforms are sampled in time. Images are sampled in space. Time-varying scenes (video) are sampled in both time and space.

To address this issue, we need to understand the basic concept of **sampling**. For simplicity, we will focus on sampling of a waveform in time. In Chapter 3, you were introduced to sampling of images in space, which is slightly more complicated, because an image has two spatial variables, whereas a waveform has only a single time variable. Sampling of a waveform refers to the process of recording values at distinct time points along a signal. Usually, samples of a signal are observed at uniformly spaced time instants. We demonstrate sampling graphically in Figure 5.2. As we have already learned, a signal, such as sound, can be plotted on a graph, such as in Figure 5.2(a). We can take measurements of this signal at regular time intervals to get the "sampled signal" shown in Figure 5.2(b). Here, the heights of the dots above or below the time axis represent the samples. In Table 5.1, we show the list of time values and the numbers corresponding to the sampled version of this analog signal. For this example, you can see that we take measurements of the original signal every 0.1 s. From this list, we see that the signal value was 2.7748 at time 1.3 s, 1.9113 at time 1.4 s, and 0.6002 at time 1.5 s. The table of sampled values does not tell us directly what the signal value was at 1.45 s, 1.376 s, or any time other than the specific times listed.

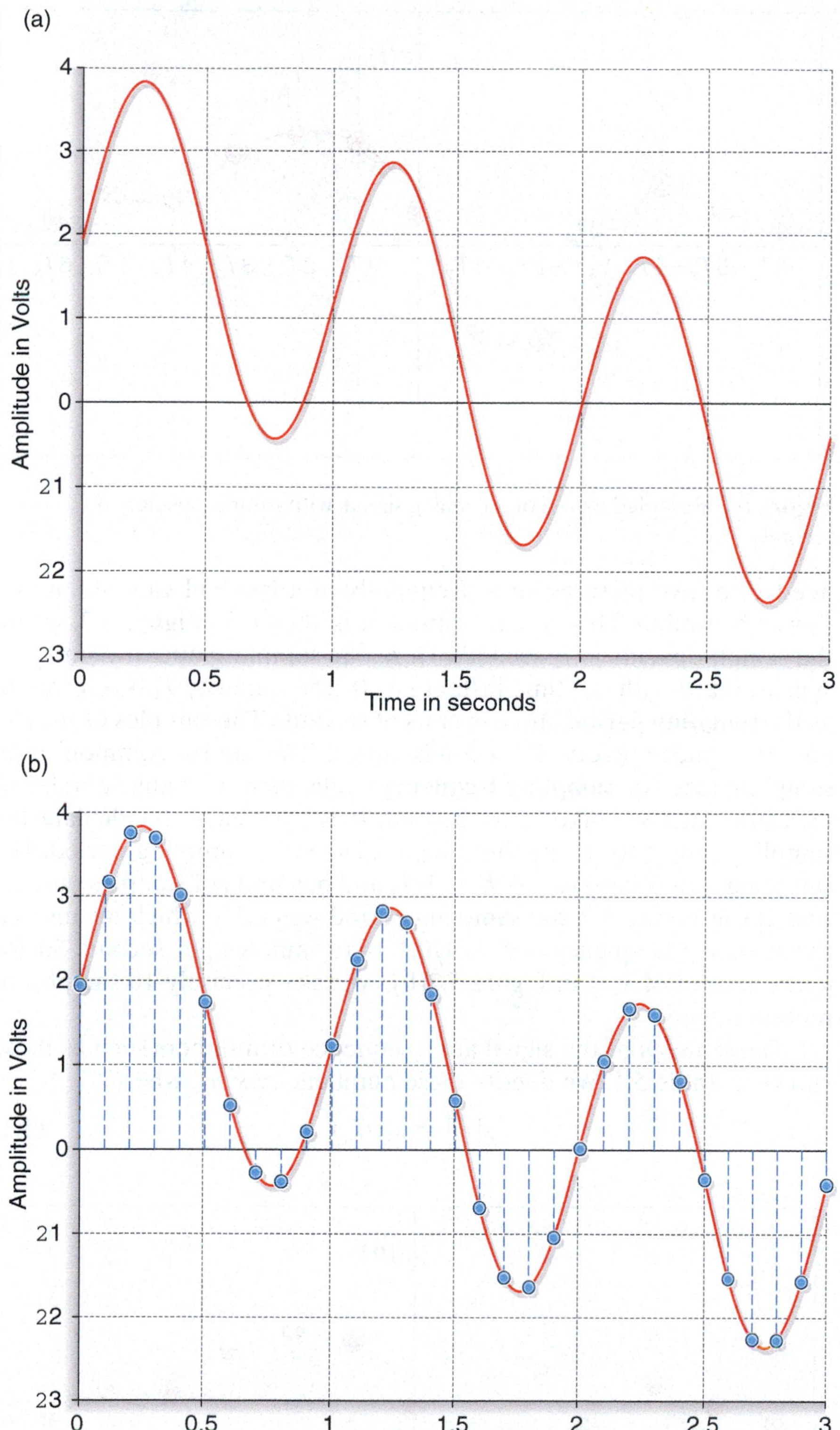

Figure 5.2 (a) An analog signal $s(t)$. (b) A plot of samples of the signal $s(t)$.

Table 5.1 Sampled Values for the Signal in Figure 5.2(a)

Time (s)	Sample Values
0.0	2.0000
0.1	3.1674
0.2	3.8694
0.3	3.8289
0.4	3.0465
0.5	1.8006
0.6	0.5412
0.7	−0.2814
0.8	−0.3885
0.9	0.2217
1.0	1.2732
1.1	2.3188
1.2	2.9112
1.3	2.7748
1.4	1.9113
1.5	0.6002
1.6	−0.7078
1.7	−1.5621
1.8	−1.6835
1.9	−1.0707
2.0	0.0000
2.1	1.0807
2.2	1.7233
2.3	1.6508
2.4	0.8637
2.5	−0.3601
2.6	−1.5718
2.7	−2.3223
2.8	−2.3346
2.9	−1.6092
3.0	−0.4244

For the signal shown in Figure 5.2(b), we used a sample spacing of 0.1 s. In general, however, the sample spacing will be something other than 0.1 s, and we will select it based on how fast or slowly the signal changes. An analog signal $s(t)$ that varies quickly must be sampled more frequently than an analog signal that varies slowly—just as we

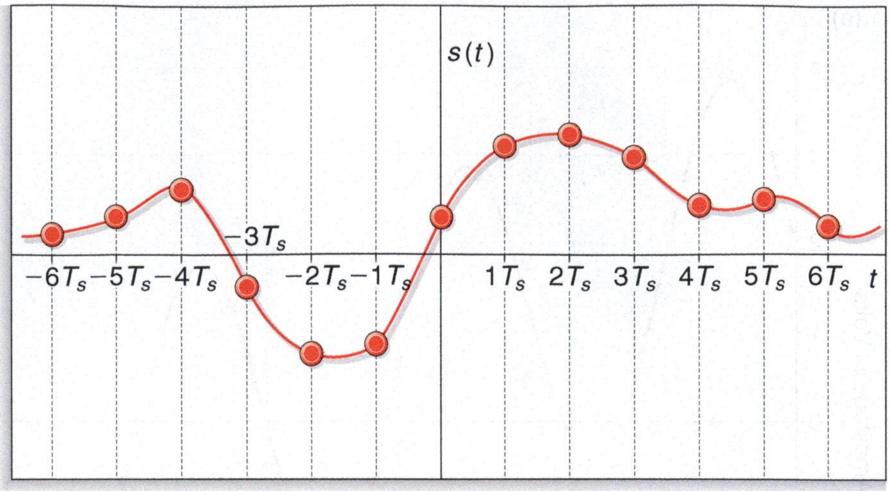

Figure 5.3 Sampled values of an analog signal, with sample spacing of T_s seconds.

needed to take pictures more frequently of a football game than of a flower blooming. This general situation is shown in Figure 5.3, where the sample spacing is T_s seconds. Here, the sampling operation may be synchronized with the time instant $t = 0$. The quantity T_s is referred to as the **sampling period** and has units of seconds. The samples of the signal are spaced every T_s seconds apart. The more common term **sampling rate** (or **sampling frequency**), tells us how many samples or measurements we take every second. It is very easy to calculate the sampling rate if we know the sample spacing, or sampling period. The sampling rate is denoted by $f_s = 1/T_s$ and has units of samples per second. So, in Figure 5.2, the sampling period was 0.1 s, which means that the sampling frequency is $f_s = 1/0.1 = 10$ samples per second. So, for every second of $s(t)$ in Figure 5.2(b), we take precisely 10 samples or measurements.

Sampling Period (T_s): Spacing in time between two adjacent samples; $T_s = 1/f_s$.

Sampling Rate or Sampling Frequency (f_s): Number of samples per second, $f_s = 1/T_s$.

The samples of the signal are a sequence of numbers such as those shown in Table 5.1. We denote these numbers by $s[n]$, where

$$s[n] = s(nT_s) \tag{5.1}$$

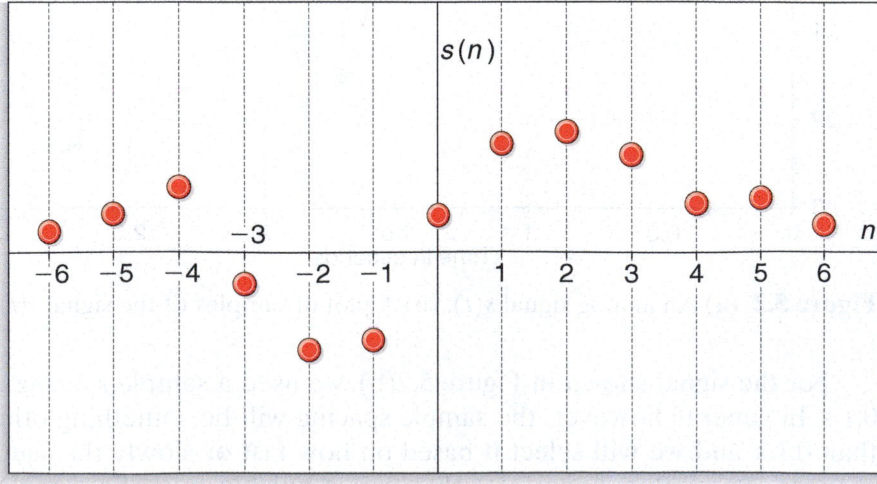

Figure 5.4 Samples of the signal in Figure 5.3.

and n can have the values 0, ±1, ±2, ±3, and so on. The use of square brackets in $s[n]$ indicates that it is a sequence of numbers, a function of a variable that is an integer like n, rather than an analog signal that is a function of a continuous variable like t. It follows that $s[0] = s(0)$, $s[1] = s(T_s)$, $s[2] = s(2T_s)$, $s[3] = s(3T_s)$, and so forth. Likewise, $s[-1] = s(-T_s)$, $s[-2] = s(-2T_s)$, and so on. We can plot samples of the signal in Figure 5.3 as the **sequence** in Figure 5.4.

Ordinarily, there is no simple mathematical function or formula describing the signal $s(t)$ (which might correspond to music) and its samples $s[n]$. However, to give us some experience in sampling, the following examples use some signals that have simple shapes.

Sequence: An ordered set of numbers. A sequence of samples is the set of numerical sample values of a signal.

INTERESTING FACT:

Music is stored on a CD as a list of numbers. These numbers are obtained by sampling, or measuring, the music 44,100 times per second. So, for each 3-minute piece of music recorded in stereo, there are 15,876,000 numbers stored on the CD.

EXAMPLE 5.1 Finding Values for $s[n]$

Suppose $s(t) = 2t - 4$ and $T_s = 0.5$. Find the values $s[n]$ of the sampled signal for $-1 \leq t \leq 1$, and determine the sampling frequency.

Solution

We have

$$s[n] = s(nT_s) = s(0.5n) = 2(0.5n) - 4 = n - 4$$

n	$t = nT_s$	$s[n] = n - 4$
−2	−1.0	$-2 - 4 = -6$
−1	−0.5	$-1 - 4 = -5$
0	0.0	$0 - 4 = -4$
1	0.5	$1 - 4 = -3$
2	1.0	$2 - 4 = -2$

Thus, $s[0] = -4$, $s[1] = -3$, $s[2] = -2$, $s[-1] = -5$, and $s[-2] = -6$. Try plotting this sequence to see that it outlines the straight-line shape of the analog signal $s(t) = 2t - 4$.

To determine the sampling frequency, we simply compute $f_s = 1/T_s = 1/0.5$, resulting in two samples per second.

EXAMPLE 5.2 Finding Values for $s[n]$

Suppose $s(t) = 7 - t^2$ and $T_s = 0.3$. Find the values $s[n]$ of the sampled signal for $-1 \leq t \leq 1$, and determine the sampling frequency.

Solution

We have

$$s[n] = s(nT_s) = s(0.3n) = 7 - (0.3n)^2 = 7 - 0.09n^2$$

Thus, $s[0] = 7$, $s[1] = 6.91$, $s[2] = 6.64$, $s[3] = 6.19$ $s[-1] = 6.91$, $s[-2] = 6.64$, and $s[-3] = 6.19$. To determine the sampling frequency for this example, we simply compute $f_s = 1/T_s = 1/0.3$, resulting in 3.33 samples per second.

Try plotting this sequence to see how well it outlines the upside-down parabolic shape of the analog signal $s(t) = 7 - t^2$.

Converting Numbers Back into Analog Information

We have learned that we can convert analog signals, such as music, into lists of numbers through the process of sampling. Sampling of a waveform is nothing more than taking measurements of the signal at regularly spaced intervals of time. However, we don't yet know how fast we need to take the measurements so that the digital signal, or list of numbers, becomes an accurate reflection of the original signal. To be more specific, we need to determine the appropriate sampling frequency so that we can build devices that can convert the lists of numbers back into the original analog signals, such as high-quality music or videos.

At first glance, it just doesn't seem possible to take a list of numbers obtained from sampling music and convert it back into the original music. The list of numbers only tells us the signal values at specific points in time. In order to be able to convert the numbers back into the analog signal, don't we need to know what the signal was doing at all points in time so we can make a continuous signal?

To help gain some intuition about good sampling rates, let's consider the sampling of a sinusoidal signal that has a frequency f and a period $T = 1/f$. After all, we know from Chapter 2 that complicated signals can be constructed as sums of sinusoids. So, if we can figure out the necessary sampling rate for a sinusoid, we also can determine the necessary rate for other signals made with sinusoids. Look at Figure 5.5. It shows a sinusoid sampled at a moderate rate of several samples per period. Here, T_s is the spacing between samples; $f_s = 1/T_s$ is the sampling frequency. Now compare Figure 5.5 with Figure 5.6. Figure 5.6 shows the same sinusoid sampled at a lower rate, which, in turn, means that the sampling interval T_s is larger.

If we were given the samples $s[n]$ in Figure 5.5 and told that they came from some sinusoidal signal having a frequency no higher than that shown, we would have little trouble perfectly re-creating the original analog signal $s(t)$ by drawing a sine wave through the samples.

However, in Figure 5.6, the sampling rate has been reduced to just two samples per period. Because $T_s = \frac{1}{2}T$, $f_s = 2f$ samples per second.

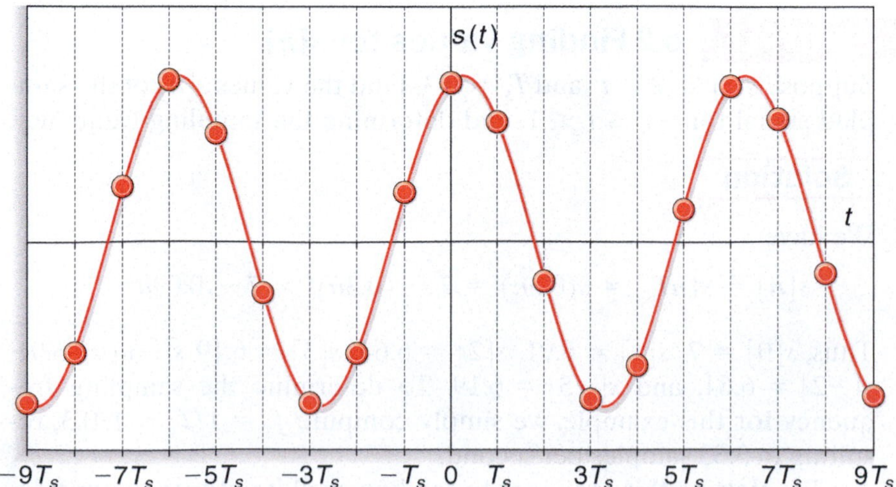

Figure 5.5 Sinusoid sampled at a moderate rate of several samples per period.

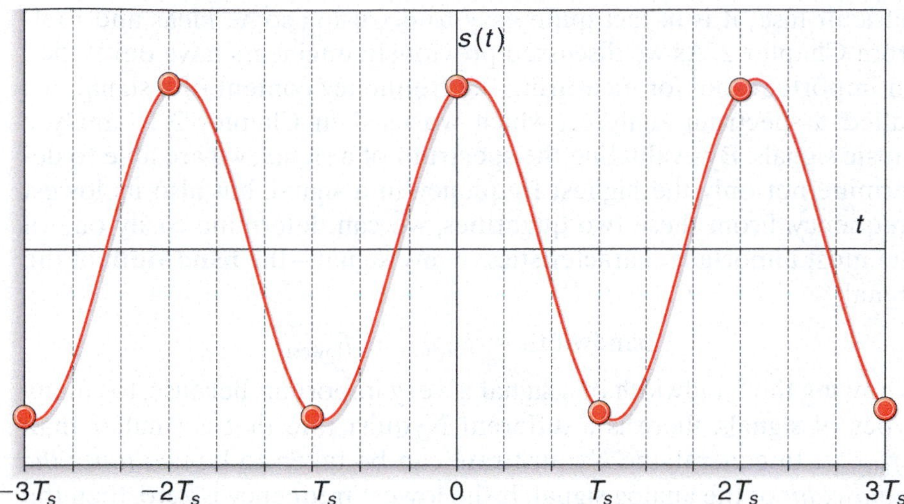

Figure 5.6 Sinusoid sampled at a lower rate of two samples per period.

Here, it looks like we might be able to come close to guessing the sine wave that goes precisely through these samples. But notice that if the sinusoid were shifted to the right slightly so that the sinusoidal signal were equal to 0 at time $t = 0$, then all of the sample times would occur at times when the signal had a value of 0. Therefore, all of the samples would be 0. In this case, we would not be able to tell the difference between the original sinusoid and a signal that was always 0.

Therefore, it seems that the sampling rate $f_s = 2f$, used in Figure 5.6, is on the borderline of the minimum number of samples per second that is necessary in order to be able to re-create the simple sine wave from a list of numbers. Thus, from our intuitive analysis of these figures, we suspect that the minimum sampling rate must be greater than two samples per period, or, equivalently, more than twice the frequency of the sinusoid, if we hope to recover the analog signal $s(t)$ from its samples $s[n]$.

What we have just observed leads us to one of the most powerful mathematical results of the digital era: the **Nyquist sampling theorem**, named after Harry Nyquist, the Bell Laboratories engineer and physicist who first discovered the result.

Nyquist Sampling Theorem: Specifies how fast we must sample a signal such as music or a movie in order to ensure that we can re-create the original signal from the samples.

The theorem is simply stated as follows: *A sampled signal can be converted back to its original analog signal without any error if the sampling rate is more than twice as large as the highest frequency of the signal.*

We can restate this result in a mathematical form as follows: If T_s is the spacing between samples and $f_s = 1/T_s$ is the sampling frequency, then the Nyquist theorem tells us that, theoretically, we can convert the samples of an analog signal back into the original signal if

$$f_s > 2 \times f_{\text{highest}}$$

where f_{highest} is the highest frequency contained in $s(t)$. The value $2 \times f_{\text{highest}}$ is called the **Nyquist rate** and can be determined for a wide range of different types of signals so that we know how fast we must sample various signals of interest.

Nyquist Rate: Minimum sampling rate needed for a signal or image. It equals twice the bandwidth of the analog signal.

How do we determine the Nyquist rate for signals such as music, speech, or even movies? We must first determine the highest frequency in the signal we are going to be digitizing. While this might seem like a difficult task, it is in fact quite easy to do, using some ideas and tools from Chapter 2. As we discussed previously, engineers have developed an important tool for measuring the frequency content of a signal. It's called a spectrum analyzer, which we used in Chapter 2 to analyze music signals. By evaluating the spectrum of a signal, we are able to determine not only the highest frequency in a signal, but also its lowest frequency. From these two quantities, we can determine easily one of the most important characteristics of any signal—the **bandwidth** of the signal:

Bandwidth: The difference between the highest and lowest frequencies contained in a signal. Often, the lowest frequency is zero. In this case, the bandwidth is just the highest frequency in the signal.

$$\text{bandwidth} = f_{\text{highest}} - f_{\text{lowest}}$$

Knowing the bandwidth of a signal is very important because, for many types of signals, there is a different Nyquist rate that is smaller than $2f_{\text{highest}}$. In general, the Nyquist rate can be taken to be *two times the bandwidth* of the analog signal. If the lowest frequency is zero, then the signal's bandwidth is equal to the highest frequency.

INTERESTING FACT:

The latest standard for high-quality sound in the home, DVD Audio, allows for a sampling rate of 96 kHz—more than twice that of CDs. While the Nyquist sampling theorem tells us that we need to sample music only 40,000 times per second, the much higher sampling rate used by DVD audio discs make the design of the player much easier and the resulting sound better than that heard on a standard CD.

EXAMPLE 5.3 Digitizing an FM Radio Signal

How can we digitize an FM radio signal?

Solution

The lowest-frequency FM station is at 88.1 MHz, while the highest frequency an FM station uses is 107.9 MHz. The FCC allocates a bandwidth of 200 kHz for each FM station. Therefore, to properly digitize an FM station, we must sample the signal 400,000 times per second.

In Table 5.2, we have listed the minimum rate at which we must sample in order to properly digitize some common signals. We can see that the easiest signal to digitize is the human voice used over telephones. Theoretically, this signal requires only 7000 samples, or measurements, every second, while the most challenging signal to digitize, a TV signal, requires 12,000,000 samples, or measurements, every second. In practice, signals usually are sampled slightly above the Nyquist rate, to account for system imperfections.

Table 5.2 also lists the bandwidths and actual sampling rates for the signals.

Extensions to Nyquist's First Result Mathematicians and engineers added to Nyquist's original sampling result by discovering precisely how to re-create the original signal from only its samples. They showed that if a signal $s(t)$ is sampled at a rate greater than two times

INTERESTING FACT:

Cell phones use sampled versions of our voices to transmit conversations.

Table 5.2 Minimum Sampling Rates for Various Signals

Signal Type	Signal Bandwidth	Minimum Sampling Rate	Rate Actually Used
Telephone-quality speech	3.5 kHz	7000 samples/s	8000 samples/s
Music	20 kHz	40,000 samples/s	44,100 samples/s
FM radio	200 kHz	400,000 samples/s	500,000 samples/s
Standard-definition television	6 MHz	12,000,000 samples/s	14,400,000 samples/s

the bandwidth, then it can be exactly reconstructed from its samples. In fact, there is a mathematical formula for reconstructing the signal. We won't worry too much about this formula here, other than to say that there *is* a formula and that it can be implemented in a very practical way. In fact, CD players contain digital-to-analog (D/A) converters that basically implement this formula.

KEY CONCEPT

Digital-to-analog (D/A) converters are part of many digital devices and serve to re-create the original signal from samples of the signal.

The implications of the Nyquist sampling theorem are nothing short of remarkable. We can go back and forth between a sound signal and its sampled representation with essentially no loss in quality. We also can go back and forth between an analog image and the pixel-by-pixel representation of the image with essentially no loss in quality. Since it is easy to process the digital versions of these signals by using simple math—for example, to remove noise or emphasize certain features—we often manipulate signals in the digital world and then produce a high-quality reconstruction, using a D/A converter. The procedure of sampling a signal, manipulating the sequence of numbers that results from sampling, and producing an analog output signal after the manipulations is called **digital signal processing (DSP)**.

Digital Signal Processing (DSP): The manipulation or processing of sampled signals.

Bad Effects of Sampling Too Slowly

We have seen that band-limited signals can be reconstructed perfectly from their samples, as long as the sampling rate is greater than twice the bandwidth of the signal that was sampled. So you might ask yourself: What happens if we unfortunately sample a signal too slowly and thereby fail to meet the requirement of the Nyquist sampling theorem?

To address this question, we need to learn a bit more about the Nyquist sampling theorem. At the heart of this theorem is the following important fact: If we have sampled the original signal at a rate higher than the Nyquist rate, then there is only one analog signal that satisfies the measurements $s[n]$ and has the bandwidth specified by $f_{\text{highest}} - f_{\text{lowest}}$. As we said earlier, engineers and mathematicians have determined the precise formula for finding that signal, which, fortunately, turns out to be the original signal! This is why your CD player is able to re-create the original music so precisely from just the samples.

However, if we don't sample fast enough, it turns out that there are many signals that pass through the samples $s[n]$. This situation is shown in Figure 5.7. In Figure 5.7(a), we see a signal sampled fast enough so that we can determine the actual original signal. However, in Figure 5.7(b), we have unfortunately sampled too slowly, and as a result, we can see that there are many possible signals that can go through these samples and still fit within our frequency bandwidth limits. The problem is that we don't know which of these various signals is the correct signal—so, in general, there will be very large errors when we attempt to re-create the original signal from the samples.

For example, when digitizing music, if we sampled slower than the Nyquist rate and then attempted to play this result back on a CD player, we would hear "junk," or, more precisely, some signal that wasn't the original music—not a very satisfying engineering solution. This problematic situation is called **aliasing** and is something that engineers attempt to avoid when digitizing all signals. Aliasing refers to one signal "pretending to be" another signal when their samples are the same.

Aliasing: An undesired effect due to undersampling whereby one signal can masquerade as another.

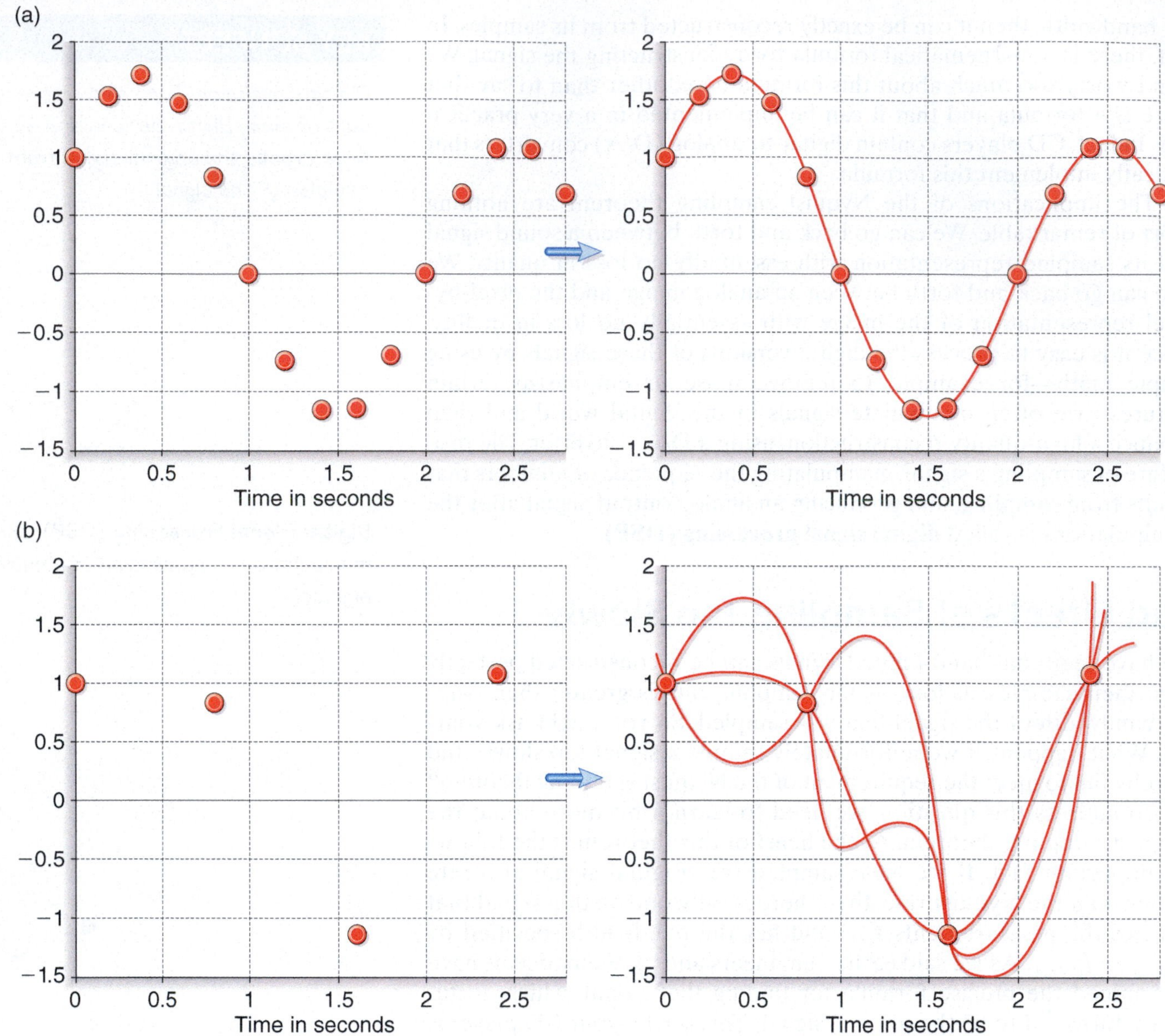

Figure 5.7 (a) Signal sampled above the Nyquist rate and reconstructed. (b) Signal sampled below the Nyquist rate. Many band-limited signals with bandwidth $> f_s/2$ that pass through the samples can be reconstructed.

EXAMPLE 5.4 Aliasing

When sampling sinusoids, the effect of sampling above the Nyquist rate is easy to describe. But what about a particular sinusoid that is sampled below the Nyquist rate?

Solution

Figure 5.8 shows 18 periods of a 720-Hz sine. The Nyquist rate for this signal is 1440 Hz (2 × 720 Hz). Suppose that we sample this signal at 660 samples per second—far below the Nyquist rate. The resulting samples are denoted by the dots in Figure 5.8. Quite surprisingly, these samples take on the shape of a sinusoid, but the frequency represented by the samples is far different from the original analog frequency of 720 Hz! And, in fact, if we attempted to re-create the original signal from these samples, we would create a 60-Hz sine wave, not a 720-Hz sine wave. This effect is called aliasing,

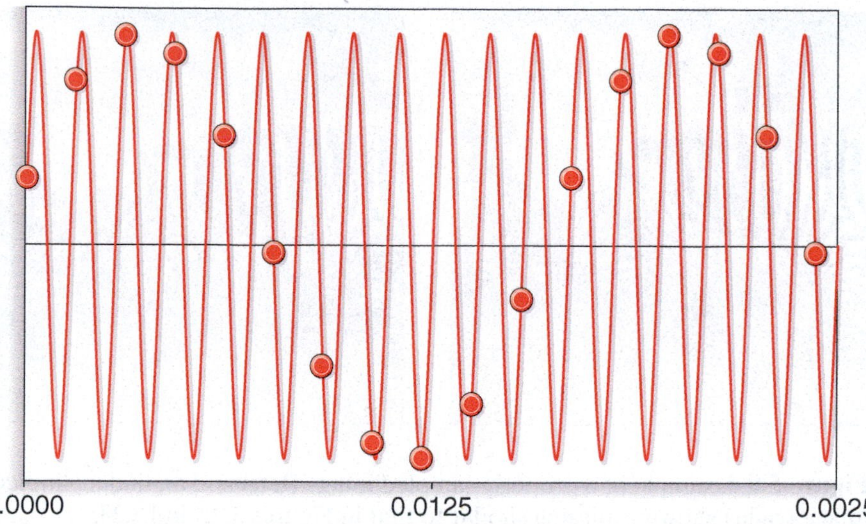

Figure 5.8 Sampling a 720-Hz sinusoid at the rate of 660 Hz. Note that the samples appear to be coming from a sinusoid at a much lower frequency.

because one frequency (720 Hz) is masquerading as another (60 Hz).

One of the best ways to experience this phenomenon is to listen to signals that have been aliased.

As we saw in Chapter 3, the bad effects of sampling too slowly sometimes occur in images and movies, except that in this case the effect is the appearance of visually distorted patterns, or artifacts. These distorted patterns are sometimes called **Moire patterns**. Because the number of lines in a TV image is a fixed standard, the sampling rate for TV images cannot be changed, so people on TV who wear narrowly striped ties, for example, might be surprised at how their ties look when the image is broadcast. Figure 5.9 shows Moire patterns caused by a sample spacing that is too large for the narrow stripes on the zebras. Not enough samples are collected to accurately represent the image.

Moire Pattern: A visual pattern or artifact caused by spatial aliasing due to insufficient sampling of an image or video.

In video, aliasing can occur both within individual image frames and from one frame to the next. If a scene contains objects that are

Infinity Project Experiment: Aliased Sinusoids, Speech, and Music

Aliasing, caused by sampling too slowly, changes the frequencies of signal components. For a sinusoidal signal, which contains a single frequency, aliasing causes that frequency to shift. For speech or music signals, which contain many frequencies simultaneously, the effect of aliasing is more complicated. In speech or music, aliasing causes the many frequencies to shift by varying amounts. The resulting error, or noise, is the sum of the aliasing errors from the individual frequencies and is often hard to predict.

Listen to a set of signals that have been sampled and then reconstructed from their samples. When the signals are sampled above the Nyquist rate, there is no aliasing and the signals sound fine. However, when sampled below the Nyquist rate, the signals are aliased. Discover what the results of aliasing sound like in the analog world. Can you hear why we must sample fast enough to avoid aliasing?

Figure 5.9 Example of a properly sampled image (left) and an undersampled image (right) showing aliasing similar to that in Figures 3.22 and 3.23.

moving, and if the scene is recorded at an insufficient frame-sampling rate, then aliasing can cause bizarre and sometimes amazing effects. As described in Chapter 3, a visual demonstration of such aliasing is present in many films that depict a rotating wheel of a wagon or car. As you know, movie cameras capture only 24 image frames per second. In most cases this sampling rate is too slow for the movie to depict a quickly rotating wagon wheel without aliasing. This effect is complicated by the fact that a wagon wheel is symmetric, with many spokes. It turns out that the necessary camera frame rate f_s needed to avoid aliasing is directly related to the number of spokes and the number of revolutions per second at which the wheel rotates. An example will help illustrate this concept.

INTERESTING FACT:

Digital cameras intentionally blur the scene prior to spatial sampling in order to limit the severity of aliasing in the form of Moire patterns.

EXAMPLE 5.5 Capturing the Image of a Rotating Wheel

Suppose a wheel with 12 spokes is rotating at 1.5 revolutions per second. Will a camera frame rate of 24 Hz cause aliasing?

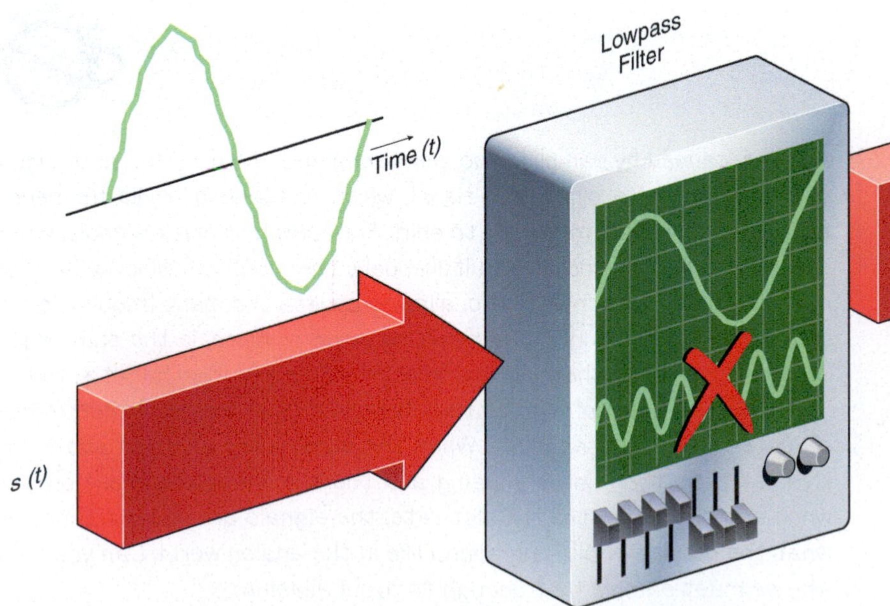

Figure 5.10 Block diagram showing an antialiasing lowpass filter that will prevent aliasing.

Solution

A single spoke of the wheel will move at a rate of $12 \times 1.5 = 18$ spoke positions per second. To capture this motion accurately, we need to sample at twice this rate, or 36 frames per second. Because the camera is not collecting enough frames per second, aliasing will occur. The resulting movie sequence may not show the wheel rotating in the right direction, and the wheel's rotation speed will be off as well.

Band-Limited Signal: A signal whose highest frequency falls below some finite value.

Lowpass Filter: A device that removes all frequencies in a signal above some selected frequency and passes all others.

Antialiasing Filter: An analog lowpass filter that precedes the sampler in order to ensure that the analog signal is band limited prior to sampling.

Limiting the Effects of Aliasing Aliasing is a serious effect that must be avoided if we are to properly digitize information. Aliasing causes frequencies to change, whether the signals are sounds, images, or video. For typical signals, images, and time-varying scenes that are composed of many frequencies, the effects are very complex, so we have to take measures to prevent aliasing. In most cases where we can measure the bandwidth of $s(t)$ and where we can choose the sampling rate, we can avoid aliasing by simply sampling faster than the Nyquist rate. However, if $s(t)$ is not **band limited**—meaning that the signal has an infinitely broad range of frequencies in it—or if we are not able to change the sampling rate—then we must first pass $s(t)$ through a device called a **lowpass filter**. This filter will remove frequency components above one-half the sampling frequency. Such a filter is called an **antialiasing filter**. This filtering operation

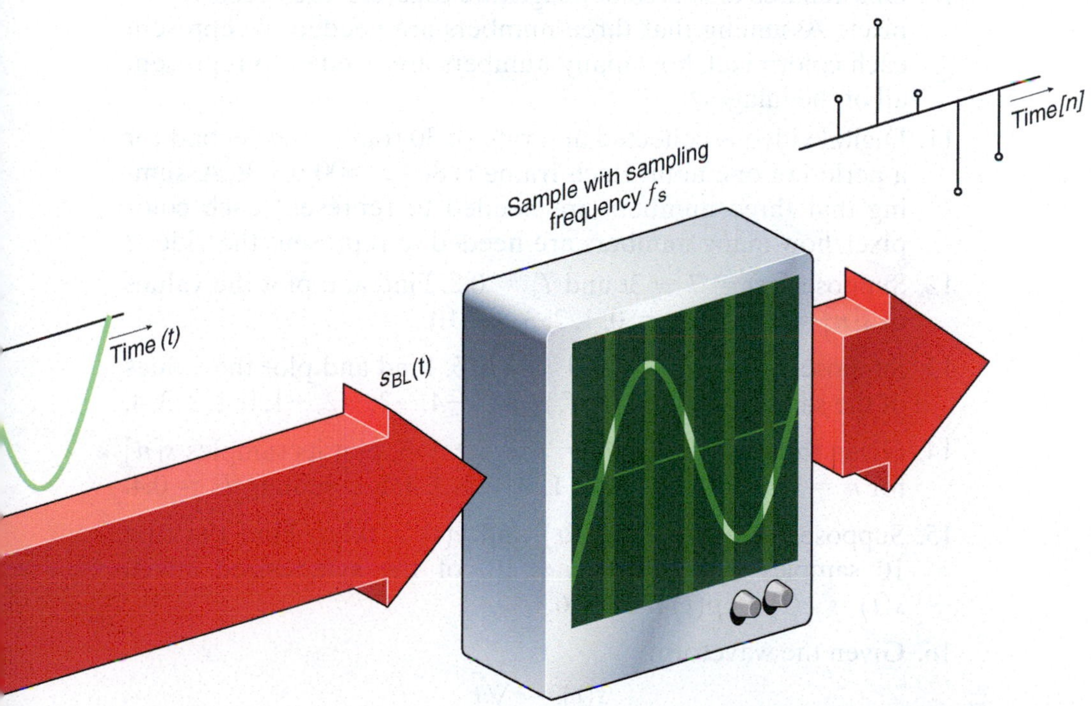

will cause the signal to be different, because it will eliminate the high frequencies in $s(t)$. However we would not be able to use these high frequencies anyway, because they are above the Nyquist limit. Removing them prevents them from corrupting the lower frequencies that we can use.

The overall scheme is shown in Figure 5.10. Here, $s(t)$ is the original analog signal, which is not band-limited. The antialiasing filter removes the high frequencies in $s(t)$, creating the new, band-limited, analog signal $s_{BL}(t)$. This signal is then sampled faster than the Nyquist rate, producing the set of samples $s_{BL}[n]$.

EXERCISES 5.2

Mastering the Concepts

1. What is the definition of "Nyquist rate"?
2. What is the definition of "band limited"?
3. What is the "Nyquist sampling theorem"?
4. What does "DSP" stand for? What is DSP?
5. What happens when a sinusoid is sampled exactly at the Nyquist rate? What happens if the sampling rate is below the Nyquist rate?
6. What is an antialiasing filter?
7. In a digital camera, how is aliasing avoided?

Try This

8. A speech signal is sampled at a rate of 8000 samples per second for a duration of two minutes. How many numbers are needed to represent the speech samples?
9. A music signal is sampled at 44,100 samples per second for a duration of 30 minutes. How many numbers are needed to represent the music samples? How does your answer change for a stereo music signal?
10. One hundred digital, color images are collected, each 1200 × 900 pixels. Assuming that three numbers are needed to represent each color pixel, how many numbers are needed to represent all of the images?
11. Digital video is collected at a rate of 30 frames per second for a period of one hour. Each frame is 800 × 600 pixels. Assuming that three numbers are needed to represent each color pixel, how many numbers are needed to represent the video?
12. Suppose $s(t) = 7 - 3t$ and $T_s = 0.2$. Find and plot the values of $s[n] = s(nT_s), n = 0, 1, 2, 3, \ldots, 10$.
13. Suppose $s(t) = t^2 + 1$ and $T_s = 0.5$. Find and plot the values of the samples $s[n] = s(nT_s), n = -4, -3, -2, -1, 0, 1, 2, 3, 4$.
14. Given the waveform $s(t) = t^2 - 4t + 3$, plot its samples $s[n]$ for $n = -5, -4, -3, -2, -1, 0, 1, 2, 3, 4, 5$, assuming $T_s = 0.4$.
15. Suppose $p(t) = 0.4\sin(0.2t)$ and $e(t) = 0.1/t$. Find the first 10 samples $s[nT_s]$, $1 \leq n \leq 10$, of the synthesized signal $s(t) = e(t) \times p(t)$ if $T_s = 6$.
16. Given the waveform
$$s(t) = \sqrt{t}$$
for $t > 0$, plot the samples $s[n]$ for $n = 0, 1, 2, 3, \ldots, 12$, assuming $T_s = 1.2$.
17. Suppose you recorded 4.5 seconds of sound, using a sampling frequency of 8 kHz. How many samples would you have measured? How about if you used a 44.1-kHz sampling frequency?
18. Telephone speech is band limited to approximately 3500 Hz. What is the minimum sampling rate that will avoid information loss for this signal?

19. In a recording studio, music is band limited intentionally to 20 kHz, since that is the upper range of human hearing for everyone but the youngest of babies. What is the minimum sampling rate for digital music?
20. The bandwidth of the audio-signal broadcast on AM radio is 5 kHz. Commercial AM radio is currently analog. If AM radio were to go digital, what would be the minimum sampling rate needed for the audio signal?
21. The bandwidth of the audio signal before being broadcast on FM radio is 15 kHz. Commercial FM radio is currently analog. If FM radio were to go digital, what would be the minimum sampling rate needed for the audio signal?
22. Suppose $s(t) = \cos(4\pi t)$ and $T_s = 0.1$ s. Find and plot the values of $s[n] = s(nT_s)$, $n = 0, 1, 2, \ldots, 10$. When sampling with this period T_s, are you sampling above the Nyquist rate?
23. Given the signal $s(t) = \sin(4\pi t)$, plot its samples, assuming $T_s = 0.3$ s. What has happened? Have you sampled above the Nyquist rate?
24. Suppose $s(t) = \cos(100\pi t)$ and $T_s = 0.003$ s. Find and plot the values of $s[n] = s(nT_s)$, $n = 0, 1, 2, \ldots, 10$. When sampling with this period T_s, are you sampling above the Nyquist rate?
25. Given the signal $s(t) = \sin(100\pi t)$, plot its samples, assuming $T_s = 0.01$ s. What has happened? Have you sampled above the Nyquist rate?
26. Suppose a sinusoid of frequency 1 kHz is sampled above the Nyquist rate at 2200 samples per second, to produce a sequence $s[n]$. Find two other analog frequencies, each higher than 1 kHz, that would alias to this same sequence.
27. You are making a movie set in the "roaring 1920s." As part of the movie, you are showing an old phonograph playing a 78-rpm (revolutions per minute) record. The record has a label in its middle. How fast must the frame rate be in order to avoid aliasing of the rotating record?
28. You are making a movie set in the 1950s. As part of the movie, you are showing a jukebox playing a 45-rpm (revolutions per minute) record. If you use a frame rate of 24 frames per second, how many frames does it take to describe one revolution of the record? Is the signal aliased?
29. You are making a television show set in the 1970s. As part of the show, you are filming a phonograph changer playing a disco record. If your 30-frames-per-second video shows one revolution of the record every 54 frames, how many revolutions per minute do you think the record is spinning? Can you tell for sure?
30. A CD player uses a variable-speed motor that initially spins at about 500 rpm (revolutions per minute) and then slows down to about 200 rpm near the end of the CD. If you are making a movie of this rotation, how many frames per second do you need in order to avoid aliasing?

In the Laboratory

31. Cut out a disc from a piece of cardboard. Using a marker, put a dot near the edge of the disc. Attach the cardboard disc to a rotating motor with a controllable speed, such as a power drill. Now, film the spinning cardboard disc, using a Web camera attached to a computer. Vary the spinning rates of the motor. What do you see? Can you find a rate that makes the disc seem nearly stationary in the video camera? Can you find a second rate that does the same thing? Repeat this experiment with a handheld video camera. Most video cameras use a frame rate of 30 frames per second to record video. How fast is your disc spinning?

Back of the Envelope

32. Suppose that you will be recording piano music. Based on the frequency of the highest note on the keyboard (which you learned how to compute in Chapter 2), and assuming that each piano tone contains significant energy in its fundamental frequency plus its first eight harmonics, what would be the minimum sampling rate needed to digitize piano music faithfully?

5.3 Binary Numbers—The Digital Choice

Bits, Bits, and More Bits

Bit: Contraction of "***bi***nary digi***t***." One of two physical states, typically thought of as a 0 or a 1. The smallest unit of information recognized by a computer or digital device.

In earlier chapters, we learned that computers and other digital devices represent information in a binary number code where the smallest unit is called a **bit** (short for "**bi**nary digi**t**"). And as we know, a single bit can represent only one of two states, which we think of as 0 and 1. If we want to represent more than just two possible states or numbers, we group many bits together.

Engineers have chosen the binary number system because the basic physical building block of nearly all digital technologies is the transistor, which can be designed to act like a switch having two distinct states that can be controlled by currents and voltages. Let's take a look at some common digital storage devices and see how they work:

- *Semiconductor memory* is sometimes called random access memory (RAM) because data can be accessed in any order at the same speed. RAM is constructed of millions of small transistor circuits on a chip (integrated circuit), where each small circuit can maintain either a high voltage or a low voltage, depending on an applied control signal. The key is that there are only two possible voltage levels—high and low, where high is about 3 volts and low is less than 1 volt (assuming a 3-volt power supply). Thus, the state of each circuit (high versus low voltage) can represent only two different numbers.
- *Magnetic disks*, such as the hard drive and floppy drive in a computer, work on principles of magnetism. Each designated location on the

disk can be magnetized in one of two directions, depending on the value of an applied control signal. This magnetization is sensed by the disk head, which rides over the spinning disk. Once again, only two states are possible at each physical location on the storage device.

- *Optical discs*, like CDs and DVDs, have designated microscopic metallic spots underneath their plastic surfaces, where each spot is either pitted or flat. Laser light is reflected off the shiny surface, and, depending on the amount of light returned to a sensor, the presence or absence of a pit is detected. As is the case with RAM and magnetic drives, these disks store only one of two possible states at each physical location on the device.

To simplify our discussion of all these devices, we refer to the two states as 0 and 1. We could call them *A* and *B*, or any other two names, but we will use 0 and 1, which is the standard nomenclature.

Since nearly all computer memory devices can store only 0's and 1's, the computer must use binary number representation and binary arithmetic. For this reason, it is important to understand the binary number system. We had a brief introduction to binary numbers in Chapter 1; however, to complete our understanding of the digitization of information, we now need to know more about them.

Binary Numbers

So far in this chapter, we have assumed that all of the samples, or measurements, of signals, such as music or pictures, can be stored as numbers in a computer. For example, a given speech sample may take on the value 4.6 or −0.5031. However, we have learned that computers and digital memories store only bits. So, how can numbers such as 4.6 and −0.5031 be represented by bits?

The answer is that computers and digital storage devices use the **binary number system** rather than the decimal number system, with which we are familiar. A binary number groups together the values of many consecutive bits, each of which is either a 0 or 1, to represent a number such as 4.6. Representing numbers as bits affects how numbers are stored and how math is done with the numbers.

Binary Number System: An arithmetic system for representing numbers as a series of bits.

As we learned in Chapter 1, the binary number system has only two symbols, unlike the decimal system, which has 10. The binary number system provides a way to represent a wide range of numerical values by using devices that can exhibit only two states at any specific location, such as high voltage or low voltage on RAM, left or right magnetization on a floppy disk, or pit versus no pit on a CD or DVD. But in all cases, a single bit can have a value of only 0 or 1.

KEEP IN MIND

The binary number system has only two symbols—0 and 1. We write larger numbers with groups of 0's and 1's.

Two bits together can represent (0,0), (0,1), (1,0), or (1,1), that is, four different possibilities. Thus, two bits can represent four different states or values. For images, these values represent the colors black, dark gray, light gray, and white. Likewise, three bits together can represent (0,0,0), (0,0,1), (0,1,0), (0,1,1), (1,0,0), (1,0,1), (1,1,0), or (1,1,1), that is, eight different possibilities. Thus, three bits can represent eight different states or values.

In general, B bits, where B is a positive integer, can represent 2^B states or values. Therefore, four bits can represent 2^4, or 16, states or values; five bits can represent 2^5, or 32, states or values; and so forth.

Base, or Radix: Identifies the type of number system being used. The weights assigned to different digits in a number are the base raised to integer powers. The base is 10 for the decimal number system and 2 for the binary number system.

KEEP IN MIND

Unlike Roman numerals, binary and decimal digits can have different values depending on their position in the number. This attribute makes the numbers much simpler to read and to use for arithmetic.

Roman	Decimal
I	1
X	10
C	100
M	1000

Positive Binary Integers

In the decimal number system, the number 10 is the **base**, or **radix**. As we know, the integer 238 takes the value 2(100) + 3(10) + 8(1), where 100, 10, and 1 represent the values of the positions of the digits 2, 3, and 8, respectively. Here, the value of each position is 10 times the value of the position to its right, with the rightmost position having a value of 1. This example is illustrated in Figure 5.11.

In the binary number system, the number 2 is the base, and each binary number is a string of bit values that can be either 0 or 1. The value of each bit position in the string is two times the value of the position to its right, with the rightmost position having a value of 1.

EXAMPLE 5.6 Converting a 4-bit Number

Convert the 4-bit number 1101_2 from binary to decimal. Here, we have used the subscript 2 to indicate a binary, or base-2, number.

Solution

The decimal values of the bit positions are 8, 4, 2, and 1, respectively. So, the 4-bit number 1101_2 takes the value 1(8) + 1(4) + 0(2) + 1(1) = 13. Figure 5.12 shows the binary representation of the decimal number 13.

EXAMPLE 5.7 Converting a 5-bit Number

Convert the 5-bit number 10101_2 from binary to decimal.

Solution

The decimal values of the bit positions are 16, 8, 4, 2, and 1 respectively. Thus, the 5-bit number 10101_2 takes the value 1(16) + 0(8) + 1(4) + 0(2) + 1(1) = 21.

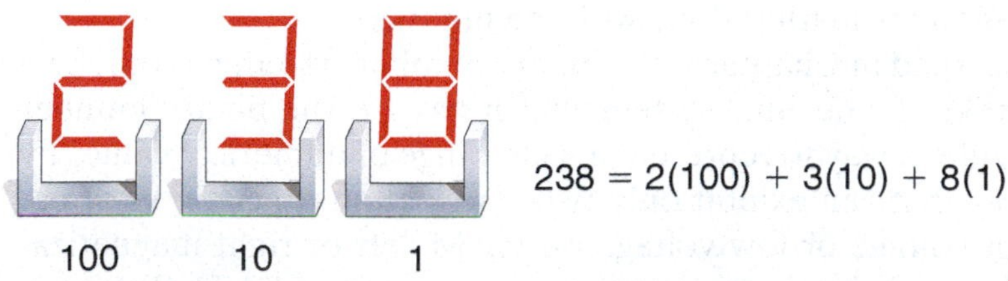

Figure 5.11 Decimal representation of the number 238.

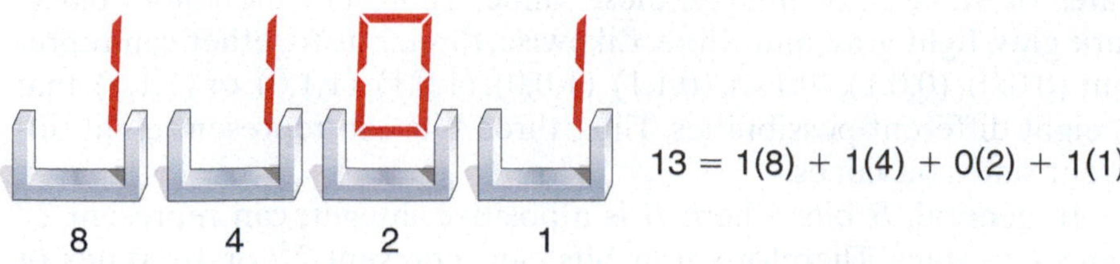

Figure 5.12 Binary representation of the decimal number 13.

EXAMPLE 5.8 Converting a 6-bit Number

Convert the 6-bit number 101100_2 from binary to decimal.

Solution

The decimal values of the bit positions are 32, 16, 8, 4, 2, and 1, respectively. So, the 6-bit number 101100_2 takes the value $1(32) + 0(16) + 1(8) + 1(4) + 0(2) + 0(1) = 44$.

How would we represent the decimal number 238 in binary form? This representation will require quite a few binary digits! As we know, each binary digit represents a power of two. We can write 238 as some combination of 128, 64, 32, 16, 8, 4, 2, and 1, which are each a power of two. As you can easily prove to yourself, the particular combination we need is $238 = 1(128) + 1(64) + 1(32) + 0(16) + 1(8) + 1(4) + 1(2) + 0(1)$, which is 11101110_2 in binary form. Figure 5.13 illustrates this binary representation.

To help with conversions between decimal and binary representations of integers, all 5-bit binary numbers and their decimal equivalents are shown in Table 5.3. Table 5.4 shows examples of some 8-bit and 10-bit binary numbers. For the 8-bit numbers in Table 5.4, the value of the leftmost bit position is $2^7 = 128$, so that the bit values are (128, 64, 32, 16, 8, 4, 2, 1). For the 10-bit numbers, the value of the leftmost bit position is $2^9 = 512$, so that the bit values are (512, 256, 128, 64, 32, 16, 8, 4, 2, 1). The largest integer that can be represented by a B-bit binary number has all bits set to 1 and has a value of $2^B - 1$.

Conversion from Decimal to Binary

As we just learned, it is quite easy to convert a binary number such as 101 into the corresponding decimal number (in this case 5). We only have to add up the powers of 2 corresponding to the bits that have a value of 1. What about the other way around? We can't always look up the values in a long table like Table 5.3. How can we convert decimal numbers into binary form, so that we may store and manipulate them in a computer or other digital device?

There are efficient algorithms for making the conversion from decimal to binary representation, based on the structure of binary representation. We also can use the simple method demonstrated in Chapter 1. We will now discuss the straightforward method presented in the first chapter in more detail and discover a way of easily converting from decimal integers to binary integers. An example will help to get us started.

Suppose we wish to convert the decimal number 213 to binary. The number 213 lies between $128 = 2^7$ and $256 = 2^8$, so we will need 8 bits to represent this number. Our job is to find the bit values B_0 through B_7 satisfying

$$B_7(128) + B_6(64) + B_5(32) + B_4(16) + B_3(8) + B_2(4) + B_1(2) + B_0 = 213$$

1 1 1 0 1 1 1 0

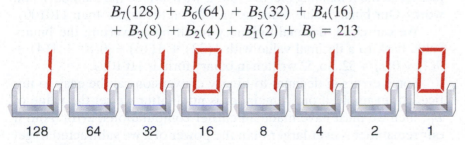

128 64 32 16 8 4 2 1

Figure 5.13 Binary representation of the number 238.

Table 5.3 The First 32 Integers Expressed in Binary and Decimal Form

Binary	Decimal
00000	0
00001	1
00010	2
00011	3
00100	4
00101	5
00110	6
00111	7
01000	8
01001	9
01010	10
01011	11
01100	12
01101	13
01110	14
01111	15
10000	16
10001	17
10010	18
10011	19
10100	20
10101	21
10110	22
10111	23
11000	24
11001	25
11010	26
11011	27
11100	28
11101	29
11110	30
11111	31

Table 5.4 Some Larger Integers Expressed in Binary and Decimal Form

Binary	Decimal
01011011	91
10000000	128
10101100	172
11010110	214
11111111	255
0111001100	460
1000000000	512
1011001010	714
1100101011	811
1111111111	1023

Most Significant Bit: In a binary number, the leftmost 0 or 1, because it has the most weight in the binary representation.

Least Significant Bit: In a binary number, the rightmost 0 or 1.

We refer to B_7 as the **most significant bit**, since it carries the most weight in the binary representation, and we call B_0 the **least significant bit**, since it carries the least weight. The position of the most significant bit, B_7, has value $128 = 2^7$, which is smaller than 213. So, 128 is "contained in" 213, and we set B_7 equal to 1. Subtracting 128 from 213 gives the remainder 85. This remainder needs to be represented by the remaining bits. The position of bit B_6 has value $64 = 2^6$, which is smaller than the remainder 85. Thus, 64 is "contained in" 85, and we set B_6 to 1. Subtracting 64 from 85 gives the next remainder, 21. Bit position B_5 has value $32 = 2^5$, which is too large to use in representing the remainder 21. Thus, we set B_5 to 0. The position of B_4 has value $16 = 2^4$, which is smaller than the remainder 21. So, we set B_4 to 1 and subtract 16 from 21 to give the remainder 5. The position of B_3 has value $8 = 2^3$, which is too large to use in the representation of the remainder 5. Thus, we set B_3 to 0. The position of B_2 has value $4 = 2^2$ which is smaller than 5. Setting B_2 to 1 and subtracting 4 from 5 gives the remainder 1. The position of B_1 has value $2 = 2^1$, which is larger than the remainder 1, so we set B_1 to 0. Finally, we set B_0 to 1 to exactly represent the remainder 1. The result is $213 = 11010101_2$. The overall procedure used in this example is illustrated in Figure 5.14.

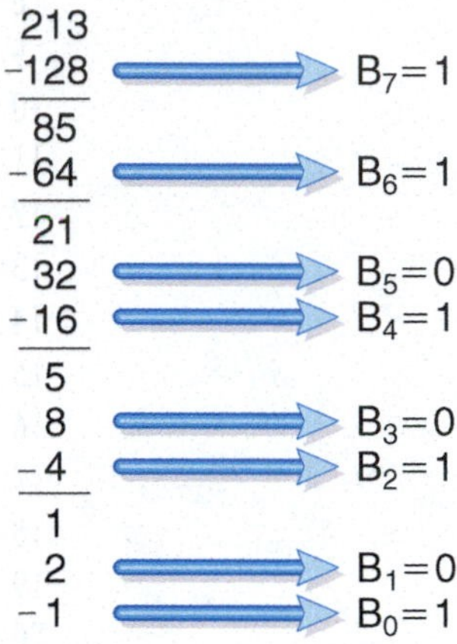

Figure 5.14 Conversion of the decimal number 213 to the binary number 11010101_2.

Let's try two more examples to gain further practice in decimal-to-binary conversion.

EXAMPLE 5.9 Converting a Decimal Number to Binary Form

Convert the decimal number 52 to binary form.

Solution

Using the algorithm described previously, we start by finding a power of 2 larger than our decimal number 52. Since $64 = 2^6 > 52 \geq 2^5 = 32$, we will need 6 bits for our binary value and we will start by finding B_5. Our calculations, for this entire process are as follows:

$$\begin{array}{rl} 52 - 32 = 20 & B_5 = 1 \\ 20 - 16 = 4 & B_4 = 1 \\ 4 - 8 < 0 & B_3 = 0 \\ 4 - 4 = 0 & B_2 = 1 \\ 0 - 2 < 0 & B_1 = 0 \\ 0 - 1 < 0 & B_0 = 0 \end{array}$$

Note that we could have stopped at B_2 when our remainder was 0, and just set all the remaining bits to 0, but if we continue the method, it still works. Our binary value, starting with B_5 on the left, is then 110100_2.

We can check decimal-to-binary result by converting the binary value back to a decimal value with $1(32) + 1(16) + 0(8) + 1(4) + 0(2) + 0(1) = 52$. So, 52 written in binary form is 110100_2.

Some errors in decimal-to-binary conversion can be easy to notice. For example, if our remainder is not 0 after we set the value of B_0, then we must have made an earlier computational error. And if our remainder is ever larger than the power of 2 we subtracted to get it, then we must have left out one of the more significant bits.

EXAMPLE 5.10 Another Decimal-to-Binary Conversion

Convert the decimal number 79 to binary form.

Solution

Using the algorithm described previously, we start by finding a power of 2 larger than our decimal number 79. Since $128 = 2^7 > 79 \geq 2^6 = 64$, we will need 7 bits for our binary value, and we will start by finding B_6. Our calculations for this entire process are as follows:

$$\begin{array}{rl} 79 - 64 = 15 & B_6 = 1 \\ 15 - 32 < 0 & B_5 = 0 \\ 15 - 16 < 0 & B_4 = 0 \\ 15 - 8 = 7 & B_3 = 1 \\ 7 - 4 = 3 & B_2 = 1 \\ 3 - 2 = 1 & B_1 = 1 \\ 1 - 1 = 0 & B_0 = 1 \end{array}$$

Our binary value, starting with B_6 on the left, is then 1001111_2. We can check our result by converting the binary value back to a decimal value with $(64) + 0(32) + 0(16) + 1(8) + 1(4) + 1(2) + 1(1) = 79$. So, 79 written in binary form is 1001111_2.

You may wonder how computers carry out arithmetic operations such as addition and multiplication with the numbers represented in binary form. There is no need (or desire!) to convert numbers back to decimal form before performing computations. The procedures for binary arithmetic are completely analogous to those for decimal arithmetic. The main difference is that, when adding a column of binary numbers, as soon as the total is 2 or greater, there must be a carry, because the only binary digits are 0 and 1. With decimal numbers, there is no carry until the total reaches 10 or greater.

Binary Fractions

In most cases, the numbers that we want to store in a computer or digital device are not integers. Usually, there is a fractional part. Fortunately, the binary representation of positive fractions, or positive integers plus fractions, is similar to the representation of positive integers.

In the decimal world, we have a decimal point that separates the fractional part from the integer part. In the binary world, we use the same point notation, but we call it the **binary point**. The binary point is similar to a decimal point. It separates the integer and fractional parts of a binary number. The binary number 1101.101_2 represents the decimal number $1(8) + 1(4) + 0(2) + 1(1) + 1(\frac{1}{2}) + 0(\frac{1}{4}) + 1(\frac{1}{8}) = 13.625$. This example is illustrated in Figure 5.15.

Binary Point: Notation used to separate the integer and fractional parts of a binary number.

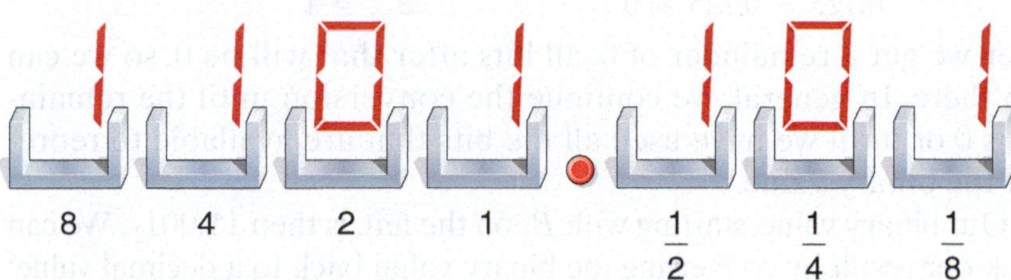

Figure 5.15 Binary representation of 13.625.

When binary fractions are involved, the value of the position to the left of the binary point is 1, and the values of the positions to the right then decline by factors of 2, giving position values $\frac{1}{2}, \frac{1}{4}, \frac{1}{8}$, and so forth. Movement of the binary point P places to the right, where P is any positive integer, corresponds to multiplying the number by 2^P. Likewise, movement of the binary point P places to the left corresponds to division by 2^P.

See Table 5.5 for several binary fractions and their decimal equivalents. Notice that the numbers in the bottom four rows are one-half of those in the top four rows, respectively, because the binary point has been shifted one place to the left.

Table 5.5 Some Fractional Numbers Expressed in Binary and Decimal Form

Binary	Decimal
100.01	4.25
011.10	3.5
110.00	6.0
111.11	7.75
10.001	2.125
01.110	1.75
11.000	3.0
11.111	3.875

EXAMPLE 5.11 Converting a Binary Number with a Binary Point to Decimal Form

Convert 11.01_2 from binary to decimal form.

Solution

$11.01_2 = 1(2) + 1(1) + 0(\frac{1}{2}) + 1(\frac{1}{4}) = 3.25.$

EXAMPLE 5.12 Another Conversion with a Binary Point

Convert 100.101_2 from binary to decimal form.

Solution

$100.101_2 = 1(4) + 0(2) + 0(1) + 1(\frac{1}{2}) + 0(\frac{1}{4}) + 1(\frac{1}{8}) = 4.625.$

EXAMPLE 5.13 Writing a Decimal Fraction in Binary Form

Write the decimal fraction 3.125 in binary form.

Solution

The method we used for converting decimal integers to binary integers works for fractions also. Using that algorithm, we start by finding a power of 2 larger than our decimal number 3.125. Since $4 = 2^2 > 3.125 \geq 2^1 = 2$, we will start with B_1. However, in this case, we don't know how many bits we will need, because we have to represent the fractional part as well as the integer part. Our calculations for this entire process are as follows:

$$\begin{array}{rr} 3.125 - 2 = 1.125 & B_1 = 1 \\ 1.125 - 1 = 0.125 & B_0 = 1 \\ 0.125 - 0.5 < 0 & B_{-1} = 0 \\ 0.125 - 0.25 < 0 & B_{-2} = 0 \\ 0.125 - 0.125 = 0 & B_{-3} = 1 \end{array}$$

After we get a remainder of 0, all bits after that will be 0, so we can stop there. In general, we continue the conversion until the remainder is 0 or until we have used all the bits that are available to represent the binary value.

Our binary value, starting with B_1 on the left, is then 11.001_2. We can check our result by converting the binary value back to a decimal value: $1(2) + 1(1) + 0(\frac{1}{2}) + 0(\frac{1}{4}) + 1(\frac{1}{8}) = 3.125$. Thus, $3.125 = 11.001_2$.

Negative Binary Numbers

Samples of speech, music, and other signals often can be negative numbers. Sinusoidal signals are negative half of the time. So, how do we represent negative numbers in the binary number system? There are numerous ways to do so. The method most similar to the way we handle decimal numbers is to add a single leftmost bit to the representation, where a 0 value for that bit indicates a positive number and a 1 indicates a negative number. Thus, if 101_2 is 2 bits plus a sign (leftmost) bit, it represents the decimal number -1, whereas 001_2 represents $+1$. Likewise, if 11010_2 is 4 bits plus a sign bit, it represents the decimal number -10, whereas 01010_2 represents $+10$. (See Table 5.6 for some more examples.) This combined representation of negative as well as positive numbers is called **sign–magnitude form**. Sign–magnitude form is just one of several ways to represent both positive and negative binary numbers.

Notice that there are two ways to represent zero in binary form, just as in decimal form, where $+0 = -0$.

Table 5.6 Three-Bit Sign–Magnitude Values

Binary value	Decimal value
000	+0
001	+1
010	+2
011	+3
100	−0
101	−1
110	−2
111	−3

Sign–Magnitude Form: A way to represent negative as well as positive binary numbers. The leftmost bit is reserved to indicate the sign of the number. Zero indicates positive, whereas 1 indicates negative.

INTERESTING FACT:

Computers often store numbers in a format called *floating-point notation*. In this case, numbers are written as $M \times 10^E$, where M is called the mantissa and E is the exponent. Both M and E are then stored using bits. This notation allows us to store a much wider range of numbers with a fixed number of bits.

EXAMPLE 5.14 Converting from Sign–Magnitude Form to Decimal

Convert 1101_2 in sign–magnitude form to decimal.

Solution

$1101_2 = -5$.

EXAMPLE 5.15 Another Conversion from Sign–Magnitude Form to Decimal

Convert 01011_2 in sign–magnitude form to decimal.

Solution

$01011_2 = 11$.

EXAMPLE 5.16 Another Conversion from Sign–Magnitude Form to Decimal

Convert 11101.11_2 in sign–magnitude form to decimal.

Solution

$11101.11_2 = -13.75$.

EXAMPLE 5.17 Writing a Decimal Number in Sign–Magnitude Form

Write the decimal number -29.25 in sign–magnitude form.

Solution

$-29.25 = 111101.01_2$.

When interpreting binary numbers, you must know or be told when the representation is in sign–magnitude form, in which case the first bit is a sign bit. Similarly, if a binary number has a fractional part, you must know the location of the binary point. The binary digits alone do not tell you what type of binary number representation is being used.

The ASCII Code

Up to now, we have focused on digitizing information such as music or images. However, there are many other forms of information that also must be digitized in order for us to be able to take full advantage of the computer-based world.

Suppose we want to share a text file or send an e-mail. How can text be represented using only binary numbers? The same question comes up in the design of our digital yearbook that we discussed at the beginning of this chapter: How can we write our words as numbers made up of only 1's and 0's? There are many ways that we can do this, but no matter which way we choose, we all must agree to use the same procedure or *code*. Otherwise, we will be expressing our words in different numerical languages, which will make it impossible to communicate! This situation is no different than if one person wrote a note in English and gave it to a person who spoke only Spanish. The note would be correct, but the code would not be common to both the writer and the reader.

ASCII Code: Binary representation of the letters and other characters on a keyboard.

To solve this problem, engineers and computer scientists have agreed upon a common standard code to represent letters of the alphabet in binary form. This code is called the **ASCII code**. Table 5.7 shows the entries of the ASCII code for the letters "A" through "Z," the lowercase letters "a" through "z," the numbers 0 through 9, and some important punctuation marks.

We will discuss the code for just a few letters and punctuation marks. The letter "A" is represented by the binary number 1000001 in ASCII. The letter "B" is 1000010. The letter "a" is represented by 1100001, whereas "b" is 1100010. To make a capital letter into a small letter, the bit next to the leftmost bit is set to 1, which is the same as adding 32 to the decimal value. A period is represented by 0101110 and a comma is 0101100.

This code may not seem too exciting, but it is almost a certainty that all the digital devices that you use—from your CD player to your calculator or computer—use the ASCII code to store letters!

From these examples, you can see that the ASCII code uses 7 bits to store its values. Hence, there are $2^7 = 128$ possible entries in the ASCII code table, enough for the letters (both upper- and lowercase) and numbers shown in Table 5.7 and for some control characters and symbols that are used less often. Using Table 5.7, you can learn to code text in ASCII and to read ASCII the way a computer does.

INTERESTING FACT:

ASCII stands for American Standard Code for Information Interchange. This code was first published in 1968. There are many national variants of ASCII for countries having languages that contain characters other than the standard English letters. We talk about this aspect more in Chapter 7.

EXAMPLE 5.18 Decoding an ASCII Message

Using Table 5.7, decode the following ASCII message: 1000111, 1101111, 1100100, 1101001, 1100111, 1101001, 1110100, 1100001, 1101100, 0011011

Solution

Go digital!

Table 5.7 Entries of the ASCII Table for Commonly Used Characters

Character	ASCII Code in Decimal	ASCII Code in Binary	Character	ASCII Code in Decimal	ASCII Code in Binary
A	65	1000001	k	107	1101011
B	66	1000010	l	108	1101100
C	67	1000011	m	109	1101101
D	68	1000100	n	110	1101110
E	69	1000101	o	111	1101111
F	70	1000110	p	112	1110000
G	71	1000111	q	113	1110001
H	72	1001000	r	114	1110010
I	73	1001001	s	115	1110011
J	74	1001010	t	116	1110100
K	75	1001011	u	117	1110101
L	76	1001100	v	118	1110110
M	77	1001101	w	119	1110111
N	78	1001110	x	120	1111000
O	79	1001111	y	121	1111001
P	80	1010000	z	122	1111010
Q	81	1010001	0	48	0110000
R	82	1010010	1	49	0110001
S	83	1010011	2	50	0110010
T	84	1010100	3	51	0110011
U	85	1010101	4	52	0110100
V	86	1010110	5	53	0110101
W	87	1010111	6	54	0110110
X	88	1011000	7	55	0110111
Y	89	1011001	8	56	0111000
Z	90	1011010	9	57	0111001
a	97	1100001	@	64	1000000
b	98	1100010	;	59	0111011
c	99	1100011	.	46	0101110
d	100	1100100	,	44	0101100
e	101	1100101	~	126	1111110
f	102	1100110	\	92	1011100
g	103	1100111	"	34	0100010
h	104	1101000	!	33	0011011
i	105	1101001	?	63	0111111
j	106	1101010	'	96	1100000

EXERCISES 5.3

Mastering the Concepts

1. What is a bit?
2. Explain why computer storage devices store information as bits as opposed to some other way, such as with numbers between 0 and 9.
3. What are the three types of storage methods commonly used in digital devices today?
4. What is meant by *most significant bit* and *least significant bit*?
5. What is the ratio of the most significant bit to the least significant bit?
6. What is the radix, or base, of a numbering system?

Try This

7. How many different numbers can be represented by (a) 6 bits, (b) 7 bits, (c) 8 bits, and (d) 10 bits?
8. How many bits would be required to store each of the following decimal numbers: 10, 100, 1000, 10,000, 100,000, and 1,000,000?
9. Convert the following positive integers from binary to decimal: 010_2, 101_2, 011_2, and 111_2.
10. Convert the following positive integers from binary to decimal: 1000_2, 1010_2, 0111_2, and 1101_2.
11. Convert the following positive integers from binary to decimal: 10000_2, 10101_2, 01111_2, and 11011_2.
12. Convert the following positive integers from binary to decimal: 100001_2, 110001_2, 011110_2, and 101101_2.
13. Convert the following positive integers from binary to decimal: 10101010_2, 11011011_2, 11100001_2, and 01110111_2.
14. Convert the following positive integers from binary to decimal: 11111111_2, 111111111_2, 1111111111_2, and 111111111111_2.
15. You are given a dozen fireplace matches, which you line up side by side. Each match head can be either up or down. The number of states or values that can be represented by this set of match heads is the number of possible orientations of the match heads. How many states can these matches represent?
16. Driving down a long street, you pass under 20 streetlights. Each light could be either working or burned out. How many possible combinations are there of burned-out and working street lights?
17. Write the following positive decimal integers in binary form: 4, 6, 8, 12, 13, and 15.
18. Write the following positive decimal integers in binary form: 16, 20, 25, and 31.
19. Write the following positive decimal integers in binary form: 32, 39, 57, and 63.
20. Write the following positive decimal integers in binary form: 64, 70, 111, and 127.

21. Write the following positive decimal integers in binary form: 128, 138, 195, and 255.
22. Write the following positive decimal integers in binary form: 256, 1024, 2048, and 4096.
23. Write the following positive decimal integers in binary form: 743, 1500, 3001, and 4104.
24. Write the following positive decimal integers in binary form: 849, 1800, 3307, and 4608.
25. Convert the following positive fractions from binary to decimal: 10.1_2, 10.0_2, 01.0_2, and 11.1_2.
26. Convert the following positive fractions from binary to decimal: 1.01_2, 1.00_2, 0.10_2, and 1.11_2. How are your answers related to the answers to Exercise 5.3.25? Do you know why?
27. Convert the following positive fractions from binary to decimal: 100.01_2, 111.11_2, 011.10_2, and 101.101_2.
28. Convert the following positive fractions from binary to decimal: 1000.1_2, 1111.1_2, 0111.0_2, and 1011.01_2. How are your answers related to the answers to Exercise 5.3.27? Do you know why?
29. Convert the following positive fractions from binary to decimal: 10101.010_2, 11011.011_2, 1110.0001_2, and 0111.0111_2.
30. Convert the following positive fractions from binary to decimal: 111.111_2, 1111.1111_2, 11111.11111_2, and 111111.111111_2.
31. Write the following decimal fractions in binary form: 5.5, 7.25, 9.125, and 12.0625.
32. Write the following decimal fractions in binary form: 8.375, 14.625, 15.875, and 19.3125.
33. Consider a binary fraction having 6 digits to the left of the binary point and 4 digits to the right. In decimal notation, how many digits to the left and right of the decimal point would be needed to store numbers covering the same range as that covered by the binary representation? Which representation, binary or decimal, requires more digits? Why?
34. Consider a decimal fraction having 6 digits to the left of the decimal point and 4 digits to the right. In binary notation, how many digits to the left and right of the binary point would be needed to store numbers covering the same range as that covered by the decimal representation? Are you surprised?
35. Convert the following integers, in sign–magnitude form, from binary to decimal: 110_2, 010_2, 111_2, and 011_2.
36. Convert the following integers, in sign–magnitude form, from binary to decimal: 1100_2, 0100_2, 1111_2, and 0111_2.
37. Convert the following integers, in sign–magnitude form, from binary to decimal: 11000_2, 01000_2, 11111_2, and 01111_2.
38. Convert the following integers, in sign–magnitude form, from binary to decimal: 11011_2, 10101_2, 01101_2, and 10011_2.
39. Convert the following fractions, in sign–magnitude form, from binary to decimal: 110.01_2, 010.10_2, 111.11_2,and 011.00_2.
40. Convert the following fractions, in sign–magnitude form, from binary to decimal: 1100.001_2, 0100.010_2, 1111.100_2, and 0111.111_2.

41. Write the following decimal fractions in sign–magnitude binary form: −5.25, +7.5, −9.375, and +12.625.
42. Write the following decimal fractions in sign–magnitude binary form: −8.625, +14.875, −15.625, and −19.5625.
43. Write the ASCII sequence representing the word *Hi*.
44. Write the ASCII sequence representing the term *MP3*.
45. Write your first name in ASCII.
46. What does the following ASCII sequence spell? 1001101, 1100001, 1110100, 1101000.
47. What does the following ASCII sequence spell? 1001001, 1101110, 1100110, 1101001, 1101110, 1101001, 1110100, 1111001. Imagine trying to read an entire book this way!

In the Laboratory

48. How many megabytes (MBs) of RAM does your lab computer have? How did you figure this out? (One byte is eight bits.)
49. How many gigabytes (GBs) of storage does your computer hard disk allow? How did you figure this out?
50. Find the complete ASCII table listed in Chapter 7. Besides the regular alphabetical characters and punctuation marks, what other types of characters are included in the table?

Back of the Envelope

51. Explain how a set of blue cards and red cards could be used to represent sequences of bits.
52. Explain how a handful of pennies could be used to represent sequences of bits.
53. Explain how a set of apples and oranges could be used to represent sequences of bits.
54. Explain three other ways that bits could be represented, other than the ways listed in the previous exercises.
55. In a standard deck of cards, using just the clubs and the diamonds, and paying attention only to color (not to value or rank), how many values or numbers can you represent?
56. The decimal number system was invented by humans because we have 10 fingers. Digital devices use the binary number system because only one of two states can be stored at any physical location. Suppose that spiders could store numbers and do math. What base would they likely use for their number system? Convert the following numbers from "spider math" to decimal: 41_8, 123_8, 321_8, 457_8, and 777_8.
57. Write the year in binary that Alexander Graham Bell first demonstrated the telephone. The Web will tell you more about this inventor than just his birth date.
58. Write the year in binary that Bardeen, Brittain, and Shockley received the Nobel Prize for invention of the transistor.
59. Write the year in binary that Jack Kilby received the Nobel Prize for invention of the integrated circuit.
60. Write the elevation of Death Valley, in feet above sea level, in sign–magnitude binary form.

61. Write the elevation of the deepest point in the ocean, in feet above sea level, in sign–magnitude binary form.
62. Suppose that a tiny, superfast memory contains 100 bits. How many states or values can this memory represent? Your answer will be a very, very large number. Approximate it by a number written in the form 10^M for some integer value of M, by using the fact that 2^{10} is about equal to 1000.
63. How does the answer to Exercise 5.3.62 compare with Avogadro's number from chemistry class? How does the number of different states that the 100-bit memory represents compare with the number of molecules in 1000 kilograms of hydrogen?
64. A new memory technology is developed where each storage location can represent five states, instead of just two. Let's call these quantities "quintary digits," or "quints." How many different values or states can be represented by (a) 6 quints, (b) 7 quints, (c) 8 quints, and (d) 10 quints? How do these answers compare with those for Exercise 5.3.7?
65. Can you use "quints" as described in Exercise 5.3.64 to count on your hands? What is the largest number you can count to? What if you used your feet as well?

5.4 Using Bits to Store Samples: Quantization

In our discussion of sampling in Section 5.2, we assumed implicitly that computers store the numerical values of samples of a signal, or pixels of an image, *exactly*, without any loss of accuracy or precision. You might be surprised to learn that computers almost always store numbers with less precision—that is, fewer decimal places—than the actual real number typically would have. (See the adjacent Interesting Fact box for an example of storing the number π.) But this situation is really no different than when you are taking measurements in a science experiment, such as noting fluctuating temperatures, and choose to write down the values using only one or two decimal places. The actual temperature, if you could measure it accurately enough, might have tens, hundreds, or even thousands of decimal places.

INTERESTING FACT:

Why can't the number pi (π) be stored perfectly on a computer? Because $\pi = 3.141\ldots$ has an infinite number of decimal places. To store it would require an infinite amount of memory. It is just simply not possible.

Real-world digital memory devices use only a fixed number of bits for the storage of each number, which implies a limited or finite precision. We encountered this idea earlier in Chapter 3 when we considered the storage of image pixels by using only binary numbers. In this section, we'll focus in more detail on how general numbers and signal measurements are coded in a finite-precision binary format.

Suppose a speech or music signal is sampled to produce the set of samples $s[n]$. The accuracy achieved in storing a sequence $s[n]$ of numbers depends on how many bits are used to store each sample. More bits mean greater precision. Of course, more bits also mean more storage space and more complex engineering designs. Figure 5.16 shows a sampled analog signal. The computer cannot store every possible value that this signal might take, because there are an infinite number of them! For example, suppose that one of the samples had the value of

Table 5.8 Sampled Values Quantized and Stored as Bits

True value	Quantized value	Binary number
2.5000	2	10
3.3783	3	11
3.9132	3	11
3.8966	3	11
3.3288	3	11
2.4183	2	10
1.4997	1	01
0.9179	1	01
0.8740	1	01
1.3712	1	01
2.2020	2	10
3.0304	3	11
3.5203	3	11
3.4644	3	11
2.8633	3	11
1.9261	2	10
0.9901	1	01
0.3928	0	00
0.3430	0	00
0.8413	1	01
1.6800	2	10
2.5228	3	11
3.0333	3	11
3.0035	3	11
2.4338	2	10
1.5323	2	10
0.6359	1	01
0.0811	0	00
0.0761	0	00
0.6205	1	01
1.5060	2	10

Quantization: The process of changing sample values to discrete levels. *Rounding,* whereby the sample values are changed to the nearest levels, is the most common form of quantization.

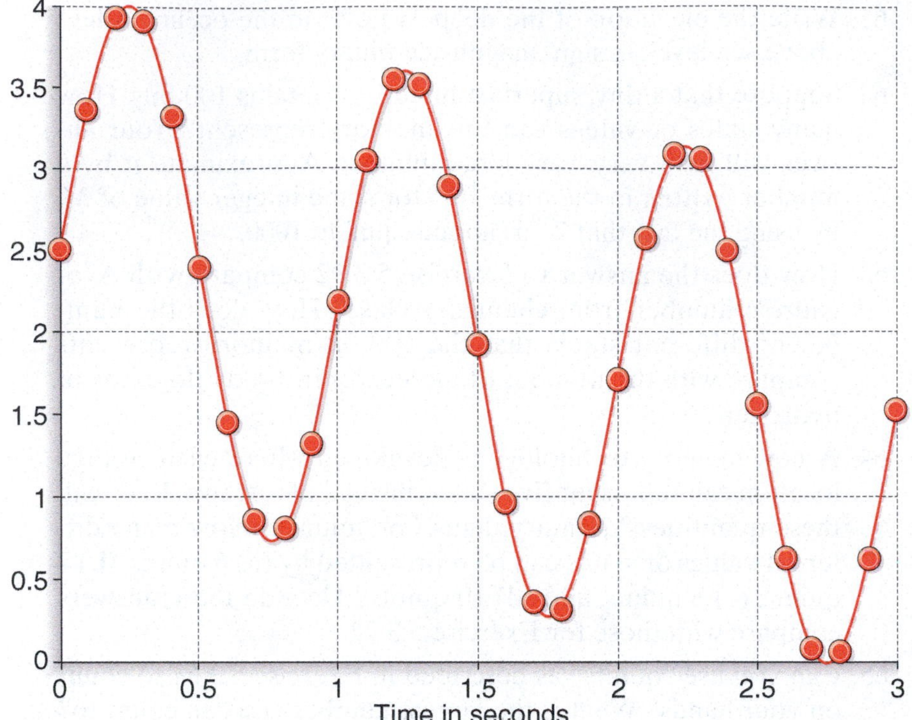

Figure 5.16 A sampled signal.

$\frac{1}{3} = 0.33333\ldots$. This number cannot be represented exactly as a binary or decimal number with a fixed number of bits or digits, respectively, because it repeats forever. Using binary numbers with a fixed number of 0's and 1's, we can store only a finite number of signal values. Furthermore, even if we could store each sample exactly, by using a very large number of bits, we might not want to, because doing so would require lots of storage. In general, we should use the smallest number of bits that will provide the accuracy we need for the given engineering application.

The signal in Figure 5.16 ranges between 0 and 4 in value. Suppose we decide to store each signal sample with only 2 bits, which would therefore correspond to only four possible stored signal values. And to make matters simple, let's set the four signal values at 0, 1, 2, and 3. They could be set differently if we like. For each signal value $s[n]$, we will then round the true value to the nearest of the four levels that we have agreed to represent. This process of signal *rounding*, shown in Figure 5.17(a), is a form of **quantization**. The dashed lines indicate whether a sample is rounded up or down. Figure 5.17(b) shows the quantized version of $s[n]$ that actually would be stored.

The values of the quantized samples are listed in Table 5.8. Since there are only four quantization levels, these levels can be represented by 2 bits. We represent the levels 0, 1, 2, and 3 by the binary values 00, 01, 10, and 11, respectively. The binary representations for the quantized values are shown in the third column of the table.

The approximation of the samples in Figure 5.17(a) is a very crude one, because we used only four quantization levels, or, equivalently, 2 bits, to store each sample, or signal measurement. We could greatly improve the accuracy of our digitized information by using more bits per sample. Most engineering applications typically do use more bits per sample.

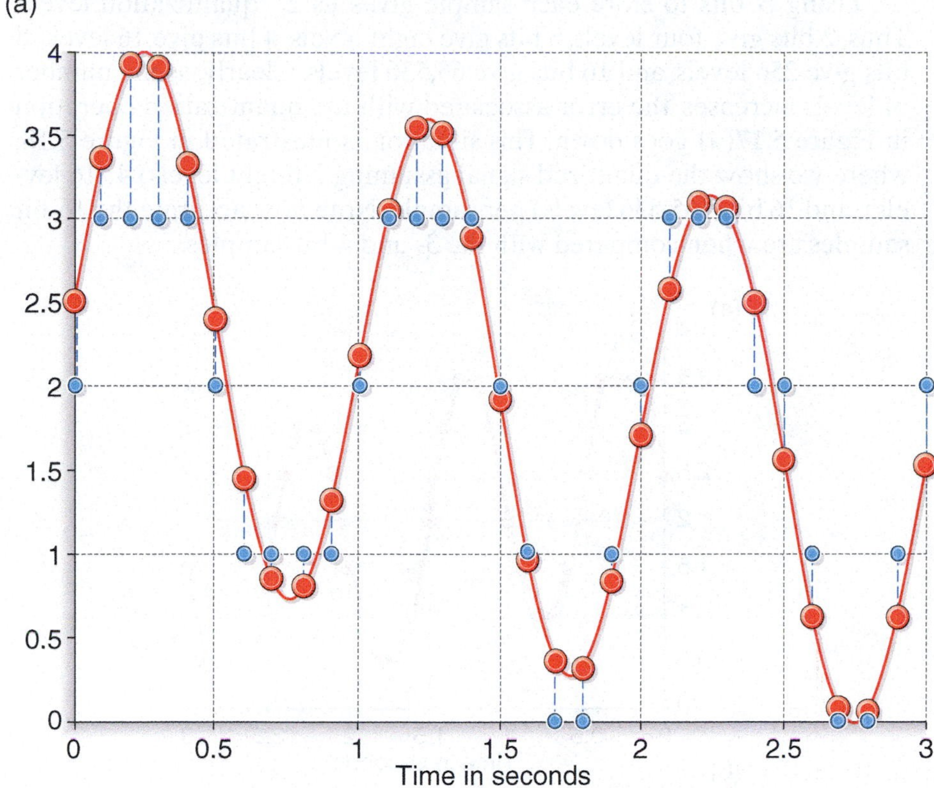

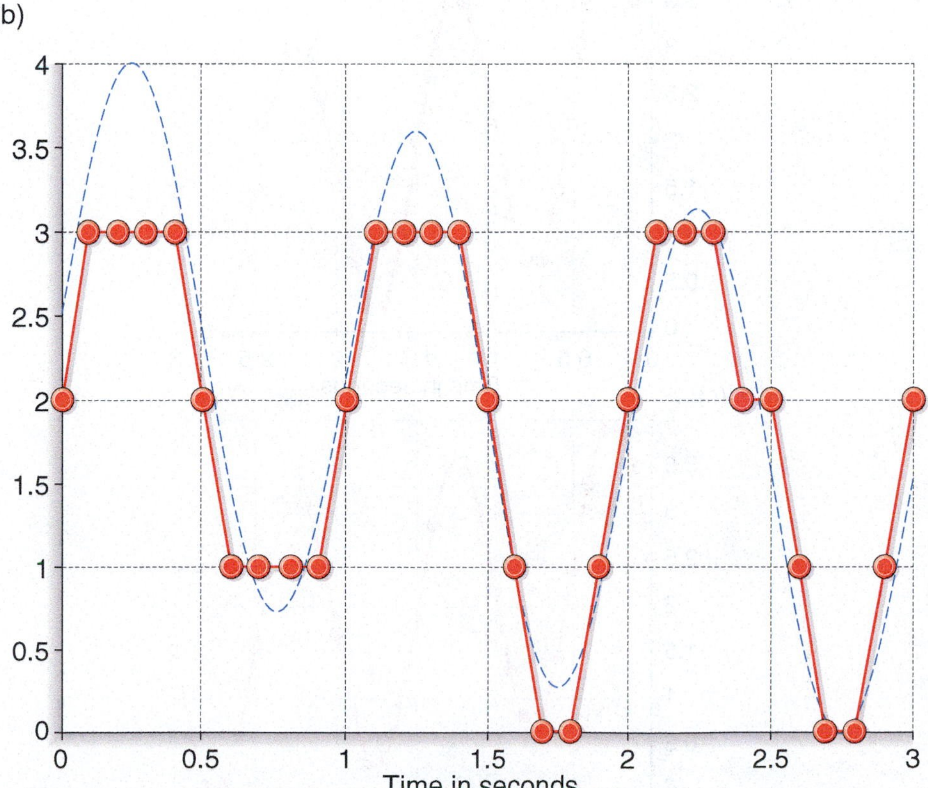

Figure 5.17 The quantization process. (a) Orange and blue dots represent original and quantized values, respectively. (b) Signal quantized using 2 bits. The dots are the quantized signal values to be stored on the digital device. The original signal is shown with a dashed line.

INTERESTING FACT:

Music on CDs is stored using 16 bits per sample. This means that with a sampling rate of 44,100 samples per second, 254,016,000 bits are used to store a 3-minute song in stereo.

Using B bits to store each sample gives us 2^B quantization levels. Thus, 2 bits give four levels, 3 bits give eight levels, 4 bits give 16 levels, 8 bits give 256 levels, and 16 bits give 65,536 levels. Clearly, as the number of levels increases, the error associated with the quantization operation in Figure 5.17(a) goes down. This situation is illustrated in Figure 5.18, where we show the quantized signal assuming 3 (eight levels) 4 (16 levels), and 16 bits (65,536 levels) per sample. Note how accurate the 16-bit samples are when compared with the 3- and 4-bit samples.

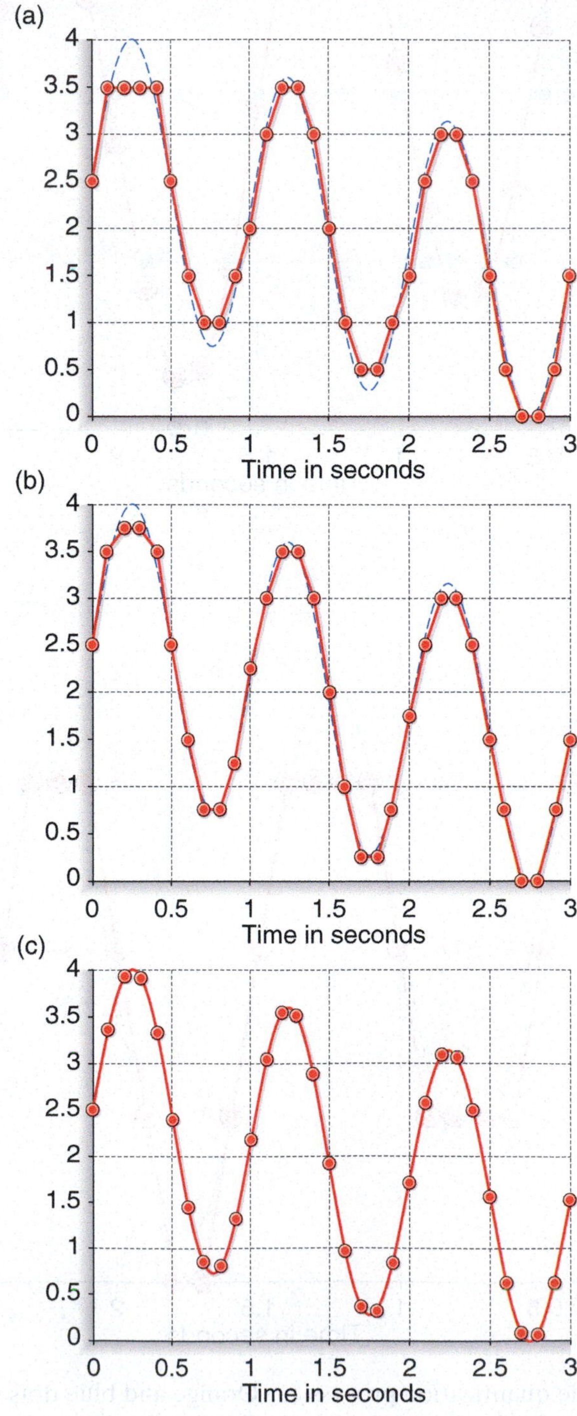

Figure 5.18 An analog signal quantized using (a) 3 bits, (b) 4 bits, and (c) 16 bits. Notice how precise the samples are when stored using 16 bits.

Unlike those in Figure 5.17, the samples in Figure 5.18 can have fractional values. Fractions are not a problem, because they can be represented in binary form, as we have discussed. Usually, the quantization interval is not equal to 1, and the signal can be both positive and negative. In general, a signal may have a so-called **dynamic range** extending from $+A$ to $-A$, in which case the total interval of size $2A$ would be divided up into $L = 2^B$ amplitude levels for B-bit binary representation.

Dynamic Range: The difference between the largest and smallest amplitudes or intensities that a signal takes on.

The levels should be set to cover the full dynamic range of the signal; otherwise, there will be amplitude **clipping**, whereby large amplitudes will be lowered in the coding process to the largest quantization level available. This situation is similar to effects described in Chapter 4 when a constant was added to all image pixels to brighten an image and when image pixels were multiplied by a constant to enhance contrast. Either operation could create new pixel values exceeding the largest amplitude that could be represented using the allotted number of bits.

Clipping: An error caused in the quantization operation if the quantizer levels do not cover the full dynamic range of the signal. Clipping results in the tops and bottoms of the signal being "clipped off."

Severe clipping is visible in the first two peaks of Figure 5.17 and in the first peak of Figure 5.18(a). We could have greatly reduced the clipping by making a different choice for the quantization levels. For example, in Figure 5.17 it would be best to set the levels higher at 0.5, 1.5, 2.5, and 3.5, rather than at 0, 1, 2, and 3, respectively, as shown. It is equally important that the quantization levels not be set too high. Otherwise, the signal being sampled may not rise above the first couple of levels, in which case the extra bits used to represent the higher levels would always be 0 and would be wasted.

KEEP IN MIND

A binary representation with B bits can represent 2^B levels of signal height or image intensity. More bits provide higher fidelity—better music and better pictures, for example. However, more bits also require more storage, more processing, and longer transmission times.

How many bits should we use to store digitized information like pictures, movies, or music? The answer depends on the application. We have seen previously that 8 bits are usually used to store individual pixels in a gray-scale digital image. However, our ears are much more sensitive to quantization errors than our eyes; therefore, engineers use 16 bits to store individual audio samples on an audio CD. This large number of bits is one of the reasons that CD music sounds so good. Thus, different applications require different levels of precision.

No matter how quantization is performed, the number of bits per sample places the ultimate limit on the accuracy of the digitized values of the information. The combined processes of sampling and quantization are called digitization or **analog-to-digital (A/D) conversion**. The electric circuit that performs the conversion is called an analog-to-digital (A/D) converter. Most A/D converters are fabricated on integrated circuits or chips. A/D converters may be found in your cell phone, computer, automobile, or any place where analog signals are to be stored, processed, or transmitted digitally.

Analog-to-Digital Conversion: Sampling of an analog signal followed by quantization of the samples to binary numbers. Also called digitization.

Quantization Errors

As we saw in the last section, the quantization operation causes errors in representing digitized information. We call these errors **quantization noise**, because the effect of quantization errors sounds like noise in a digitized music signal and looks like noise in a digital image.

Quantization Noise: The error or noise introduced into signal samples through the process of quantization.

Clearly, the more accurately we store the signal samples or measurements, the smaller the errors or quantization noise will be. As we have seen, we can reduce the errors by using more bits to represent each numerical value on a computer or digital system. The amount of noise decreases as the number of bits increases and the corresponding number of

> **KEEP IN MIND**
>
> When applied to signals, **digitization** refers to the combined operations of sampling and quantization.

quantization levels increases. This phenomenon was observed in Figure 5.18. In practice, the amount of noise also depends on the dynamic range of the signal, since a wider dynamic range implies that the quantization intervals must be spaced more widely. But the general rule still holds: Using more bits to represent information means greater accuracy, which translates into better sound, richer colors, and more realistic graphics.

Since engineers want to be able to determine the impact of quantization noise in various applications, we must construct a way of measuring the amount of noise. If we look back at Figure 5.17(a), the quantization error is simply the difference between the original signal samples, represented by the big dots, and the rounded or quantized values, represented by the small dots.

As in all engineering applications, the size of an error in any given measurement or signal is always relative to the measurement itself. For example, a 1-inch error when measuring something as long as a foot is very large. However, a 1-inch error when measuring something that is a mile long is very small. Therefore, when constructing our definition of the size of the quantization noise, we need to make sure that we take into account the size of the signal we are measuring.

> **Signal-to-Noise Ratio (SNR):** The ratio of maximum signal level to maximum noise level, usually expressed in a logarithmic form in decibels (dB). For quantization noise measured in dB, SNR $= 6B$, where B is the number of bits per sample, and where best use is made of the quantizer's dynamic range.

Engineers have come up with a very important measure of the relative size of any type of noise, even quantization noise. This measure is called the **signal-to-noise ratio (SNR)**. There are many similar definitions of the signal-to-noise ratio. In this book, we use the simplest, which is the ratio of the largest signal amplitude to the largest noise amplitude:

$$\text{SNR} = \frac{\max\langle|\text{signal}|\rangle}{\max\langle|\text{noise}|\rangle}$$

Here, the vertical bars indicate the absolute value, or magnitude, of the numbers, and "max" means the maximum over all time for an audio signal or over all space for an image. While this equation might look a bit difficult to use, it is actually very simple.

For the signal in Figure 5.17(a), it is easy to see that the maximum, or the largest value of the signal, is 4. This value gives us the numerator in the SNR equation. Next, we have to determine the denominator, or the maximum value of the noise. As we have learned, the noise is the difference between the original, or true, values of the signal and the quantized values. The maximum value of the noise occurs near the first peak of the original signal and has a value approximately equal to $4 - 3 = 1$. Thus, the SNR of the signal in Figure 5.17(a) is approxi mately $\frac{4}{1} = 4$, meaning that the quantization noise or error is one-fourth the size of the signal being measured.

SNR is measured easily from plots of the noise-free signal and the noise corrupting the signal. Let's illustrate this concept with an example. Figure 5.19(a) shows samples of a sinusoidal signal $s(t) = 0.8 \cos(30t)$, sampled at every $T = 0.01$ s. Clearly, the maximum absolute value for the signal sample is

> **INTERESTING FACT:**
>
> Digital signal processors are high-speed integrated circuits or computer chips that are used in a wide variety of applications involving digital music, pictures, and movies. They are even used in antilock brakes.

$$\max\langle|s[n]|\rangle = 0.8$$

Figure 5.19(b) shows the signal samples that have been quantized using 4 bits. We set the $2^B = 16$ quantization levels to be spaced uniformly with spacing of $\frac{1}{8}$ covering a range from about -1 to $+1$. The quantized signal is represented using only the levels $\left[-\frac{15}{16}, -\frac{13}{16}, -\frac{11}{16}, -\frac{9}{16}, -\frac{7}{16}, -\frac{5}{16}, -\frac{3}{16},\right.$

$-\frac{1}{16}, \frac{1}{16}, \frac{3}{16}, \frac{5}{16}, \frac{7}{16}, \frac{9}{16}, \frac{11}{16}, \frac{13}{16}, \frac{15}{16}]$, which provides uniform coverage of signal levels from -1 to $+1$.

Because we are using quite a few quantization levels, the original and quantized signals look nearly the same in Figure 5.19(a) and (b), respectively. The signals are different, though, because the quantization operation has caused some error. Figure 5.19(c) shows the quantization noise, which is the difference between the original signal and its quantized

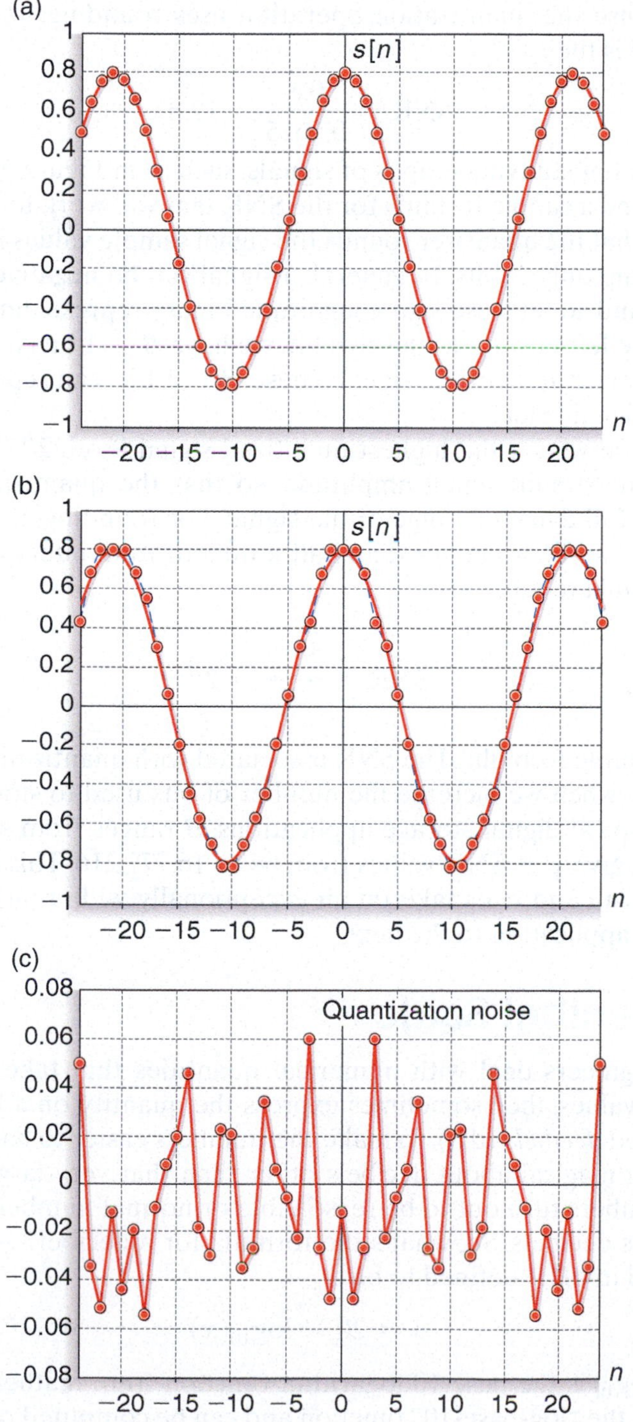

Figure 5.19 (a) Sampled sinusoid with amplitude 0.8. (b) Quantized version of the sinusoidal signal in (a). (c) Quantization noise in the signal in (b).

version. We see that the maximum absolute value of the quantization noise is

$$\max\langle|\text{noise}[n]|\rangle = 0.06$$

for the range of n shown in the figure. If we plot the quantization noise for a broader range of n, we will see that the maximum noise is just a little larger, namely,

$$\max\langle|\text{noise}[n]|\rangle = 0.0625 = \tfrac{1}{16}$$

This value is half the distance between the levels in the quantized version because the quantization operation uses rounding. The SNR for this signal is then

$$\text{SNR} = \frac{0.8}{0.0625} = 12.8$$

Instead of studying graphs of signals, such as in Figure 5.19, we can actually find a simple formula for the SNR that will work for all signals. Suppose that the quantizer rounds the signal sample values to numbers stored using only B bits. In general, a signal can be negative as well as positive, and so let's use sign–magnitude binary representation. By reserving the leftmost bit as the sign bit, we have $B - 1$ bits remaining to store the magnitude of the signal. These $B - 1$ bits can represent 2^{B-1} possible signal values.

Suppose we set the highest quantized signal level, 2^{B-1}, to represent the maximum signal amplitude, so that the quantization levels cover the full dynamic range of the signal. For rounding, the quantization error can be no larger than half a quantization interval. Thus, the SNR formula reduces to

$$\text{SNR} = \frac{2^{B-1}}{\frac{1}{2}} = 2^B$$

What a simple formula! The SNR associated with quantization increases quickly when we increase the number of bits used to store the samples. In typical digital storage applications, B ranges from about 6 bits to 24 bits, giving us SNR values from 64 to 16,777,216. This means that the SNR can and does take on an exceptionally wide range of values from one application to the next.

The Decibel Scale

When engineers deal with numerical quantities that take on a wide range of values, they sometimes express the quantity on a logarithmic scale called *decibels* (dB), to make the numbers easier to handle. There is nothing magical about decibels other than that very large and very small numbers turn out to be reasonable or normal numbers when expressed as decibels. So, what's the formula for a decibel? A number x expressed in dB is defined to be

$$x = 20 \times \log_{10}(x)$$

where $\log_{10}$ is the base-10 logarithm function from mathematics. It is known as the "log-base-10" function and can be computed on nearly all calculators. The $\log_{10}$ function increases very slowly as x increases. Table 5.9 lists some values of $\log_{10}(x)$ for different x values.

Table 5.9 Table of Values of $\log_{10}(x)$

x	$\log_{10}(x)$
0.01	−2
0.1	−1
1	0
10	1
100	2
1000	3
10,000	4
100,000	5
1,000,000	6

EXAMPLE 5.19 Converting a Large Number to Decibels

What is the number 5,432,276 in decibels?

Solution

$20 \log_{10}(5{,}432{,}276) = 134.69$ dB.

EXAMPLE 5.20 Converting a Small Number to Decibels

What is the number 0.0000235 in decibels?

Solution

$20 \log_{10}(0.0000235) = -92.57$ dB.

So, what do these two examples teach us? First of all, decibels can be both positive and negative numbers. By the basic properties of the logarithm, the decibel value will be positive when the original number is greater than one, negative when the original number is less than one. Secondly, through the decibel calculation, very big and very small numbers get converted into numbers with which we are more familiar. Because of this attribute, engineers often express wide-ranging numbers in decibels.

We have seen that the signal-to-noise ratio can range from very small to very large; as a result, engineers generally calculate the SNR in terms of decibels. Table 5.10 includes a few examples of the signal-to-noise ratios of various audio systems—typical SNRs for sources from AM radio to DVD audio systems. As we have learned, the higher the SNR value, the lower is the noise level in the signal, while the lower the SNR value, the higher is the noise level. We can see from the table that AM radio has the lowest SNR value and, therefore, the most noise, while DVD audio has the highest SNR value and thus the least noise. Anyone who has listened to both immediately will recognize the difference!

Calculating the SNR for Quantization Noise As it turns out, it is quite easy to determine on dB scale the signal-to-noise ratio associated with digitizing signals. Expressing SNR in decibels, we have

$$\begin{aligned} \text{SNR} &= 20 \times \log_{10}(2^B) = 20 \times B \log_{10}(2) \\ &= 6.02B \end{aligned}$$

Thus, on a dB scale, the SNR grows by about 6 dB for every bit we use to represent the samples. As we have already learned, the audio samples stored on a CD use 16 bits. Using our equation, we easily can calculate that the SNR for CD music is 16×6.02, or approximately 96 dB, as listed in Table 5.10. This value is substantially higher than for both AM and FM radio, but not as good as for DVD audio. Why? In DVD audio, each audio sample is stored using 24 bits, which translates into an SNR of 144 dB. Try the next: Experiment to hear signals at a variety of SNRs, where the associated noise is caused by quantization or clipping.

KEEP IN MIND

Decibels are used to measure signal levels that cover a wide range. Decibels are used to measure sound volume because the human ear can hear sounds ranging from a quiet whisper, which is very soft, to a rock concert, which is very loud.

KEY CONCEPT

A basic property of a logarithm function is that log(x) is positive when x is greater than one and negative when x is less than one.

Table 5.10 Signal-to-Noise Ratio (SNR) of Various Audio Sources (calculated in decibels (dB), the higher the dB value, the better is the sound quality)

Music Source	Typical Signal-to-Noise Ratio in dB
AM radio	50 dB
FM radio	75 dB
CD audio	96 dB
DVD audio	144 dB

INTERESTING FACT:

The SNRs for AM and FM radio vary depending on transmitter power, distance from the transmitter, type of receiver antenna, and other factors.

INTERESTING FACT:

Audio sound accuracy, as measured by its signal-to-noise ratio, increases by 6 dB for every additional bit we use Infinity Project to represent the signal samples.

Infinity Project Experiment: Quantization and Clipping

Listen to speech and music stored at many bits per sample and at few bits per sample, corresponding to high and low SNRs, respectively. Then listen to speech and music that have been **clipped** due to the quantization levels not covering the full **dynamic range** of the signals. Describe what you hear.

EXERCISES 5.4

Mastering the Concepts

1. Define the operation of quantization.
2. What is meant by clipping?
3. What is a digital signal?
4. What is analog-to-digital conversion?
5. What is quantization noise?
6. What is the definition of SNR on a linear (not logarithmic) scale?

Try This

7. The sampled signal $s[n] = 1.5 + 2\cos(0.6n)$ is quantized to the four levels 0, 1, 2, and 3. Find and plot the original samples $s[n]$ and their quantized values for $n = 0, 1, 2, \ldots, 12$. Assign 2-bit binary values to the quantized samples.
8. The sampled signal $s[n] = 1.75 + 2\cos(0.6n)$ is quantized to the eight levels 0, 0.5, 1, 1.5, 2, 2.5, 3, and 3.5. Find and plot the original samples $s[n]$ and their quantized values for $n = 0, 1, 2, \ldots, 12$. Assign 3-bit binary values to the quantized samples.
9. The sampled signal $s[n] = 0.035n^2$ is quantized to the eight levels 0, 0.5, 1, 1.5, 2, 2.5, 3, and 3.5. Find and plot the original samples $s[n]$ and their quantized values for $n = 0, 1, 2, \ldots, 10$. Assign 3-bit binary values to the quantized samples.
10. Repeat Exercise 5.4.9, except with 4 bits and the levels 0, 0.25, 0.5, 0.75, 1, 1.25, 1.5, 1.75, 2, 2.25, 2.5, 2.75, 3, 3.25, 3.5, and 3.75. What is the dynamic range of the quantized signal?
11. Suppose that the SNR in a particular application can range from 1 to 1,000,000 when measured on a linear scale. What would be the range of SNR when measured in dB?
12. Suppose a speech signal is quantized with 8 bits per sample. Assuming that the range of quantization amplitudes is matched to the signal amplitude, what is the SNR, measured in dB, of the speech signal?
13. Rework Exercise 5.4.12, except make the quantization levels match the amplitudes from −1 to 1, and make the actual maximum signal amplitude 0.7. How does your answer compare with that for Exercise 5.4.12?
14. An audio signal to be stored on a DVD is quantized to 24 bits per sample. How much higher is the SNR (again measured in dB) as compared with that of an audio CD?

15. A system used to process electronic signals can handle voltages up to 5 V without clipping. The amount of noise produced by the circuitry of the system is 1 millivolt (0.001 V). What is the SNR of this system in dB?

In the Laboratory

16. Most portable audio, home audio, and car audio systems have an FM tuner, an AM tuner, and an amplifier. Each of these components can add noise to the signals that you listen to from your speakers. Find an owner's manual for an audio system and look up the SNRs for the FM tuner, AM tuner, and amplifier. How many bits would be needed to process signals in digital versions of these systems? Which of the three systems—FM, AM, or amplifier—handles signals that would require the most bits in order to sample accurately?

Back of the Envelope

17. There is some clipping in the quantization of the signal in Figure 5.17(a). In the text, we explained how the clipping can be reduced by resetting the quantization levels. Figure 5.18(a) also shows some clipping. How should the eight levels used in Figure 5.18(a) be reset to minimize the maximum quantization error? Each pair of adjacent levels should be the same distance apart.

Master Design Problem

We have learned a great deal about digitizing information ranging from pictures and movies to sounds, music, and text. So, now let's return to the design of the digital yearbook. In starting our design, let's specify a few design parameters. Let's assume that our hypothetical school has about 1500 students and teachers. We will assume that the "conventional" part of the yearbook—the text and images—occupies about 250 pages.

If we consider direct storage of the digitized images and video, we will soon see that our hope for a digital yearbook is unrealistic, because too much storage would be required. Digital imagery and, in particular, video signals require large amounts of storage in their "raw" formats. Fortunately, we can use a technology called *compression* to reduce the amount of storage space required for these signals. You will learn more about compression and how it is done in Chapter 6. For now, let's assume that we can compress binary text files by a factor of 2. In other words, we have ways to halve the amount of storage needed for the text from the number of bits we need in a standard ASCII representation. We also will assume that by using compression technology, we can compress speech by a factor of 5, music by a factor of 3, images by a factor of 10, and video by a factor of 100.

After summing the required storage for the compressed text, speech, music, images, and video, we will add in an extra 10% for "overhead" to account for page layouts, background graphics, means of indexing the digital material, and so forth.

KEEP IN MIND

Compression is the process of reducing the size of a computer file by removing redundancy. Different amounts of compression are possible for text, speech, music, images, and video.

Text

We need to decide how much text our yearbook will contain. Let's assume that each page contains 25 lines of text on average (most of the space will be used by images), with 10 words per line and 6 letters per word on average. We will ignore punctuation. These assumptions give (250)(25)(10)(6) = 375,000 letters. We typically store a single letter in ASCII by using 8 bits, or 1 byte. Thus, we project that the text in our yearbook will require 375 kilobytes (KB) of storage, prior to compression. Assuming a compression factor of 2, the required amount of storage for text would be 187.5 KB.

Byte: A group of 8 bits.

Speech

We will assume that 1600 sound clips are collected, with an average length of 1 minute per clip. Speech can be band limited using an anti-aliasing filter to less than 4000 cycles per second (Hz) without significant loss in sound quality. Thus, let's assume that the speech is sampled at 8000 samples per second, which is the Nyquist rate. Furthermore, assume that each sample is represented in 8 bits. This gives a total of (1600)(60)(8000)(8) = 6,144,000,000 bits = 768 megabytes (MB), prior to compression. A compression factor of 5 would reduce this amount to 153.6 MB. We see that the speech will require nearly 1000 times as much storage as the text!

Music

The music to be stored might be recordings of the orchestra or jazz band or might be from musical theater productions. We also will lump any other audio, such as nonmusical theater productions, into this category if they are to be recorded with high fidelity. Let's assume that 12 hours of high-fidelity stereo will be recorded. The human ear can hear frequencies up to about 20,000 cycles per second (20 kHz). Thus, let's assume that the audio signal is band limited to 20 kHz prior to sampling and then is sampled at a rate of 40,000 samples per second. Each of the two stereo channels will be coded with 16 bits per sample. This gives a total of (12)(60)(60)(40,000)(2)(16) = 55,296,000,000 bits = 6,912,000,000 bytes, prior to compression. A compression factor of 3 would reduce this amount to 2304 MB. The music will require about 15 times as much storage as the speech.

Images

Assume that each of the 1500 students and faculty will have a medium-sized portrait photo of size 800 × 600 pixels. Each pixel will consist of three colors represented in 8 bits per color. Furthermore, let's assume that there will be 500 other photos of all sorts of student activities. Some of these photos will be larger than the student and staff portraits, so we will assume image sizes of 1200 × 900 in order to enhance image quality. The total amount of storage required for images will be (1500)(800)(600)(3)(8) + (500)(1200)(900)(3)(8) = 30,240,000,000 bits = 3,780 MB, prior to compression. A compression factor of 10 would reduce this amount to 378 MB. The images will require more storage than the speech, but far less than the music.

Video

We may wish to incorporate long video segments from athletic contests or theater productions, as well as many much shorter segments. Video will require lots of storage, so we will try to limit the amount of video in our yearbook. We will plan on storing 100 short video clips, with an average duration of 1 minute each, plus 10 longer video segments having an average duration of 15 minutes each. The total number of minutes of video will be 250. We will assume a frame rate of 30 frames per second and a frame size of 640 × 480 pixels. As with the still-frame images, each pixel will be represented by three colors, with 8 bits per color. This leads to (250)(60)(30)(640)(480)(3)(8) = 3,317,760,000,000 bits = 414,720 MB, prior to compression. A compression factor of 100 would reduce this amount to 4147.2 MB. Despite the high compression factor, this amount of video requires more storage than any other component of our yearbook—about twice the storage required for the 12 hours of music.

KEEP IN MIND

Speech requires far more storage than text. Music requires more storage than speech. Video requires more storage than music. A standard CD can store 650 MB of information, and a standard DVD can store 4.7 GB. Therefore, one DVD can store the information found on seven CDs.

Total Storage and Iteration on the Design

The sum of the storage required for the compressed text, speech, music, images, and video is 0.1875 + 153.6 + 2304 + 378 + 4147.2 = 6983 MB = 6.983 GB. Adding in an extra 10% overhead (multiplying 6.983 GB by 1.1) gives a total storage figure of 7.68 GB. How does this amount compare with the amount of storage available on a CD?

The answer is, "not too well"! A traditional CD can hold only 650 MB of information. Thus, our digital yearbook will not fit onto a single CD. Twelve CDs would be required. The problem is that both the music and the video require a tremendous amount of storage.

A single DVD could do the job, though, depending on the type of DVD. A one-sided, single-layer DVD can hold 4.7 GB. A double-sided, single-layer DVD can hold over 9 GB. Thus, we are in luck; one double-sided, single-layer DVD easily would hold our digital yearbook. Of course, if we had hoped to store more hours of video, even the double-sided DVD would not do the job.

Suppose we wish revise our yearbook design so that the digital yearbook can fit onto one single-sided DVD. What information from the yearbook should we cut out? Figure 5.20 shows the percent of total storage occupied by the various types of information. Since the music and video dominate the storage requirement, let's reduce the amount those. Reducing the music from 12 hours to 10 hours will cut the required music storage from 2304 MB to 1920 MB, giving a savings of 2304 − 1920 = 384 MB. Completely eliminating the longer video segments will reduce the required video storage from 4147.2 MB to 1658.9 MB, giving a savings of 4147.2 − 1658.9 = 2488.3 MB. These two reductions, plus the corresponding overhead savings, will be (1.1)(384 + 2488.3) = 3.16 GB. Subtracting this savings from the original total required storage gives 7.68 − 3.16 = 4.52 GB, which is just about right for a single-sided DVD. Good engineering requires the ability to be constantly modifying or improving your design!

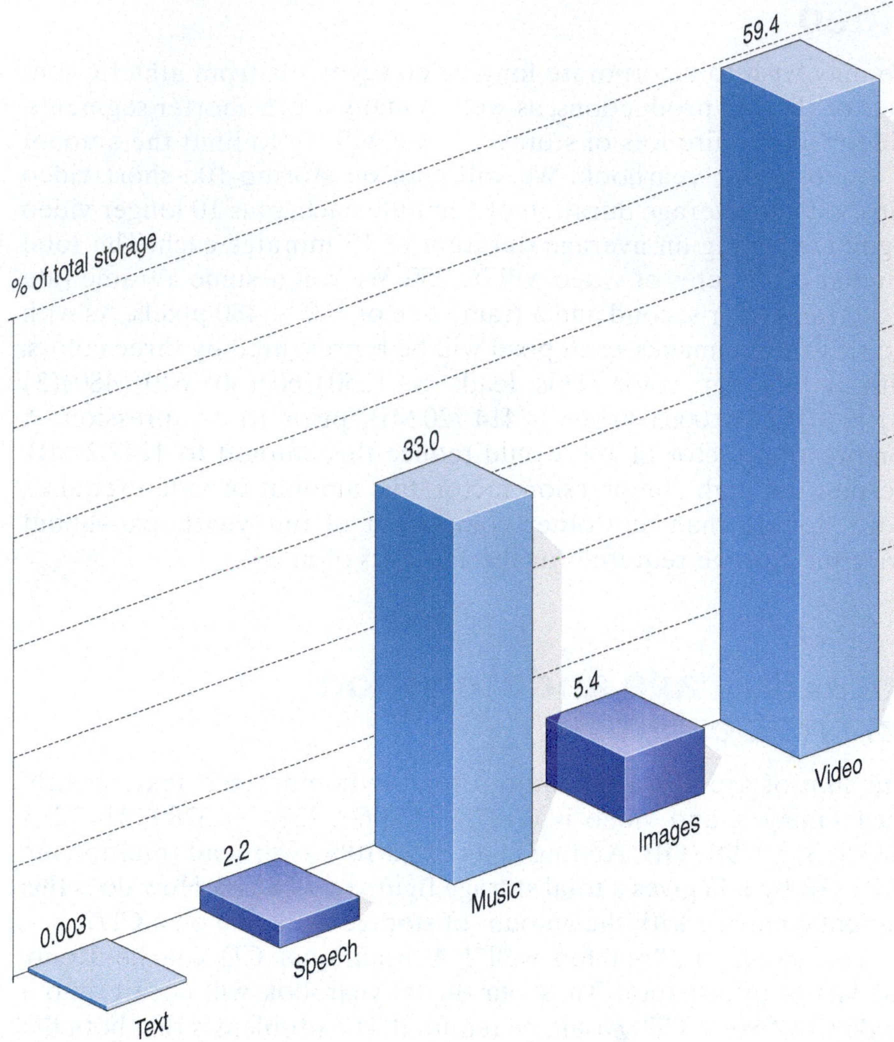

Figure 5.20 Percentage of total storage occupied by the various media comprising the digital yearbook.

Big Ideas

Math and Science Concepts Learned

Our goal in this chapter was to discover how computers and other digital devices represent various forms of information. Specifically, we were interested in digitizing text, speech, music, images, and video to create a digital yearbook. We encountered several key concepts in our study.

Information can be represented by numbers. In the case of signals, such as audio or images, these numbers are samples in time of the audio signal's height or samples in space of the image's intensity. In the case of text, the ASCII code assigns a binary number to each letter and punctuation mark.

When sampling signals or images, the samples must be spaced closely enough to prevent loss of information. In particular, the sampling rate, measured in samples per second for signals and samples per meter for images, must be at least twice the frequency bandwidth of the signal or image. This minimum acceptable sampling rate is called the Nyquist rate.

Sampling below the Nyquist rate causes aliasing, which makes it impossible to recover the original analog signal from its samples. For sinusoidal signals, aliasing causes a shift in the sinusoid's frequency.

Each physical location in computer memory, on a CD, or on a DVD can store only one of two states—for example, high voltage or low voltage in RAM, or pit versus no pit on a CD or DVD. Thus, computer memory, a CD, or a DVD can represent at each physical location only one of two possible values. For simplicity, we call these values 0 and 1. Each 0 or 1 is called a bit.

Bits can be grouped together to represent numbers other than 0 or 1, including larger integers, fractions, and negative numbers, by using the binary number system.

The number of bits in a binary number determines the number of amplitude levels that can be represented. A B-bit binary number can represent $L = 2^B$ signal values. Because samples of signals or images can take on a wide range of amplitude values, these samples must be quantized, or rounded, to one of the levels that can be represented by a B-bit binary number.

The value of B is chosen by the system's designer. A larger value for B assures less quantization error or noise. When measured on a decibel scale, the signal-to-noise ratio associated with quantization is approximately $6B$. Thus, storing samples by using $B + 1$ bits instead of just B bits will increase the SNR by about 6 dB. However, increasing the number of bits used to store information comes at the expense of increasing the size of the memory and increasing the transmission time in a communication system.

Important Equations

The Nyquist sampling rate is

$$f_s > 2f_{\max}$$

where f_s is the necessary sampling rate, measured in samples per second, and $f_{\max}$ is the bandwidth of the signal being sampled, measured in cycles per second or Hz.

The signal-to-noise ratio for a general signal and noise is

$$\text{SNR} = \frac{\max<|\text{signal}|>}{\max<\text{noise}|>}$$

The signal-to-noise ratio, measured in decibels (dB), for a quantized signal, where B is the number of bits used to represent the samples, is

$$\text{SNR} = 6B$$

This equation assumes that the maximum quantization level is matched to the maximum signal level.

Building Your Knowledge Library

W. Aspray, *John von Neumann and the Origins of Modern Computing*, MIT Press, Cambridge, MA, 1990.

M. Brain, ed., *Marshall Brain's How Stuff Works*, Hungry Minds, New York, 2001.

J. H. McClellan, R. W. Schafer, and M. A. Yoder, *DSP First: A Multimedia Approach*, Prentice Hall, Upper Saddle River, NJ, 1998.

K. Steiglitz, *A Digital Signal Processing Primer*, Addison-Wesley, Menlo Park, CA, 1996.

Coding Information for Storage and Secrecy

chapter 6

From the last chapter, we know that in our world of information—from text and music to images and video—everything is going digital. When converting information into a digital form, we know that we have some choices to make, such as how we represent the information in binary form, and how many bits we use. But what if we want to do more with our bits? Consider the following:

1. We could have so much information that a simple binary conversion would produce too many bits to conveniently store or manage. For example, we might have a whole music library that we would like to carry around with us wherever we go. Can we reduce the number of bits so that the overall data are easier to work with?
2. In the process of using our information, errors might be introduced. For example, a scratch on a CD or DVD might make some parts of the disc unreadable. How can we place information on the disc so that these errors can be detected and corrected?
3. The information that we are encoding might be sensitive and private, such as a digital cellular telephone call to a close friend. How can we alter the information so that others can't easily read and understand it?

OUTLINE

Code: The rules that define the meanings of a set of bits. For example, the ASCII code assigns 7-bit binary numbers to letters, spaces, punctuation, and other computer characters.

Encoding: The act of translating meaningful symbols into a set of bits.

Decoding: The act of recovering meaningful symbols through the translation of a set of bits.

INTERESTING FACT:

The word *code* originates from the Latin word *codex*, which literally means "tree trunk" or a piece of wood in which important words could be carved.

All of these possibilities are realized through the use of codes. A **code** is a set of rules that defines the meanings of groups of bits. **Encoding** describes the act of translating symbolic information into the bit language of a digital device. Conversely, **decoding** describes the act of interpreting the bits as meaningful symbols to the user of the device. Devices and programs that encode and decode information are digital "translators" that change digital information from one useful form to another. In this chapter, we are going to explore ways to design clever codes so that we can save, use, and protect important digital information.

Design Objective: A Digital Backpack

You do a lot of things at school, such as take notes, listen to lectures, read books, hand in homework, eat lunch, and socialize with your friends. To keep your school life going, you probably have used a bag or backpack to hold important information and items. A traditional backpack is just a container for your stuff. But what if the backpack could do much more? Consider the following:

1. Suppose it could "take notes" for you by recording the lecture and taking pictures of the chalkboard or whiteboard. Given that you are in school every day, the amount of audio and images that you would need to store could get huge. How would you store all that information efficiently?
2. It could hand in your homework for you when you walk into the classroom by "beaming" your assignment to the teacher's backpack. How would you make sure that others wouldn't be able to copy your work before or when you hand it in?
3. If you lose your backpack, all your private information might not be so private anymore. How could you keep your personal information safe in case it falls into the wrong hands?

In this design, we assume that we already have a way to digitize the information that we need in a simple manner, so that we can use all the ideas that we learned in Chapter 5 to help us. Digitizing information, if done right, gives us a nearly perfect copy of the original. It also has the advantage of being easy to transport, because new generations of digital devices continually are getting smaller and more efficient. For this design, we must ask ourselves the following questions:

- **What problem are we trying to solve**? We want to design a digital scheme by which large amounts of information, in binary form, can be stored efficiently and securely so that other individuals can access the information only when we want them to.
- **How do we formulate the underlying engineering design problem**? Some of the important features of the digital backpack include the following:
  1. It should preserve our information in an accurate and efficient manner, using the fewest possible number of bits.
  2. It should have some way of releasing our personal information only when we need to use that information.
  3. It should enable the user to define a level of security by which the information can be accessed or used.

- **What are the rewards if our design satisfies our needs**? If we do a good job in designing our digital backpack, we can expect the following:
  1. We won't have to carry large amounts of paper and books around to class anymore.
  2. All the information that we care about—text, pictures, and sounds—will be stored for us in one place.
  3. We won't have to worry about remembering phone numbers, addresses, bank account numbers, or other personal information; all of this information will always be available from our digital backpack.
  4. Even if we lose the digital backpack, we won't have to worry about ordinary people getting access to our personal lives.
  5. We might find some other important uses for the backpack. For example, the secure method we come up with for handing in homework could be used for financial transactions such as buying lunch, recorded music, or movie tickets. That way, our digital backpack could replace our wallet or purse.
- **How will we test our design**? The security level with which we protect our information should be measurable in some way. Also, we should make sure that we have a way of knowing how efficiently our storage schemes work. Finally, we can test our digital backpack by making one available for others to use and seeing how many of them find it useful in their daily lives.

6.1 Introduction

Classes of Coding Methods

All codes can be put into one of the following four classes:

- **Formatting**—A formatting code describes how information will be translated into binary patterns so that it can be read, understood, heard, or seen later.
- **Compression**—A compression code attempts to reduce the number of bits needed to represent a set of information.
- **Error correction** or error control—An error-control code adds extra bits to a set of binary information so that errors introduced into the information can be detected and, in some cases, corrected.
- **Encryption**—An encryption code modifies the bits that represent information so as to conceal their true values from unauthorized or unwanted viewers.

Compression: An encoding method that tries to reduce the number of bits needed to store information.

Error Correction: An encoding–decoding procedure that tries to detect and fix errors when they occur.

Encryption: An encoding method that tries to hide information from others.

While the objectives of these four classes of codes are different, all four of the classes are based on mathematical principles of codes. In addition, most real-world applications use more than one type of code. For example, in our digital backpack, we might take a history paper and convert the letters and punctuation into digital form, such as an ASCII representation. The ASCII characters could be compressed to reduce the number of bits needed to store the paper. Then we might use an error-correction code so that the information can still be read even if

INTERESTING FACT:

If we want to store N symbols in our codebook, at least some of the codewords will be $\log_2 N$ bits long or longer. If N is a power of two and all of the codewords are chosen to have exactly the same length C, then the length of all of them is given by $\log_2 N$ bits.

INTERESTING FACT:

As you know, time is counted out using multiples of 60. Sixty seconds equals one minute, 60 minutes equals one hour, and even 12 hours is a simple fraction (one-fifth) of 60. The reason for this choice is that the ancient Babylonians used base-60 to count all numbers. Their counting system is also used to represent angles in degrees: There are 360° in a circle.

some of the bits are lost due to memory errors made by the digital device. Finally, we might want to encrypt the paper so that no one else can read what we wrote until we hand it in.

Design of Simple Formatting Codebooks

We begin by defining the concept of a codebook. A codebook is a table that translates important symbols into the strings of binary bits that we will use to represent them. The string of bits we use for each symbol is called a codeword, and the length of the codebook is called the codebook size. An important example of a codebook is the ASCII code in Table 5.7. It translates every letter of the Roman alphabet into a different 7-bit string. The length of each of the codewords is seven bits, and the codebook size, the total number of symbols represented, is 128.

In many cases, the codeword length and the codebook size are directly related. Suppose that C denotes the codeword length and N is the codebook size. In the case of ASCII, C and N are related by

$$N = 2^C$$

In other words, the number of unique codewords corresponds to the number of different forms that the binary codewords can take. In fact, every possible 7-bit binary sequence is represented in ASCII, and we could not represent an additional codeword without repeating the codewords that have already been used. This fact immediately shows us that

$$N \leq 2^C$$

for codebooks made up of binary codewords that are all of equal bit lengths. Example 6.1 shows us how to use this fact to design simple codebooks for formatting information.

EXAMPLE 6.1 A Codebook to Format the Whole Numbers Zero to Nine

In many digital systems, we need to store only the whole numbers zero through nine. In such cases, we can assign binary codewords to each of the individual numbers. Suppose we use the same number of bits in each codeword. How many bits in each codeword do we need in order to represent these numbers?

Solution

The answer to this question can be reasoned from the relationship between N and C. Since we need to represent 10 numbers (0, 1, 2, 3, 4, 5, 6, 7, 8, and 9), we set $N = 10$. Thus, we need a value of C such that

$$10 \leq 2^C$$

Since C must be an integer greater than zero, we can simply start calculating the possibilities for the right-hand side of the inequality as follows:

$$2^1 = 2$$
$$2^2 = 2 \times 2 = 4$$
$$2^3 = 2 \times 2 \times 2 = 8$$
$$2^4 = 2 \times 2 \times 2 \times 2 = 16$$

Since 16 is greater than 10, we see that $C = 4$ gives a valid codeword length to represent the whole numbers between zero and nine. We can design the code as follows:

"0" = 0000 "1" = 0001 "2" = 0010 "3" = 0011 "4" = 0100
"5" = 0101 "6" = 0110 "7" = 0111 "8" = 1000 "9" = 1001

We've used the binary-to-decimal number conversion relationship described in the last chapter to design this code. This assignment is not the only possible choice, however. In fact, any unique assignment of 4-bit codewords to numbers works.

INTERESTING FACT:

When using 4 bits to represent 10 numbers, we have the capability to make $2^4 = 16$ different codewords. What can we do with the six unused codewords? The *hexadecimal code* assigns the six letters "A" through "F" to the last six codes in this codebook as

"A" = 1010,
"B" = 1011,
"C" = 1100,
"D" = 1101,
"E" = 1110, and
"F" = 1111.

EXAMPLE 6.2 Designing a Formatting Code for the Time on a Digital Clock

Figure 6.1 shows the face of a typical digital clock. On the clock, there are four different types of information:

1. the hours display (from 1 to 12);
2. the minutes display (from 00 to 59);
3. the seconds display (from 00 to 59);
4. the AM–PM symbol.

A digital clock uses a binary representation to store the time in its internal memory. How can we represent this information digitally?

Solution

When solving engineering problems, we have many choices as to the overall design. We will look at several and pick the best.

Design #1: Suppose we use a method that we already understand: Represent each decimal digit on the clock readout with a binary codeword. Since there are six such digits, we would need $6 \times 4 = 24$ bits. How would we handle the AM–PM designation? There are just

Figure 6.1 The face of a digital clock, using six decimal digits and a one-bit AM–PM indicator to show the time of day.

two possibilities, so a single-bit codeword would do, with "AM" = 0 and "PM" = 1. In sum,

$$\text{Total number of bits} = (4 \times 6) + 1 = 25$$

Design #2: We could pair up the digits on the clock as hours, minutes, and seconds, respectively. The possible number of hours is 12, so we will need 4 bits, because $2^4 = 16$ is greater than 12. For the minutes and seconds values, each of these units ranges from 0 to 59. We need at least 6 bits for each, because $2^6 = 64$ is a value greater than or equal to 60. We can code the AM–PM designation as before. So, we have

$$\text{Total number of bits} = 4 + (2 \times 6) + 1 = 17$$

We've saved 8 bits per codeword over our previous design.

Design #3: We can measure the time of day in number of seconds past 12 midnight—for which the maximum value is a single very large number—and encode this number with a single codeword. How many seconds are in a day? We can calculate this value as

$$60\ \text{s/min} \times 60\ \text{min/hr} \times 24\ \text{hr/day} = 86{,}400\ \text{s/day}$$

So, we would need a value of C such that 2^C is greater than or equal to 86,400. Some trial and error with a calculator shows that the smallest value of C is 17, because

$$2^{17} = 131{,}072 \quad \text{and} \quad 2^{16} = 65{,}536$$

Thus,

$$\text{Total number of bits} = 17$$

It turns out that Designs #2 and #3 are both equally good at representing the time of day in digital form, using the fewest possible number of bits.

EXAMPLE 6.3 Designing a Formatting Code for the Dates of a Digital Calendar

Figure 6.2 shows the face of a digital calendar that keeps the date in its memory.

Figure 6.2 The face of a digital calendar. It uses six decimal digits to specify which day it is within a century.

There are three different types of information within this calendar display:

1. the month display (from 1 to 12);
2. the day display (from 1 to 31);
3. the year display, which we'll assume needs to store only numbers between 2000 and 2099. In the 22nd century, you might want to get another digital calendar!

How can we represent this information digitally? As in the previous digital-clock example, we have many possible designs:

Design #1: Representing each decimal digit with a 4-bit codeword would require $6 \times 4 = 24$ total bits.

Design #2: We could use a 4-bit codeword to store the month value, because $2^4 = 16$ is a number greater than or equal to 12. Similarly, a 5-bit codeword would work for the day value, because $2^5 = 32$. Finally, we have 100 different possible years to represent, so we could use 7-bit codewords for the year ($2^7 = 128$). The total number of bits we would use in this design is thus $4 + 5 + 7 = 16$.

Design #3: How many days are in a century? A simple calculation shows that

$$(365 \text{ days/year} \times 100 \text{ years/century}) + 25 \text{ leap days/century} = 36{,}525 \text{ days}$$

The smallest number of bits required in order to represent 36,525 unique dates is 16. Thus, this method is as efficient as that in Design #2. Both schemes represent the information on the digital calendar by using the fewest possible number of bits.

These examples show how digital encoding can be used to format information. In general, an encoder takes information—such as characters, letters, or signal samples—and turns these symbols into bits for storage, transmission, or display. Both the input to the encoder and the output of the encoder are digital; that is, the encoder simply changes the representation of the discrete-valued information, not its overall content. A decoder reverses the encoding process, taking the encoded bits and turning them into useful information or signals.

Both of these processes essentially are identical to the procedure you used to decode the ASCII messages in Chapter 5. These procedures involve scanning the entries of a table for the proper translation from symbols to bits, and vice versa. This process is very similar to how you might use a language dictionary to translate words from one language to another. For this reason, the codebook is sometimes called the **codebook dictionary**.

Codebook Dictionary: The table that describes the translation of symbols to their binary representation.

An important feature of the coding process is that it is reversible: You can recover the desired information without any loss of useful content. In other words, the correct decoder turns an encoded message back into an apparent copy of the original message before it was encoded.

EXERCISES 6.1

Mastering the Concepts

1. What are the four reasons for which information is encoded into digital form? Give an example of how each code class would be used in a particular application.
2. What is the difference between codebook size and codeword length? What is the relationship between them?
3. Which is usually larger, the codebook size or the codeword length?

Try This

4. A store has 45,835 different items in its inventory. You are asked to design an efficient scheme for indexing all of the items, using binary codewords of equal length C. How many bits per codeword should you use?
5. An audio CD uses 16 bits per sample to record information. The sampling rate of a CD is 44.1 kHz.
   a. If you were to use a single audio sample to index a piece of information—that is, if $C = 16$—how large could the codebook size N be?
   b. How many audio seconds would it take up on the CD to represent all the different possible codewords in the codebook? Assume that both channels of the two-channel (stereo) recording store the same codeword.
6. Repeat Exercise 6.1.5 for an audio DVD in which the audio signal is recorded using 24 bits per sample at a sampling rate of 96 kHz.
7. Design a codebook dictionary for the seven colors of the rainbow (red, orange, yellow, green, blue, indigo, and violet). Suppose you want to store the color white. Is it possible to change the codebook without lengthening the number of bits per codeword? How would you do it?
8. You are designing a digital compass for a vehicle.
   a. Suppose you want to index the four main map orientations (north, south, east, and west) and the combined directions of northeast (NE), northwest (NW), southeast (SE), and southwest (SW). How many bits will you need in your codeword?
   b. Suppose instead that you want to give a readout in degrees. How many bits do you need in your codeword now?
   c. Write out your codebooks and all of their codewords for parts (a) and (b).
9. Astrophysicists from NASA estimate that there are 10^{21} stars in the universe. Determine the codeword length that would be required in order to index all of them if all codewords are of the same length C (*Hint:* Remember that $2^{10} \approx 10^3$.)

Back of the Envelope

10. ASCII encodes English text by using 1 byte per symbol (either letter, space, or punctuation mark). Take any book that you've

read recently and pick a page at random. Take any full line on that page and count the number of symbols in the line of text. Don't forget to count spaces and punctuation marks. Next, count the number of lines on the page. Now multiply the number of symbols per line by the number of lines, and then multiply the result by the number of pages in the book. About how many bytes would it take to store all of the characters and punctuation in this book, using ASCII code as a formatting method? A CD-ROM holds around 700 MB of information. How many similar books could be stored on a CD-ROM, using ASCII?

11. Suppose you could record everything that you see out of both of your eyes, using digital video cameras. Let the digital video cameras record video at 30 frames per second. Assuming that you live to your 110th birthday, how many frames of video would you record? How many bits would you need in order to index (not store) your stereo "life movie," using one unique index per frame? Does the answer surprise you?
12. Take a look at any car license plate in your vicinity. What is the format of the license plate? (That is, what characters or symbols are allowed in each position? How many characters are there?) Using this information, figure out how many cars can be identified uniquely using the current license-plate format. How many people live in your state? Does the license-plate format allow for more cars than there are people currently living in your state?

6.2 Principles of Compression

Compression of Matter versus Information

After we drink a can of soda and return the empty can to a recycling plant, the aluminum cans are crushed or compressed. When transporting aluminum cans, a crushed can saves physical space. An intact can and a crushed can are shown in Figure 6.3. If you were exceptionally talented with such tasks, you might try to "uncrush" a crushed can, putting it back into its cylindrical shape. Most of us would agree that expanding a can back into its original form is hard. We would also agree, though, that the compressed can

INTERESTING FACT: Launched in April of 1997, DVD (digital versatile disc) video has been the most successful consumer electronics product in history. DVDs contain high-quality video images in a digital format accompanied by digital multichannel surround sound. Many discs for movies contain additional information about the movie, such as running audio commentaries by the directors or actors, behind-the-scene featurettes on the movie's production, and still artwork. This presentation is made possible by the DVD's digital format, which allows convenient storage of many types of information on a single disc.

Figure 6.3 An aluminum can in its original and compressed shape.

contains the same amount of metal as the original can; the physical makeup of the can hasn't changed in any way.

Compression describes the act of reducing the number of characters, symbols, or bits used to represent any given signal. This is shown conceptually in Figure 6.4. Compression of information is like compression of something physical: By compressing information, the amount of physical resources (e.g., space) that we use to represent it is reduced. Unlike a crushed soda can, however, a compressed signal may be reconstructed to closely resemble its original form.

There are many physical technologies used to store information—optical discs such as CDs and DVDs, magnetic discs such as computer floppy and hard disks, and memory

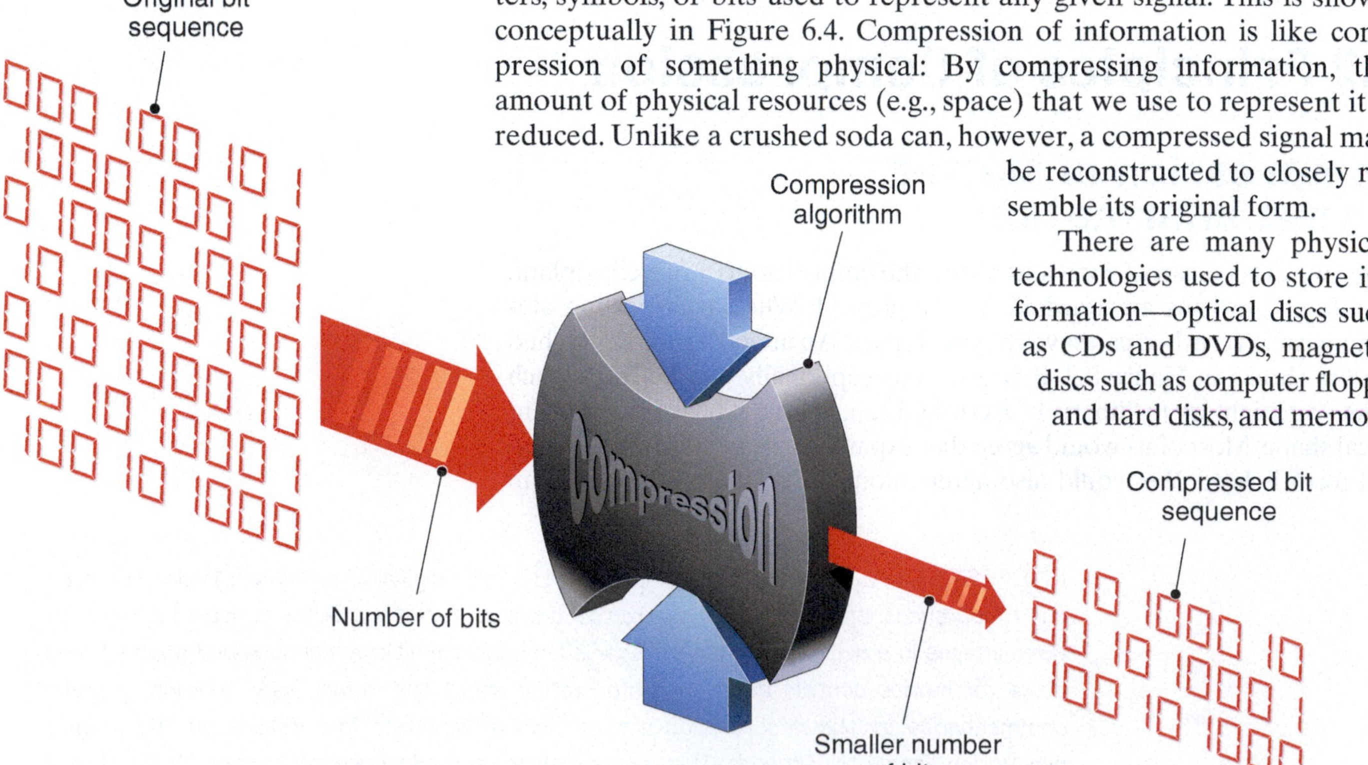

Figure 6.4 Compression is the act of reducing the number of symbols or bits needed to represent a signal or other data.

chips. Some of these technologies can store an amazing amount of information in a small physical space. If we are getting so good at storing larger and larger amounts of information in the same amount of material, why do we need compression? Here are some reasons:

- For many applications, the raw-data format requires too many bit locations to store easily in existing media. A good example is digital video, as we will discuss later in this chapter.
- As consumers, we have an ever-greater desire to store more information digitally. Digital data forms are easier to use, easier to manipulate, and easier to communicate to other digital devices. Compression gives us a way to make better use of our existing storage capabilities.
- Compression gives us the ability to transmit data more efficiently using fewer resources such as energy or battery power, transmission time, and communication bandwidth. We'll learn more about these issues in Chapter 7.

With all these advantages, it would seem that compression would and should be a natural part of every system. However, there are two drawbacks to compression:

- Since compression and decompression are both coding steps, they require more computer software and processing time to complete, making the overall system more complex than if the raw-data format were used exclusively.
- Almost all compression methods delay the progress of the signal through the system, because they usually need to examine a group of symbols or bits before they can perform a compression or decompression step. For this reason, the resulting compressed or decompressed signal can be made available only after a delay. In many systems, the delay is small (e.g., 24 milliseconds for digital audio that is compressed using the MP3 audio compression standard), but it can be a limiting factor in some cases.

EXAMPLE 6.4 The Need for Video Compression

DVDs can store about two hours of high-quality video and multichannel surround sound. They use clever audio and video compression technologies to get this information onto one single-layer DVD, which has a capacity of about 4.7 GB. How many DVDs would we need to store the uncompressed information of a two-hour movie?

Solution

For simplicity, we will consider only the video portion of the data stream, as the audio portion requires a much lower storage capacity in its raw format. To determine the amount of storage needed to represent two hours of video, we can use Equations (3.5) and (3.7) from Chapter 3. We need to determine the number of pixels required for each frame, the number of bits for each color pixel, and the number of frames needed for a two-hour movie. The number of pixels, N, in a single frame within a digital version of a television signal with 480 horizontal rows and 640 vertical columns of information is

$$N\frac{\text{pixels}}{\text{frame}} = 480\frac{\text{rows}}{\text{frame}} \times 640\frac{\text{pixels}}{\text{rows}} = 307{,}200\frac{\text{pixels}}{\text{frame}}$$

If we use 24 bits to store each color pixel in each digital image of our movie, the total number of bytes needed to store each TV frame is

$$B\frac{\text{bytes}}{\text{frame}} = N\frac{\text{pixels}}{\text{frame}} \times \frac{3\text{ bytes}}{\text{pixel}} = 307{,}200\frac{\text{pixels}}{\text{frame}} \times \frac{3\text{ bytes}}{\text{frame}} = 9.21 \times 10^5\frac{\text{bytes}}{\text{frame}}$$

All we need to do now is multiply the number of bytes per frame by the number of frames of video that there are in a two-hour movie. This calculation is as follows:

$$\frac{\text{bytes}}{\text{movie}} = B\frac{\text{bytes}}{\text{frame}} \times \frac{30\text{ frames}}{1\text{ s}} \times \frac{3600\text{ s}}{1\text{ hr}} \times \frac{2\text{ hr}}{1\text{ movie}} = 1.99 \times 10^{11}\frac{\text{bytes}}{\text{movie}}$$

In other words, to store the video for a two-hour movie in its digital uncompressed format would take 199 GB, which is *over 40 times* the storage capacity of one single-sided DVD! Using DVDs in this format would test both your strength and patience. Not only would you have to carry home 40 discs in order to play one movie, but also you would have to change the disc roughly every three minutes!

The video compression method used in the DVD standard (called MPEG-2) reduces the size of this information to less than 2.5% of its original bit size, allowing a movie to be saved in a much smaller package. Isn't it amazing how well it works?

Compression Ratio

Compression Ratio: The ratio of the number of bits in the original representation of a signal to the number of bits in a compressed version of the signal.

The **compression ratio** of a compressed signal is defined as the ratio of the number of bits in the original representation of the signal to the number of bits of the compressed version of the signal, or

$$R_{\text{compression}} = \frac{(\#\text{ of bits in original signal})}{(\#\text{ of bits in compressed signal})}$$

The compression ratio is usually a number greater than one, with larger values corresponding to higher levels of compression. For example, in digital video, a compression ratio of

$$R_{\text{compression}} = 40$$

means that, on average, every 40 bits of the original video signal are reduced to 1 bit in the compressed video signal.

Another way to express the compression ratio is by using the notation $B_{\text{original}} : B_{\text{compressed}}$, where B_{original} is the number of bits in the original signal and $B_{\text{compressed}}$ is the number of bits in the compressed version of the signal. For example, the value of $R_{\text{compression}} = 40$ corresponds to a compression ratio of 40 : 1.

Compression methods can also be applied to symbols that have already been encoded using a formatting code. Remember ASCII? It

INTERESTING FACT:

The compression ratios for images are not as high as those for video. Good video compression schemes use the similarity between image frames in the video stream to reduce the size of the compressed signal further than if each image were compressed separately. That is why image compression ratios are typically only about 8 :1 or so.

uses 7 bits per character. It turns out that, with some cleverness and effort, ASCII-formatted English text can be compressed to roughly 1.35 bits per character on average—a compression ratio of about 5:1. How we can compress text in this way will be considered later in this chapter.

Lossy versus Lossless Compression

There are two general types, or "flavors," of compression technology. While both achieve the overall goal of compression—reducing the number of bits needed to store or transmit a set of numbers—they do it in different ways.

Lossy compression describes a class of compression techniques that throw away information about the sets of numbers or symbols being compressed. Lossy compression literally amounts to giving up certain features of the set of numbers or symbols. The goal in lossy compression is to lose those features that wouldn't be noticed or that aren't considered important in the application. By its very nature, lossy compression is not reversible; that is, you cannot get back the exact original set of numbers from the compressed version.

Lossy Compression: A class of compression techniques that selectively throw away information about the sets of numbers or symbols being compressed.

Lossless compression describes a class of compression techniques for which nothing about the original set of numbers or symbols is lost in the compression process. Lossless compression is akin to letting the air out of a balloon; nothing about the balloon has been changed in the process except its size and shape. By putting air back into it, the balloon

Lossless Compression: A class of compression techniques for which nothing about the original set of numbers or symbols is lost in the compression process.

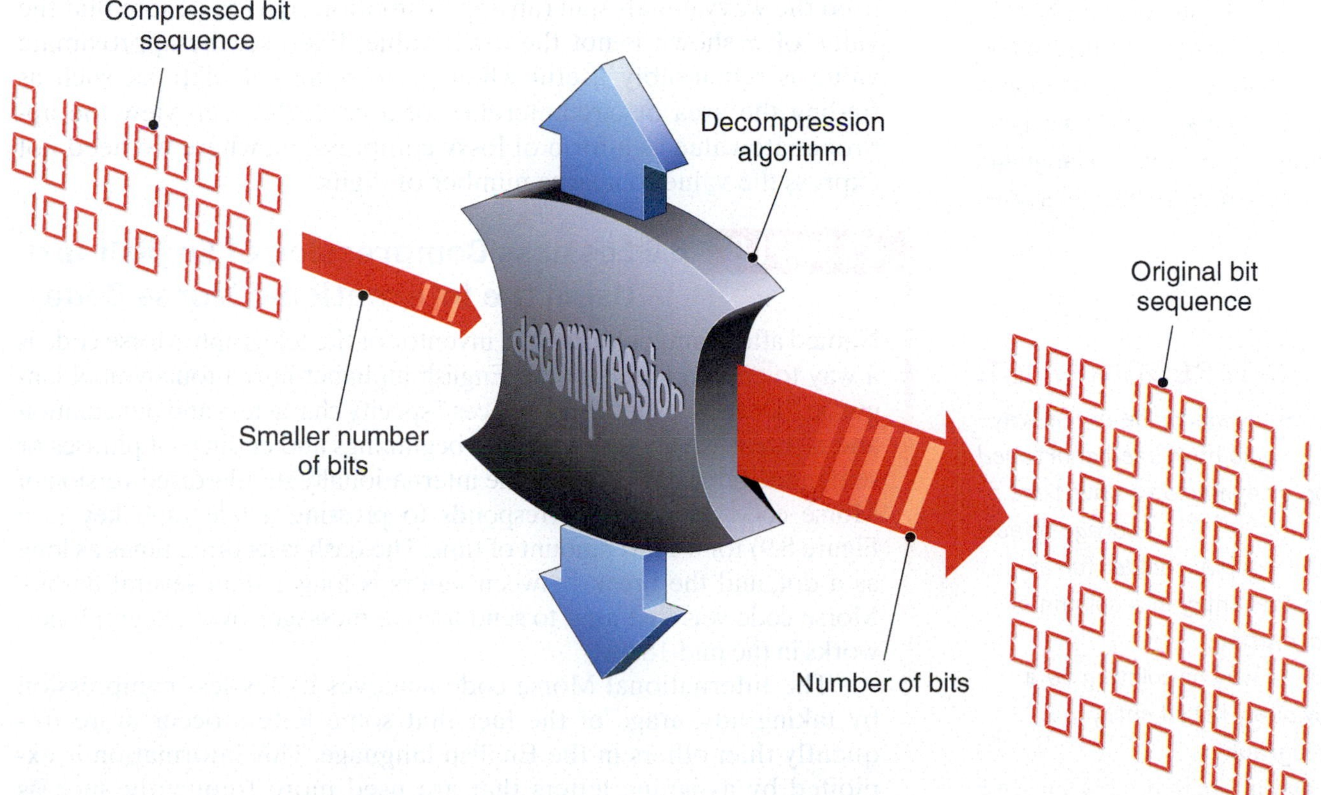

Figure 6.5 Decompression is the act of reconstructing the original signal, or something close to it, from a compressed version.

recovers its original size and shape. Unlike deflating a balloon, however, compressing a set of numbers requires a mathematical process or algorithm. With lossless compression, the process is reversible and the reverse process is called decompression (Figure 6.5). In this case, the exact set of numbers can be recovered by undoing the compression process. Because of this requirement, however, lossless compression does not achieve as high a compression ratio on most signals as lossy compression does.

The applications of lossy and lossless compression in Examples 6.5 and 6.6 will help illustrate some of their features.

EXAMPLE 6.5 Lossy Compression by Rounding Off a Numerical Result

There are many situations throughout life where being overly precise ends up wasting effort and resources. In this text, we have tried to save your time and resources by simplifying the way we represent numbers. In most places, we have rounded all calculations to some useful set of digits—usually three to four significant digits. Numbers such as the value of pi,

$$\pi = 3.141592653589793\ldots$$

have been rounded to a suitably smaller set of significant digits, such as

$$\pi \approx 3.1416$$

so that we can perform our calculations with ease. Although your calculator shows more digits of precision than that above, it, too, approximates the true value to some number of digits. Here, we have used the wavy equals sign (an approximation sign) to denote that the value of π shown is not the exact value. Even so, the approximate value is remarkably useful when performing calculations, such as finding the area or circumference of a circle. We can view the approximate value as a form of lossy compression, where we need not express the value to a large number of digits.

INTERESTING FACT:

Samuel F. B. Morse, who was born on April 27, 1791, and died on April 2, 1872, was the inventor of the electric telegraph (1832–1835) and Morse code (1838). Morse code is a way to encode letters of the English language into a four-symbol language.

EXAMPLE 6.6 Lossless Compression of English Text Using the International Morse Code

Named after Samuel Morse, the inventor of the telegraph, Morse code is a way to encode letters of the English alphabet into a four-symbol language. "Dots," "dashes," and "spaces," specify characters and punctuation and a "break" symbol denotes the beginnings and endings of phrases or sentences. Table 6.1 illustrates the internationally standardized version of Morse code. Each dot corresponds to pressing a telegraph key (see Figure 8.9) for a short amount of time. The dash lasts three times as long as a dot, and the break between letters is longer than several dashes. Morse code was first used to send textual messages over telegraph networks in the mid-1800s.

The International Morse code achieves its lossless compression by taking advantage of the fact that some letters occur more frequently than others in the English language. This information is exploited by assigning letters that are used more frequently, such as "E" and "T," to dot-and-dash patterns that are short and use more dots than dashes. These letters take less time and effort to send. Conversely, infrequently used letters, such as "Z" and "Q," are assigned

INTERESTING FACT:

There are many examples of the ways in which people have used shortened names or symbols to save time or storage space:

- *Ditto*—a placeholder for an identical entry in a column of information;
- *Ibid*—a way of pointing to a previously listed entry in a bibliography.

And how about the last time you said "Me too?" It saved you from having to say everything all over again, didn't it?

Table 6.1 The Codebook for the International Morse Code

Letter	Morse	Letter	Morse	Letter	Morse
A	·–	N	–·	Ä	·–·–
B	–···	O	–––	Á	·––·–
C	–·–·	P	·––·	Å	·––·–
D	–··	Q	––·–	Ch	––––
E	·	R	·–·	É	··–··
F	··–·	S	···	Ñ	––·––
G	––·	T	–	Ö	–––·
H	····	U	··–	Ü	··––
I	··	V	···–		
J	·–––	W	·––		
K	–·–	X	–··–		
L	·–··	Y	–·––		
M	––	Z	––··		

Digit	Morse	Punctuation Mark	Morse
0	–––––	Full stop (period)	·–·–·–
1	·––––	Colon	––– ···
2	··–––	Comma	––··––
3	···––	Question mark (query)	··––··
4	····–	Apostrophe	·––––·
5	·····	Hyphen	–····–
6	–····	Fraction bar	–··–·
7	––···	Brackets (parentheses)	–·––·–
8	–––··		
9	––––·		

long sequences that have more dashes than dots. The use of variable-length symbols helps to minimize the amount of transmission time it takes to send English text via Morse code.

Choosing a Compression Method

Lossy and lossless compression methods are used in two different ways in modern technology:

1. Lossy compression methods are most useful for signals and information that we perceive, such as sounds and images. The quality of what we see and hear is closely tied to the limitations of our eyes and ears, which, while being remarkable measuring devices, do have limits in their ability. Compression exploits these limits by throwing away "unnecessary" bits. As stated previously, lossy compression generally gives bigger compression ratios than does lossless compression.
2. Lossless compression methods are most useful for signals and information that we count on for accuracy. Financial and medical records, computer programs, and text documents are some examples of data that are often compressed by lossless compression methods.

EXERCISES 6.2

Mastering the Concepts

1. Name two different lossy compression methods and two different lossless compression methods.
2. Can one compress a text file by simply throwing away bits from the ASCII representation of each letter? Why or why not?
3. Which compression method (lossy compression or lossless compression) preserves a numerically perfect signal in compressed form?

Try This

4. Suppose a particular set of images can be compressed from their raw 8-bits-per-pixel format by a factor of 10:1. You have 100 of these black-and-white digital images that are 640×480 pixels in size. How much disk space do you need in order to store all 100 images in a compressed form?
5. Suppose a particular single-sided, single-layer recordable DVD holds 4.2 GB of information. You have 100 GB of digital video to store in its raw format. What compression ratio do you need in order to store all of your video on a single DVD?
6. The MP3 audio encoding standard often allows CD-quality stereo audio to be compressed with a negligible loss of audio quality with an 8:1 compression ratio. How many four-minute songs can fit on an MP3-encoded CD? (A standard audio CD holds about 74.5 minutes of stereo music.)
7. Rounding often is used in engineering calculations to save storage space. How many bits are required in order to store the fractional part of a number when the fractional part is rounded to four significant decimal digits?

Back of the Envelope

8. How efficient would it be to transmit random sets of letters such as "WEOFIJC VHKJSDHFKASUDY RWERUIOLGHS MVN" using Morse code? Do you think the Morse coding scheme is any better than what would result if we were to scramble the character assignments in the codebook table in Table 6.1 and use the new table instead? Why or why not?
9. Slavic languages, such as Russian, use more consonants per word in a sentence than English does. Which do you think would take longer, using Morse code to send an English novel or using Morse code to send a Russian version of the same novel? Give reasons for your answer.

6.3 More on Lossless Compression

As mentioned in Section 6.2, one way to reduce the amount of storage we need for information is to throw unimportant information away using lossy compression. Lossy compression makes a lot of practical sense; why bother saving what you don't need? In many situations, however, there are good reasons to store information without any degradation or loss:

- For data records that must remain error free, such as bank records, medical records, and text documents, lossy compression cannot be used. Imagine the uproar that would be created if banks started using lossy compression to store amounts of money in personal bank accounts; even small errors in account balances would not be tolerated by the public.
- In some situations, we might not know which component of the signal is unimportant. A good example of this situation is in data collection for scientific experiments. Recently, scientific organizations from several countries have taken digital images of most of the surface of the Earth, using cameras on orbiting satellites such as the International Space Station (Figure 6.6). This information is expected to be useful for many future studies about our planet (e.g., studies on weather patterns, crop yields, and other phenomena). When storing these images, these organizations want to keep as much of the information intact as possible. Even a slight degradation might limit the usefulness of the data for as-yet-unknown studies.

INTERESTING FACT:

JPEG stands for Joint Pictures Experts Group, an international standards body that makes decisions on ways to store images digitally.

Figure 6.6 The International Space Station, a platform designed for the gathering of scientific measurements about the Earth, its atmosphere, and outer space.

Run-Length Coding

Run-length coding is a lossless compression technique that we'll use to reinforce the basic concept of lossless compression. Run-length coding is applicable to streams of bits that have several 1's or 0's appearing in succession, such as those produced by office facsimile machines. A simple example illustrates the technique.

Run-Length Coding: A coding method for losslessly compressing binary signals that contain long strings of contiguous 1's and 0's.

EXAMPLE 6.7 Run-Length Encoding of a Bit Sequence

Suppose we have the 30-bit sequence

010000000111111000111111100000

This sequence could be described in the following way:

"A single 0, followed by a single 1, followed by seven 0's, followed by six 1's, followed by three 0's, followed by seven 1's, followed by five 0's." We could use shorthand to represent the foregoing sentence as

(1,0) (1,1) (7,0) (6,1) (3,0) (7,1) (5,0)

INTERESTING FACT:

In many compression methods, the location in which a number appears is often as important as the value of the number. For example, what bit sequence is produced from the run-length-encoded representation [0] (1) (1) (6) (7) (3) (7) (5)? How is it different from the sequence with the run-length-encoded representation [0] (1) (1) (7) (6) (3) (7) (5)?

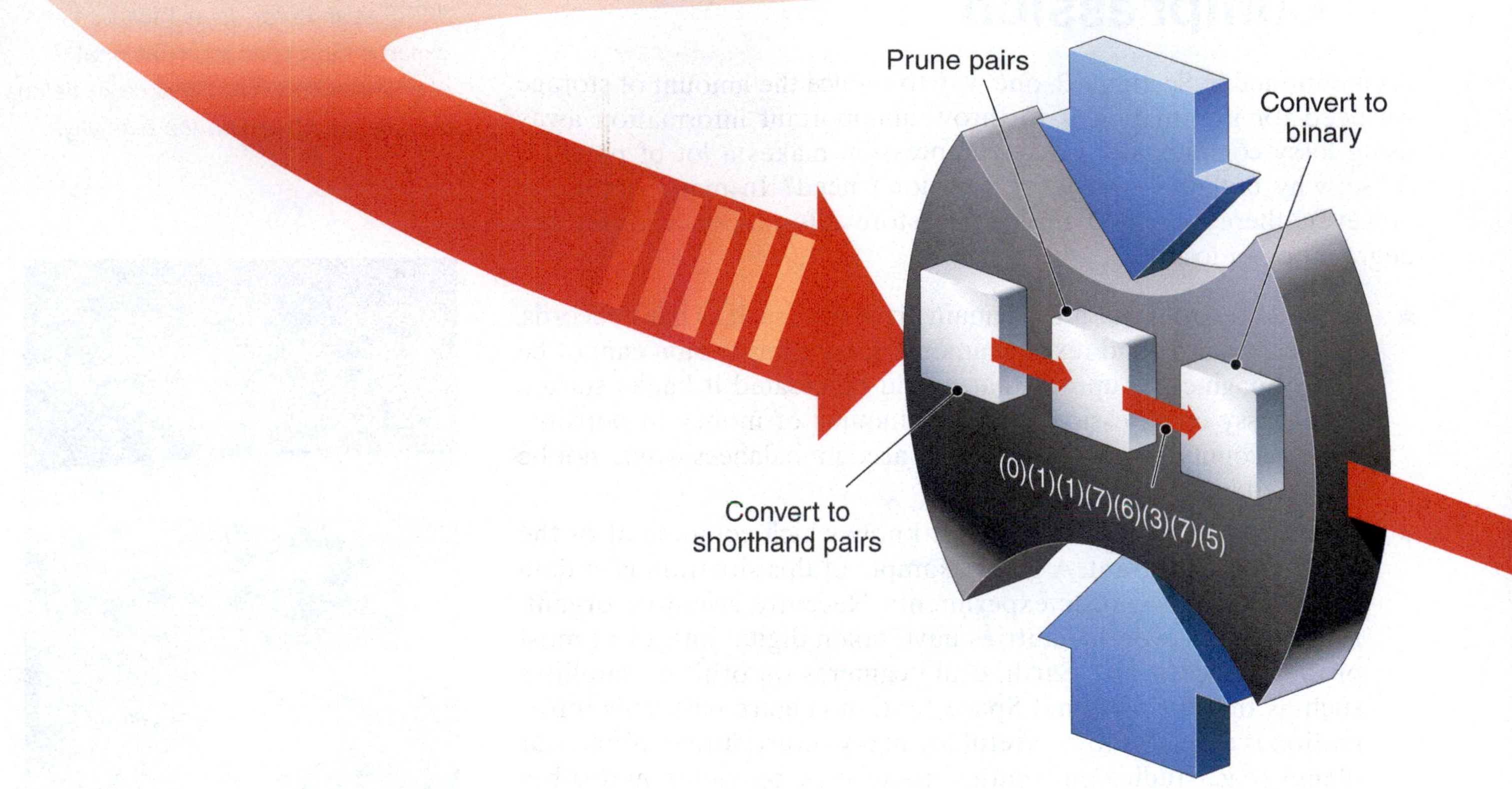

Figure 6.7 Example of run-length coding where a string of 30 bits is losslessly compressed into a string of 22 bits.

> **INTERESTING FACT:**
> Office facsimile machines use a more sophisticated form of run-length encoding. By compressing the black-and-white information they scan from a page, they can reduce the amount of time required to send a fax by a factor of 5 to 10.

The first number in each pair gives the number of bits to be represented, and the second number gives the bit value. Since the bit values are alternating, we could shorten this list of numbers even further as

[0] (1) (1) (7) (6) (3) (7) (5)

In this new representation, the first bracketed symbol is the first bit value, and the remaining numbers are the numbers of 0's and 1's that should be repeated.

We still cannot store this sequence of numbers in binary form, however, because the numbers are not in the form of bit values. To get these numbers to bit values, we convert the decimal numbers to their binary representation, noting that all numbers are between 0 and 7, a 3-bit range. So, we have

[0] [001] [001] [111] [110] [011] [111] [101]

or, more compactly, 0001001111110011111101. This sequence of 22 bits can be used to reconstruct the original 30-bit sequence without any errors. Since 22 is less than 30, we have performed compression without any loss of information. All the steps of this compression process are shown as a block diagram in Figure 6.7.

In Example 6.7, the compression ratio (15:11, or 1.36) is not that great, because we do not have many long sequences of successive 1's and 0's. In other situations, the number of successive 0's and 1's is much greater, and thus the compression ratio can be higher. Of course, more bits would be needed to represent each number of successive 0's. Three bits per sequence would not be enough.

Example 6.7 illustrates the general features of lossless compression:

- Lossless compression maps bits to bits; that is, a series of bits is turned into another, hopefully shorter, series of bits.
- There is a clearly identifiable encoding procedure that creates the shortened bit sequence from the original bit sequence.

In addition, there must be a clearly identifiable decoding procedure that expands the shortened bit sequence out to the original bit sequence. Otherwise, there is no way to get the original bits back!

Lossless compression schemes work well for certain types of signals, but not others. We might ask when lossless compression can be done.

INTERESTING FACT:

Many people use compression methods when ordering special license plates (vanity plates) for their cars, such as

U R GR8

and

FULL 2 MT

These plates use the phonetic sounds associated with letters and numbers to sound out words. The decoding of these license-plate messages, however, is not always easy, because the rules for decoding the messages are not well defined.

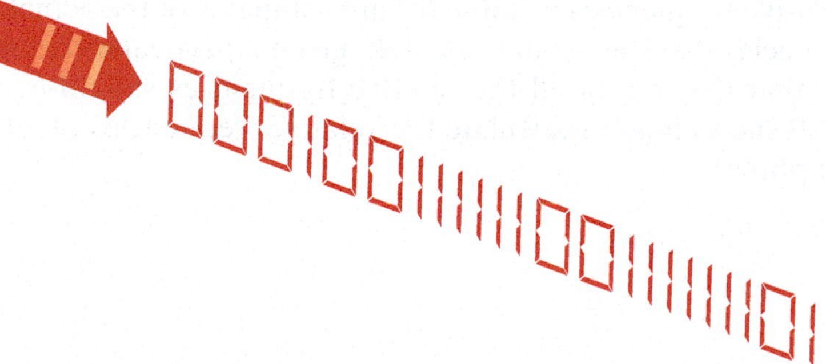

Relative Frequency

The ability to compress any set of data in a lossless fashion is tied to the concept of relative frequency. **Relative frequency** of occurence is defined as

$$f_R(E) = \frac{\text{Number of observations of an event } E}{\text{Number of total observations}}$$

where f_R is the relative frequency and E is an event of interest.

Relative frequency is tied to our notion of probability and chance. For example, if the weatherperson tells us that there is an "80% chance of rain today," we understand this phrase to mean, "For days with weather conditions like today, it rains on four out of every five days on average."

In the weatherperson's case, the event of interest, E, is daily rain. While this simple example illustrates the concept of relative frequency, the actual measurement of relative frequency in compression is much more direct than weather prediction. In fact, it is almost always the case that the data we're compressing give us both the number of observations of the event E and the total number of observations. So, we can calculate relative frequency directly before we encode the information.

Relative Frequency: A measure of how frequently an event occurs, computed by taking the ratio of the number of observations of the particular event and the total number of observations of all events.

INTERESTING FACT:

Table 6.2 shows that a space occurs on average once every 1/0.1859 = 5.3 symbols. This means that the average length of any word in an English sentence is about 4.3 letters.

Table 6.2 Relative Frequencies of Letters in the English Language

Letter	Relative Frequency
A	0.0642
B	0.0127
C	0.0218
D	0.0317
E	0.1031
F	0.0208
G	0.0152
H	0.0467
I	0.0575
J	0.0008
K	0.0049
L	0.0321
M	0.0198
N	0.0574
O	0.0632
P	0.0152
Q	0.0008
R	0.0484
S	0.0514
T	0.0796
U	0.0228
V	0.0083
W	0.0175
X	0.0013
Y	0.0164
Z	0.0005
Space	0.1859

EXAMPLE 6.8 Relative Frequency of Letters in the English Language

Anyone familiar with the English language knows that certain letters are used more often than others. For example, the letter "E" is used in many words, whereas the letters "Q" and "Z" aren't used very much. If we were to take a look at a large amount of English text, how often would each of the letters appear?

Table 6.2 shows the relative frequencies of the letters in English text as calculated by counting the number of each in a large number of books. In this table, all punctuation has been ignored, and only the nonletter spaces between words have been considered. The table shows that the letter "E" is used over 10% of the time, whereas the letter "Z" is used only 0.5% of the time. In other words, about 1 out of every 10 symbols in English is an "E," whereas only 1 out of 200 symbols is a "Z."

Figure 6.8 shows a bar graph of the relative frequencies of these letters, with the most frequently used letters shown on the left.

The relative frequencies in Table 6.2 are estimates of the actual relative frequencies that one would calculate given a particular text message. Note that the sum of all the relative frequencies should be 1.0. Example 6.9 shows how to calculate the relative frequencies of letters for a given phrase.

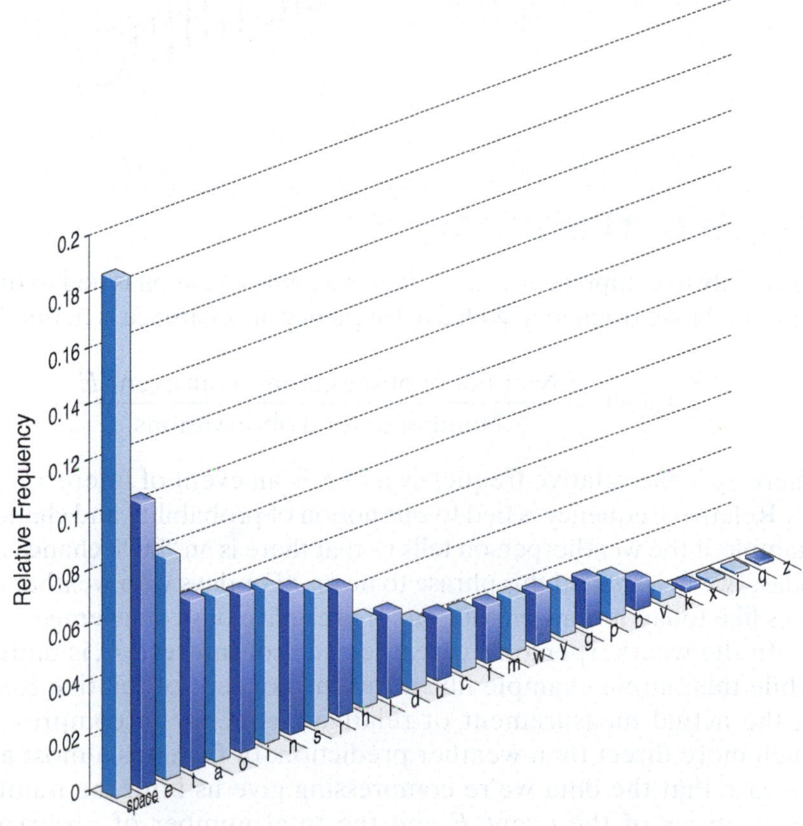

Figure 6.8 A bar chart showing the relative frequencies of letters in the Roman alphabet when used in English text.

EXAMPLE 6.9 Calculating Relative Frequency of Text

Consider the following popular children's phrase:

"I scream,you scream,we all scream,for ice cream"

In this phrase, the words are separated by either a space or a comma but not both. This phrase contains six "e"s, five "a"s, five "c"s, five "r"s, four "m"s, three "s"s, two "l"s, two "o"s, three commas, and one each of capital "I," small "i," "f," "u," "w," and "y". Including the six spaces in between words, the phrase contains a total of 47 characters. What is the relative frequency of each of these letters and symbols?

Solution

Using the definition of relative frequency, we find that some of the values are

$$f_R(\text{letter "e"}) = \frac{6}{47} = 0.1277$$

$$f_R(\text{letter "a"}) = \frac{5}{47} = 0.1064$$

$$f_R(\text{letter "m"}) = \frac{4}{47} = 0.0851$$

$$f_R(\text{letter "s"}) = \frac{3}{47} = 0.0638$$

$$f_R(\text{letter "l"}) = \frac{2}{47} = 0.0426$$

$$f_R(\text{letter "i"}) = \frac{1}{47} = 0.0213$$

The relative frequencies of the other symbols are equal to one of these values. For example, $f_R(\text{letter "c"}) = f_R(\text{letter "a"})$, because there are five "c"s and five "a"s in the phrase.

Using Relative Frequency to Compress Information

We can use the relative frequencies of a set of symbols or characters to come up with an efficient way to represent the symbols or characters. The bits we save then can be used to our advantage, allowing us to store more information in the same amount of space. To do this task, we must establish a table relating each symbol to a particular unique bit sequence. Unlike in the ASCII table (Chapter 5), however, we will allow an important modification: Not all of the bit sequences that we use need to be the same length. For example, some letters could be assigned short bit sequences, whereas others might have long sequences assigned. Of course, we would need to provide the table along with the bit sequence itself if we want to read back the original information.

KEY CONCEPT

One of the basic ideas of lossless compression is to use the shortest bit sequences for those events that happen most frequently.

The basic ideas that we'll use in representing the information are as follows:

1. Symbols that happen more frequently should be described using as short a name or bit sequence as possible.
2. Rare symbols, on the other hand, can be described using a long name or bit sequence, since they occur less often.

If done correctly, our bit assignment procedure will save us valuable storage and transmission resources. A good code will do the job of reducing the amount of storage space needed to record a given set of information. But how do we figure out the quality of our code in compressing information?

Average Codeword Length: The average number of coded bits it takes to store an original symbol sequence, using a particular code.

Average Codeword Length One way is to determine the **average codeword length**, or the average number of coded bits it takes to store or send an original symbol sequence. The average codeword length can be calculated as

$$L_{\text{average}} = f_R(s_1) \times L_1 + f_R(s_2) \times L_2 + f_R(s_3) \times L_3 + \cdots + f_R(s_N) \times L_N$$

which can be expressed using summation notation as

$$L_{\text{average}} = \sum_{i=1}^{N} [f_R(s_i) \times L_i]$$

where

$$L_i = \text{code length of } i\text{th codeword}$$

and

$$f_R(s_i) = \text{relative frequency of } i\text{th symbol}$$

In these equations, N is the number of different symbols that appear in the signal or message.

The average codeword length depends on the way labels are assigned to symbols in a set of data. If short bit names are assigned to symbols that occur more frequently, then the average codeword length will be small.

INTERESTING FACT:

The Greek Letter Σ

Engineers often use the Greek letter sigma (Σ) to represent a sum of a series of terms, especially if there are a lot of terms to be added together. In Greek Σ is an S.

EXAMPLE 6.10 Long-Distance Area Codes and Rotary-Dial Telephones

Virtually all modern telephones use Touchtone® dialing, in which a push-button keypad creates tones to denote the 12 numbers and characters on the telephone keypad. Each of the digits takes about the same amount of time to send. Prior to the invention of Touchtone dialing, telephones had rotary dials (Figure 6.9). This dial had a label containing the numerals 0 to 9. Attached above the dial was a disk with finger-sized holes over each of the numerals. To dial a telephone number, one inserted her or his finger into the holed disk at the digit to be dialed, rotated the disk clockwise to the metal hook stop, and let go. The dial would rotate back to its original position while creating a series of electrical pulses that were sent along the telephone line to the telephone company's central office. The time required to send each dialed digit was directly proportional to its value. Sending a 0 (which was really a 10) took 10 times as long as sending a 1. Larger numbers, such as 8 and 9, took much longer to dial than smaller numbers, such as 1 or 2.

As the population of the United States grew, the number of available telephone numbers across the country had to increase in order to support more distinct users of the telephone network. Before the invention of long-distance area codes, the number of digits per tele-

Figure 6.9 A telephone with a rotatory dial.

phone number had grown from three or four to seven. Direct-dial long distance was not possible, as all telephone calls to places outside the user's home region had to be operator assisted. By adding three numbers at the front of all telephone numbers, the convenience of local calling without an operator was extended to long-distance telephone calls. We might ask how the long-distance area codes were assigned.

Solution

The engineers at American Telephone and Telegraph (AT&T) figured that the cities with the highest population would receive the most long-distance telephone calls. To make dialing to these big cities faster, these engineers gave area codes with the fastest-dialing three-digit numbers to the largest populated cities (for example, area code 212 for New York, area code 213 for Los Angeles, area code 312 for Chicago, area code 313 for Detroit, and so on). Less populated areas, such as Alaska, were expected to receive far fewer calls and therefore were given area codes that took a long time to dial. Choosing area codes in this way made a lot of sense. It reduced wear and tear on telephones and on the telephone user's fingers by minimizing the amount of rotation of the pulse dial for frequently called numbers.

With the advent of Touchtone dialing, the advantage of smaller-numbered area codes disapperead. New area codes no longer needed to be designed with this efficiency in mind.

INTERESTING FACT:

The original set of area codes all had a "0" or a "1" as the middle digit. Now any one of the ten digits on a telephone dial can be used as the middle digit of a North American telephone area code.

Example 6.10 illustrates how the careful assignment of labels in a code can help save time and effort in an application. This form of convenience is used in many aspects of our everyday lives. Sometimes, we employ such conveniences without directly thinking about them.

EXAMPLE 6.11 Average Codeword Length for a Phrase

Consider the following phrase from our previous discussion in Example 6.9:

"I scream,you scream,we all scream,for ice cream"

In this phrase, there are 16 different characters.

Solution

We could set up a code that assigns 4 bits to each of the 16 different characters. For example, we could have

space = 0000, "e" = 0001, "a" = 0010, "c" = 0011, "r" = 0100,
"m" = 0101, "s" = 0110, comma = 0111, "l" = 1000, "o" = 1001,
"I" = 1010, "i" = 1011, "f" = 1100, "u" = 1101,
"w" = 1110, and "y" = 1111

Then we could save the sequence of bits, as well as the codebook that translates these bits to the individual characters. The average codeword length per symbol is clearly $L = 4$ bits, since the individual codewords are each four bits long. Assuming that the codebook were available to whomever received or read the compressed data, the number of bits needed to store or send the phrase would be $47 \times 4 = 188$ bits.

We can do even better than this value of 188 bits, however, since certain characters such as "e" and "a" appear more frequently than others. Suppose we use the code given by

space = 100, "e" = 101, "a" = 000, "c" = 001, "r" = 010,
"m" = 1100, "s" = 1101, comma = 0110, "l" = 0111,
"o" = 11100, "I" = 111010, " i" = 111011, "f" = 111100,
"u" = 111101, "w" = 111110, and "y" = 111111

This codebook uses codewords with fewer bits to name characters that appear more often in the message. Using the relative-frequency values from our example, we find that the average codeword length for this new code is

$$L_{\text{average}} = \sum_{i=1}^{N} [f_R(i\text{th letter}) \times L_i] =$$

$$\frac{1}{47}[6(3) + 6(3) + 5(3) + 5(3) + 5(3) + 4(4) + 3(4) +$$

$$3(4) + 2(4) + 2(5) + 1(6) + 1(6) + 1(6) + 1(6) + 1(6) + 1(6)] =$$

$$\frac{175}{47} = 3.72 \text{ bits/symbol}$$

This number might seem strange to you: What defines 72 hundredths of a bit anyway? Remember, however, that this number is an *average* value. It makes sense only over a block of symbols. What it states is that we can encode 100 symbols (or letters, in this case) by using only 372 bits for groups of sentences having a similar structure to that of our example.

The sequence of bits for this coded version of the phrase would be 175 bits long, not including the codebook information, which is shorter than the 188 bits needed for the previously coded version. Of course, the codebook has been designed to store just this message, so it is likely that it would be less useful for other phrases. For long text messages (say, a book), however, the size of the codebook would be much shorter than the coded message, so we would get a greater savings in storage by using this method instead of one that uses codewords that have identical bit lengths. Morse code described in Example 6.6 was designed in a manner similar to this example.

The Limits of Lossless Compression

Lossless compression seems like an answer to all our compression needs. We can represent information that we care about with fewer bits without any loss. But how many bits can we get rid of and still have a faithful representation of our information? Can we continue to reduce the number of bits without end? Example 6.12 suggests that a fundamental limit exists for all lossless compression methods.

EXAMPLE 6.12 A "Great" Compression Method?

The president of a large Internet company has an idea one day. He is barely familiar with compression technologies, but he believes that after a compression algorithm has been applied to a signal or a piece of data, the number of bits needed to store or transmit the signal is always less than the number of bits in the original. So, he thinks, why not take the compressed signal and run it through the compression process again and again? See Figure 6.10. There should be fewer bits remaining after the second compression step, and then fewer yet after the third step—yielding an even better compression ratio. The president immediately calls his corporate lawyers, who start discussing the idea and a possible patent.

Before long, word of this idea gets to the company's chief engineer. She listens to the idea, but realizes that it cannot work. If it could, then, as shown in Figure 6.10, the original signal or data file could be repetitively compressed until it is just 1 bit, a 1 or a 0. It's impossible to decompress the complete signal if all you have left is 1 bit. She explains to the disappointed president that compression algorithms don't always reduce the number of bits, particularly when the algorithm in use counts on some characteristic of the signal (like long run lengths) that isn't there anymore after the first step of compression has been completed. Now, more educated, the president asks her just what can be expected: What is the very best that a great compression scheme ought to be able to do? "Good question," she says. "I bet it has something to do with a concept called **entropy**."

Entropy: In terms of information, the fewest number of bits per symbol required in order to store and recover a data set.

Example 6.12 suggests that there is a fundamental limit in the mathematical theory of coding that states the lossless compressibility limits of any data set. This fundamental limit is known as the **entropy**. Suppose we have a set of N unordered symbols s_i, each with relative frequency $f_R(s_i)$. Then the entropy of the symbol set is defined as

$$H = -\sum_{i=1}^{N}[f_R(s_i) \times \log_2(f_R(s_i))]$$

The units of the entropy H are bits. The entropy of a symbol set is a measure of its randomness and therefore how much information is needed to describe it. A set of N symbols in which each of the symbols has roughly the same value of relative frequency—that is, each relative-frequency value is around $1/N$—has high entropy. By contrast, a symbol set in which certain symbols occur much more often than other symbols has lower entropy.

INTERESTING FACT:

The term "frequency" is used in two ways in this book. Both meanings are related to how frequently something measurable occurs or happens. In the case of relative frequency of occurence, we measure how many times a symbol is observed in a particular sequence of symbols and divide this value by the total number of symbols in the sequence. For sinusoidal frequency, we count how many times the periodic sine function repeats every second.

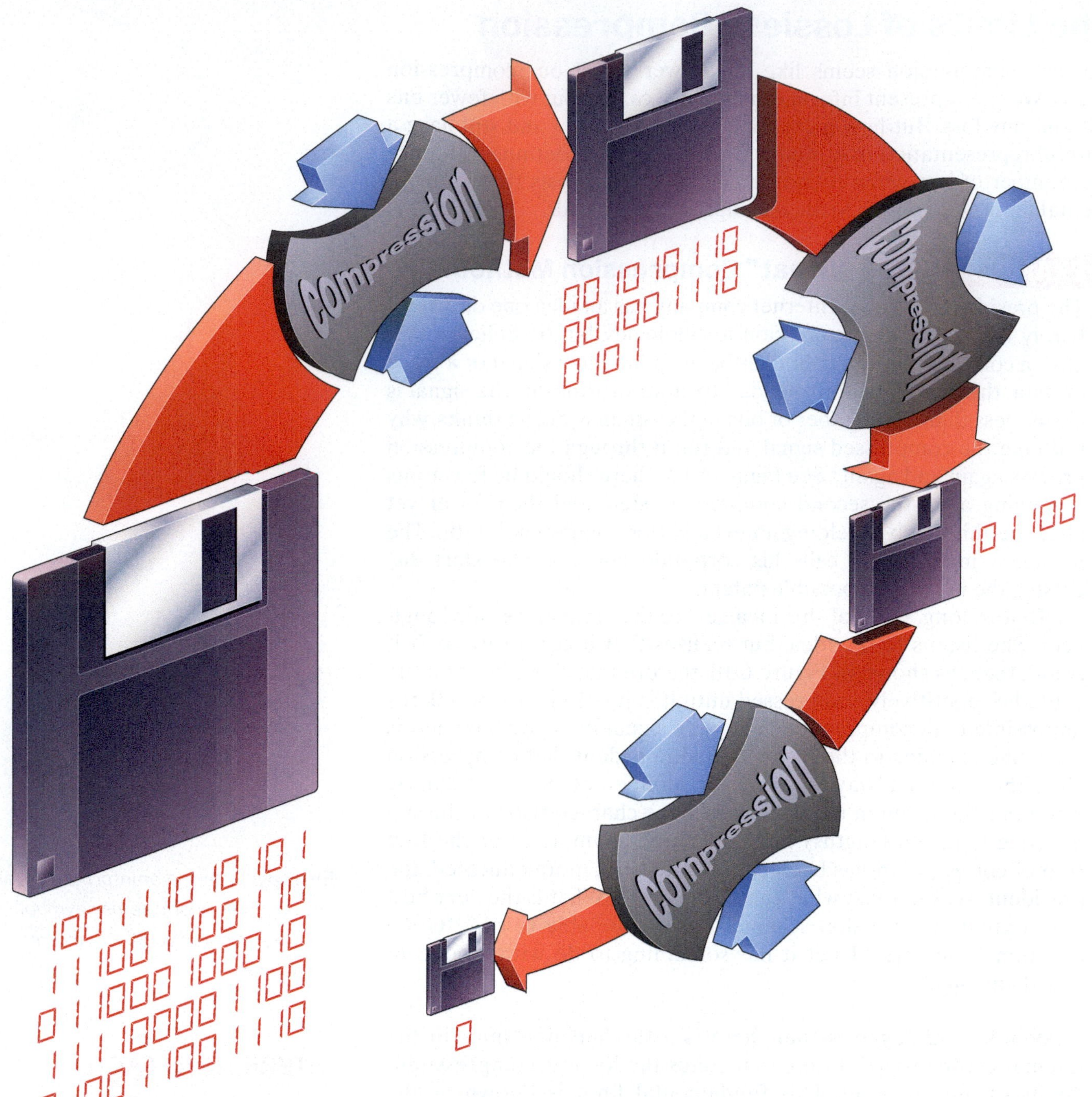

Figure 6.10 The procedure envisioned by a company's CEO in Example 6.12 for obtaining a huge compression ratio by repetitively compressing the input data.

The entropy of a symbol set defines a lower limit on the average codeword length that can be achieved when losslessly compressing the information in the set. This fundamental boundary is

$$L_{\text{average}} \geq H$$

In other words, we can compress a data set no further than its entropy value times the number of elements in the data set. Example 6.13 shows how this limit works.

EXAMPLE 6.13 Fundamental Limit to Compressing a Phrase

Recall again the phrase from the Examples 6.9 and 6.11:

"I scream,you scream,we all scream,for ice cream"

If we treat each letter as a separate symbol and ignore the spatial placement of letters, we can compute the entropy of the symbol sequence quite easily. We use the formula given previously for H, using the relative frequencies that we calculated for each of the $N = 16$ different letters in our 47-symbol phrase. This calculation results in the value

$$H = 3.70 \text{ bits}$$

This value is less than the 4 bits per codeword that we would need to store these symbols if we gave each one of the codewords the same length. The value is remarkably close to the average code length of the proposed code in Example 6.11, which achieved a value of $L = 3.72$ bits per coded symbol. The entropy limit tells us that our proposed code is quite good and near the best it can be under the assumptions we've used to compute the code.

Coding General English Text We can calculate the minimum number of bits required in order to efficiently code the letters of English text from the entries of Table 6.2. This coding assumes that every letter in English appears in a random order in words, but still occurs with the relative frequency given in Table 6.2. The value we obtain for the entropy is 4.08 bits per character, including spaces. What does this mean? Well, we have already seen that standard ASCII coding of English text uses 7 bits for upper and lowercase letters, numbers, and punctuation. This means that by taking advantage of the relative frequencies of the letters in English text, we might save about 20% in storage or transmission time.

We can go even further if we also realize that the letters are not random from one to the next. For example, consider the short phase "I love ice c." We have a pretty good idea that the next letter should be an "r." So, when taken together within sentences, letters are not completely random from one letter to the next. Using the structure built into the spelling and grammar of English, we can encode English text by using as few as 1.3 bits per character. This value is lower than the 4.08 bits per character that we computed previously because we are now also using the ordering of the letters to code the information, not just their relative frequencies. This means that a book with approximately 2 million characters (80 characters per line, 50 lines per page, and 500 pages) can be stored using only 2.6 Mbit or 325 kB (kilobyte) of data. This capability would allow us to store over 2000 books on a single 700-MB CD-ROM. Compare this storage capacity with that for standard ASCII stored as 1 byte per character. We would have to store 2 MB of data for each book! We could store only 350 books on a CD-ROM when using ASCII.

What does all this tell us? We can use the natural redundancy in English text to efficiently store books, magazines, or anything else with lots of English words in it. The same concept can, of course, be applied to other languages as well.

EXERCISES 6.3

Mastering the Concepts

1. A call to which of the following three cities would take the longest time to dial, using a rotary telephone?
   a. Honolulu, Hawaii (area code 808)
   b. Santa Barbara, California (area code 805)
   c. El Paso, Texas (area code 915)
2. A call to which of the following three cities would take the shortest time to dial, using a rotary telephone?
   a. Manhattan, New York City (area code 212)
   b. Cleveland, Ohio (area code 216)
   c. Beverly Hills, California (area code 310)
3. Are all the letters in the English language equally likely to appear in a sentence? If not, which letter is most likely to appear? Which letter is least likely to appear?
4. What are the units of entropy?
5. What is the smallest codeword length (in bits) that a codeword can have?
6. Suppose you have two *N*-bit sequences. One sequence has roughly the same number of 1's and 0's, whereas the other sequence is mostly 1's with only a few 0's. Which sequence is more likely to have a smaller entropy?

Try This

7. Encode the following bit sequences, using run-length coding, and calculate the compression ratio for each:
   a. 000000111001000001111110
   b. 11000000000011111000000111111111110000010
8. Decode the following run-length-coded sequences:
   a. [1] (4) (5) (6) (3) (7)
   b. [0] (2) (15) (11) (5) (7)

   What is the compression ratio that run-length-coding provides in each case? (Assume 4 bits are used to store all lengths.)
9. Consider the sequence of 22 bits that is produced by Example 6.7.
   a. Compress this sequence of bits again, using run-length coding. How many bits are needed to represent the recompressed signal?
   b. Is the value you obtain in part (a) greater than or less than the original run-length-coded sequence? Explain.
10. Suppose you have a set of *N* symbols in which there is only one of each symbol.
    a. What is the relative frequency of each symbol in terms of *N*?
    b. What is the entropy of this symbol set?

In the Laboratory

11. Get any bag of your favorite colored candy and calculate the entropy of its contents by color. If you could design a code that is as efficient as theoretically possible, what would be the average length of the codewords?

12. Everybody's birthday falls in one of the 12 months of the year. Calculate the relative frequencies of birthday months for all the students in your classroom. What is the entropy of this information?
13. Computers typically have a "ZIP" utility that allows one to compress any ordinary file, using a lossless compression technique. Take some files and, through experimentation, calculate the compression ratios provided by such a utility. You can usually look at the "before" and "after" size by highlighting the original and compressed files, respectively, without clicking on them. See how much, if any, you can compress the following types of file:

    a. a text file (.txt extension)
    b. a JPEG image file (.jpg extension)
    c. a document file (.doc extension)

    Which of the three is most compressed by the ZIP utility?

Back of the Envelope

14. Run-length coding works well for bit strings that have long runs of 1's or 0's. Do you suspect that ASCII-coded text would be compressed a lot by using run-length coding? Why or why not?
15. In the summertime, the weatherperson predicts that, on any given day in a particular area, the probability of rain is 10%, the probability of sunny weather is 89%, and the probability of snow is 1%. What is the entropy limit of the weather, assuming that the weatherperson is accurate? Suppose you were to write a program to store the weather over the entire 92-day summer period. About how many bits would you need to store one summer's weather report, assuming that the weather changed randomly from one day to the next? Compare this value with the nominal value of two bits per day that one would use if no compression were possible.

6.4 More on Lossy Compression

Is Being Lost Always Bad?

The word "loss" has a bad feeling about it. "Being lost" means not knowing where one is. Lost money or possessions may be gone forever. Even the phrase "You lost me" usually means that one is confused. But sometimes losing something can be good, especially when coding information for compression.

Compression methods that throw away information for the sake of efficiency are called lossy compression methods. Like other things that are lost, the information removed during the compression step cannot be regained. The goal in performing lossy compression is to remove the portion of the signal or information that we can't perceive anyway or that doesn't matter to us. If done well, lossy compression can save lots of storage space or transmission time, with only a slight degradation in the quality of the signal. It may surprise you to learn that MP3 audio

MP3: An acronym for Audio Layer 3 of the MPEG1 Multimedia Coding Standard.

and DVD-quality video signals are lossily compressed versions of the true audio and video sources, respectively. These signals still sound and look remarkably good. We can store a lot more audio and video signals in the same memory space when these signals are compressed using lossy compression methods.

MP3: One Example of Lossy Compression

MP3 (or, more accurately, Audio Layer 3 of the MPEG1 Multimedia Coding Standard) is the name used for a group of popular audio compression standards. MP3 is popular right now because it is economically useful. By encoding a CD-quality audio track into one of the MP3 formats, one reduces the file size required to store the track onto a computer hard disk, CD-R, or flash RAM card.

Since the file size is reduced, one can store more songs onto the corresponding storage medium, allowing hours of music to be stored in a portable MP3 player, for example. The quality of the sound in the compressed format is not as good as that of the original CD, but the difference is not that noticeable to most people in casual listening situations. The MP3 format is also helpful when one wants to download music off of the Web, because the reduced file size effectively reduces the amount of download time needed.

The Role of Perception in Lossy Compression

Most lossy compression methods rely on our own perceptual limitations. To really understand how lossy compression methods work, we need to understand how human sensory systems work. In this section, we will briefly discuss the abilities and limitations of the human hearing system as it pertains to lossy audio-coding methods. A similar understanding of the human visual system is used by engineers to design lossy video-coding methods.

Perceptual Characteristics of the Human Ear As you probably are aware, human hearing abilities differ from those of other animals. A dog whistle, for example, produces a loud, high-frequency sound that is beyond our own hearing abilities, but that can be heard clearly by dogs. The fact that we cannot hear certain types of sounds means that we need not store all sounds when saving digital representations of audio signals. This concept is at the root of all lossy audio-coding methods.

Figure 6.11 shows a plot of the range of sounds that we can hear in terms of their frequency content and intensity. The left-to-right scale is the frequency of the sounds in cycles per second, or Hertz (Hz), and the vertical scale is the intensity of the sound, also known as the sound pressure level (SPL), measured in decibels (dB). The borders of the shaded area are the boundaries of human hearing and are worth exploring further:

Threshold of Quiet: The lower intensity limit of the human hearing system. Describes the softest sounds that humans can possibly hear.

- At the bottom of the graph, we find the fundamental lower limit of human perception, the **threshold of quiet**. Sounds that fall below this threshold are too soft for humans to hear. The threshold of quiet is different at different frequencies; in particular, it is higher at low frequencies and lowest at middling frequencies. Thus, it is the bass sounds in a signal that are hardest to hear when the volume of the sound is low.

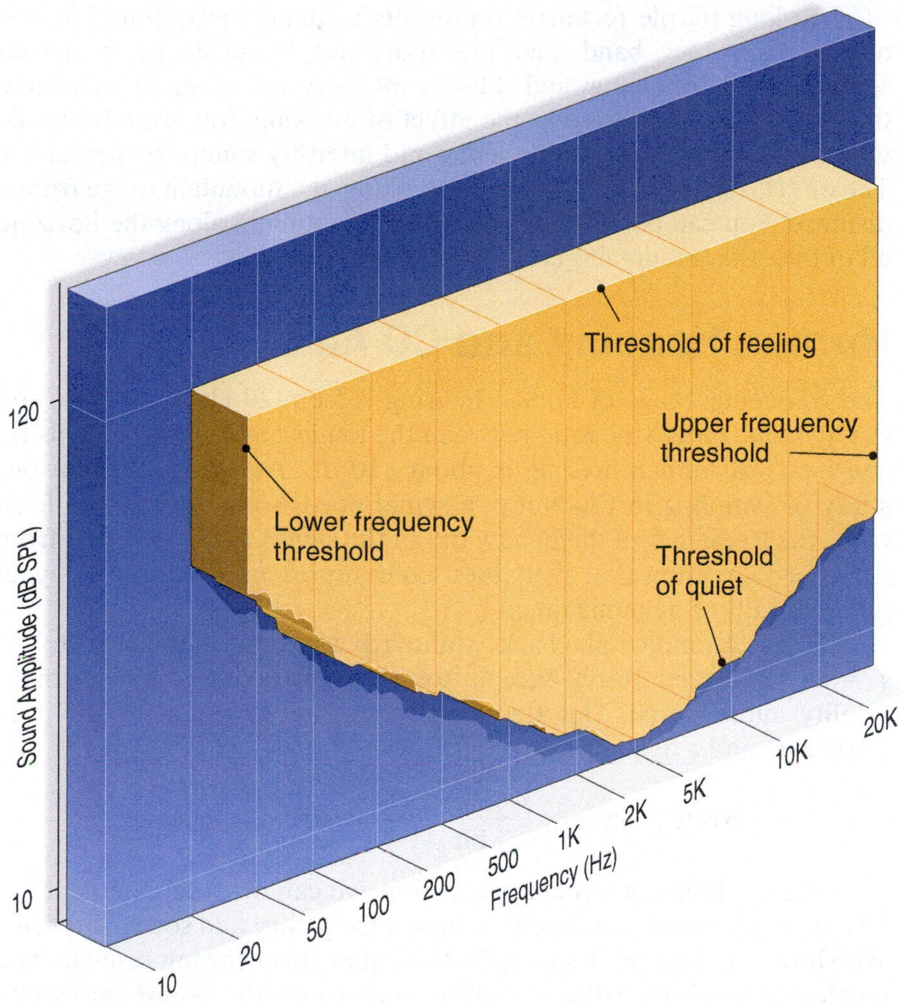

Figure 6.11 Range of audibility for humans in terms of both the range of frequencies that can be perceived and the range of audio power levels. (Note that the frequency scale is logarithmic, not linear.)

- At the top of the graph, we have the fundamental upper limit of human perception, the **threshold of feeling**. The threshold of feeling is beyond the point at which one's ears can be damaged permanently.
- At the left of the graph is the **lower frequency threshold** of human hearing. Sounds below 20 Hz are not perceived as signals with pitch; rather, they are perceived more as vibrations by the human body.
- At the right of the graph is the **upper frequency threshold** of human hearing. Sounds higher than about 20,000 Hz (or 20 kHz) cannot generally be heard. These sounds are the ones that your dog and other animals can hear, although these animals have limits to their high-frequency listening capabilities as well.

There is another important aspect of the human hearing system that is useful for perceptual coding systems. When two sounds of a similar frequency, but different intensity, are played together, humans perceive only that the louder sound is playing. This effect is known as **masking**. Figure 6.12 shows how masking works for signals that contain only a narrow range of frequencies.

INTERESTING FACT:

It is known that long exposures to sounds as soft as 75 dB SPL in some frequency ranges can cause permanent hearing damage. For this reason, it is always a good idea to turn down your audio device whenever you have been listening to it for more than a few minutes—your ears will thank you later!

Threshold of Feeling: The upper intensity limit of the human hearing system. Describes the loudest sounds that humans can possibly hear without pain.

Lower Frequency Threshold: The frequency below which humans cannot hear sounds. Such sounds are felt as vibrations and not heard.

Upper Frequency Threshold: The frequency above which humans cannot hear sounds.

Masking: A psychoacoustic effect of the human hearing system whereby soft sounds are made inaudible by louder sounds that are close in frequency.

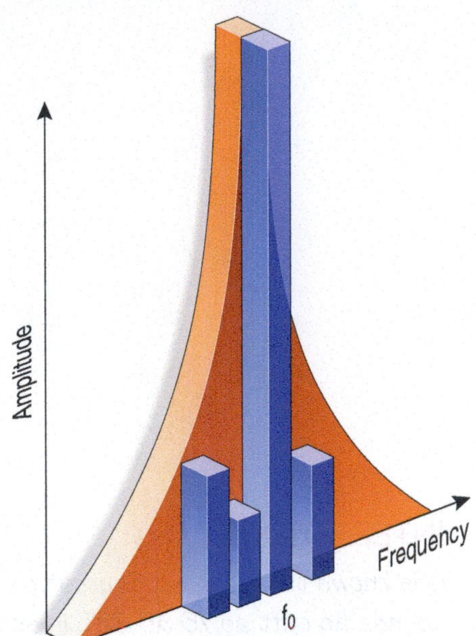

Figure 6.12 An illustration of the masking effect. When humans hear the strong signal at frequency f_0, they can't hear weaker ones with frequencies close to f_0. The strong one is said to mask the weaker ones.

The long purple rectangle represents a sound concentrated in one narrow frequency band, and the short purple rectangles represent sounds close in frequency and of lower intensity. The orange area around the long rectangle represents the effect of masking. Any sounds that lie within the orange area in frequency and intensity cannot be perceived. The effect is like what you see when looking at a mountain range from a distance: You can perceive only the tallest mountain along the horizon; all of the other mountains are lost in the tall mountain's shadow.

Human Hearing and CD Audio

The frequency range of human hearing is from 20 Hz to 20 kHz, and the *dynamic range* (the ratio between the loudest and softest sounds we can hear) of human hearing is about 110 dB. As we know from our study of sampling in Chapter 5, a signal needs to be sampled at least twice the frequency of the bandwidth of the signal. Moreover, also from Chapter 5, we know that every bit used to represent values of the signal gives us 6 dB of dynamic range.

The CD audio playback standard uses a sampling rate of $f = 44.1$ kHz and 16 bits/sample to represent a single-channel high-fidelity audio signal. This translates to an upper frequency limit of 22.05 kHz and a dynamic range of

$$16 \text{ bits/sample} \times \frac{6 \text{ dB SPL}}{\text{bit}} = 96 \text{ dB SPL/sample}$$

Like the limits of human perception, we can draw the limits of the CD audio playback standard in terms of frequency and sound intensity. We show this plot in Figure 6.13. Note that this drawing represents a particular amplifier volume setting that places the 96-dB maximum range at 96 dB SPL. It is possible to either increase or decrease the volume setting, which would shift the position of the yellow rectangle up or down, respectively, on the plot.

As can be seen in Figure 6.13, the usable range of CD audio is greater than that of the human hearing system in terms of frequency range and loudness level at this amplifier setting. This fact is the main reason that CDs sound so good; the standard defines a quality that is *better* than even the best human ear can sense for moderate volume levels. The fact that the CD audio standard exceeds human hearing abilities means that we can reduce the number of bits further and still get good-sounding audio. This fact is even more apparent when masking is considered, because CD audio doesn't even take advantage of masking in its operation. Perceptual coding systems exploit masking and other aspects of the human auditory system.

Transparent Compression

Transparent Compression: Any signal or data compression method that introduces no perceptual distortion into the encoded signal or data.

Transparent compression (also called *transparent coding*) describes any signal or data compression method, lossy or not, that introduces no perceptual distortion into the encoded signal or data. Transparent compression is the goal of every lossy compression method. We don't care if the signal that we're watching or listening to is different from the original signal; we just don't want to see or hear the differences.

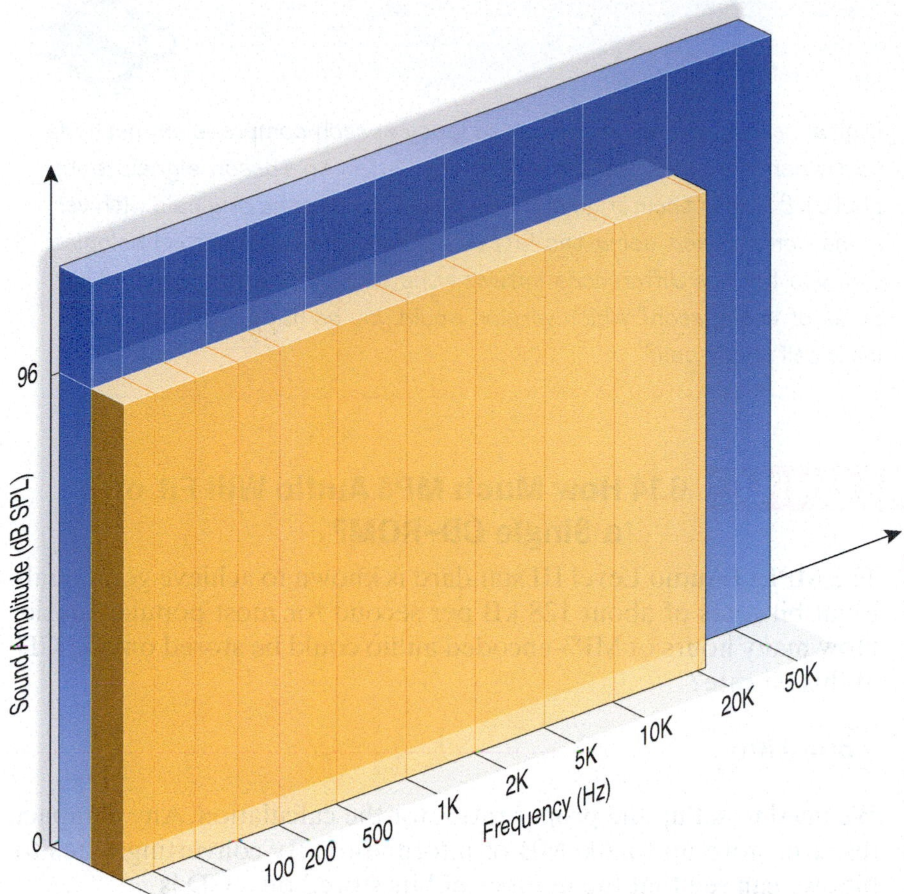

Figure 6.13 The range of audio frequencies and power levels a compact disc (CD) was designed to accurately reproduce.

MP3 Audio Encoding: An Example of Transparent Compression The popular MP3 coding format is actually three different formats whose long names are MPEG Audio Layers I, II, and III, respectively. The common component of the MPEG Audio Layers I and II coding methods uses a form of compression called **subband coding**. This compression method breaks up the sound into particular frequency bands in much the same way as the structure of our ear enables us to analyze the frequency content of sounds. Masking thresholds based on the ear's ability to detect sounds in adjacent frequency bands are then used to pick the bit accuracies used to encode each band's information. The individual encoding the audio information can select one of several bit rates from 32 kB per second for the lowest-quality monaural sounds to 384 kB per second for high-quality stereo sounds. CD-quality audio signals require approximately 1.41 MB per second in their uncompressed stereo form, so MP3 files offer compression ratios of about four for the most bit-costly encoding method, and even higher compression ratios for lower levels of quality. Moreover, extensive tests with listeners indicate that MPEG Layer II encoded audio at 192 kB is similar to CD-quality audio at 1.41 MB. This difference represents a compression ratio of about 7 : 1, a significant saving in storage needs.

Subband Coding: A compression method whereby the sound signal is broken up into different frequency bands, and each frequency band signal is assigned its own bit sequence.

Infinity Project Experiment: Speech Compression

Digital cellular phones make use of lossy speech compression methods to transmit speech signals efficiently. Listen to speech signals sampled at 8 kHz in their original forms, and compare the originals with versions compressed using the latest speech compression technologies. Can you tell the differences between the original and compressed versions of the speech? Which version would you be happy with when making a cell phone call?

EXAMPLE 6.14 How Much MP3 Audio Will Fit on a Single CD-ROM?

The MPEG Audio Level III standard is known to achieve good quality at bit rates of about 128 kB per second for most popular music. How many hours of MP3-encoded audio could be stored on one CD at this bit rate?

Solution

We need to set up the proper ratios for the calculation. Any compact disc can store up to 700 MB of information. By converting bytes to bits, we can see that the number of bits stored on a CD is

$$N\frac{\text{bits}}{\text{CD}} = 700 \times 10^6 \frac{\text{bytes}}{\text{CD}} \times \frac{8\text{ bits}}{1\text{ byte}} = 5.6 \times 10^9 \frac{\text{bits}}{\text{CD}}$$

The rate at which the MP3 music is to be read off the encoded CD is

$$r_{\text{MP3}} = 128 \times 10^3\text{ bits/s} = 1.28 \times 10^5\text{ bits/s}$$

The number of seconds of audio that can be stored can be found by dividing these two numbers:

$$T_{\text{MP3}}\frac{\text{s}}{\text{CD}} = N\frac{\text{bits}}{\text{CD}} \times \frac{1}{r_{\text{MP3}}\text{bits/s}}$$

Plugging in the values, we get

$$T_{\text{MP3}}\frac{\text{s}}{\text{CD}} = 5.6 \times 10^9\frac{\text{bits}}{\text{CD}} \times \frac{1}{1.28 \times 10^5\text{ bits/s}} = 4.375 \times 10^4\frac{\text{s}}{\text{CD}}$$

Finally, converting the time units into hours, we obtain

$$T_{\text{MP3}}\frac{\text{s}}{\text{CD}} = 43{,}750\frac{\text{s}}{\text{CD}} \times \frac{1\text{ hour}}{3600\text{ s}} = 12.15\frac{\text{hour}}{\text{CD}}$$

In other words, a CD can hold over 12 hours of audio in MP3 format at near-CD quality. If you repeat the calculation for a single-layer single-sided DVD, the amount of music that such a DVD can hold is about 82 hours!

EXERCISES 6.4

Mastering the Concepts

1. Is the ability of the human ear to hear sounds better or worse than the sonic capabilities of a CD-quality audio system?
2. Suppose you are listening to the car radio at such a high volume that you could not hear the warning siren of a nearby ambulance. What psychoacoustic effect are you demonstrating?

Try This

3. A particular digital audio recorder uses 20 bits to store each sound sample. Calculate the dynamic range that could be obtained with such a scheme.
4. Suppose you are designing the audio hardware for a new handheld video game. Most users of the video game will be playing in places where there is background noise. Only a 40-dB dynamic range is needed for the device. How many bits per sample should you use?

Back of the Envelope

5. Suppose that a computer hard disk can store 1000 GB of information. How many hours of MP3 audio can be stored on this hard disk?

6.5 Coding to Detect and Correct Errors

When you listen to sound from a CD or watch the action-packed scenes from a DVD, you do not notice that these devices are constantly fixing mistakes caused by small imperfections on the discs. Large memories in laptop computers can also operate correctly even when some parts of the memory produce errors. The ability to function correctly in spite of some errors is really important for extending the useful life of products such as CDs and allowing reasonable manufacturing tolerances for complex systems and components such as large digital memories. Both of these factors reduce the cost of the devices we use everyday.

Special coding methods that provide **error detection** and **error correction** capability ensure the high quality of signals from these and other digital devices. Error detection refers to the ability to figure out whether there is an error somewhere in a sequence of bits. If we detect a single error, we do not know which bit is wrong. We know only that one of the bits in the sequence is wrong. If we can only detect errors, then we can avoid using bad data, but we have no way to know how to correct it. Error correction refers to the ability to fix erroneous symbols or bits when they are detected by using some mathematical procedure on the symbol or bit string.

Error Detection: The process of analyzing a bit sequence to see if an error has been made somewhere in the sequence.

Error Correction: The process of analyzing a bit sequence to detect which bits are wrong and then replacing them with the correct values.

Error detection and correction are used in many applications:

1. When a digital device is reading information from a physical storage medium, such as the shiny surface of a CD or DVD, physical defects or debris such as dust, scratches, or oily fingerprints can obscure

some of the bits. Error correction capability allows the device to provide high-quality audio and video output despite these problems.

2. When a wireless cellular telephone is communicating with a base station, the connection between the two can become corrupted by interference from other electronic devices, power lines, and atmospheric phenomena. Error correction enables the conversation to continue in spite of the corruption.
3. Large digital memory chips used in computers can be used even if a few of the bits have failed. Otherwise, a problem with a single bit could make an entire gigabyte of memory useless for a general computer applications.
4. When storing information for a long time on magnetic disks, environmental effects such as a stray magnetic field from another electronic device can introduce some errors on the disk. Special coding used when the data are stored allows errors to be detected and possibly corrected.

All error detection and correction are based on designing codes that use only some of the possible bit strings to represent data, making all other strings invalid. If all bit strings are valid, there is no way to know if an error has been made when the data are read or received. Consider the simple case of reading a single bit. It can be a 0 or a 1, and if an error has been made, there is no way to detect it, because either a 0 or a 1 is a valid bit. If we wanted to store the bit more reliably, we might store it twice so that a 0 would be stored as 00 and a 1 would be stored as 11. Then we would have to read 2 bits in order to get one bit of information, but if a single error occurred we would detect it. The error would cause us to read either 01 or 10, and neither is valid. Unfortunately, we would not be able to correct the error, because we would not know which bit to change.

INTERESTING FACT:

Error detection and correction for an n-bit string are possible only if the code designer makes some of the 2^n possible bit combinations invalid.

We could store the bit even more reliably if we stored it three times, so a 0 would be stored as 000 and a 1 would be stored as 111. Then we could easily detect an error, because anything other than 000 or 111 is not a valid pattern for the 3 bits. Now we could also correct a single error by a "majority rules" vote whenever an invalid group of 3 bits is read. If we read 001 or 010 or 100, the bit would be declared a 0, and if we read 110 or 101 or 011, the bit would be declared a 1. Methods that are similar to this example, but more efficient, can be applied to storing and transmitting bit strings instead of single bits. The number of errors that can be successfully detected and corrected will depend on the percentage of the bit patterns that are defined to be invalid by the maker of the code.

INTERESTING FACT:

The number of errors that can be successfully detected or corrected will depend on the percentage of bit sequences that are made invalid by the code designer.

A Simple Method of Error Detection: Parity

A good example of error detection uses the concept of **parity**. Parity means "equivalence" or "of equivalent classes or types." If two things have parity, they are similar to one another. Let's consider a simple example of parity, using the number of bits in a sequence that are 1's. Suppose you have a string of 8 bits such as

Parity: In error detection, classifies bit sequences by whether the number of 1's in the sequence is an even or an odd number.

$$S_1 = 00011010$$

This 8-bit string has three 1's and five 0's. We see that the bit string has an odd number of 1's in it, so we say this string has odd parity. Now, if we had another bit string

$$S_2 = 00110101$$

we would count four 1's and four 0's. Since this bit string has an even number of 1's, we say it has even parity. If all 8-bit strings are valid, then half will be even strings and half will be odd strings. If an error occurs in one of the bits in an even string, it will change it into an odd string. Similarly, a bit error in an odd string will change it into an even string. But we will have no way to know the error occurred, because both even strings and odd strings are valid.

We can use parity for error detection by making sure that all valid strings have even parity. Then any single-bit error will cause the string to have odd parity, and we can easily detect that an error has occurred by counting the 1's and noting if the number is even or odd. One way to make all valid 8-bit strings S have even parity is to add a ninth bit p, called the **parity bit**. The value of this parity bit is selected to make an even number of 1's in the new 9-bit string Sp. We have to read 9 bits for every 8-bit string of data we want, but the additional bit gives us the ability to detect a single error.

Parity Bit: A bit added to a bit string to guarantee that the string will have an even number of 1's for even parity or an add number of 1's for odd parity.

Consider the four 8-bit strings in the table below. The first and fourth strings each have three 1's, which is an odd number of 1's, so the added parity bit is a 1 to make the 9-bit pattern have four 1's for even parity. The third string, S_3, has five 1's, which is also an odd number of 1's, so its parity bit is also 1, and the 9-bit string S_3p has six 1's. The second string has four 1's, which is an even number of 1's, so its parity bit is 0 to keep the number of 1's in the 9-bit string even.

8-bit data string S	9-bit even-parity string Sp
$S_1 = 00011010$	$S_1p = 000110101$
$S_2 = 00110101$	$S_2p = 001101010$
$S_3 = 10110101$	$S_3p = 101101011$
$S_4 = 00110001$	$S_4p = 001100011$

Suppose now that an error occurs in one of the bits in bit string S_1, creating the new bit string

$$S_{1X} = 00010010$$

If we had only the 8-bit string, we could not detect that an error had occurred. However, if that same error occurred in the 9-bit string with parity, we would have

$$S_{1X}p = 000100101$$

We can immediately detect that an error has occurred because there are three 1's in this string, which is an odd number of 1's.

In the foregoing example, we could detect that an error had occurred, but we could not tell which bit changed. We could not even know whether

INTERESTING FACT:

ASCII codewords for characters are 7 bits long, but to allow error detection, they are often sent as 8-bit codewords with a parity bit added to each character code. This 8-bit code is conveniently stored in one byte in a computer's memory.

the bit that changed was the parity bit or one from the 8-bit string. Now suppose we have read the 9-bit string

$$S_p = 001101011$$

which has five 1's. We know that an error has occurred, but we cannot tell if the string was supposed to be S_2, S_3, S_4 or one of six other possibilities, because we do not know which bit changed.

There are many other ways to code for error detection and error correction. For example, we could have added a parity bit that would guarantee an odd number of 1's for all valid strings instead of an even number of 1's. Or we could have added several parity bits for different parts of the strings so that we could also correct an error. The parity-bit example and the simple example of storing the same bit two or three times illustrate some general concepts about coding for error detection and correction:

1. We have to change the signal or information in a special way in order to recognize later if errors have been made.
2. Generally, the change in the information increases the amount of memory that we use to store the information or the time needed to send the signal over a communications link.
3. There is no new information contained in the new coded version, which uses more bits. We have increased the length of the message without increasing its content in order to make it tolerant to errors.

The foregoing concepts work for any type of information, not just bit strings, and for any added "clues" that we append to that information. For example, UPC codes on product labels have a 10-digit code for the product, followed by a check digit, so that if a single digit is scanned incorrectly, an error will be detected. Bank account numbers and credit card numbers may also have at least one "check digit" so that if a single digit is entered incorrectly, the account number will be rendered invalid. That way, you are less likely to deposit your savings into someone else's account or charge your purchases to another person's credit card.

EXERCISES 6.5

Mastering the Concepts

1. What is the difference between error detection and error correction?
2. What is the difference between even parity and odd parity?
3. Why can't we detect an error in an 8-bit string if all bit strings are valid?
4. What is a parity bit?
5. The system of storing each bit twice can be thought of as a parity system. Is it even parity or odd parity?

Try This

6. If each data bit is stored twice for error detection, what percentage of the possible bit patterns is not valid? If each data bit is stored three times for error correction, what percentage of the possible bit patterns is not valid?
7. Find the value of the 5-bit even-parity strings for the following 4-bit strings:

4-bit data string	5-bit even-parity-string
S	Sp
0011	
0111	
0101	
1000	

8. The following 5-bit strings were read: Each 5-bit string represents a 4-bit string with a parity bit added to guarantee even parity. Find the strings that have an error.

   00101 01001 10101 00000 11111 11000 11011 11001
   00110 00001 10111 01001

9. The 9-bit string 001101011 is read and an error is detected because it does not have the expected even parity. Find the single bit that changed if the string should have been S_2p given previously. Repeat for S_3p and S_4p.
10. Each ASCII character is identified by a 7-bit code. If a parity bit is added to each character, what is the percentage increase in the storage required for text with parity compared with that for text without parity?

Back of the Envelope

11. The system of storing each bit three times allows us to detect a single error and to correct a single error. Can we also detect a double error in which 2 bits have changed? Would we be able to distinguish a double-bit error from a single-bit error, using only the 3 bits read?
12. The idea of a parity bit applies to strings of all lengths. What factors would you consider when deciding whether to use a parity bit for each bit, each byte, or each megabyte?

6.6 Coding for Security

"Your Secret Is Safe with Me"

Privacy is important. Bank accounts, medical records, and the contents of telephone calls are all types of digital information that many of us want to be secure and private. So how is this information actually protected from the prying eyes of others? In the remainder of this chapter, we'll take a quick look at ways of keeping data private. From our discussion, we'll see how information security is both an important and challenging task.

The Internet and Information Security

The recent increase in popularity of the Internet has made information security one of the most important challenges facing modern society. The Internet has the ability to affect almost every type of human social and commercial interaction, so maintaining secure communications across the Internet is a critical issue. The leaders of software companies, which design the programs and the infrastructure required to access the Internet, are aware of the public's desire for privacy. Most software programs have built-in safeguards to protect a Web surfer's data. Many people, however, are not aware of these safeguards or of the many challenges involved with Internet security.

Cryptography: A form of coding whose goal is to make information difficult to read or be understood by others.

Decryption: In cryptography, a generic name for the decoding process.

Key: In cryptography, a unique numeric or symbolic sequence used to decrypt or encrypt important information.

Codes, Keys, and Cryptography

Whenever we format, compress, or add error correction capabilities to information, we assume that everyone knows the rules by which we did our job and can recover the data for themselves if they wish. Sometimes, however, we would like to selectively make information available to the intended user, but make it impossible to read or be understood by others. This form of coding is known as **cryptography**. We call the particular type of coding done for this purpose encipherment or encryption, and we call the corresponding decoding processing decipherment or **decryption**.

Cryptography has a strong parallel to physical security. When you use a key to lock a car door, a locker, or your front door (Figure 6.14), you are trying to prevent others from accessing the inside of the car, the locker, or your house or apartment. Similarly, when you encrypt data or information, you are trying to prevent others from accessing those data or that information. Unfortunately, the word "trying" is critical in both contexts. It is almost always the case that a locked door or locker can be broken into by a malicious person with enough time, money, or a big-enough crowbar. Similarly, with enough time, money, or computing resources, an encrypted signal can be decrypted and read by someone else. For these reasons, you should always view such data security methods as forms of protection that are not perfect.

Figure 6.14 Traditionally, metal keys have been the way to control entry to spaces and information that people would like to protect.

Cryptography also shares another element with physical security: the concept of a **key**. A physical metal key is a device that helps you gain access to something you own. Similarly, a data key is a sequence of numbers that helps you gain access to information that you encrypted. A data key generally doesn't store any information in and of itself, just like a physical metal key doesn't hold your physical possessions inside of it.

The main difference between a physical key and a data key is that a data key can be copied more easily. This difference is what makes data security so hard to maintain. Imagine if anyone could copy an apartment or car key just by looking at it and remembering it!

Passwords, Access Codes, and PINs

A **password** is the name given to an alphanumeric key used to access computer and network systems. Most on-line Internet account services are accessed by a password that the user chooses when he or she sets up the account.

Password: A set of numbers or letters that a person uses to protect an account from being accessed by others.

Access codes, also called **personal identification numbers (PINs)**, are numerical sequences that many financial institutions and telephone companies rely on to secure a user's account access and services. These numerical sequences are digital keys that are usually 4 to 20 numbers long. If you have a bank account, you most likely have chosen a PIN in order to withdraw your money at an automated teller machine (ATM).

Access Codes and Personal Identification Numbers (PINs): A password made up of numerical digits that is often used by financial institutions to secure an individual's account from unwanted access.

The same issues that surround data security also apply to passwords, access codes, and PINs. Since these numbers can be communicated easily, they can only discourage, rather than prevent, others from accessing your computer account or monetary savings. That is why you should never choose a simple or easily identifiable password or PIN for your account access. A good guesser could cause trouble for you. In fact, you should get into the habit of changing your password or PIN from time to time to prevent others from gaining access to your "digital life."

EXERCISES 6.6

Mastering the Concepts

1. What is the difference between encryption and decryption?
2. How is a data key similar to a physical key? How are they different?
3. When picking access codes for their ATM bank cards, many people use a well-known personal number, such as their birthday. How secure is this choice?

Back of the Envelope

4. Write down five ways you could use cryptography to secure information in your life.
5. Several modern luxury cars have a numeric keypad on the car door that will open the door without a key if one knows the right numeric sequence to punch. Discuss the benefits and risks of this feature.

6.7 Simple Encryption Methods

Rotational Encoding

Perhaps the simplest text-encrypting method is **rotational encoding**. You might have seen this type of encoding method before, as it is sometimes used in puzzles in the entertainment section of a daily newspaper. Here's how it works: Consider the following sentence, written in all capital letters to avoid capitalization issues:

Rotational Encoding: An encryption method for encoding letters in English text by using a rotational shifting operation.

SALLY USUALLY WEARS SNEAKERS WHEN
SHE IS RUNNING OUTSIDE

INTERESTING FACT:

Rotational encoding is one of the first known encryption methods to be used in the Western world. It was used by Julius Caesar around 50 B.C. In Caesar's code, every letter was shifted by three places (that is, the letter "A" became the letter "D"). At the time, this code would have provided a high degree of information security.

A rotational encoder substitutes a letter for every letter appearing in the sentence. Thus, every letter "A" will be replaced by a letter different from "A," every letter "B" will be replaced by a letter different from "B," and so on. We have to be careful, however. We cannot assign the same replacement letters to different original letters, because we won't be able to get back the original words. Also, we need a simple way to describe the substitution of letters so that we can define a useful key to help us decode the message. In general, any letter can be used to replace any other. In rotational encoding, however, we'll do it in a very simple way.

To make the assignment unique and easy to describe, we'll use the following rule: Whatever letter we choose to replace "A," we'll use the next letter to replace "B," and the next letter after that to replace "C," and so on. If we get to the end of the alphabet, we'll just "wrap around" the assignment, starting with "A." For example, if we assign an "F" for every "A," then we'll assign a "G" for every "B," an "H" for every "C," and so on, up to assigning a "Z" for every "U." Then we'll assign an "A" for every "V," a "B" for every "W," and so on. To make things even easier, we'll leave the spaces between words exactly where we found them and not change them into some other letter or character. Figure 6.15 shows this type of assignment, where we have used a pair of concentric letter wheels

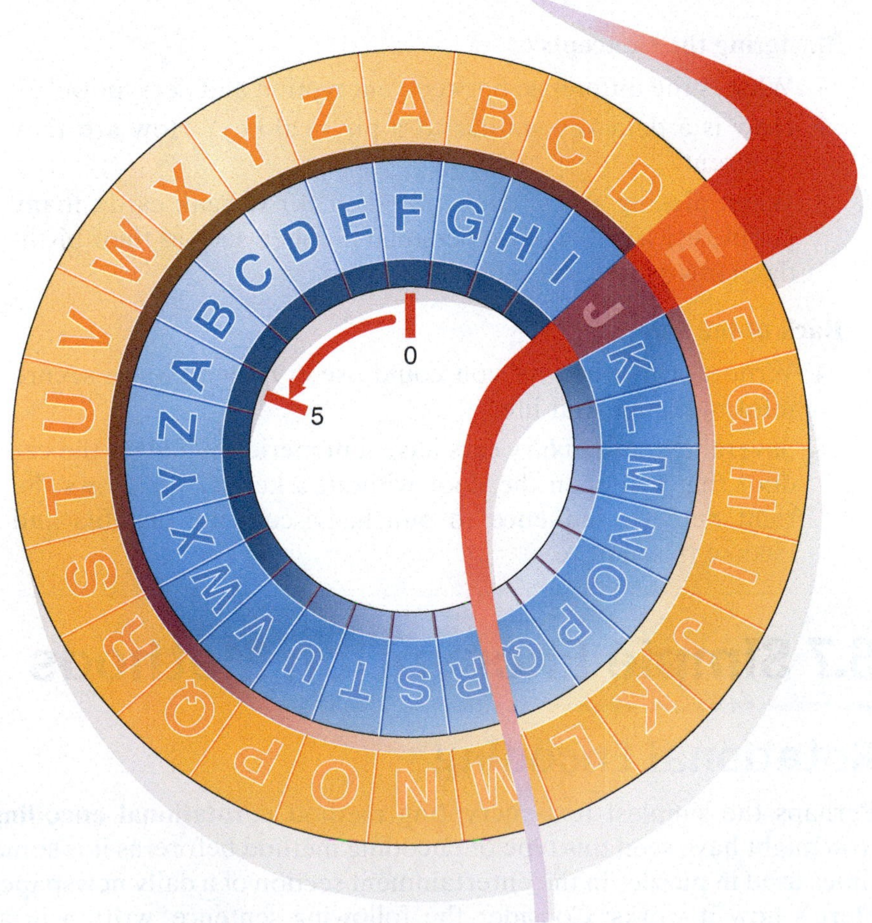

Figure 6.15 The nested coding wheels of a rotational encoder. The inner ring is rotated 5 positions counter clockwise relative to the outer ring.

to show the letter assignments. The outer ring of letters corresponds to the letters in the original sentence, and the inner ring of letters corresponds to the encrypted sentence. The difference between these two letter wheels is a rotation, which is how the encoder gets its name.

With this substitution in place, our sentence becomes

XFQQD ZXZFQQD BJFWX XSJFPJWX BMJS
XMJ NX WZSSNSL TZYXNIJ

We call an enciphered string of text a **cryptogram**. At first glance, the cryptogram we've created here doesn't make any sense. So, we've achieved our objective of preventing casual readers from understanding the sentence.

Cryptogram: An encrypted string of text or binary data that results from enciphering a message or file.

What is the key for this encoder? We can use the number of counterclockwise shifts of letters between the outer and inner letter wheels of Figure 6.15 to define the key. In this case, since the wheels are offset by five positions counterclockwise, the key for this encoder is $K = 5$. Knowing the key and the encoding method immediately allows decoding of the message.

Identity Coding: A coding scheme that produces the same symbol or bit sequence as the original symbol or bit sequence.

To decode the sequence, we simply rotate the inner wheel by K positions clockwise from the so-called **identity coding**, in which the inner and outer letter wheels match. This decoder setting is shown in Figure 6.16. Now, we use the letters in the encrypted sentence with the

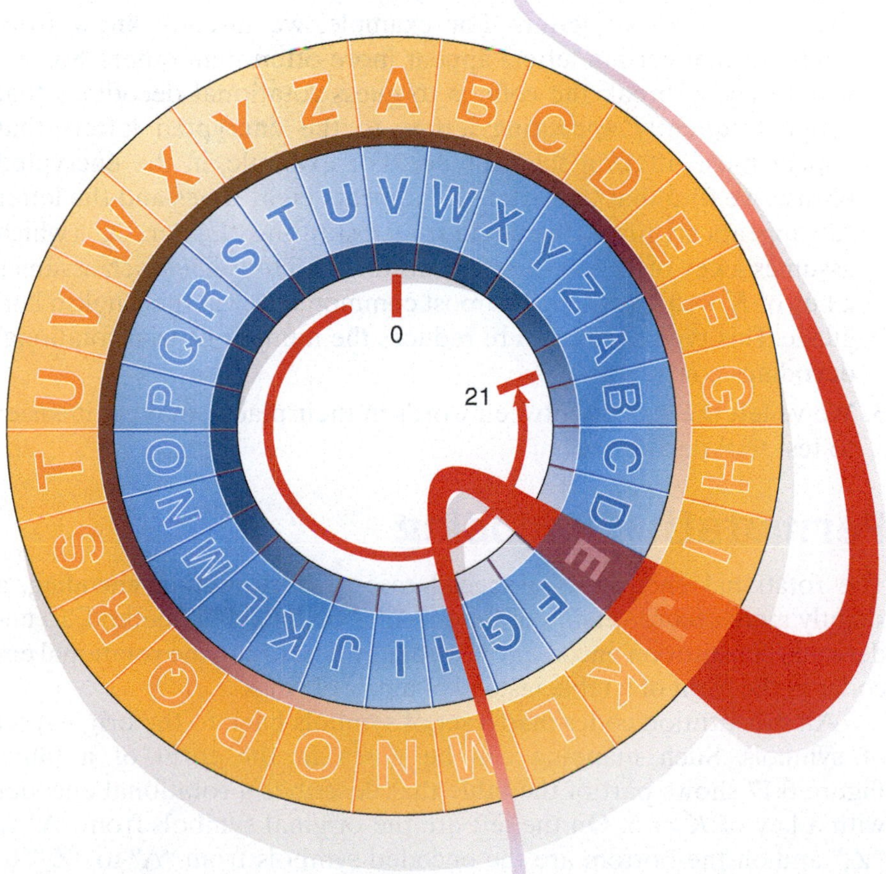

Figure 6.16 The nested rings of the rotational decoder. The inner ring is rotated 5 positions clockwise, which is the same as 21 positions counter clockwise.

INTERESTING FACT:

Because of the circular nature of the shift of a rotational encoder for the English alphabet, any shift larger than 25 is identical to a shift of between 0 and 25. For example, a shift of 30 is equivalent to a shift of 4 (because 4 + 26 = 30). A shift of 65 is equivalent to a shift of 13 (because 13 + 26 + 26 = 65). To get the equivalent small shift value for these larger shifts, just divide the larger shift by 26 and take the remainder. This type of calculation is called "modulo-26" arithmetic.

outermost letter wheel to find the corresponding decrypted letters on the innermost letter wheel. We see that "X" maps to "S," "F" maps to "A," for example.

The rotational encoder illustrates three important concepts of an encryption method:

1. Symbols in a message are mapped back to the same symbol set used for the message; only the assignments of the symbols change.
2. A well-defined procedure exists to both encrypt the original message and decrypt the encrypted message. The encryption key is used for these procedures.
3. A correct key exists that decrypts the encrypted message properly and without loss. An erroneous key produces a useless result.

Breaking a Rotational-Encoded Cryptogram

The rotational encoder is a simple encryption method—so simple, in fact, that it doesn't provide a lot of security. Consider the following facts:

1. The encryption algorithm allows 1 of only 25 different keys. It is a simple task to try all possible rotational decodings to see which one produces a reasonable message at the output.
2. The English language has a structure that can be used to figure out the assignments of letters. For example, we already know from Table 6.2 that certain letters appear more often than others. So, one way to try to break the code is to guess rotational decodings that assign frequently appearing letters to the encrypted letters that appear most often in the sentence. For example, in the encrypted phrase, we find that the letter "X" appears eight times and the letter "J" appears six times. We could try substituting "E" for "X" (which assumes a key of $K = 19$) or substituting "E" for "J" (which assumes a key of $K = 5$), as "E" is the most common letter in the English language. This type of procedure reduces the number of trial rotational decodings considerably.
3. We've left the spaces between words in their places, making it easier to test trial decodings.

INTERESTING FACT:

Testing the quality of a code often involves the public. Sometimes, a code designer uses his or her code in a message in a public place and asks others to try to decipher the coded message to test how well the code works. Companies and organizations have even set up contests with prizes for successful code breakers.

Permutation Encoding

Permutation Encoding: An encryption method of encoding letters in English text by using a shuffling operation, which allows more codes than simple rotations.

The rotational encoder is a special case of **permutation encoding**, a slightly more secure version of the substitution encoding method. To understand substitution encoding, we begin by describing the rotational encoder in the form of a table instead of a pair of rings.

All substitution codes map a set of symbols back to the original set of symbols. Such mapping can be given in the form of a table. Figure 6.17 shows part of the table that describes a rotational encoder with a key of $K = 5$. On the left are the original symbols from "A" to "Z," and on the bottom are the encoded symbols from "A" to "Z." To express the mapping, we put a single 1 in every row and column. Filling in the table in this way defines the code.

To see what encrypted letter will be used for each original letter, we find the letter to be encrypted on the left, and we follow along the row

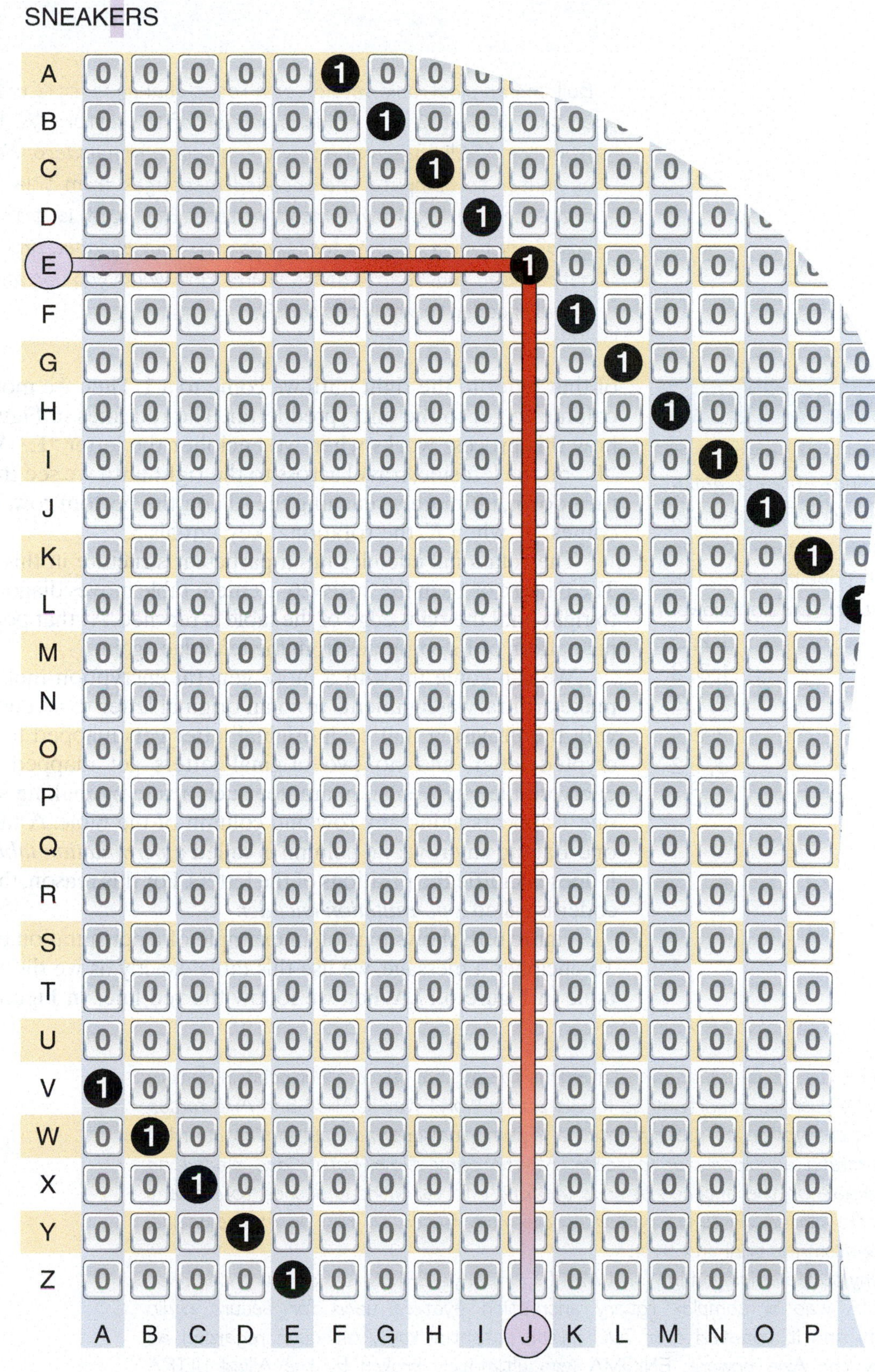

Figure 6.17 The permutation table corresponding to the nested rotational coding wheels of Figure 6.5 when the key, K, is 5.

Infinity Project Experiment: Rotational Encoder and Decoder

Building an electronic version of a rotational encoder is not very hard; all that is needed is a numeric version of the letters "A" through "Z" from the ASCII code and a way to calculate remainders. We can use a computer to compute the encoded messages from this information and a chosen key. Now comes the test: How easy is it to break this code? Try typing in some messages into your encoder, and look at the output. Can you "read" what is there? How would you use another rotational encoder to break the code?

KEY CONCEPT

Permutation is the process of shuffling things so as to get a different arrangement. For example, the set of numbers 1, 2, and 3 can be permuted in six possible ways. The possible permutations are as follows:

1, 2, 3
1, 3, 2
2, 1, 3
2, 3, 1
3, 1, 2
3, 2, 1

of that letter to the right until we come to a 1. Then we move down that column to get to the encrypted letter. The red lines in Figure 6.17 show how this process works when we encipher the letter "E." We start with "E" on the left side, travel across to the right until we see the 1, and then turn down, ultimately reaching the "J" on the bottom row. Thus, "E" becomes "J" when K, the rotational key, equals 5.

The rotational encoder has a particular structure in this table format. The pattern of 1's in the 26 × 26–element table moves diagonally from left to right until the right edge of the table is reached. At that point, the diagonal pattern of 1's continues from the left edge again.

We can come up with a more general encryption method. It would make arbitrary assignments from an original letter to an encrypted letter, with the provision that each original letter gets mapped to only one encrypted letter, and no two original letters get mapped to the same encrypted letter. We can guarantee such a rule by making sure that only one 1 appears along any row and column of the table. A table with only one 1 along any row and column is called a *permutation table*, because all it does is shuffle the positions of the letters. For this reason, this generic encoder is called a *permutation encoder.*

Figure 6.18 shows part of an example of a permutation encoder table. To encipher a message, we use this table exactly as we did the rotational table in Figure 6.17. When we follow the red lines in Figure 6.18 we see

INTERESTING FACT: It is not an overstatement to say that at no other point in history has cryptography been as important as it was during World War II. Allied and Axis forces relied heavily on encryption to protect internal communications, and all parties guarded their coding secrets as closely as possible. At the same time, both sides placed an enormous emphasis on the success of code-breaking efforts. Without access to efficient digital computers, these wartime cryptographers were forced to rely on mathematics, intuition, and systematic trial and error.

Of the many codes employed in World War II, Germany's ENIGMA is certainly the best known. ENIGMA was a complex rotary encryption system used for secure radio communication, and it boasted over 700 million potential keys. Although regarded as unbreakable by the Axis powers, ENIGMA was ultimately broken by the Allied ULTRA program. Great care was then taken to ensure that Germany and other members of the Axis didn't know that the Allies had broken the code, so that the Axis communications could be monitored surreptitiously.

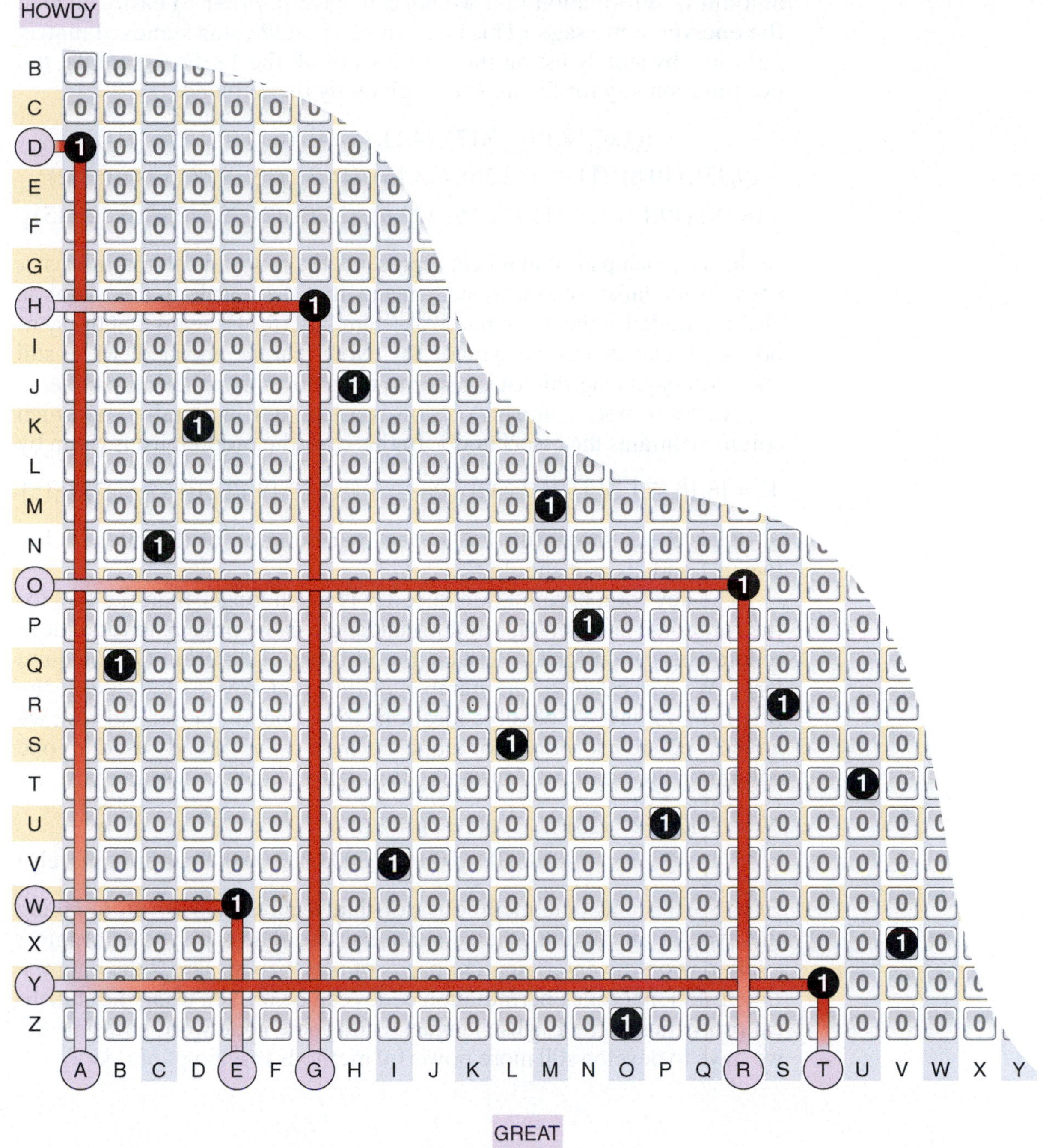

Figure 6.18 A permutation encoder table. Such tables can provide many more keys than the simpler rotational codes can.

that the letters in "HOWDY" become the letters "GREAT" in our cryptogram. In this specific table, we've assigned five letter–letter pairs such that the phrase "HOWDY" becomes "GREAT" in the encrypted message, but we have a lot of freedom to choose the permutation encoder table. The only restriction is that each row and each column have only one 1 value.

An immediate question to be asked is, what is the key for a permutation encoder? Since the encoding procedure employs a shuffling, we need to save the shuffling information as the key. Thus, the key must contain the positions of all 1's inside of the table. (This is the smallest amount of information that we need to save in order to easily decode the encrypted message.) This key can be created using standard matrix notation, by simply listing the locations of all the 1's. For example, the permutation key for Figure 6.18 is given by the sequence

$$K = [(1,6), (2,10), (3,17), (4,1), (5,24), (6,11), (7,25), (8,7), (9,23), (10,8), (11,4), (12,26), (13,13), (14,3), (15,18), (16,14), (17,2), (18,19),(19,12), (20,21), (21,16), (22,9), (23,5), (24,22), (25,20), (26,15)]$$

In this list, each pair of numbers represents the row and column of each 1 value in the table. The (4,1) entry means that the fourth original symbol, "D," is encoded as the first symbol, "A." Similary, (5,24) means that the symbol "E" is encoded as the symbol "X." Note that the encoded "M" is still "M." Although long, this list is easier to write down than the entire table.

An even more compact version is a list, by row number, of which column contains the associated 1. In this case, the ordered list is given by

$$K = [6,10,17,1,24,11,25,7,23,8,4,26,13,3,18,14,2,19,12,21,16,9,5,22,20,15]$$

A quick check shows this list to be a permutation of the integers between 1 and 26, as it should be.

The key for a permutation encoder is much longer than the key for the rotational encoder. This additional length is the price paid for a more secure encryption format. In fact, the number of different keys is quite large. We can calculate the number of different keys as follows: In the first row, we have 26 possible locations for a one. In the second row, we have 25 possible locations for a one, because we've already used up one column in the first row. Continuing through all the rows, we arrive at

$$N = 26(25)(24)(23)(22)\ldots(2)(1) = 26!$$
$$= 40{,}000{,}000{,}000{,}000{,}000{,}000{,}000{,}000 \text{ (approximately)}$$

Even so, this encoding method is not that secure. Remember, breaking this type of encryption scheme is an exercise that is played out almost every day on the entertainment pages of newspapers across the country. Like rotational codes, permutation codes can be broken using the statistical properties of the English language and some good word guessing. So, we have to develop still more powerful methods to encrypt our data.

INTERESTING FACT: Throughout much of history, information security has often been ensured by keeping the means of encrypting one's data a secret. Up until fairly recently, this approach made sense. Preventing others from seeing the mechanics of an encryption scheme presents an additional barrier to those who seek to crack that scheme. These days, however, "security through obscurity" doesn't carry as much weight as it used to. Computers are so adept at finding patterns that modern cryptography has had to adapt accordingly. Many modern encryption schemes derive their security from their key size and the sheer number of possible keys that can be created. In most cases, the way these codes work is not kept secret at all. Any encryption scheme that relies on keeping its inner workings secret should be viewed with suspicion.

EXERCISES 6.7

Mastering the Concepts

1. Which is a more secure encryption method, rotational encoding or permutation encoding? Why?
2. How many entries with a value of 1 does a permutation table have when encoding and decoding letters?
3. A permutation table has all but one row already specified. Can you figure out the last row? Try it and see!

Try This

4. ENCODE THIS SENTENCE USING A ROTATION OF TWELVE.
5. Suppose you wanted to encode the numbers 0 through 9 by using rotational encoding. Draw the rotational-encoder wheels for a shift of four values, and then write an equation for it.
6. ENCODE THIS SENTENCE USING THE FOLLOWING KEY:

$$K = [(1,24), (2,8), (3,21), (4,25), (5,15), (6,26), (7,2), (8,22), (9,16), (10,19), (11,10), (12,7), (13,4), (14,3), (15,11), (16,14), (17,6), (18,12), (19,23), (20,9), (21,5), (22,20), (23,18), (24,13), (25,17), (26,1)]$$

7. Give the five-by-five permutation table that describes the following permutation: GREAT becomes TEAGR.
8. Consider a slightly modified rotational encoder that reverses the order of the letters in the inner wheel of Figure 6.15. That is, the outer ring of letters goes around clockwise, but the inner ring of letters goes around counterclockwise. Draw this rotational encoder for shifts of 5, −5, and 12 characters. Then, use this rotational encoder to encode a simple message, and compare the encoded version with that of the original rotational encoder (which doesn't reverse the letter order). Which one of the two encoding methods provides more protection of the message?
9. It is possible to develop permutation encoders that are easier to describe compared with the completely random encoding table in Figure 6.17. Consider the key given by

$$K = [(1,4), (2,8), (3,12), (4,16), (5,20), (6,24), (7,2), (8,6), (9,10), (10,14), (11,18), (12,22), (13,26), (14,3), (15,7), (16,11), (17,15), (18,19), (19,23), (20,1), (21,5), (22,9), (23,13), (24,17), (25,21), (26,25)]$$

   Draw the encoding table for this permutation encoder. Just write the 1's in the table to save time. Do you notice any structure to the encoding method?
10. Consider the effects of double encoding.
    (a) If a rotational encoder is used with key K_1, and then the encoded result is passed through a second rotational encoding

using key K_2, write the permutation table for the double encoding. How would you describe the total effect of the double encoding? (*Hint:* Write the two permutation tables separately as shown in Figure 6.17. Then rotate the second table so that its inputs are aligned with the outputs of the first. Then start with the inputs to the first table and follow the path of each through both tables to the doubly encoded output.)

(b) Repeat part (a) for doubly encoding with two different permutation encoders instead of two rotational encoders. Use tables such as the one shown in Figure 6.18.

(c) Repeat part (a) for a rotational encoding followed by a permutation encoding.

In the Laboratory

11. Make a rotational encoder by using cut-out paper circles and a pencil. Punch the pencil through the center of the paper circles to hold the two discs together, or use some other center fastener. Then encode a message. Hand the message to a friend (or foe?) and see if he or she can decode the message without your encoder.

Back of the Envelope

12. Suppose you want to express the key for a rotational encoder in a binary representation. How many bits would you need? How many bits would be required in order to express the key for a permutation encoder?
13. Is it possible to develop a rotational encoder that swaps letter sets, so that a message that is rotation encoded twice results in the same message back? If so, what is the rotation value?
14. A rotational encoder uses a single angle value to represent a rotational offset. How many degrees does a rotation by a single letter represent?
15. Work out the details of a two-dimensional rotational encoder, where letters appear on the surface of two concentric spheres and one sphere is offset with respect to the other.

6.8 Encrypting Binary Sequences

Cryptography was first developed to protect short textual messages, such as military orders or diplomatic communications. One of the first uses of cryptography in Western history was by Julius Caesar, who encrypted messages to his military leaders during the height of the Roman Empire in ancient times. Many things have changed since then. Computers are available to help us create ciphers, break ciphers, and do the coding and decoding for us rapidly. Many people and companies rely on information held in computer databases and would like that information to be protected from unauthorized viewing or harm. The development of high-speed electronic and optical communications has brought with it the desire for privacy. In many cases, though, the information to be protected is not text. It can be anything in digital form, such as images, audio, and video signals. All of these considerations have led the developers of cryptography systems toward methods that permit the protection of large amounts of binary information. Let's examine how this encipherment is usually accomplished.

The Exclusive-OR Operation and Cryptography

The **exclusive-OR operation** is a way to combine two sequences of bits into one. Suppose you have an important bit sequence that you want to keep private. As an example, we'll take the sequence

$$S = 101100010100100001010010$$

To keep this sequence private, you generate a bit sequence that is the same length as the original sequence. We'll choose this sequence to have no relationship to the data S, and in fact we'll choose it to be as random as possible. We'll call it the key K:

$$K = 001001101001110100110101$$

The exclusive-OR operation combines these two sequences, bit by bit, by comparing their values. If each bit in the pair is the same, a 0 is generated. If the bits are different, a 1 is produced. Table 6.3 shows the result of the exclusive-OR operation. In this table, the first bit corresponds to a bit from the original sequence, and the second bit corresponds to a bit from the same position in the key. Using this operation, we have

$$S = 101100010100100001010010$$
$$K = 001001101001110100110101$$
$$E = 100101111101010101100111$$

where E is the encrypted bit sequence. Notice how E looks nothing like S and K, implying that the signal is encrypted. We therefore can make E known to everyone, but we keep the key K private.

To decrypt the encoded signal E, we need to use the key K again. It turns out that the exclusive-OR operation can be used to get back the original message from the encrypted one by combining E and K. Because E is 1 where K and S are different, it tells us which bits of K to change to make it look like S. Employing the exclusive-OR again, we have

$$E = 100101111101010101100111$$
$$K = 001001101001110100110101$$
$$S = 101100010100100001010010$$

So, the original secret message S is found. This form of encryption works, but it has some disadvantages:

- The key K must be exactly the same length as the signal S to be encrypted. This requirement puts demands on storage capabilities, not to mention the fact that we have to generate long sequences of random-looking bits for the key.
- Security is lost if the key K becomes available.

In the next two subsections, you'll see how both of these difficulties were overcome by clever engineers, mathematicians, and computer scientists.

Exclusive-OR Operation: A means of combining two sequences for bits into one, used for both encryption and decryption of information. This operation is the same thing as modulo-two arithmetic—adding the two inputs, dividing by two, and keeping the remainder. The exclusive-OR is 1 when the two bits are different and zero when the two bits are the same .

Table 6.3 Exclusive-OR Operation

First Bit	Second Bit	Exclusive-OR
0	0	0
0	1	1
1	0	1
1	1	0

Pseudo-Random-Number Generators

The cryptographic method discussed in the previous subsection uses long sequences of random-looking bits for the key. Generating truly random bit sequences is actually a hard task. We could generate random numbers between 0 and $N - 1$ and then translate each number into its corresponding bit representation, where N is a power of two. That way, each bit representation is exactly $\log_2 N$ bits long. But then we would need a random sequence of numbers. If you were to just bang on a numeric keypad and try to generate random numbers, you might fall easily into a pattern after a while. Fortunately, we only need random-*looking* numbers. We'll call these numbers *pseudorandom,* because they only appear to be random to us. The method that we'll now describe produces long sequences of pseudorandom numbers. Hence, it is called a **pseudo-random-number generator**.

Pseudo-Random-Number Generator: A mathematical device for generating long strings of random-looking bits.

The pseudo-random-number generator employs four different numbers: A, B, N, and a starting value $X(0)$ called the **seed** of the generator. The generator then produces random numbers $X(n)$, $n = 1, 2, 3, \ldots$ using these four numbers. It effectively generates a long and apparently random sequence of numbers, using only four numbers. The numbers are generated using the equation

Seed: The starting value in a pseudo-random-number generator.

$$X(n+1) = [AX(n) + B]\,\text{mod}(N)$$

where $\text{mod}(N)$ denotes the **modulo-*N* operation.** The modulo-N operation takes the remainder of the quantity in the brackets after dividing this quantity by N. Thus, the pseudorandom values stored in $X(n)$ will be integers between 0 and $N - 1$. The pseudo-random-number generator described produces integer numbers only in the range from 0 to $N - 1$, because of the modulo-N operation.

Modulo-N Operation: An operation that takes the remainder of a number when divided through by N. For example the modulo-12 value of 77 is 5. (Seventy-seven divided by 12 is 6 with a remainder of 5.)

Let's see how this type of pseudo-random-number generator creates numbers.

EXAMPLE 6.15 Pseudo-Random-Number Generator

Generate a 24 bit key with a pseudo-random-number generator using $A = 533$, $B = 227$, $N = 64$, and $X(0) = 125$.

Solution

The first four numbers from the pseudo-random-number generator are

$$X(1) = [533(125) + 227]\,\text{mod}(64) = [66{,}852]\,\text{mod}(64) = 36$$
$$X(2) = [533(36) + 227]\,\text{mod}(64) = [19{,}415]\,\text{mod}(64) = 23$$
$$X(3) = [533(23) + 227]\,\text{mod}(64) = [12{,}486]\,\text{mod}(64) = 6$$
$$X(4) = [533(6) + 227]\,\text{mod}(64) = [3425]\,\text{mod}(64) = 33$$

To produce bit values from this sequence, we know that 6 bits are required to represent integers from 0 to 63, because $64 = 2^6$. Therefore, we have

Decimal	Binary
36	100100
23	010111
6	000110
33	100001

Putting the bits next to each other, we generate the 24-bit key

$$K = 100100010111000110100001$$

We easily could use this method to generate more numbers in the sequence and thus make longer and larger keys for K.

Infinity Project Experiment: Pseudo-Random-Number Generator

Modulo operations are calculated easily using digital devices, so we can have a computer make pseudo-random-number sequences given the values of A, B, N, and X(0). Try different values of these numbers and see what is generated. What combinations of A, B, N, and X(0). make the number sequences look most random? What combinations produce not-so-random results? How might you determine how random a generated sequence is?

An important feature of the pseudo-random-number generator is the number of pseudorandom numbers it can produce. It turns out that, because of the modulo-N operation, at most N random numbers are produced before the sequence *begins to repeat.* Therefore, the choice of N is critical. We need to choose N large enough so that we produce a long-enough key. When N is a power of 2, $\log_2(N)$ is the number of bits needed to represent N values from 0 to $N - 1$. Then the maximim number of bits that can be generated from a pseudo-random-number generator is

$$L \text{ bits} = N \text{ numbers} \times \log_2(N) \frac{\text{bits}}{\text{number}}$$

Therefore, in Example 6.15, where $N = 64$, we can generate a bit sequence that is $L = 64(6) = 384$ bits long before the sequence repeats.

The sequence of numbers produced by a pseudo-random-number generator depends on the four values $A, B, N,$ and $X(0)$. In practice, the values of A, B, and N are usually made available to everyone, and only the seed value $X(0)$ is kept secret. Thus, $X(0)$ represents our key K. Here again, we can see the connection between N and the randomness of the sequence. As it turns out, the number of different "random" sequences that a pseudo-random-number generator can produce is equal to N (because a seed value outside the range $[0, N - 1]$ is equivalent to some seed value in this range through the modulo-N operation). So, good data security is achieved only when large N values are chosen. Most commercial encryption methods today use 128-bit keys or seeds, corresponding to an N value of

$$N = 2^{128} \approx 10^{38}$$

Searching through all possible 128-bit seeds is a lengthy process even for a very fast computer. So, if one doesn't know the seed, one has a difficult time decoding a bit sequence encoded using the pseudo-random-number generated from the seed.

A Simple Cryptography System A simple cryptography system would use the pseudo-random-number generator in the following way:

1. To communicate a message, both the sender and recipient of the information need to know $A, B, N,$ and the seed $X(0)$. Typically, A, B, and N are built into the encryption hardware or software, and $X(0)$ is communicated by some other secure method to the receiver.
2. The sender generates the pseudorandom sequence of numbers, using the values of $A, B, N,$ and $X(0)$, and converts this sequence into bits.
3. The sender then performs the exclusive-OR operation on successive pairs of bits from the message to be sent and the pseudorandom number. This operation makes the binary *cryptogram*.

KEEP IN MIND

Usually we want to choose A, B, and N so that they do not have any common factors. Then we are more likely to get a sequence that has the maximum length of N before repeating.

INTERESTING FACT:

When N is a power of 2, for example, $N = 2^k$, digital computers can perform the modulo-N operation without any division. After the multiplication and addition are performed to compute $X(n + 1)$ from $X(n)$, the remainder after division by N will be the same $k = \log_2(N)$ rightmost bits of the result. The computer simply discards all the bits to the left and keeps only these k rightmost bits to get $X(n + 1)$ modulo N.

KEY CONCEPT

10^{38} is big! Remember there are fewer than 10^{10} humans on earth. This means that there are 10^{28} (or a one followed by 28 zeros) keys for every human using 128-bit keys!

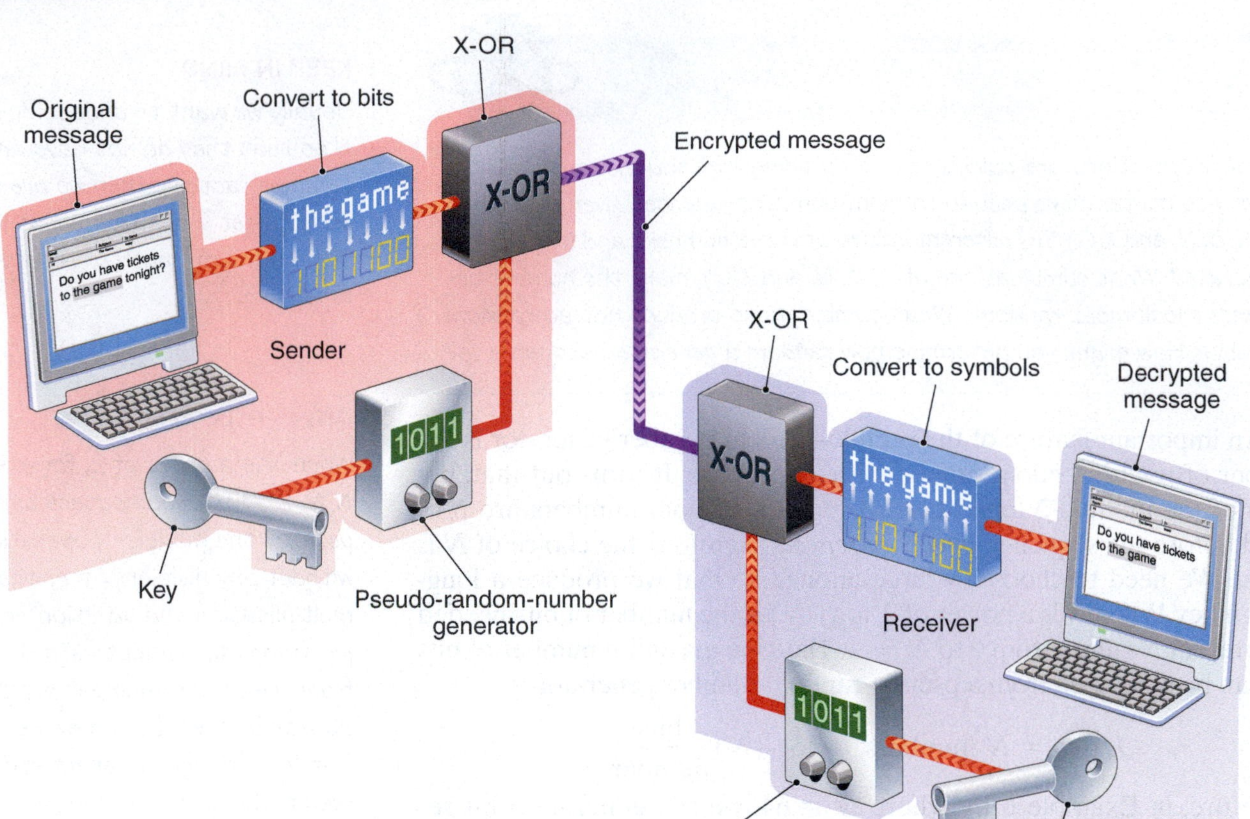

Figure 6.19 A cryptographic transmission system that uses binary pseudo-random streams as keys to encrypt binary multimedia data.

4. The sender transmits the encoded message to the recipient.
5. The recipient generates her or his own version of the pseudo-random-number sequence, using the values of A, B, N, and $X(0)$, and then converts these numbers into bits.
6. The recipient then performs the exclusive-OR operation on successive pairs of bits from the encrypted message and the pseudo-random bit sequence, recovering the original message. This entire process is shown in Figure 6.19.

While appearing to be simple, this basic scheme is at the heart of many practical cryptographic systems used by industry and governments. The digital cellular telephones used by many people all over the world rely on the automatic generation of a pseudo-random binary key stream that is $2^{42} - 1$ bits long. At normal transmission rates, it takes 41.5 days to repeat. Your digitized and compressed voice is then exclusive-Or-ed with this key stream to produce the information actually sent out of your cellular phone on a radio signal.

Sharing Keys with Everyone

The scheme for secure communications shown in Figure 6.19 is widely used in practice. Versions of it are also used to protect the data held on some computer disk drives. This technique can secure any information that we've been clever enough to code into a binary format. The most important issue with these schemes is *key management*, which is the problem of securely providing the key information to the intended recipient so that he or she can read the decrypted message. Getting the key to the desired recipient is a hard problem. In some cases, the keys to important data are transported by hand, as in the military, where armed guards are used. When they are not in use, the keys are put away for safekeeping, such as in a strong locked safe.

In many situations, we would like to have a simple way to share a key with another person—for example, to encrypt an e-mail message to send to a friend's Internet account. Armed guards and locked safes aren't of much use to us here. Fortunately, engineers in the 1970s came up with a clever mathematical approach to the solving the problem. Their technique is called **public-key cryptography**, based on the fact that everyone can share a private key in a public way with everyone else. Don't be fooled though. There are still secrets in there somewhere!

Public-Key Cryptography: An encryption method that allows the sending of secret information without requiring the sender and receiver to share a secret password or key.

Public-Key Cryptography

The coding system described in the previous section and shown in Figure 6.19 requires a secret seed $X(0)$ for the number generator. From this seed and the knowledge of A, B, and N, anyone can generate the full pseudorandom number and decode your encrypted message with the exclusive-OR operation. What public-key cryptography provides is a method to generate a secret seed that only the sender and an intended recipient know. In this technique, no single person has the capability to make a seed. Instead, two people work together to build a seed whenever they want to share information. Here's how it works.

Suppose each of these users has a different *public key* that he or she shares with everyone openly. Each public key is generated from a *private key* in such a way that no one knows anyone else's private key. In public-key cryptography, a simple mathematical function is used to combine any one public key with any one private key to generate the secret seed $X(0)$ used to make the bit sequence for K. The trick is to find a mathematical function that makes the *same* secret seed when each person in a pair of people combines her or his private key with the other person's public key. Of course, the form of this mathematical function is not secret, and the number of different seeds that it can calculate is finite. If we make the number of possible public keys enormously large, however, it becomes nearly impossible for anyone to use two public keys to figure out the secret seed $X(0)$.

Many mathematical functions could be used for generating public keys and secret seeds in public-key cryptography. One example is described in the following subsection.

A Method for Sharing Binary Keys in Public-Key Cryptography

One way to generate public keys and secret seeds for public-key cryptography combines the monomial function C^J, where C and J are both integers, with the modulo-N operation $\text{mod}(N)$ that was introduced in our discussion of rotational encoding. When used properly, this method generates a seed $X(0)$ that is easily decodable only by the sender and the recipient of a set of information. Here's how the system works.

Both the sender and the recipient choose a common integer value C between 0 and $N - 1$. Generally, this value is kept secret, although it doesn't have to be. Next, the sender chooses a large integer I as a private key that only he knows, and the recipient chooses a large integer J as a private key that only she knows. Neither I nor J

are communicated publicly or transmitted. Then the sender computes his public key

$$P = [C^I]\,\text{mod}(N)$$

and the recipient computes her public key

$$Q = [C^J]\,\text{mod}(N)$$

Both sender and recipient send the public keys P and Q to one another, using any open and public method of communication. Now the sender receives Q from the recipient, and he computes the value

$$[Q^I]\,\text{mod}(N)$$

The recipient, having received P, computes the value

$$[P^J]\,\text{mod}(N)$$

It can be shown that

$$[Q^I]\,\text{mod}(N) = [P^J]\,\text{mod}(N) = [C^{IJ}]\,\text{mod}(N)$$

In other words, these two numbers are identical. The sender uses this number as the seed $X(0)$ to generate a pseudorandom bit sequence to encode the message with the exclusive-OR operation as described previously. The sender then sends the encrypted bit sequence to the recipient, who then uses this number to generate the same pseudo-random bit sequence and decode the message with the exclusive-OR operation. Since no one knows $X(0)$, the information is kept secret. Here are a couple of examples to help illustrate how the secret seed is generated.

EXAMPLE 6.16 Computing Powers of C modulo N

Suppose $C = 5$. Compare the first 6 powers of C modulo N when $N = 16$ to the first 6 powers when $N = 17$.

Solution

The value $C^{J+1}\text{mod}(N)$ can be computed in two ways. We can raise C to the power $(J + 1)$ and then perform the modulo N operation. However, if we already know $C^J\text{mod}(N)$, we can simply multiply that value by C and then perform the modulo N operation. This allows us to work with much smaller numbers when J gets large. The second method is used in this solution. On each line of the table we compute $C^{J+1}\text{mod}(N)$ from $C^J\text{mod}(N)$ and then enter that result into the next line of the table.

$N = 16$

J	C^Jmod (N)	$C \times (C^J$mod $(N))$	C^{J+1} mod (N)
0	1	5	5
1	5	25	$25 - 16 = 9$
2	9	45	$45 - 2 \times 16 = 13$
3	13	65	$65 - 4 \times 16 = 1$
4	1	5	5
5	5	25	$25 - 16 = 9$
6	9	45	$45 - 2 \times 16 = 13$

$N = 17$

J	C^Jmod (N)	$C \times (C^J$mod $(N))$	C^{J+1} mod (N)
0	1	5	5
1	5	25	$25 - 17 = 8$
2	8	40	$40 - 2 \times 17 = 6$
3	6	30	$30 - 17 = 13$
4	13	65	$65 - 3 \times 17 = 14$
5	14	70	$70 - 4 \times 17 = 2$
6	2	10	10

When $N = 16$, the powers of $C = 5$ after the modulo N operation have only four distinct values. Since we would not want P or Q to be 0 or 1, we only have three values that we could use. When $N = 16$, the values shown in the table are all different. If the table were extended, we would see that the powers of 1 go through all values from 1 to 16 before repeating.

EXAMPLE 6.17 Generating another Secret Seed

Suppose $N = 32, I = 5$, and $J = 7$ are chosen. The value of C chosen is 11. Compute $X(0)$.

Solution

Then

$$\begin{aligned} P &= [C^I]\,\mathrm{mod}(N) \\ &= [11^5]\,\mathrm{mod}(32) \\ &= [161{,}051]\,\mathrm{mod}(32) = 27 \end{aligned}$$

and

$$\begin{aligned} Q &= [C^J]\,\mathrm{mod}(N) \\ &= [11^7]\,\mathrm{mod}(32) \\ &= [1{,}9487{,}171]\,\mathrm{mod}(32) = 3 \end{aligned}$$

So, the numbers 27 and 3 are passed publicly. Then, processing these values, we get

$$[Q^I]\,\mathrm{mod}(N) = [3^5]\,\mathrm{mod}(32) = 19$$
$$[P^J]\,\mathrm{mod}(N) = [27^7]\,\mathrm{mod}(32) = 19$$

The seed value of 19 is the same in both cases. This number would be used to generate the sequence $X(n)$ from $X(0)$ given A, B, and $N = 32$.

The important feature of this seed-generating method is that the values of I and J *are never communicated.* Therefore, it is very difficult for anyone else to generate the seed $X(0)$, because it depends on both of the unknown values I and J. If N is large enough, the number of possibilities of any seed is so large that a supercomputer is needed to search through all of them in order to decrypt the encrypted information.

What is the main disadvantage of this public-key cryptographic system? The two people must send the values of P or Q to each other before a private connection is made. This requirement might be inconvenient in some situations, such as in broadcast communications, where the desired information need flow only "one way." Where two-way communication of information is possible, however, public-key encryption is a very useful method. In fact, it is widely used on the Internet to make on-line financial transactions, such as credit card purchases from a website, more secure.

EXERCISES 6.8

Mastering the Concepts

1. What is the purpose of a pseudo-random-number generator?
2. What is a seed? How do we use it to generate pseudorandom numbers?
3. Can A and B of a pseudo-random-number generator be the same value? Why or why not?
4. What happens when $B = N$ for the pseudo-random-number generator? What happens if $A = N$?

Try This

5. Determine the exclusive-OR of the following pairs of bit sequences to generate the encrypted sequence E:
   a. $S = 010001010$, $K = 110011001$
   b. $S = 001000011111$, $K = 111100001100$
6. From your answers in parts (a) and (b) of Exercise 6.8.5, apply the key K to the encrypted sequences E, using the exclusive-OR operation. What sequences do you obtain?
7. Suppose a pseudo-random-number generator is created using the values $A = 10$, $B = 9$, $N = 23$, and $X(0) = 15$. What is the maximum number of different values that this pseudo-random-number generator can make?
8. Find the first five numbers from the following pseudo-random-number generators:
   a. $A = 13, B = 7, X(0) = 2$, and $N = 16$
   b. $A = 25, B = 14, X(0) = 152$, and $N = 128$
   c. $A = 151, B = 39, X(0) = 305$, and $N = 256$
9. Determine the bit sequences that result from the number sequences produced by the pseudo-random-number generators in Exercise 6.8.8.
10. In a public-key cryptography system, the given values are chosen to create the seed. Find the resulting values P and Q as well as the common secret seed.
    a. $N = 16, I = 5, J = 9$, and $C = 11$
    b. $N = 32, I = 11, J = 15$, and $C = 7$
    c. $N = 128, I = 13, J = 15$, and $C = 19$

Back of the Envelope

11. You, too, can encrypt messages to communicate with a friend! First, generate a seed value, using public-key encryption. Then generate a bit sequence, using a pseudo-random number generator, and use this bit sequence to encode a secret message to your friend, using ASCII for the letter encoding and the system in Figure 6.19. Since each codeword is 7 bits long, use $N = 128$ for generating both the seed value and the pseudo-random-number sequence. Have your friend decode your secret message and read it back to you.

Master Design Problem

Garage sales, yard sales, "spring cleaning" sales—all of these events bring people together for weekend commerce and bargain hunting. The joy of finding a collectible item at one of these sales is always matched by the happiness of the seller, who gets paid to rid himself or herself of old stuff.

The rise of the popularity of the Internet has created a new opportunity: global commerce on a per-individual scale. Now, "virtual garage sales" go on at all hours of the day and night. The companies that establish the websites where items are bought and sold often collect money for each transaction that is carried out. Realizing this, and with your newfound understanding of compression and encryption, you decide to design a "virtual garage" where people can buy and sell personal items.

What will your virtual garage need in order to be successful in carrying out its daily business?

- Customers who use the site must be confident that their personal information—their identity, financial account numbers, and ordering history—is not divulged to others.
- When a transaction is carried out, both buyer and seller need a secure way for exchanging information without interference from other individuals.
- Your virtual garage sale should be able to lists millions and millions of items for sale—so that anyone who visits the site at any time will find something of interest to buy.

Clearly, compression and encryption technologies are needed in order to make your website a success.

1. Describe the overall functionality of your virtual garage sale website. How will sellers post information on the items they wish to sell? How will buyers choose what they want to buy? Sketch out on paper the basic interfaces for the buyer and the seller, respectively.
2. In order to efficiently store all of the items and their listings on your website, you intend to provide common listings for identical items listed by multiple sellers. How should you index the information to make the best use of your storage space?
3. When a seller and a buyer wish to exchange financial information, such as bank accounts, how should the information be encoded so as to ensure privacy? Give the complete details of the method.
4. Once your website is up and running, you discover that some participants are attempting to get account information of other users without the users' knowledge. How can you change your encryption scheme to better protect your clients' information? How can you make your encryption secure even if most of the details of the encryption method are known to all?

With the success of your virtual garage sale website, you are now ready to develop other business ideas for other would-be entrepreneurs. What other new ideas do you have for using bits and bytes to the world's advantage?

Big Ideas

Math and Science Concepts Learned

This chapter has discussed ways of changing digital information either to reduce the number of bits required to represent the information or to keep the information private.

Compression describes the act of reducing the number of characters, symbols, or bits used to represent any digital signal. Compression usually lowers the cost of storing information, because fewer resources, such as hard disk space, are needed. We have to decompress the information when we need it, however, so extra processing steps and time are required in order to use the compressed information.

We measure the improvement produced using the compression ratio, the ratio of the number of bits in the original data to that which remains after we've compressed the data. This quantity is usually expressed as a number (hopefully greater than one) or as a classic ratio, such as 4 : 1. The larger the compression ratio, the better, in most applications.

There are two general classes of compression methods. Lossy compression techniques throw away certain portions of the data or information that wouldn't be noticed or aren't considered important. Lossless compression techniques represent the information so that fewer bits are required without losing any of the information contained in the original.

Run-length coding is one example of a lossless compression method. Run-length coding works best on binary signals where there are long stretches of identical 1's or 0's in the data stream. The concept of relative frequency can also be used to assign codewords to symbols for compression. Longer codewords are associated with symbols that occur less frequently.

The entropy of a symbol set is a mathematical quantity that measures the randomness of the symbol set. The units of entropy are bits. The entropy of a symbol set is a lower limit on the average number of bits needed to losslessly store each symbol in the set. The entropy can be used to figure out how good a particular lossless code is for a given set of information.

Lossy compression uses our limits in human perception to reduce the number of bits needed to store a signal. For example, our ears can hear a wide range of sounds in terms of their amplitudes and frequency content, but these ranges are still limited in mathematical terms. Masking of sounds allows us to encode audio signals selectively so that fewer bits are required. Lossy compression can often obtain higher compression ratios than lossless compression on the same data.

Codes for error detection and error corrections require additional bits so that some data sequences are not valid. Using a parity bit is a simple method of error detection.

Cryptography describes a set of methods that are designed to make information private. By encrypting information, we encode the information to hide its contents from others. We use decryption to undo the encryption process and read the information again. Passwords and access codes are examples of keys, sequences of symbols or bits that are used to "unlock" an encrypted message.

Rotational encoding and permutation encoding are two examples of simple encryption methods that are not very secure when used on written messages. We can use features of the language, such as the relative frequency of letters, to "crack" the code. A more secure encryption method is public-key encryption.

Modern "streaming" encryption of digital data uses the exclusive-OR operation to combine the bits to be encrypted with another bit sequence of the same length. The exclusive-OR operation can be used both to encrypt and decrypt the data if this other bit sequence is known. To generate long random bit sequences, a pseudo-random-number generator is often used. A pseudo-random-number generator creates a long sequence of seemingly random numbers, using the modulo operation. The key of the pseudo-random-number generator is the initial value that generates the sequence, which is also known as the seed.

Traditional encryption methods require that a common digital key be secretly provided to both the sender and recipient of the data to be protected. The practical difficulty of providing these keys in a secure fashion threatened to limit the availability of encryption to those who needed it for modern communication and data storage. Fortunately, a simple way, termed "public-key cryptography," has been developed to create a private seed value that is communicated publicly between two individuals. This private seed is then used by a pseudo-random number generator to create long bit sequences to encode messages. Web portals often use this encryption method to perform secure financial transactions.

Important Equations

Design of simple formatting codebooks:

$$N \leq 2^C$$

Compression ratio:

$$R_{\text{compression}} = \frac{\text{\# of bits in original signal}}{\text{\# of bits in compressed signal}}$$

Relative frequency:

$$f_R(E) = \frac{\text{Number of observations of an event } E}{\text{Number of total observations}}$$

Average codeword length:

$$L_{\text{average}} = \sum_{i=1}^{N} [f_R(s_i) \times L_i]$$

$$L_i = \text{code length of } i\text{th codeword}$$

$$f_R(s_i) = \text{relative frequency of } i\text{th symbol}$$

Entropy of a symbol set:

$$H = -\sum_{i=1}^{N} [f_R(s_i) \times \log_2(f_R(s_i))]$$

Lower bound on average codeword length:

$$L_{\text{average}} \geq H$$

Pseudo-random-number sequence generator:

$$X(n+1) = [AX(n) + B]\,\text{mod}(N); \quad n = 0, N-1$$

Length of a bit sequence from a pseudo-random-number generator:

$$L = N \log_2(N)$$

Equations for public-key cryptography:

Sender's public key: $P = [C^I]\,\text{mod}(N)$

Recipient's public key: $Q = [C^J]\,\text{mod}(N)$

Calculation of the shared key, or key seed:

$$[Q^I]\,\text{mod}(N) = [P^J]\,\text{mod}(N) = [C^{IJ}]\,\text{mod}(N)$$

Building Your Knowledge Library

Kahn, David. *The Codebreakers*, Scribner, 1996.

Tuchman, Barbara, *The Zimmerman Telegram*, Ballentine Publishing Group, 1996.

These are both good texts for learning more about codes.

C
OK
2 ABC

Communicating with Ones and Zeros

chapter 7

In earlier chapters, we learned that voices, music, images, and other types of signals and media can be represented digitally as a list of symbols or bits. We saw that each of these media can be converted into symbols and that these symbols can be converted back into a form that humans can recognize easily and interpret. In addition, we learned that those symbols can be stored and recovered and that they may be manipulated or "processed" to improve some aspect of how we interpret them.

In this chapter, we examine the communication of digital information. We define "communication" as the movement of digital or symbolic representations of information from one location to another, physically separate, location that may be millimeters or millions of miles away. At the new location, the information might be stored, reconstructed for human use, or sent to yet another destination. A wide variety of communications systems, such as the telephone, electronic mail, and broadcast radio, is used routinely in everyday life.

OUTLINE

Design Objectives for Digital Communications Systems

Digital Communications: The movement of binary or symbolic representations of information from one location to another.

INTERESTING FACT:

Sound travels at 341 meters per second at sea level, at a temperature of 15°C. As the altitude increases, the atmospheric pressure drops, and so does the velocity of sound. The velocity is also affected by humidity, temperature, and pressure. For example, at the top of Mt. Everest at a temperature of −25°C, the speed of sound would be 310 meters per second.

INTERESTING FACT:

Engineering design is often driven by economic objectives.

INTERESTING FACT:

The largest long-distance telephone communications network in America typically handles more than 300,000,000 telephone calls per business day and has handled as many as 450,000,000 calls in one day.

Before we begin our investigation of **digital communications**, let's first define our objectives from the perspective of engineering design. We start by asking and answering the following questions:

- **What problem are we trying to solve?** We want to move multimedia information from one location to another. In principle, we would like to move the information as quickly as possible, receive it as accurately as possible and execute the process as cheaply as possible.
- **How do we state the underlying engineering design problem?** The potential user of a communications system usually will specify the system's requirements; that is, the customer, through user feedback, will define the set of characteristics the system must have in order to make it worth his or her money. These characteristics typically include connection speed, connection accuracy, and cost. Other characteristics may include such features as the time delay between transmission and reception, the security of the communications system, and its ease of use. A common additional requirement is that the communications system must work in accordance with government regulations or industry standards.
- **What are the rewards if the designed communications system satisfies all of the user's needs?** Engineering design is often driven by economic objectives. Who will benefit from a well-designed communications system? First, you, as the designer, will benefit from producing a communications system that meets all of your and the customer's needs, because the customer will pay you for the job. If you do a good job, you also could benefit from the project by having the opportunity to work with the same happy customer again on a future project. Future projects may be more challenging, interesting, and profitable. Once you have established a good relationship with this customer, he or she may pass your name on to other customers, thus growing your business. Developing better technology for multimedia communications can have even greater rewards, as certain advances could have a global benefit, giving people around the world the ability to work together on a far broader scale than was ever possible before. But there may be other less desirable consequences to certain advancements, which are unintended. For example, electronic mail is now so inexpensive that many users receive numerous unwanted messages each day.
- **How will you test your design?** As we mentioned previously, most customers will provide a list of specifications. An important part of the engineering design process is to turn those specifications into a list of tests. Both the designer and the customer must agree that successful completion of the tests means that the system works as desired. In the case of multimedia communications systems, these tests typically include assessment of the following:
  1. *Peak data rate*—the maximum rate at which a short message sequence can be sent. The rate is often specified in units such as characters per second, words per minute, or bits per second.
  2. *Average data rate*—the sustainable rate at which long messages can be sent. This quantity will always be lower than the peak data rate, because it includes the time spent setting up and maintaining the connection.

3. *Error rate*—the maximum number of errors as a fraction of the total number of bits transmitted.
4. *Error-free seconds*—the number of seconds in which no data are received with an error.
5. *Delay*—the maximum and minimum delay between transmission and successful reception.

A carefully written **test plan** is usually part of the initial contract between the designer and the prospective customer. This plan should describe exactly how each of the measurements will be made so that both parties have confidence in the test's outcomes.

So, what do we need to learn in order to understand communications systems and then to design them successfully? In this chapter, we explore the common elements of all communications systems. We also explore design goals for simple communications systems and the possible methods of achieving those goals. We will start with simple solutions using basic technology and then watch them evolve into higher-performance systems that employ state-of-the-art technology.

7.1 Introduction

Basic Concepts and Definitions

Although there are many different ways to communicate information, all methods rely on the same basic principles and use the same three basic components to successfully transfer information from an originating source to a destination. This simple communications system is shown in Figure 7.1. On the left-hand side of the figure, you see that a user (the sender) provides information to the **transmitter**. The transmitter accepts the information from the user and turns it into a signal that can be conveyed over the **communications channel**, which is the physical medium that carries the signal. The medium might be an optical fiber, a pair of telephone wires, or the air itself. The medium carries the signal from the transmitter to the **receiver**, shown on the right-hand side of the figure. At this receiver, which is the intended destination of the transmission, the signal

Transmitter: A device or circuit that converts a communication signal into a form that can be conveyed to a distant physical location. For example, a signal might be converted into sound vibrations in the air or an electrical signal on a wire.

Communications Channel: The physical medium that carries a signal from the transmitter to the receiver. Examples include telephone wires, optical fibers, and the air between you and the person listening to you speak.

Receiver: A device that recovers transmitted information from the transmitted signal and converts it into a form that the recipient can use.

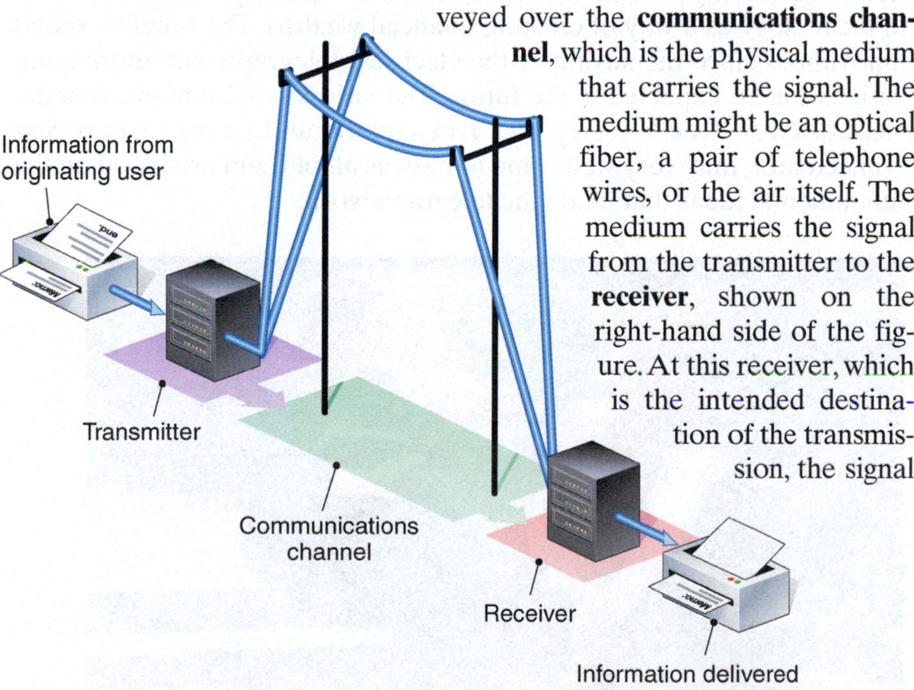

Figure 7.1 Block diagram of a simple communications system.

INTERESTING FACT:

The digital version of a two-hour movie requires about 400 billion bits of information. If it were to be sent from the movie studio directly to the movie theater, the digital transmission system would have to carry more than 50 million bits per second. In contrast, your digital cellular phone requires only about 8,000 bits per second to carry each of your calls.

INTERESTING FACT:

Samuel F. B. Morse (1791–1872), the inventor of the telegraph, was a portrait painter by profession.

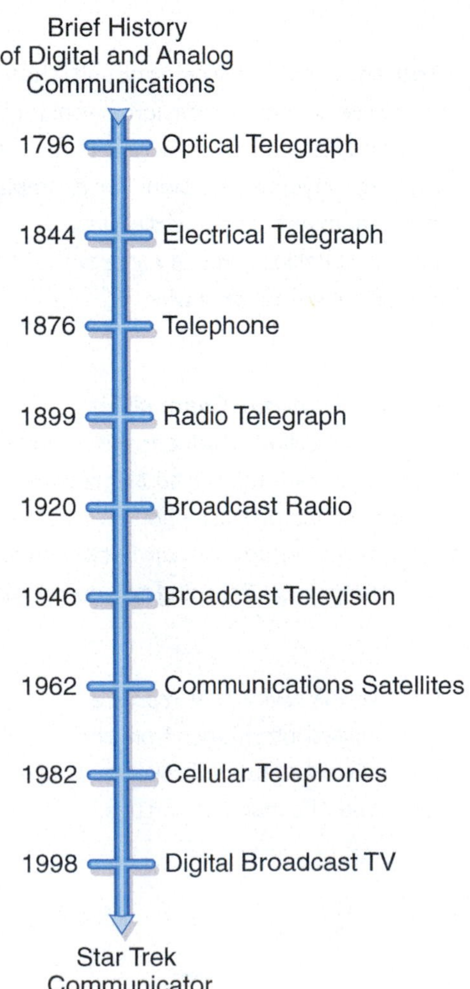

is captured with an appropriate device, such as an antenna. The receiver attempts to recover the transmitted data as accurately as possible and then delivers these recovered data to the user at the distant location.

As a simple illustration of this process, we can interpret the communication between two people, as shown in Figure 7.2, in terms of the three basic components in Figure 7.1. The information to be communicated exists in the mind of the speaker. The "transmitter" is the speaker's vocal cords and the parts of her brain that convert mental information into the mechanical motion of her vocal cords and mouth to generate speech sounds. The "communications channel" is the air between the two people. The speech sounds travel through the air from the mouth of the speaker to the ears of the listener. The "receiver" consists of the listener's ears, which capture the speech sounds, and her mental processes that convert those sounds into meaningful ideas.

Brief History of Digital Communications

The timeline in the margin highlights some important events in the evolution of digital communications. While people have been communicating with each other since before recorded history, the invention of the first practical electrical telegraph by Samuel Morse during the 1830s, culminating in his invention patent filing in 1838, marks the beginning of modern digital electronic communications. This invention, which was first demonstrated publicly in 1844, freed communication from mechanical conveyance by horses, trains, or ships and allowed information to be sent almost instantaneously. The telegraph was the first commercially successful application of electricity. It made it possible to govern large countries from a central capital, to conduct business on an international scale, and to operate diplomatic and military forces across the continents and oceans from a single headquarters. With it also came the field of electrical engineering and the concept of venture capital (private investment in technology as a way of creating financial wealth). The timeline shows innovations since the advent of the electrical telegraph, and more innovations can be expected in the future. The visionary communication devices of Gene Roddenberry, *Star Trek* creator, and George Lucas, *Star Wars* creator, may very well come to pass, as all of them are based on the fundamental ideas that made the telegraph work.

Figure 7.2 Spoken communication between two people.

EXERCISES 7.1

Mastering the Concepts

1. For the following communications systems, identify and describe the transmitter, the channel, and the receiver:
   a. a cellular telephone
   b. broadcast radio
   c. cable television
   d. a facsimile (fax) system
   e. a television remote control
2. Discuss the performance of these specific communications systems in terms of how much it costs to send a message, how long it takes you (on average) to send a message, how many words are in an average message, and how many errors are in a message. Which is fastest? Which is the most expensive per word? Which is most reliable?
   a. e-mail
   b. telephone calls
   c. letter sent by U.S. first-class mail
   d. letter sent by Federal Express

Try This

3. A horse on the old Pony Express routes could travel about 20 miles per hour. The Pony Express was replaced by the telegraph, which could carry messages at about 180,000 miles per second. How many times faster could the telegraph carry a message than a horse?
4. How many telegraph lines operating at 5 bits per second would it take to carry a 50-Mbit-per-second digital television signal?

Back of the Envelope

5. Let's try to determine the rate of communication for speaking, using a simple experiment. Have three of your friends read a selected paragraph of this book out loud, and measure the time it takes each of them to say the words. Count both the number of words and the number of characters in the paragraph.
   a. Compute the communication rate in units of words per minute. Do so for each person, and find the average rate.
   b. Compute the communication rate in units of characters per second, as you did in part (a).
   c. Which unit, words per minute or characters per second, is the most natural measure of spoken language?
6. Repeat the previous exercise for typing instead of speaking by measuring the rates at which different students can type the same paragraph on a keyboard. Repeat the computations in parts (a) and (b). For for part (c), determine which unit, words per minute or characters per second, is the most natural measure of typing on a keyboard.

7.2 A Simple Communications System

INTERESTING FACT: Modern digital communications systems use signals based on simple mathematical functions. Using these simple signals allows us to design and build inexpensive, yet reliable, systems to carry audio, video, and data.

Two people communicating by speaking to each other seems simple enough. It is intuitive and commonly practiced, and very young children can do it. But the audio speech signals themselves are very complex. The physical, cognitive, and physiological processes of transmitting and receiving such apparently simple communications are so complicated that scientists today do not understand them fully. These are interesting and current research areas for psychologists and biologists.

In contrast, electronic communications systems, designed and built by humans, in which the receiver is not a human listener, rely on simpler, well-understood signals. These systems use signals with simple mathematical representations not designed for the human ear. Using these simpler signals allows us to design and build inexpensive, yet reliable, systems that can communicate audio, video, and data over great distances a lot faster than we can talk.

Design of a Simple Communications System

Let's look at a simple digital communications system. The objective of this system is to send information wirelessly over a short distance, such as across the room. We will have to design each of the three basic components of a communications system illustrated in Figure 7.1.

Tones: Signals of constant frequency and amplitude represented by a sine or cosine function.

We first consider a method of sending textual information from one side of the room to the other by turning the individual letters of the text into audible **tones** instead of using verbal speech. The text characters will be transmitted by spelling each word via a set of audible tones of different frequencies. An audio loudspeaker will be the transmitting device instead of vocal cords. The communications channel will be the air between the transmitting and receiving sites. The receiver will capture the audio tones transmitted through the air by listening with a microphone instead of with ears. If this system works correctly, the receiver will recognize the frequencies of the transmitted tones and will be able to turn them back into the original transmitted text.

Although there are many possible ways to transmit information wirelessly, we consider an acoustical system first because it is most similar to our direct experience of talking and listening. There are also many possible ways to represent letters with audio tones. One simple approach would be to associate each letter with a unique tone, a sinusoidal signal of constant amplitude and frequency. Figure 7.3 shows how this system could work. This figure shows the specific details of our design for each of the three basic components of a communications system from Figure 7.1. The receiver in Figure 7.3 must determine what characters were transmitted based on the sound patterns it receives. Then the receiver must make a list of the identified characters and display them to the recipient of the message. For the receiver to work properly, it needs to identify each character by its unique sound pattern.

KEEP IN MIND

A tone is specified completely by three parameters. The frequency of the sinusoid or cosinusoid is represented by f, the maximum amplitude by a, and the phase angle by ϕ.

Other possible sound patterns for this system might also be considered. A person could simply choose to spell the word, saying each letter in turn so that the signal associated with each letter would be the sound of someone saying that letter. However, human speech creates a very complex signal that usually communicates far more information than just the text of the message. A listener might also learn the identity of the speaker, the mood of the speaker, or the regional dialect used by the speaker. Because of these factors and the similarity of many letter sounds, such as "b" and "p," speech signals would be more difficult for a simple electronic receiver to interpret correctly.

To make it easier for the electronic receiver, simpler signals—in this case, unique sinusoidal signals—are chosen to represent the letters. Although sinusoids are unfamiliar to people as a way of communicating information, they are far easier for an electronic system to interpret correctly. When we analyze this method later, we will also find that it can transmit data much faster than a human speaker and listener.

INTERESTING FACT:

The human ear is not sensitive to the phase, ϕ, of a single tone, and therefore cannot tell the difference between a sine and a cosine of the same frequency and amplitude. Electrical communications equipment can tell the difference, however, and can use that ability as one method for sending and receiving digital data.

Operation of a Simple Communications System

Assume that the information to be transmitted is a list of capital letters chosen from the Roman alphabet. Examples of messages might be "MEET ME AT FOUR PM" or "HOUSTON WE HAVE A PROBLEM." Designing the system shown in Figure 7.3 requires the selection of a different tone to represent each character. We'll discuss how this is done later, but for now we'll use the mapping of letters to tones shown in Table 7.1. Sending the letter *A* requires the creation of a sinusoid with a frequency of 300 Hz. The letter *B* would be indicated by a sinusoid with a frequency of 400 Hz. The other letters would be generated in the same way, with a 100-Hz difference between each adjacent pair of character tones. With 27 tones, we can send all 26 capital letters and the space character needed to separate words. If punctuation or numbers were also needed, more tones would have to be added.

Table 7.1 Mapping of Capital Letters to Frequencies in Hertz (Hz) of Audible Tones, with 100-Hz Separation between Adjacent Tones

Letter	Frequency (Hz)
A	300
B	400
C	500
D	600
E	700
F	800
G	900
H	1000
I	1100
J	1200
K	1300
L	1400
M	1500
N	1600
O	1700
P	1800
Q	1900
R	2000
S	2100
T	2200
U	2300
V	2400
W	2500
X	2600
Y	2700
Z	2800
space	2900

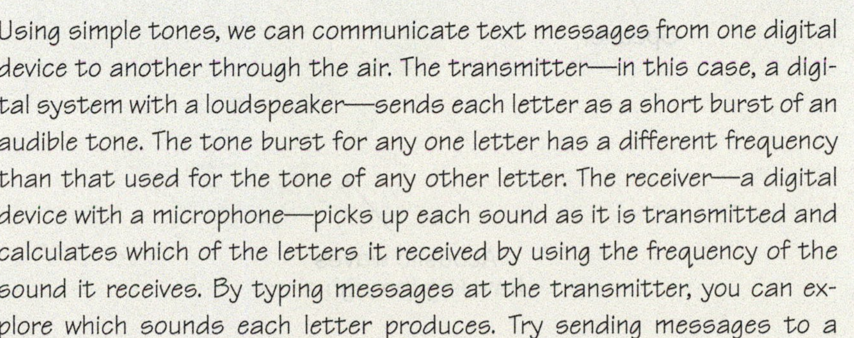

Infinity Project Experiment: Audio Communication of Messages, Using One Tone per Letter

Using simple tones, we can communicate text messages from one digital device to another through the air. The transmitter—in this case, a digital system with a loudspeaker—sends each letter as a short burst of an audible tone. The tone burst for any one letter has a different frequency than that used for the tone of any other letter. The receiver—a digital device with a microphone—picks up each sound as it is transmitted and calculates which of the letters it received by using the frequency of the sound it receives. By typing messages at the transmitter, you can explore which sounds each letter produces. Try sending messages to a friend to see how this system works. What happens if you type too fast? What if someone else is talking in the room at the same time?

The general mathematical representation of a tone is given by Equation (7.1),

$$s(t) = a \times \cos(2\pi ft + \phi) \qquad (7.1)$$

where $s(t)$ represents the strength of the signal to be transmitted as a function of the time variable t. The tone is completely specified by three parameters. The frequency of the sinusoid is represented by f, the maximum amplitude by a, and the phase by ϕ. Using Table 7.1, we find that the specific signals for the letters A and B are given by $s_A(t)$ and $s_B(t)$, respectively, in Equation (7.2). For the simple system we are designing, the phase will not be important, so it is set to zero. The amplitude a will always be the same at the transmitter.

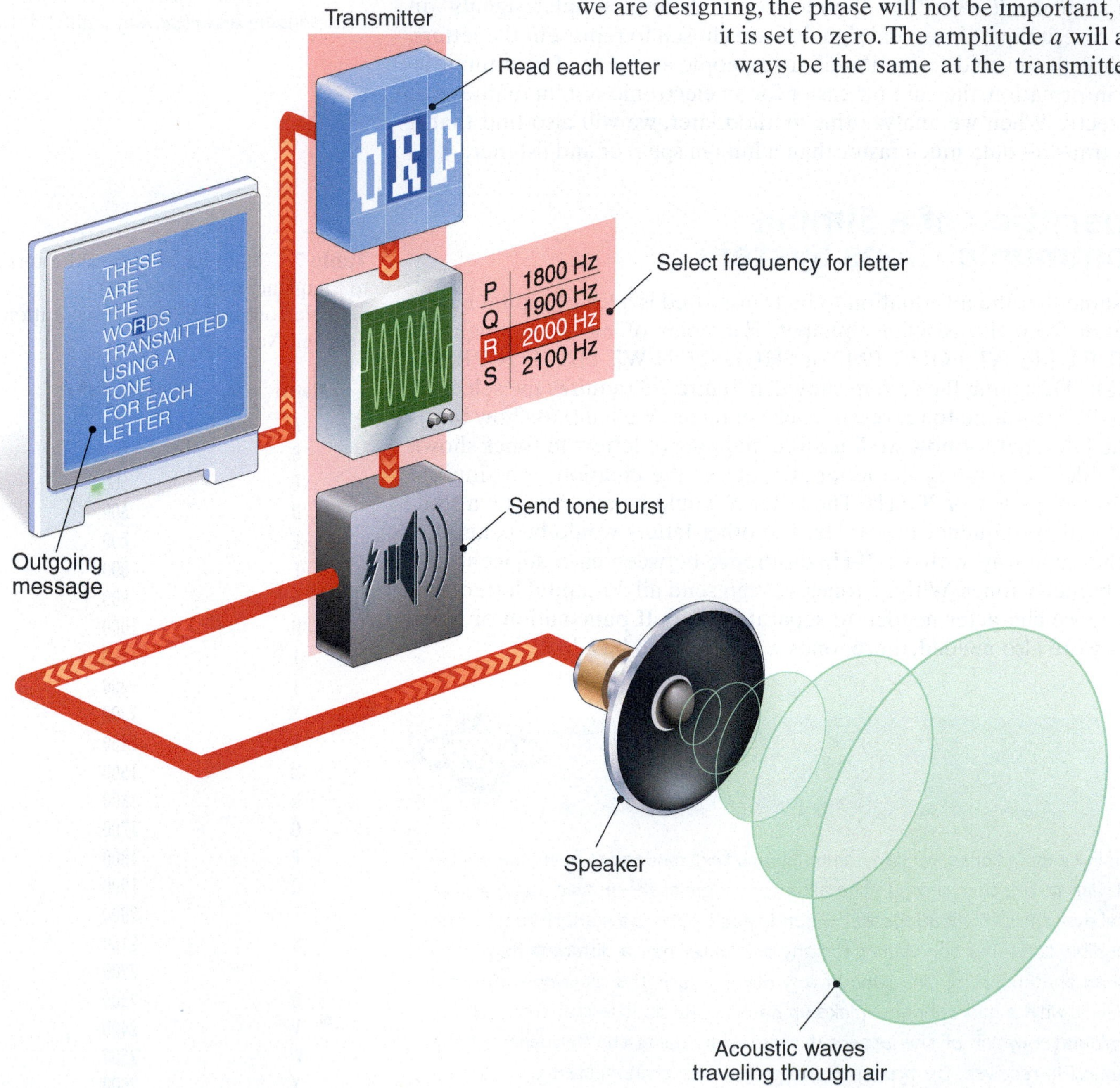

Figure 7.3 A communications system for transmitting text by using distinctive tones for each letter.

For the receiver, the amplitude will vary depending on the distance between the transmitter and the receiver so we will assume that a wide range of amplitude values will be acceptable:

$$s_A(t) = a \times \cos(2\pi \times 300t)$$
$$s_B(t) = a \times \cos(2\pi \times 400t) \quad (7.2)$$

Operation of the Transmitter How, then, would this communications system send the word MEET? Upon receiving the list of the characters in the message from the originating user, the transmitter would send a sequence of four tone bursts for the four-character word MEET. A "burst signal" is a signal that is present for only a short interval of time. A "tone burst" is a burst signal that has a constant frequency and amplitude during the short time interval. For the message MEET, the first tone burst, corresponding to the letter *M*, would have a frequency of 1500 Hz. The next two tone bursts would be at 700 Hz, and the last would be at 2200 Hz. There would be some quiet time between each burst, so that the two *E*s would be distinguishable

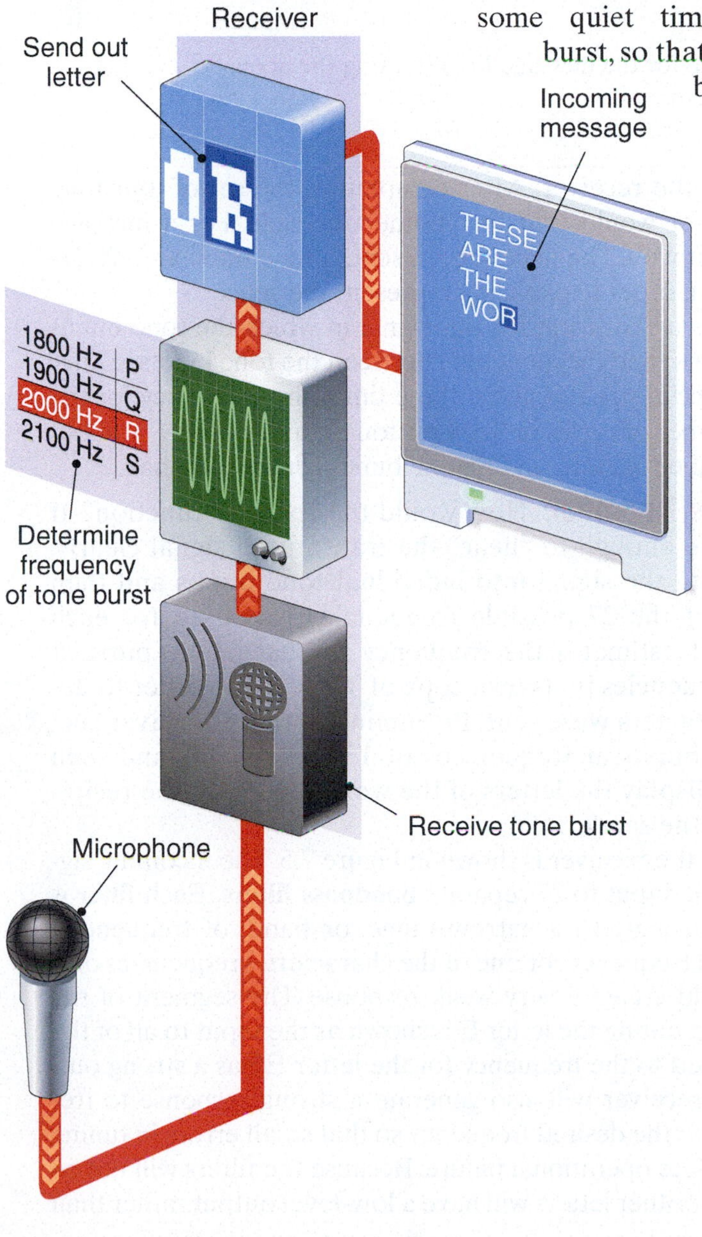

KEEP IN MIND

The human auditory system can normally hear sounds in a range from 15 Hz to 20,000 Hz, but it is most sensitive to sounds in the range from 1000 to 4000 Hz. Most speech falls in this range, and engineers have designed the telephone and cellular phone to take advantage of this situation. They were designed to carry a frequency range of 300 to 3400 Hz.

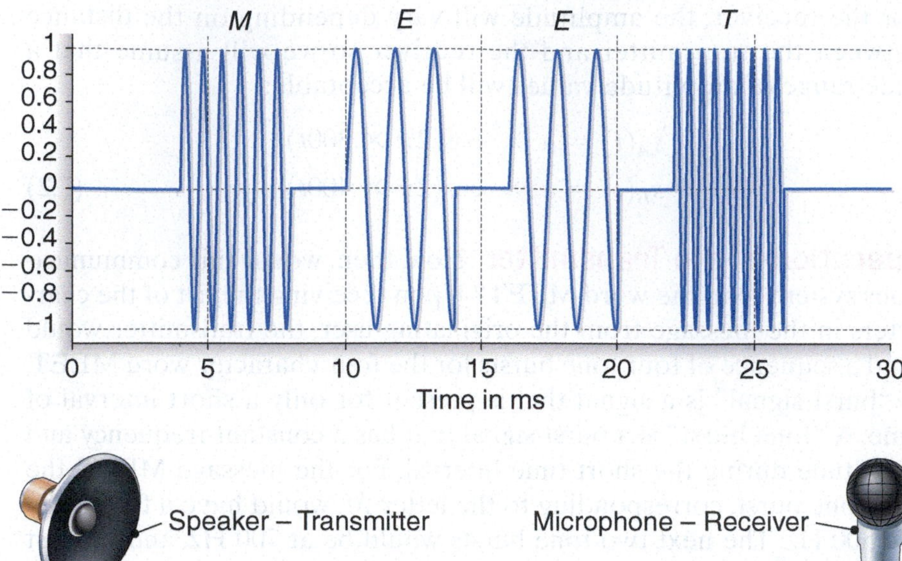

Figure 7.4 The signal for the message MEET, using the acoustic communications system.

from each other at the receiver. After completing the set of four tone bursts, the transmitter would wait for another message. For a message with several words, it would be necessary to send a tone burst at 2900 Hz between the words in order to explicitly represent the space.

Figure 7.4 shows a waveform for this signal in which the tone bursts last 4 milliseconds (ms) and the quiet time between the tone bursts is 2 ms. The frequency for the *T* signal is about three times the frequency for the *E* signal, so the number of cycles in the 4-ms tone burst for the *T* is about three times the number of cycles in the tone bursts for the *E* signal.

Operation of the Receiver How would the receiver function? If the receiver is close enough to "hear" the transmitted signal clearly, its job is to separate the signal into individual tone bursts and then determine which of the 27 possible frequencies was used for each tone burst. Once it estimates the frequency for each tone burst, it "looks up" the frequencies in its own copy of Table 7.1 in order to determine which characters were sent. Presuming that the receiver successfully detected bursts at frequencies of 1500, 700, 700, and then 2200 Hz, it would display the letters of the word MEET to the recipient as intended by the sender.

Filter: A simple system (digital or analog) that allows only prespecified frequencies to pass from its input to its output.

Bandpass Filter: A filter that only "passes," or responds strongly to, a specific range, or "band," of frequencies.

A closer look at the receiver is shown in Figure 7.5. The incoming signal is provided as the input to 27 separate **bandpass filters**. Each **filter** is tuned to respond strongly to a narrow range, or band, of frequencies around the specified frequency for one of the characters. Frequencies outside this range would create a very weak response. The segment of the message signal representing the letter E is shown as the input to all of the filters. The filter tuned to the frequency for the letter E has a strong output signal. A good receiver will also generate a strong response to frequencies very close to the desired frequency so that small errors in tuning will not cause complete operational failure. Because the filters will not be perfect, the filters for other letters will have a low-level output rather than a zero-level output for a signal associated with the letter *E*. As shown in Figure 7.5, the two adjacent filters for *F* and *D* have very weak output signals, and all other filters have a flat output. In this case, the decision logic has a relatively easy task to determine which character was sent.

The block diagram of the receiver structure shown in Figure 7.5 is similar to a block diagram for human hearing, although the actual physical receivers are quite different. The structures in the human ear that respond directly to sound are shown in Figure 7.6. Sound travels into the human ear and causes vibrations in the fluid of a coiled tube inside the ear. This tube, which looks like a snail shell, is called the **cochlea**.

Cochlea: A fluid-filled coiled tube in the ear that responds to different frequencies at different points along its length.

Due to the shape of the cochlea, with its varying stiffness and diameter, different locations along the tube respond best to different frequency ranges. The length of the tube in the cochlea is about 30 millimeters (mm). The maximum response for the lowest frequency that humans can

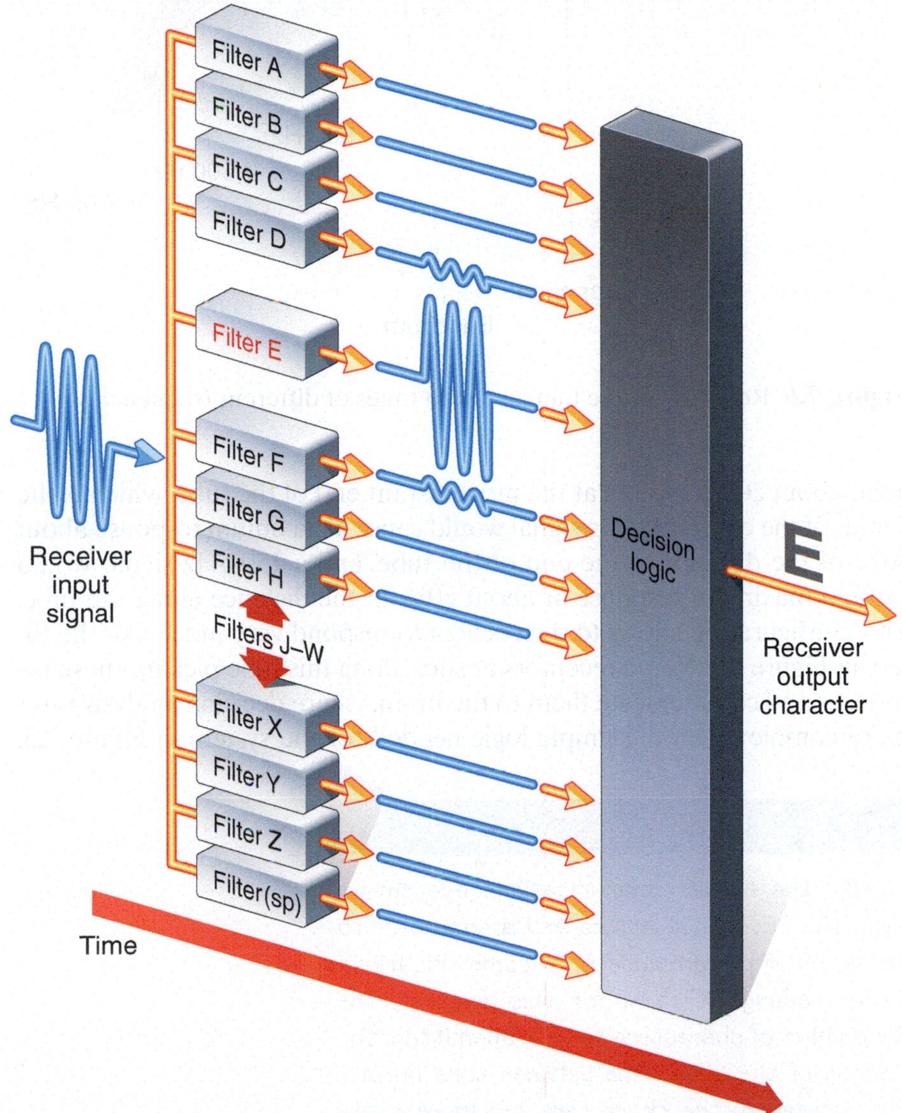

Figure 7.5 The receiver structure identifies tone bursts and determines which letter has been sent. In this case, the tone for the letter *E* has been transmitted and detected. The filters for *F* and *D* have a low response, and all other filters have no noticeable response.

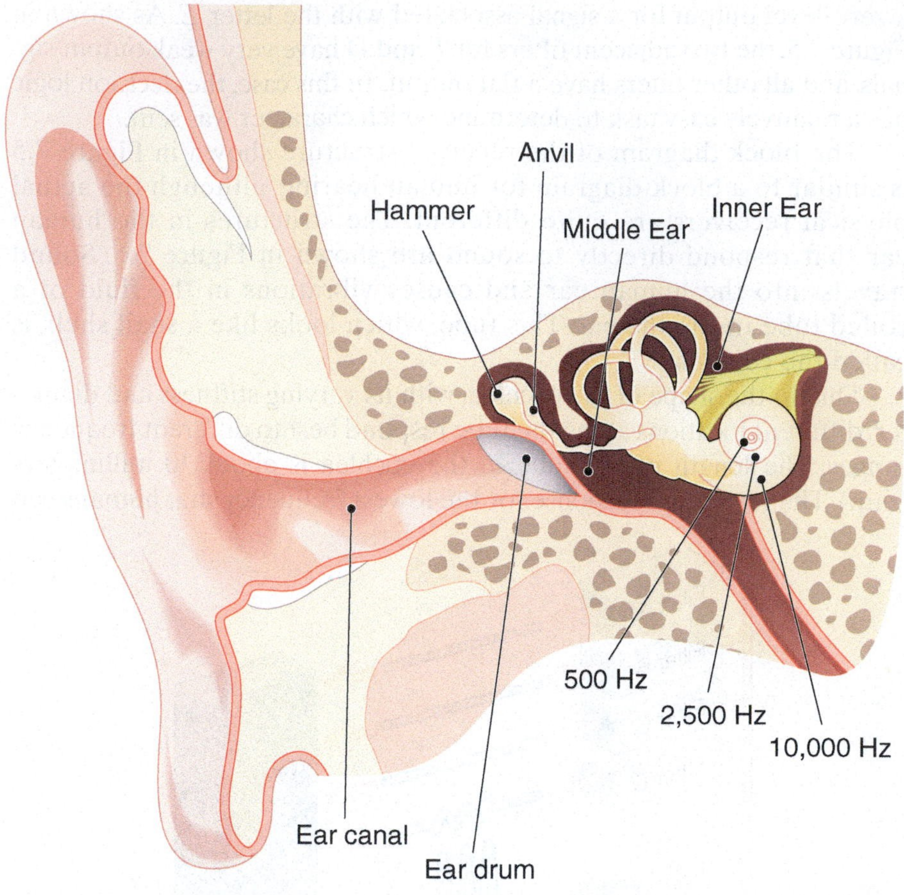

Figure 7.6 Response of the human ear to tones of different frequencies.

hear, about 20 Hz, occurs at the most distant end of the tube, which is the center of the coil. A 500-Hz signal would cause a maximum response about 80% of the distance to the end of the tube, and a 1500-Hz signal would cause a maximum response at about 60% of the distance along the tube. This configuration allows to the cochlea to respond very much like the filters in Figure 7.5. Nerve receptors at sites along this tube pick up these responses and communicate them to the brain, where decision analysis is far more complex than the simple logic needed for the system in Figure 7.5.

KEY CONCEPT

A message of any length can be sent using the simple communications system we designed as long as the message only uses the 26 capital letters and a character to indicate a space between words. One measure of performance of a communications system is how long it takes to send a message. Clearly, for this system, the transmission time depends directly on the number of characters to be transmitted, the durations of the tone bursts, and the length of the quiet time between tone bursts. When the duration of the tone bursts is increased, the characters are more easily distinguished by the receiver, but the overall transmission rate is reduced. If a system can successfully receive 10 bursts per second, then the transmission rate is 10 characters per second, or 10 symbols per second.

EXAMPLE 7.1 Determining Transmission Rate

A communications system uses 15-ms tone bursts and a 5-ms quiet period between each tone burst. How many characters per second (char/s) can be sent? How many words per minute can be sent if each word is about six letters long (including spaces)?

Solution

Each character needs 15 ms for the tone burst plus 5 ms quiet time, for a total of 20 ms/char. The transmission rate will be

$$\frac{1000 \text{ ms/s}}{20 \text{ ms/char}} = 50 \text{ char/s}$$

If a word is assumed to be six characters long, then the transmission rate is

$$\frac{50 \text{ char/s} \times 60 \text{ s/min}}{6 \text{ char/word}} = 500 \text{ words/min}$$

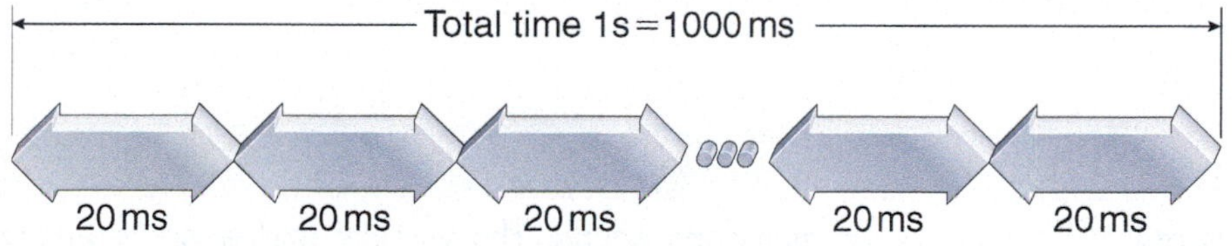

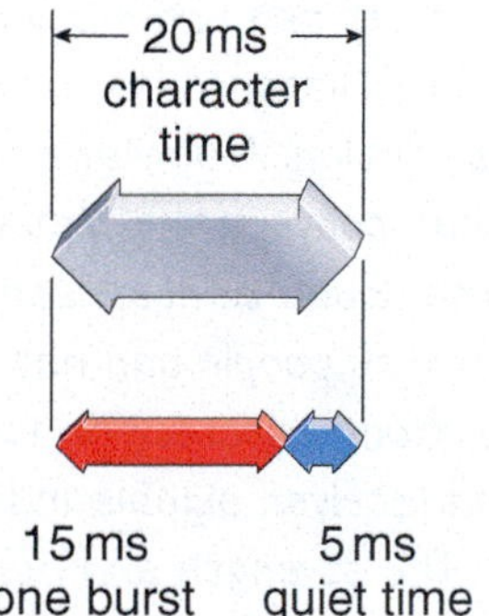

EXAMPLE 7.2 Determining the Length of the Tone Burst

Suppose a text communications system sends letters at a rate of 2000 characters per minute. If the quiet period between each tone burst is 10 ms, how long must the tone burst be for each letter?

Solution

The number of minutes/character is 1/2000 characters/minute. Converting minutes to ms, the number of milliseconds per character for this system is

$$\frac{1000 \text{ ms/s} \times 60 \text{ s/min}}{2000 \text{ characters/min}} = 30 \text{ ms/char}$$

This 30 ms time interval includes the quiet period. Therefore, the length of the tone burst must be 30 ms − 10 ms = 20 ms.

The fundamental concept that makes this system work is that the transmitter sends an acoustic signal for each character that is *sufficiently different* from the acoustic signals for all other characters, so that, under normal operating conditions the receiver can distinguish that character's signal from the signals of other characters. This fundamental concept applies to all communications systems no matter what channel, message coding, or transmission is used. The specifications for what makes the signals sufficiently different depend on how complicated and expensive the receiver can be, the characteristics of the communications channel, and the conditions under which the communications system is expected to operate. If the signals associated with all possible transmitted characters are sufficiently different, then the receiver should be able to accurately distinguish and interpret all of them. Therefore, at the signal's destination, the receiver should be able to accurately report the whole message to the user without making any errors.

INTERESTING APPLICATION

Using Tones to Help the Visually Impaired Read Printed Text and Line Drawings

There are some communications systems that do use audio tones to communicate directly to a human listener rather than to an electronic receiver. A device developed to assist a blind person in reading printed text, called a stereotoner was developed by Harvey Lauer. It converts printed character images into sequences of combinations of 10 tones. Horizontal strips of printed text are sliced vertically as shown in Figure 7.7(b). Each vertical strip is divided into 10 squares, and each row of squares is assigned one of the 10 tones. If a square is mostly black, then the corresponding tone is turned on and the pattern of the vertical strip creates a particular multitone sound. The vertical strips are scanned from left to right, creating a sequence of multitone sounds that can communicate the image information to a trained listener. Using more tones can improve vertical resolution. A similar concept is used in navigation aids that convert the distance of objects into sound patterns. Such devices could transmit an ultrasonic signal, which people can not hear, and receive reflected signals from objects. The time between the transmitted and received signals indicates the distance of the object. The strength and duration of the reflected signal provides information about the object's surface. This information can be encoded in audio tones for the user.

Figure 7.7 Using tones to help the visually impaired read printed text. (a) An example of printed text to be read with a "stereotoner." (b) Text with a grid superimposed to indicate the horizontal and vertical strips. (c) Text image divided into vertical slices on 10 squares each. (d) Multitone sounds created by each vertical slice expressed in standard scale notation. Note that the stereotoner frequencies are all four times as large as those shown in the music score notation, with the lowest frequency at 440 Hz and the highest at 3520 Hz.

HELP !
(a)
HELP !
(b)
HELP !
3520 Hz
1760 Hz
880 Hz
440 Hz
(c)
(d)

EXERCISES 7.2

Mastering the Concepts

1. What is the role of a transmitter in a communications system? What is the role of a receiver? When one person is talking to another, who is the transmitter, and who is the receiver?
2. Use Equation (7.1) to write the expression for a tone with a frequency of 1000 Hz, a peak voltage of 4 V, and an initial phase ϕ of zero degrees.
3. What is the peak amplitude of a tone given by the equation $s(t) = 7\sin(300\pi t + 1.5)$? What is its phase? What is its frequency in Hz if the units of time are seconds?
4. An audio communications system similar to that in Figure 7.3 is used to communicate the numbers 0 through 9. The system uses tones with frequencies of 500 Hz for 0, 600 Hz for 1, and so on, up to 1400 Hz for 9. Finally, a tone with frequency 1500 Hz is used for a decimal point (.).

   a. What sequence of frequencies would you use to communicate the number 478.65?
   b. Write down the sequence of frequencies that corresponds to the first 10 digits of π. (Don't forget the decimal point!)
5. DNA strands are defined by their sequence of amino acids. There are only four amino acids in DNA, and they are designated by the letters *A*, *G*, *C*, and *T*. Suppose that we want to send information about a DNA string from one location to another and choose to do so using a tone burst of a different frequency for each of the four different amino acids. Suppose further that each tone burst will last for 0.9 seconds followed by a quiet interval of 0.1 seconds.

   a. How many different tone frequencies must the transmitter be able to send?
   b. How many tone filters are needed at the receiver?
   c. How does the receiver decide if a tone is being sent?
   d. How does the receiver decide which tone is being sent?
   e. How long would it take to send information about a DNA sequence consisting of a string of 10,000 amino acids?

Try This

6. Assume that the receiver in an audio communications system is capable of distinguishing tone bursts that are only 25 Hz apart instead of 100 Hz apart and that the letter *A* is assigned the frequency 300 Hz. Make a mapping of letters to frequencies similar to Tables 7.1 and 7.2, with 25-Hz spacing between adjacent pairs. What frequency range would be needed for the character set if your table were used?
7. You are designing an expanded communications system that still uses a single frequency to represent each character, but the characters will include both upper- and lowercase letters, all 10 decimal digits, and a space.

a. How many different tones are needed?
b. If the lowest acceptable frequency is 300 Hz and the highest acceptable frequency is 3400 Hz, what frequency separation between the tones should be used?

8. If each tone burst lasts 4 ms and the bursts are separated by 1-ms quiet intervals, how long would it take to transmit the message "MEET ME AT FOUR PM FOR COFFEE"? (*Hint*: Don't forget that the space between words must also be represented by a tone burst.)

9. An audio communications system is being designed to transmit telephone numbers in the format 8005551212, with no dashes or spaces between the digits of a single telephone number. Furthermore, each phone number is separated from every other by a single space.
   a. How many tones would be required in a single-tone-per-symbol scheme?
   b. Suppose each tone burst lasted 5 ms, with a 2-ms quiet period. How many telephone numbers could be transmitted per minute with this scheme?

10. One way to estimate the frequency of a signal is to count the number of times the signal changes sign from positive to negative. This procedure is often called counting "zero crossings." A sinusoidal signal should have two zero crossings for every cycle.
   a. How many zero crossings would be counted for a 300-Hz sinusoidal tone burst of 10 ms? Of 5 ms?
   b. Repeat part (a) for a 400-Hz sinusoidal tone burst.
   c. If a tone bursts lasts 10 ms, how many zero crossings would distinguish an *A* at 300 Hz from a *B* at 400 Hz? If the tone bursts lasts 5 ms, how many zero crossings would distinguish the two characters?

In the Laboratory

11. What causes a single-tone-per-symbol audio communications system to fail? Outline all the ways that errors might be introduced using this communications system.

Back of the Envelope

12. Pick up any paperback book and look at a single page.
   a. How many different symbols are represented on the page? Include all spaces and punctuation marks.
   b. How many different tones would be required to communicate this book, using an audio transmission system similar to that in Figure 7.3?
   c. Suppose you wanted to transmit the entire page via this transmission scheme. Assuming that you could transmit five symbols per second, how long would it take? About how long would it take to transmit the entire book this way, assuming that each page has a similar number and type of characters?
   d. Do you read faster or slower than the answer to part (c)?

7.3 Sources of Error in a Communications System

Causes of Errors in Communications Links

Although we try to prevent it, sometimes a communications system makes errors. An error happens in the communication of a character if

1. no character is detected at the receiver;
2. an unwanted character is produced at the receiver when the transmitter is silent; or
3. the wrong character is detected at the receiver.

It might surprise you, but such errors occur in practical systems more often than you think! To understand how these errors happen, let's take a look at a specific example using our audio communications system in Figure 7.3. Suppose the receiver cannot distinguish between a tone burst at 700 Hz and one at 800 Hz. In this case, the receiver will not be able to accurately reconstruct the transmitted message if it contains either *E* or *F*. Similarly, if the receiver is unable to "hear" bursts with frequencies above 2500 Hz, then it will simply not acknowledge that the letters *X*, *Y*, and *Z* and the space have been transmitted. Clearly, both of these cases must be avoided, because they degrade the accuracy of the received message.

The system's designers are responsible for making sure that the receiver can distinguish all of the possible tones and that all of the tones are within the receiver's "hearing range." There are practical circumstances, however, that can limit the system's performance in spite of the care used by the designers. Three common causes of degraded performance in a communications system are **weak signals**, **noise**, and **interference**. Figure 7.8 shows examples of these conditions, using the MEET signal shown in Figure 7.4.

Weak Signal: To *attenuate* a signal is to make it weaker. Typically, the further a signal travels in a communications system, the more attenuated, or weaker, it becomes.

Weak Signals Suppose the transmitter is far enough away from the receiver that the acoustical signal becomes inaudible. At some point, it will become too weak to be detected accurately. This situation is illustrated in Figure 7.8(a). If this weak signal is the input to the bank of filters, as shown in the block diagram in Figure 7.5, then even the strongest outputs will be very weak, and the decision logic will not be able to determine which filter output is strongest.

Noise: Disturbances in signals that can make it difficult for the receiver to operate correctly.

Noise A transmitted signal is always received with some "noise." If the noise power is high, it may mask the signal enough to make it difficult to detect the presence of each tone burst or to accurately determine the frequency of a tone. If either of these problems exist, then characters will be missed altogether by the receiver, or they may be interpreted as the wrong character. Figure 7.8(b) demonstrates this problem.

Interference: Signals or disturbances unrelated to the signal you want to receive that can make it difficult for the receiver to operate correctly.

Interference In some cases, the receiver may be able to "hear" from two or more transmitters simultaneously. If the frequencies used by the various transmitters overlap to the extent that the receiver can't separate them, then, just as with the case for noise, the receiver may make errors or even miss the desired tone bursts altogether. In Figure 7.8(c),

Infinity Project Experiment: Effects of Weak Signals and Noise

The receiving system can make mistakes when using tones to communicate characters from one device to the next. The receiver has a hard time making a good decision about what letter was sent if it hears noise at the same time. In this experiment, we can simulate the effect of noise in the room by generating its sound artificially and adding it to the transmitted signals. How well can the receiver detect the right letters when a little noise is present? What happens as the level of noise is increased? Can you tell with your ear what is being transmitted when noise is present?

White Noise: Noise received along with the desired signal in a communications system can have approximately the same amount of energy at all of the frequencies used by the system. When this happens, the noise is called *white*—by an analogy to white light, which is composed of a mixture of all of the colors in the visible spectrum.

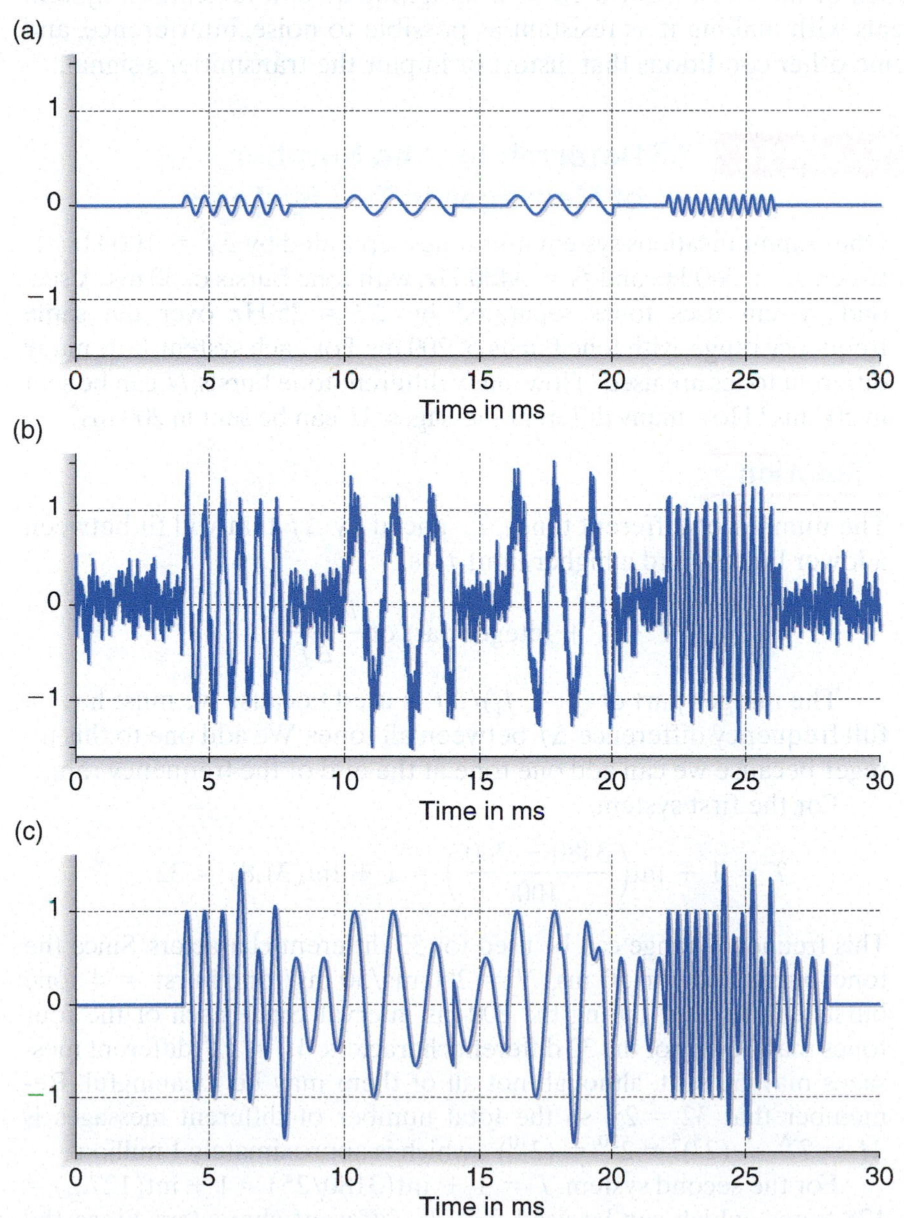

Figure 7.8 Three causes of signal degradation and errors demonstrated with the signal for the message MEET: (a) weak signal; (b) noisy signal; and (c) interference from weaker message HELP.

KEY CONCEPT

When receiving a signal in the presence of noise or interference, more confident detection and more accurate determination of frequency will be possible if the burst carrying each letter is allowed to remain on for a longer time. While this approach will usually improve the accuracy of reception, it also slows down the rate of transmission. This trade-off between speed of transmission and accuracy of reception is a fundamental one in telecommunications systems.

the MEET signal and a weaker HELP signal, which begins 2.0 ms later, are received at the same time. This problem is similar to trying to listen to one person talk while another is also talking.

Sometimes the interfering signal is a delayed version of the transmitted signal. This situation frequently happens in radio and television transmission systems when the signal from the transmitter takes different, or "multiple," paths to the receiver. Often, one of the signals reflects off a building or mountain. The receiver can become confused when confronted with the sum of these two signals instead of just one. This situation is known as *multipath reception*.

From these examples, you see that a number of problems can occur in the communications channel that will make a communications system prone to make errors or miss signals altogether. We will find that much of the work associated with designing a communications system deals with making it as resistant as possible to noise, interference, and some other conditions that distort or impair the transmitter's signal.

EXAMPLE 7.3 Determining the Number of Messages in Two Systems

One communications system uses tones separated by $\Delta f = 100$ Hz between $f_1 = 300$ Hz and $f_2 = 3480$ Hz, with tone bursts of 50 ms. A second system uses tones separated by $\Delta f = 25$ Hz over the same frequency range, with tone bursts of 200 ms. For each system, how many different tones are used? How many different tone bursts, N, can be sent in 200 ms? How many different messages, M, can be sent in 200 ms?

Solution

The number of different tones, T, spaced by Δf that will fit between a lower limit f_1 and a higher limit f_2 is

$$T = 1 + \text{integer part of } \frac{f_2 - f_1}{\Delta f}$$

The integer part of $(f_2 - f_1)/\Delta f$ is used because we must have a full frequency difference Δf between all tones. We add one to this integer because we can add one tone at the end of the frequency range.

For the first system,

$$T = 1 + \text{int}\left(\frac{3480 - 300}{100}\right) = 1 + \text{int}(31.8) = 32$$

This frequency range can be used for 32 different characters. Since the tone bursts last for 50 ms, $N = 200$ ms/50 ms/tone burst $= 4$ tone bursts can be sent during the 200-ms interval. Since each of the four tones may be any of the 32 different characters, $M = 32^4$ different messages may be sent, although not all of them may be meaningful. Remember that $32 = 2^5$, so the total number of different messages is $M = T^N = (2^5)^4 = 2^{20} = (2^{10})^2$, which is approximately 1 million.

For the second system, $T = 1 + \text{int}(3180/25) = 1 + \text{int}(127.2) = 128$ tones, which can be used for 128 different characters. Since the tone bursts last for 200 ms, $N = 200$ ms/200 ms/tone burst $= 1$ tone burst can be sent during the 200-ms interval. The total number of different messages is $M = T^N = 128^1 = (2^7)^1$.

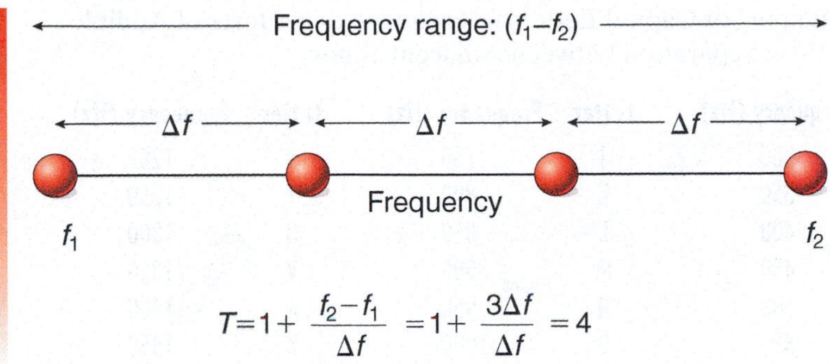

Coordination between Sender and Receiver

Although it might seem obvious, reliable reception of messages also requires that the transmitter and receiver have a common agreement about how the characters are to be represented by the tones. Many different ways of representing the characters might be used. In particular, for the one-tone-per-character system to work, there must be a prearranged agreement on

1. what tone frequencies are to be used for each character;
2. how long each tone burst will last; and
3. how much quiet time there will be between the end of one tone burst and the beginning of the next tone burst.

Using this information, the receiver can "listen" at the right times and at the right frequencies. This arrangement will allow it to be less sensitive to noise and interference as well.

The importance of the requirement for **prearrangement** can be illustrated by again assuming that the transmitter sends the letters MEET, using the tone frequencies specified in Table 7.1 and the signal shown in Figure 7.4. Suppose, for some reason, that the receiver uses Table 7.2 instead of Table 7.1 to interpret the meaning of the received tone bursts. Even if the frequencies of the tone bursts are received and identified perfectly, the receiver will decide that the letters NVVG have been sent, which is obviously not the desired result. Another receiver,

Prearrangement: For proper reception to occur, the transmitter and receiver need to have agreed on the meaning of each of the signals to be sent.

Table 7.2 A Second Mapping of the Capital Letters to Frequencies in Hertz of Audible Tones, with 100-Hz Separation between Adjacent Tones

Letter	Frequency (Hz)	Letter	Frequency (Hz)	Letter	Frequency (Hz)
A	2800	J	1900	S	1000
B	2700	K	1800	T	900
C	2600	L	1700	U	800
D	2500	M	1600	V	700
E	2400	N	1500	W	600
F	2300	O	1400	X	500
G	2200	P	1300	Y	400
H	2100	Q	1200	Z	300
I	2000	R	1100	space	2900

Table 7.3 Mapping of Capital Letters to Frequencies in Hertz of Audible Tones, with 50-Hz Separation between Adjacent Tones

Letter	Frequency (Hz)	Letter	Frequency (Hz)	Letter	Frequency (Hz)
A	300	J	750	S	1200
B	350	K	800	T	1250
C	400	L	850	U	1300
D	450	M	900	V	1350
E	500	N	950	W	1400
F	550	O	1000	X	1450
G	600	P	1050	Y	1500
H	650	Q	1100	Z	1550
I	700	R	1150	space	1600

using Table 7.3, would interpret the signal as YII. This situation is similar to a German speaker trying to communicate with one listener who understands only Chinese and another listener who understands only Arabic.

For the single-tone-per-character communications system, one might assume that this type of error is not a big problem. Since the received letters NVVG or YII are clearly not English words, it would seem obvious that a mistake had been made somewhere. Now consider instead the case where a third receiver is erroneously using the mapping table shown in Table 7.4. When the letters MEET are sent as tone bursts with frequencies of 1500, 700, 700, and 2200 Hz, respectively, the receiver will decode them as FOOL. This is an English word, and it might convey the wrong message to the recipient.

One way to guarantee that the sender and receiver are using the same lookup table and the same timing of the communications signals is to design the system to allow only one option. However, if a communications system must function correctly under different conditions of signal level, noise, and interference, it would be wise to include some options so that performance can be optimized when the operating environment changes. If timing and frequency mapping options are allowed, either the receiver must know the current characteristics of the transmitter before communications start or the receiver must be designed to be "clever" enough to figure out the transmitter characteristics for itself, based on signals that it receives.

Table 7.4 A Third Mapping of the Capital Letters to Frequencies in Hertz of Audible Tones, with 100-Hz Separation between Adjacent Tones

Letter	Frequency (Hz)
A	300
B	400
C	500
D	600
E	1400
F	1500
G	1600
H	1000
I	1100
J	1200
K	1300
L	2200
M	2300
N	2400
O	700
P	800
Q	900
R	1700
S	1800
T	1900
U	2000
V	2100
W	2500
X	2600
Y	2700
Z	2800
space	2900

Infinity Project Experiment: Different Codes

The way in which letters are assigned to each tone frequency is up to the user. In communications systems, the association of a letter to a tone frequency is called a *mapping*. Mapping is used by the transmitter to select the tone frequencies to be transmitted. At the receiver, the same mapping is used to convert a measured tone frequency back into a letter. If the same mapping isn't used in both places, then the message will get jumbled. Try different mapping tables and verify, by their sound and spectrum, that the tones used are different for each one. Then see what happens when the mapping tables at the transmitter and receiver are different. What problem occurs? How do you fix it?

EXERCISES 7.3

Mastering the Concepts

1. Assume that the maximum frequency your receiver can "hear" is 3000 Hz. What potential advantage does a system using Table 7.3 have compared with a system using Table 7.1? What potential disadvantages does it have?
2. If the zero-crossing method described in Exercise 7.2.10 is used to estimate frequency, explain how the following signal degradations might cause problems:
   a. weak signal;
   b. added noise;
   c. interference.
3. Suppose that the single-tone-per-character system were to be used in a room with an echo. How might that affect the receiver's performance? What characteristic of the transmitted signal might be changed to reduce the effect of reverberation and echo? How would changing this parameter affect the transmission rate that can be achieved by the system?
4. Why is it important to have the tables used for sending and receiving characters be the same at the transmitter and receiver? What happens if they aren't the same?
5. Why is there a fundamental trade off between speed of transmission and accuracy of reception?

Try This

6. Using the mapping in Table 7.1, determine the frequencies for the tone bursts that we would use to transmit the word MUSIC.
7. Repeat Exercise 7.3.6, using Table 7.4 to determine the tone frequencies needed to send the word MUSIC.
8. Suppose that the word MUSIC is sent according to the mapping defined in Table 7.1, but the receiver mistakenly uses Table 7.4. What "word" does the receiver interpret?
9. Repeat Exercise 7.3.8 for the case where the receiver mistakenly uses Table 7.3.
10. If the receiver mistakenly used Table 7.4 and then later one finds out that Table 7.1 was the correct one, can the receiver's errors be corrected? If yes, how can the errors be corrected? If no, explain why they cannot be corrected.
11. Repeat Exercise 7.3.9 for the case where Table 7.3 is mistakenly used. Explain the basic difference between this case and the case of Exercise 7.3.10.
12. Suppose the word MUSIC is sent by the transmitter, using Table 7.1, and that Table 7.1 is also used by the receiver, but the receiver mistakenly detects the frequency of the fourth burst as 1000 Hz instead of the correct value. What word will

the receiver send to the user? What word is reported if the receiver doesn't "hear" the fourth tone burst at all?

13. Use your calculator to execute the following steps:
    a. Use Equation (7.1) to draw a tone burst of peak amplitude 2, frequency 400 Hz, initial phase 0, and duration 10 ms.
    b. Now draw an echo of the burst from part (a) that has been attenuated by a factor of four and delayed by 4 ms.
    c. Add the two, demonstrating the effects of interference from multipath reception, which can cause an echo on a telephone line or a ghost image on a monitor screen.
    d. What makes it so difficult to determine the frequency of the received tone burst with multipath or echo interference?

14. Suppose we use a piano as a transmitter by mapping a piano key to each textual character that we might want to send.
    a. How many different characters could be sent in this way?
    b. Is this quantity enough for all of the letters, numbers, and punctuation in our alphabet?
    c. A telephone circuit reliably conveys tones between 300 Hz and 3400 Hz. Is this frequency range large enough to accurately transmit all the signals from our piano modem?
    d. *Extra credit*: If we use the piano at a transmitter, what are the lowest and highest frequencies that the communications channel will have to carry in order to faithfully convey the full range of frequencies to the receiver? (See Chapter 2 for a discussion of the frequencies used by a piano.)

15. It initially might seem that the number of different symbols transmitted could be increased indefinitely by reducing the frequency spacing Δf between the symbols. (For example, Table 7.1 uses $\Delta f = 100$ Hz, while Table 7.3 uses $\Delta f = 50$ Hz.) If we could continue to decrease Δf, we could continue to increase the data transmission rate. However, the duration T_d of a burst must be equal to $C/\Delta f$ for accurate reliable detection of tones, where Δf is the minimum frequency separation of the symbols and C is a constant that depends on the particular filter implementation. Assume that $C = 4$. In addition, assume that the quiet time between bursts is $1/\Delta f$, that the lowest frequency is 300 Hz, and that the highest frequency that is 3480 Hz.
    a. For $\Delta f = 100$ Hz, how many different symbols can be used? How many different messages can be sent in 80 ms?
    b. Repeat part (a) for $\Delta f = 50$ Hz.
    c. Repeat part (a) for $\Delta f = 200$ Hz.

16. In Table 7.1, the letter A is represented by a cosine signal at 300 Hz and the letter B is represented by a cosine signal at 400 Hz. A receiver might distinguish bursts of these two signals by simply subtracting one from the other.

a. Use a calculator to plot the *difference* of a cosine at 300 Hz and a cosine at 400 Hz over the interval from 0 ms to 0.5 ms. What is the maximum value of the difference?
b. Repeat (a) for the interval from 0 ms to 10 ms.
c. If you were a communications system designer, which time interval would you choose for a tone burst? Why?

17. Repeat Exercise 7.3.16 for the letters *Y* and *Z*.

In the Laboratory

18. You and your friend decide to set up your own "secret" communications system using a special table. Design the entries of your table, and set up a single-tone-per-character system in the laboratory. What choices have you made? How hard would it be for someone else to figure out what messages are being communicated?

Back of the Envelope

19. You are in charge of designing a receiver such as the one shown in Figure 7.5 for an application in which Table 7.1 is used. The main cause of potential errors is assumed to be interfering tone bursts from other sources that can occur at any frequency. Suppose further that the transmitter may not generate the specified tone frequencies perfectly, so that all its transmitted frequencies may be as much as 20 Hz too high or 20 Hz too low. Discuss the following trade-offs for the design of the 27 bandpass filters:

a. If each bandpass filter gives a strong response over a range of 40 Hz (for example, from 380 to 420 Hz for the letter *B*), then the full range of the transmitted signals can be received correctly. What percentage of the interfering tone bursts will also be interpreted as valid characters?
b. If each bandpass filter is designed to give a strong response over a narrower frequency range of 10 Hz (for example, from 395 to 405 Hz for the letter *B*), then what percentage of the interfering tone bursts will be interpreted as valid characters?
c. What impact would the narrower filters have on the reception of messages from the desired transmitter?

7.4 The Craft of Engineering—Improving the Design

In the last section, we discussed ways of sending information from one location to another, and we defined the basic components required by a communications system in order to accomplish our goal. A specific example demonstrated the basic concepts by sending text messages for short distances through the air, using a sequence of audible tones to represent a sequence of characters. The range of this communications system is limited by the ability of the microphone receiver to "hear" the transmitting speaker as the distance between them grows.

KEY CONCEPT

In choosing the best design for a practical application, engineers must compare alternatives and make judgments about the cost and technical performance of each alternative.

There are many other ways to represent the characters of a message, and there are many more ways to transmit a message. In this section, we'll look at two other ways to represent data, and we'll develop performance criteria to compare these new methods.

The performance and cost comparisons we will do are part of the engineering design process. *Engineering*, by definition, is the application of scientific knowledge to practical uses that satisfy human needs. In choosing the best design for a practical use, engineers must compare alternatives and make judgments about the cost and technical performance of each alternative within the context of the specific application.

Improving the Design of a Digital Communications System

The communication method developed in Section 7.2 and illustrated in Figure 7.3 is called a one-tone-per-character method, since each of the 26 letters and the space is assigned to a unique frequency or tone. The receiver implementation for this system with the best performance has 27 narrow-band filters (Figure 7.5). Each filter is tuned to respond to exactly one of the frequencies used to represent a character, so that each filter has a maximum response at the frequency associated with the character it was designed to detect. The outputs of these filters are examined constantly by the receiver's decision logic. When the energy from one filter substantially exceeds the energy from any of the other filters for a long-enough time, the receiver declares that a tone burst has been transmitted. The character transmitted is identified as the character corresponding to the filter that responds most strongly to the tone burst.

KEY CONCEPT

Cost is an important consideration for any design. The first cost we encounter is the cost of implementing a particular design. Engineers must also be concerned with other costs, such as the costs of maintenance, training personnel to use a system or device, and keeping the system or device in operation. These continuing costs must be balanced against the one-time design costs when choosing the final design.

While the 27-filter receiver design works well, it can be quite expensive compared with other approaches. The overall cost and performance of a receiver will depend on several factors, including the number of filters and the quality and cost of each individual filter. Factors affecting the cost of individual filters will include frequency selectivity and the minimum time duration needed to determine which tone was transmitted. Often, the filters are the most expensive components used in a receiver. If we wanted to build a cheaper receiver, we would explore strategies to reduce both the number of filters needed and the cost of each filter.

We must first ask if it is possible to design a transmitter and receiver pair that needs fewer tone frequencies, so that fewer expensive receiver filters are required. This is possible, but when we redesign to reduce cost, other attributes of the communications system must also change. The engineer's objective of finding the "best" design will come down to comparing the alternative designs and their attributes and finding the one best suited to both the customer's desires and his or her financial resources.

In this section, we present two alternatives to the communications system defined in Section 7.2. They both are cheaper than the design in Section 7.2 in the sense that fewer tone frequencies are used and fewer expensive filters are required. After each of these alternatives is described, we will analyze all of the designs in terms of implementation cost, transmission speed, and accuracy.

Binary Representation for Each Character

The design from Section 7.2 can be viewed as extravagant, because only one of the 27 receiving filters is used at any time. Specifically, the transmitter sends one tone burst for each character in the message, and one of the 27 filters at the receiver is expected to detect it. The other 26 "reject" the tone and have no significant energy output.

Could we send more than one tone at the same time? This approach might reduce the implementation cost of the receiver by requiring fewer costly filters, as long as the cost of interpreting a signal with several tones is less than the cost of the filters removed.

Using More Than One Tone Consider a system that uses a small number of tones to represent characters such that any combination of any number of the tones will represent some valid character. Let P be the number of tones. This system would allow 2^P combinations of tones, since each of the P tones could independently be on or off, just like the way binary numbers are represented. Actually, the P tones will be used to represent only $(2^P - 1)$ characters, since we must exclude the case where all tones are off. (Why would we do this?)

If $P = 5$, then $2^P - 1 = 31$, and we can send our 26 Roman letters, a space, and a few control characters, using a combination of up to five tone frequencies for each letter. The method would work as shown in Figure 7.9. Each character from the originator's list would be converted into a burst in the same manner as before, but the multitone burst would consist of the sum of between one and five sinusoids. The choice of the number of sinusoids, and their frequencies, would be determined by a table of the type shown in Table 7.5, where an "X" in a column indicates that a tone is on for the letter specified.

For example, the transmission of the letter A would be sent via a burst consisting of two sinusoids, one at frequency F_1 and the other at frequency F_2. Similarly, the letter B is represented by a burst containing three sinusoids, while the space character is represented by a burst with only one sinusoid, at frequency F_3.

An example of such a composite burst is shown in Figure 7.10. Suppose that we want to send the letter D using the representation in Table 7.5, and that our choice for frequency F_1 is 500 Hz and our choice for frequency F_4 is 1700 Hz. The top two traces of Figure 7.10 show bursts of 500 and 1700 Hz, while the bottom trace shows the composite burst containing the sum of the upper two traces.

KEY CONCEPT

The simple air modem used in the earlier experiments uses a single burst consisting of a single sinusoidal signal to carry a single letter of text. It is possible to use more complicated bursts, consisting of the sum of one or more sinusoids, to carry textual traffic as well. These more complicated methods usually bring the benefit of improving the rate at which messages can be sent or decreasing the implementation cost of the transmitter or receiver.

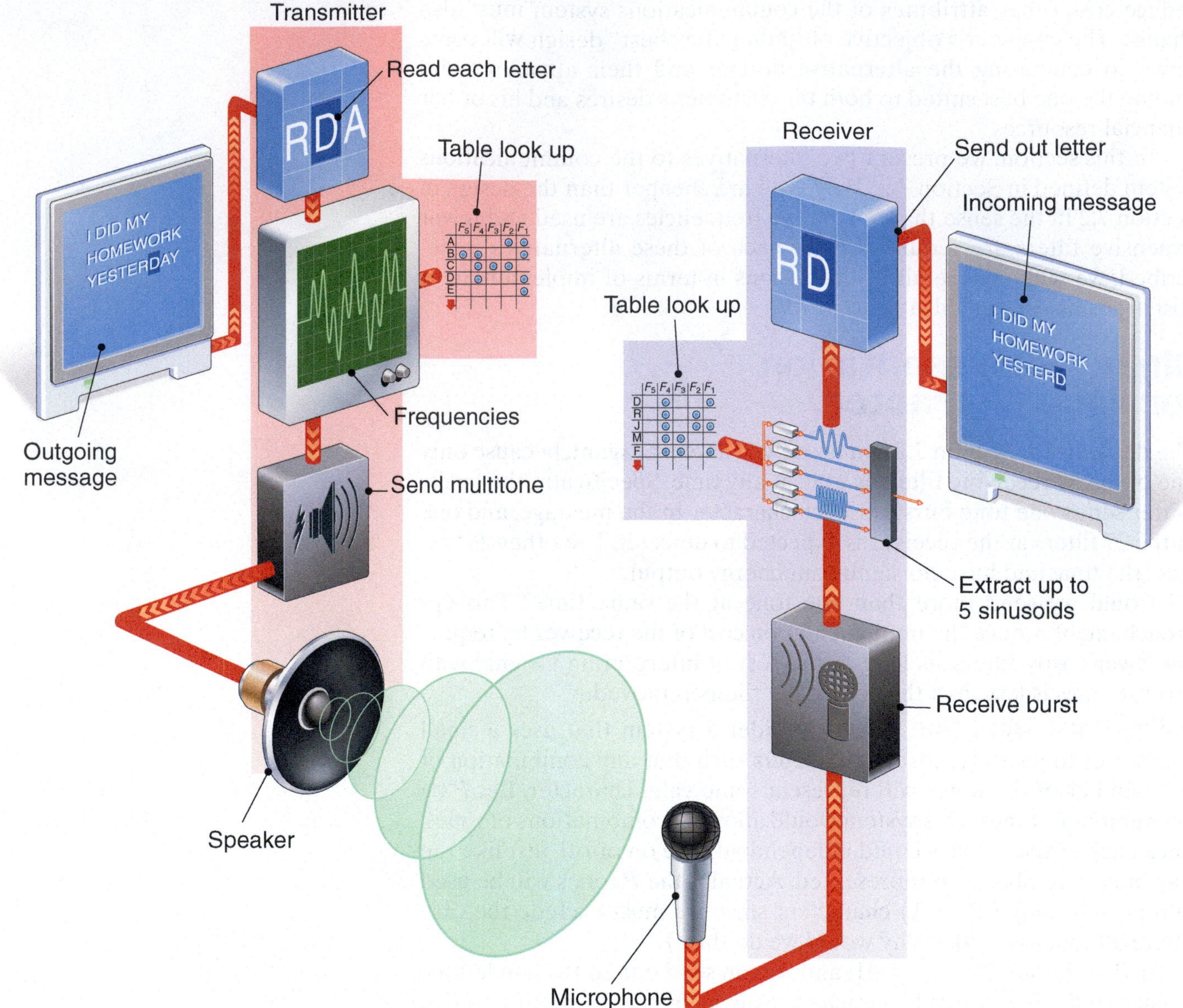

Figure 7.9 Communications method based on using any combination of five possible tone frequencies.

Table 7.5 A Possible Mapping Table Used by the Transmitter for Sending Characters as the Combination of up to Five Sinusoids

	A	B	C	D	E	F	G	H	I	J	K	L	M	N	O	P	Q	R	S	T	U	V	W	X	Y	Z	sp
F_1	X	X		X	X	X				X	X						X		X		X		X	X	X	X	
F_2	X		X				X		X	X	X	X				X	X	X			X	X	X				
F_3			X			X		X	X		X		X	X		X	X		X		X	X		X	X		X
F_4		X	X	X		X	X			X	X		X	X	X			X				X		X			
F_5		X					X	X				X	X		X	X	X			X		X	X	X	X	X	

The tone bursts have a duration of 5 ms, so we see 2.5 cycles in the 500-Hz signal and 8.5 cycles in the 1700-Hz signal. The formula for the composite signal for the letter D is similar in form to Equation (7.2), but in this case, the two signals are added as shown in Equation (7.3):

$$s_D(t) = \cos(2\pi\, 500t + \phi_1) + \cos(2\pi\, 1700t + \phi_4) \qquad (7.3)$$

The two phases are chosen to give each cosine a time delay of 4.5 ms.

Infinity Project Experiment: Audio Communication of Messages, Using Several Tones per Letter

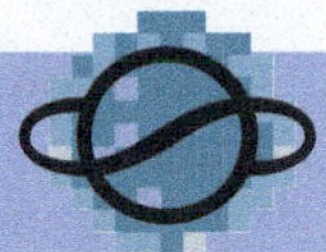

All of the audio communications systems in previous experiments use one tone to communicate a single letter of text. We can use sums of tones to send text as well. Try typing letters at the transmitter of such a system, and listen to the output; how is it different from that of the single-tone-per-letter method? Up to how many different frequencies are used to send each letter? By experimentation, find the five letters that use only one tone per letter, and use these letters to test the receiver. Now try sending text. How well does this system work? Is there a letter that uses all five tones played simultaneously to represent a character?

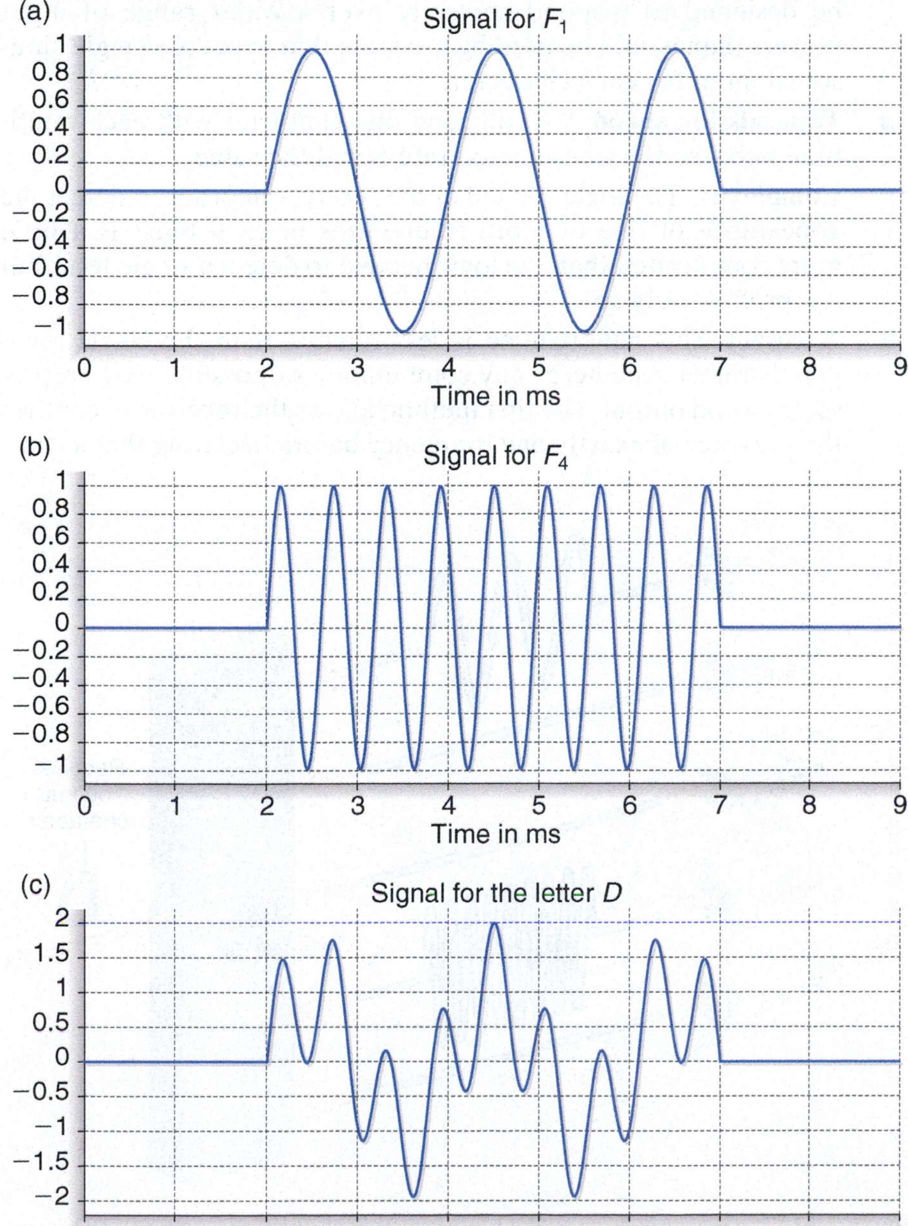

Figure 7.10 Two tone bursts, one at 500 Hz and the other at 1700 Hz, are added to represent the character D as in Table 7.5.

As shown in the block diagram in Figure 7.9, the receiver is based around a set of five filters, each tuned to one of the five frequencies F_1 to F_5. A more detailed block diagram of the receiver is shown in Figure 7.11. The receiver watches the outputs of all five filters. The receipt of a transmitted character is indicated by the presence of sufficient energy in any of the five filters. The receiver then examines all five filter outputs to determine which are responding to a transmitted sinusoid and which are not. By referring to Table 7.5, the receiver can determine which of the characters were sent.

The expected cost and performance of the new alternative we have just developed can be evaluated in terms of four important factors:

Transmission Speed: The rate at which information can be communicated, measured in communication units per time interval such as words per minute, characters per second, or bits per second.

- **Implementation cost** We now need only five filters, making this method substantially cheaper to implement than the 27-filter, single-tone system. However, since from one to five sinusoids may be transmitted simultaneously, the microphone input system must be designed to respond correctly over a wider range of input powers than would be used by a system that expects a single sinusoidal input for each character.
- **Transmission speed** We still send one character with each multitone burst, so the transmission rate is still the same.
- **Complexity** The logic needed at the receiver in order to detect the appearance of one or more frequencies in each burst is a little more complicated than the logic needed to detect a single tone, but not prohibitively so.
- **Accuracy** This new scheme is less reliable than the single-tone-per-character scheme, as any combination of possible tones represents a valid output. The first method allows the receiver to confirm the presence of exactly one frequency before declaring that a char-

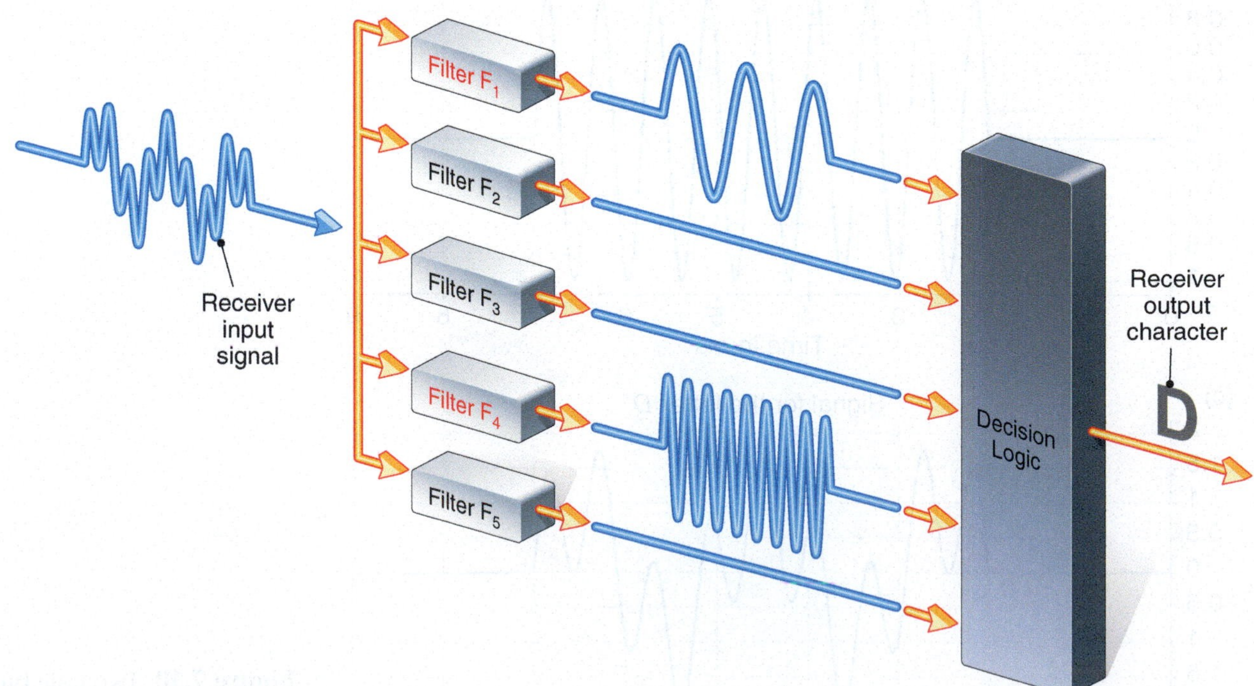

Figure 7.11 Receiver for the parallel binary system shown in Figure 7.9. The input signal is the combination of two tones for the letter *D*, shown in Figure 7.10.

acter has been received. It is common for engineers to specify the reception accuracy required for a system by defining the maximum acceptable **error rate**.

Error Rate: The ratio of the number of bits received incorrectly to the total number of bits sent by the transmitter.

Binary Interpretation of Character Codes—The Parallel Binary Method We can interpret this five-tone technique as a method of sending characters via a five-bit binary code. Each letter and control character is represented by a unique set of five bits. Each of the bits controls one of the tone frequencies that might be present in the burst. If only one bit with a value of one is used for a particular character (such as the space, the E, or the T), the burst carrying the character to the receiver consists of only one sinusoid at the designated frequency. If two of the bits have a value of one, then the corresponding two frequencies are sent. Other characters may have three or four bits with a value of one, and will respectively have three or four frequencies used in the transmitted tone burst. This system is perhaps more apparent when Table 7.5 is reorganized as Table 7.6. The contents of the tables are the same, but by reordering the letters, you can see that there is a clear binary counting pattern present. Only 27 of the 31 possible patterns are identified, allowing four additional control characters to be assigned to the four unnamed patterns, if desired. This order makes it much easier for the receiver to identify the character.

INTERESTING FACT:

The mapping from bits to characters as shown in Tables 7.5 and 7.6 was first suggested in the 1800s by a French scientist named J. M. E. Baudot. The unit of transmission speed in a data communications system, the *baud*, was named for him.

When viewing the table this way, we see that we assign to each character a binary pattern that is different from all the other patterns. We then use the values of the bits to determine whether a sinusoid at the frequency corresponding to each bit is a part of the character's multitone burst. The receiver observes the burst via five filters, each one looking for one of the five sinusoidal frequencies. When a sinusoid is detected at the filter's output, a one is reported for that bit. If no sinusoid is detected by a filter, a zero is reported for that bit. The pattern of five ones and zeroes is then matched against Table 7.6 in order to determine which character was transmitted. Since the bits are sent together in a single burst, and therefore in "parallel," we call this approach the **parallel binary method**.

Parallel Binary Method: Sending communication units such as bits or tones all at the same time.

We can show mathematically that if the transmitter is limited to turning sinusoidal bursts on and off for each character, then the parallel binary method minimizes the number of expensive filters in our receiver. However, if that limitation is not present and a single character can be represented by a sequence of tones, then there are ways of using even fewer filters. The next method we examine will reduce the number of filters to two. There is a price to pay, however, for the reduced number of filters, and the engineer must evaluate these performance and cost factors to determine the best choice for a particular application.

Table 7.6 A Reordering of Table 7.5 to Reveal the Binary Counting Progression of the Transmission Patterns for Use by the Receiver

F_1	X		X		X		X		X		X		X		X		X		X		X		X		X		X		X		X
F_2		X	X			X	X			X	X			X	X			X	X			X	X			X	X			X	X
F_3				X	X	X	X					X	X	X	X					X	X	X	X					X	X	X	X
F_4								X	X	X	X	X	X	X	X									X	X	X	X	X	X	X	X
F_5																X	X	X	X	X	X	X	X	X	X	X	X	X	X	X	X
	E		**A**	**sp**	**S**	**I**	**U**		**D**	**R**	**J**	**M**	**F**	**C**	**K**	**T**	**Z**	**L**	**W**	**H**	**Y**	**P**	**Q**	**O**	**B**	**G**		**M**	**X**	**V**	

Serial Binary Representation for Each Character

Our objective is to find a method of conveying the capital letters and a few control characters with just two frequencies and two corresponding filters at the receiver. A common technique is a variation on the parallel binary scheme we just examined. In that method, Table 7.5 was used to convert each character we wanted to transmit into a binary pattern of ones and zeros. Each bit position was identified by one of five frequencies, and the bit value was identified by whether or not the frequency was present in a single burst.

To transmit "serial binary," however, we use a sequence of five separate tone bursts for each character, with each burst containing one of two sinusoids. The sinusoid at frequency F_1 is used if a one bit is to be sent, and the frequency F_0 is used if a zero bit is to be sent. In this case, the bit value is identified by one of two frequencies, and the bit position is identified by the tone's position in the sequence of five tones. We can rewrite Table 7.5 as Table 7.7 to show this is configuration.

Serial Binary Method: A method of representing a character by a series of bits, each of which can take on a value of zero or one, and transmitting the character by sending the bits one at a time.

Frequency Shift Keying (FSK): Sending serial binary data by using one frequency to represent a zero and a different frequency to represent one. An alternative, PSK, uses different phases for ones and zeros.

We call this technique the **serial binary method**, since binary representations are used for each character, but we send the bits out serially, the ones and zeros are represented by bursts of one or the other of the two frequencies. This method is also known as the **frequency shift keying (FSK)** method, since the transmitter's frequency is shifted from one frequency to another to indicate the shift from a zero to a one, and vice versa. Since we now use only two frequencies to send sequences representing our alphabet, our receiver needs only two filters. A block diagram of this receiver is shown in Figure 7.12. Clearly, this receiver is simpler than the one shown in Figure 7.11 for the parallel binary method, because the former has only two filters instead of five. An additional benefit of reducing the number of tones is that the frequency difference between tones used by the receiver can be increased. The duration of a tone burst can be reduced if the frequency difference between the tones is increased. Using shorter tone bursts can increase the system's transmission rate.

The difference between the serial and parallel binary transmission methods can be visualized by looking at the patterns of the transmitted signals. Figure 7.13 shows the waveforms for the word IF when trans-

Table 7.7 A Table for Mapping Characters into a Sequence of Bursts, Using the Baudot Code

Burst 1	F_1	F_1	F_0	F_1	F_1	F_1	F_0	F_0	F_0	F_1	F_1	F_0	F_0
Burst 2	F_1	F_0	F_1	F_0	F_0	F_0	F_1	F_0	F_1	F_1	F_1	F_1	F_0
Burst 3	F_0	F_0	F_1	F_0	F_0	F_1	F_0	F_1	F_1	F_0	F_1	F_0	F_1
Burst 4	F_0	F_1	F_1	F_1	F_0	F_1	F_1	F_0	F_0	F_1	F_1	F_0	F_1
Burst 5	F_0	F_1	F_0	F_0	F_0	F_0	F_1	F_1	F_0	F_0	F_0	F_1	F_1
	A	**B**	**C**	**D**	**E**	**F**	**G**	**H**	**I**	**J**	**K**	**L**	**M**

Burst 1	F_0	F_0	F_0	F_1	F_0	F_1	F_0	F_1	F_0	F_1	F_1	F_1	F_1	F_0
Burst 2	F_0	F_0	F_1	F_1	F_1	F_0	F_0	F_1	F_1	F_1	F_0	F_0	F_0	F_0
Burst 3	F_1	F_0	F_1	F_1	F_0	F_1	F_0	F_1	F_1	F_0	F_1	F_1	F_0	F_1
Burst 4	F_1	F_1	F_0	F_0	F_1	F_0	F_0	F_0	F_1	F_0	F_1	F_0	F_0	F_0
Burst 5	F_0	F_1	F_1	F_1	F_0	F_0	F_1	F_0	F_1	F_1	F_1	F_1	F_1	F_0
	N	**O**	**P**	**Q**	**R**	**S**	**T**	**U**	**V**	**W**	**X**	**Y**	**Z**	**sp**

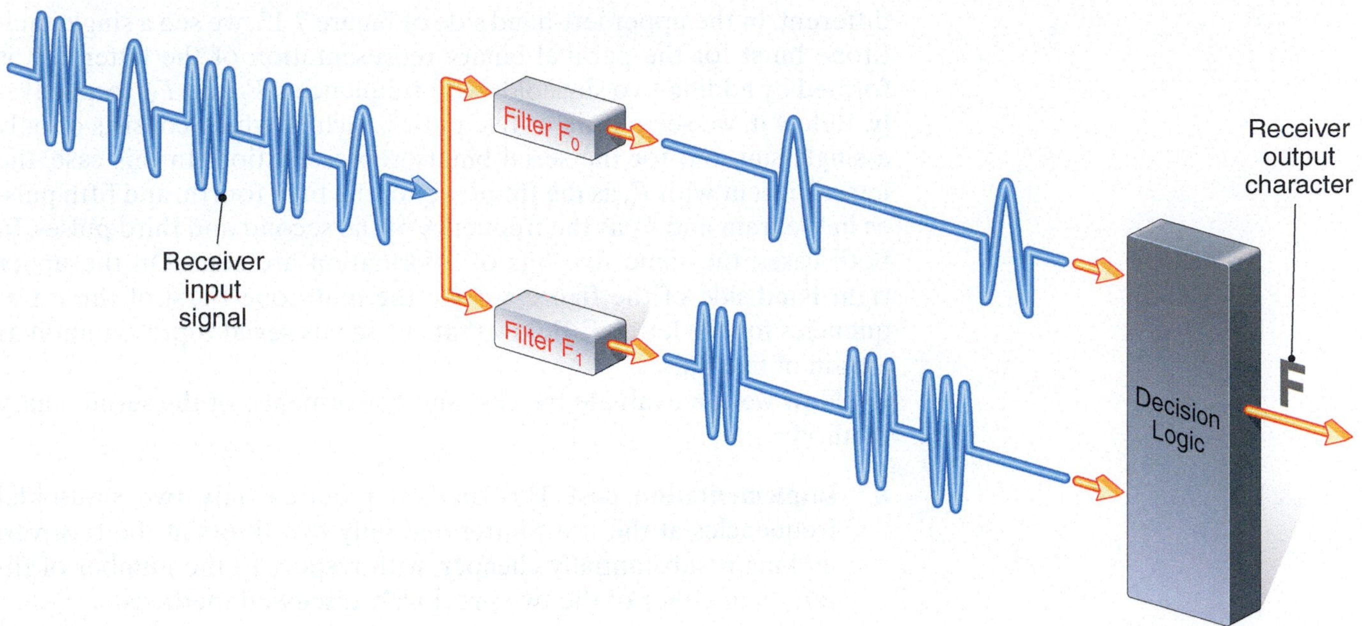

Figure 7.12 Block diagram of the receiver for a serial binary communication system. A 500-Hz tone burst is used to represent a zero bit, and a 1200-Hz tone burst is used to represent a one bit.

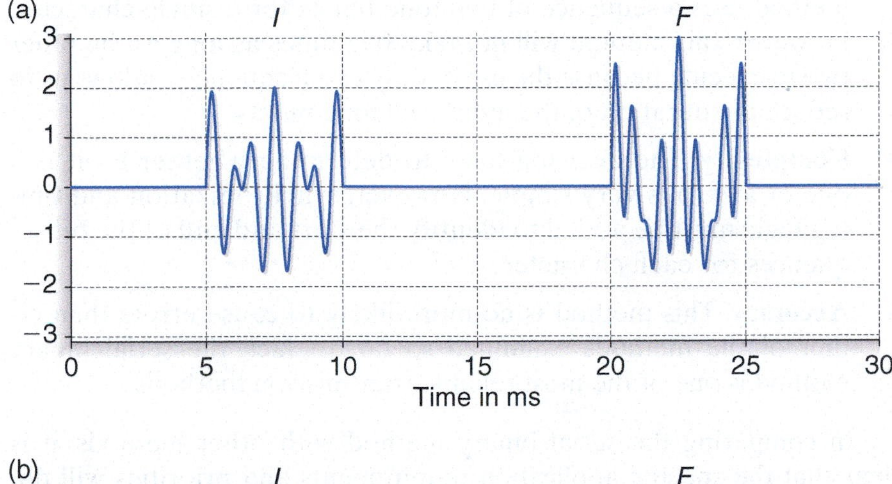

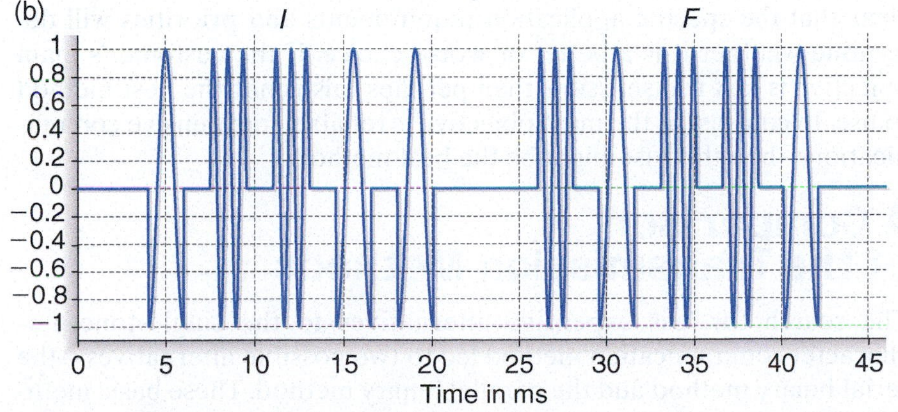

Figure 7.13 A view of the waveforms used to carry the letters *I* and *F*. (a) A single multitone burst for each character, using the parallel binary method with frequencies of 500, 900, 1300, 1700, and 2100 Hz. (b) A sequence of five tone bursts for each character, using the serial binary method with 500 Hz representing a zero and 1200 Hz representing a one.

mitted in both ways. We assume that Table 7.5 is used for the parallel method and Table 7.7 for the serial method. Recall that the mapping from each character to the binary pattern is the same in both cases, but the way in which the bit positions and bit values are represented is different. In the upper left-hand side of Figure 7.13, we see a single multitone burst for the parallel binary representation of the letter *I*. It is formed by adding two sinusoids with frequencies F_2 and F_3, respectively. Below it, we see a train of five pulses, each of which consists of only a single sinusoid for the serial binary representation. In this case, the letter *I* is sent with F_0 as the frequency of the first, fourth, and fifth pulses in the train and F_1 as the frequency of the second and third pulses. In both cases, the same five bits of information are sent. On the upper right-hand side of the figure, we see the multitone burst of three frequencies for the letter *F*. Below that, we see its serial representation as a train of five pulses.

Now we can evaluate the cost and performance of the serial binary method:

- **Implementation cost** This method requires only two sinusoidal frequencies at the transmitter and only two filters at the receiver, making it substantially cheaper, with respect to the number of filters, than either of the two previously discussed methods.
- **Transmission speed** It takes longer to transmit a character with this method if the tone burst is of the same duration as the tone bursts of the other methods. The two previous methods needed only one burst of tones to send a single character, while this method uses a sequence of five tone bursts for a single character. However, this method will not take five times as long as the other two to execute because the use of only two frequencies allows us to reduce the duration of the individual tone bursts.
- **Complexity** The decision logic to determine whether a bit is a one or a zero is very simple. However, synchronization and timing logic must be added to identify the start and end of the bit sequences for each character.
- **Accuracy** This method is no more likely to cause errors than either of the methods examined so far. In fact, the serial binary method is one of the most reliable transmission methods.

In comparing the serial binary method with other methods, it is clear that the specific application requirements and priorities will determine whether it is a better or worse choice. If the customer's main objective is fast transmission, then perhaps this is not the best method to use. In contrast, if the main objective is reliable, inexpensive communications, then this just might be the best method.

A Comparison of the Transmission Methods

The search for less expensive alternatives to the single-tone-per-character communication method led to two possible alternatives—the serial binary method and the parallel binary method. These basic methods can be compared in terms of cost and performance. Such a comparisons for a system designed to transmit 30 distinct codes is provided in Table 7.8.

Infinity Project Experiment: Audio Communication of Messages, Using Serial Binary Transmission

We've seen how text messages can be communicated using an audio transmitter and receiver with either (1) a single tone for each letter or (2) up to five tones per letter. What if we use only two tones per letter? Each tone in this scheme could correspond to a binary one or zero, and we would send a sequence of five tones for each letter. Listen to the output of the transmitter of such a scheme, and verify that five successive tones are being sent. Also check to see what two tone frequencies are being used. How long does it take to communicate information by using this "serial binary" method? How does this method's speed compare with that of the "parallel binary" method? Examine the displayed bit pattern for each character as you listen to it being transmitted. Do you think you could learn to recognize the letters by ear?

Both alternatives to the single-tone method use fewer filters and therefore are presumably cheaper to implement than the single-tone-per-character method. However, a lower implementation cost means that the speed of transmission may decline. These different methods have different attributes, some of which might be more attractive to a particular user than others. Some users might want speed, others the lowest cost, and still others the best possible accuracy when attempting to communicate through noise and interference. There is no absolute best choice among these methods. The most appropriate method for a particular use will depend on the application's communication needs and priorities.

Using Pairs of Tones to Communicate

The Touchtone® telephone keypad shown in Figure 7.14 uses a representation method that is different from the three discussed so far. Seven possible tone frequencies are used to send any of 12 possible numbers and symbols on the Touchtone keypad.

A different frequency is used for each of the three columns of buttons, and four more frequencies are used to indicate each row. Pressing a button sends a burst containing exactly two frequencies—the appropriate one for the row and the appropriate one for the column for the button that is pressed. (To verify this configuration for yourself, using a

Table 7.8 A Comparison of the Basic Attributes of Three Alternatives for Designing an Acoustical Communications System for Textual Messages Using 30 Characters

Method	Number of Filters	Relative Speed	Reliability	Automatic Detection
Single tone per character	30	5	Medium	Very easy
Parallel Binary	5	5	Medium to low	Medium
Serial Binary (FSK)	2	1	Good	Easy

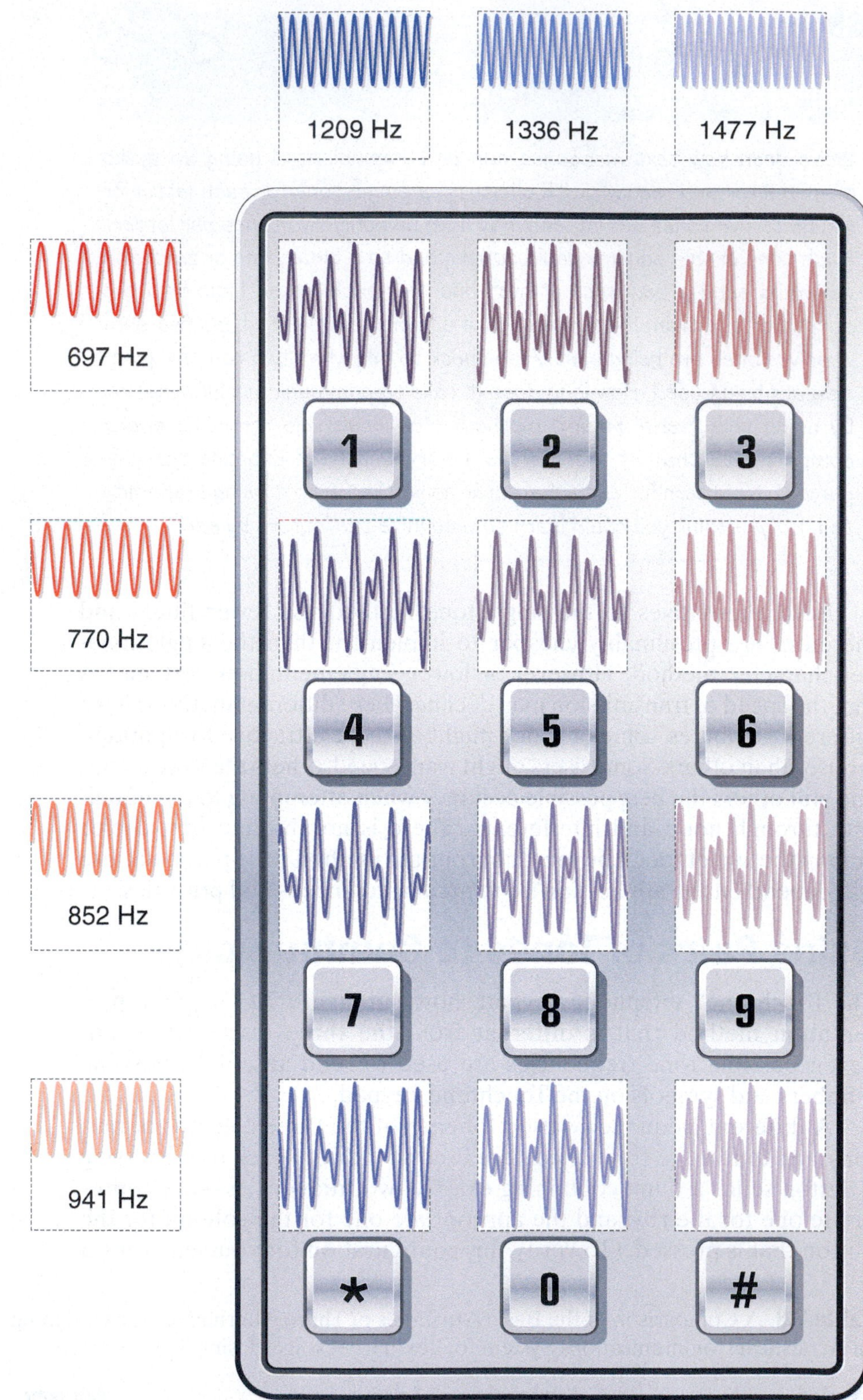

Figure 7.14 The combinations of tone frequencies used in a Touchtone telephone keypad.

Touchtone telephone, simultaneously press two buttons in the same column and then two buttons in the same row. Doing this produces the single tone associated with that row or column.)

Consider now a scheme for sending textual messages by using an extended version of the Touchtone concept (technically called *dual-tone multifrequency,* or DTMF). Suppose that we use the sets of frequencies

Infinity Project Experiment: Touchtone® Telephone

There are hundreds of millions of wired and cellular telephones that use Touchtone to "signal" the telephone number that the user wants to call. Each of the digits and characters on the dial is conveyed using a combination of exactly two tones. We easily can re-create the sound of Touchtone dialing by adding the appropriate two tones together for each button on a telephone keypad. From the spectrum of each Touchtone sound, verify that only seven different frequencies are used to send all 12 characters.

shown in Figure 7.15 to transmit each of the capital Roman letters. For example, the letter *Z* would be carried by a burst consisting of tones at 1100 and 1700 Hz. This method actually has been used to transmit textual messages over shortwave radio communications systems.

Infinity Project Experiment: Audio Communication of Messages, Using Two Tones per Letter

The two-tones-per-letter transmission scheme is yet another way to communicate letters via a digital audio communications system. Explore how the transmitter in such a scheme works by typing in messages and listening to the output. Verify, using the spectrum of each sound, that only two tones are being used at any one time.

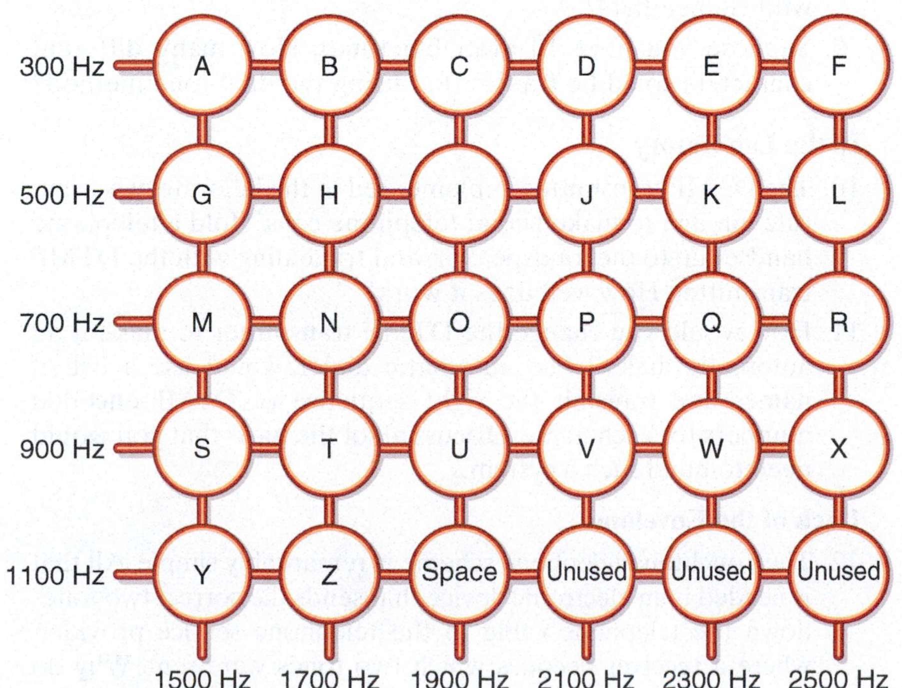

Figure 7.15 A method for transmitting characters by using the combination of two tones.

EXERCISES 7.4

Mastering the Concepts

1. How many bits does the Baudot code use to specify each of the letters of the alphabet?
2. Which is the more reliable transmission method, FSK using serial binary representations of Baudot, or the parallel binary method using bursts consisting of many tones with many frequencies? Why?

Try This

3. Redraw Figure 7.11 to show the operation when the "I" of Figure 7.13 is received. Repeat for the "F" of Figure 7.13.
4. Suppose we want to send the word MUSIC using the DTMF acoustical transmission method. What tone pairs from Figure 7.15 would be used?
5. Characterize the DTMF acoustical transmission system just as we have done the others, in terms of
    1. transmission speed,
    2. implementation cost,
    3. complexity, and
    4. expected accuracy.
6. Fill in the appropriate entries in Table 7.8 for the DTMF method.
7. Provide a general mathematical rule for determining the number of distinct tones necessary with any dual-tone method to transmit N different characters.
8. Determine the minimum number of different tone frequencies needed with the dual-tone method to convey any of 64 characters. What is the minimum number for a character set with 78 members?
9. Suppose you have 10 available tones. How many different characters could be transmitted using the dual-tone method?

In the Laboratory

10. The DTMF transmitter implemented in the laboratory is accurate enough to make actual telephone calls. Hold a telephone handset up to the loudspeaker, and try dialing with the DTMF transmitter. How well does it work?
11. How would you change the DTMF transmitter to make it an automatic dialer? The automatic dialer would use a list of names and transmit the right sequence of DTMF-encoded numbers for each name. Discuss all of the parts that you would need to build such a system.

Back of the Envelope

12. The Touchtone telephone scheme is remarkably simple. All that is needed is an electronic device that sends the correct two tones down the telephone cable to the telephone service provider, where a receiver decodes which two tones were sent. Why do

you think this transmission scheme was used? (*Hints*: How easy would it be for a person to sound like a Touchtone telephone? Could you "dial" a telephone number accidentally by talking? How would the situation change if there were only a single tone per number?)

13. Design a communications system that uses three tones per character to transmit information. How many different tones are required to send all of the letters of the alphabet plus a space?

7.5 Extensions

From Morse Code to ASCII and Beyond

The data representation methods described in this chapter have been demonstrated with a limited character set of 27 codes in order to emphasize the differences between the methods and the design process of selecting a method. However, in modern communications systems, a much larger set of characters typically is used. In order to represent upper- and lowercase Roman alphabet characters and the 10 digits, 62 character codes would be needed. Adding punctuation and control codes further increases this number. There are two basic approaches to extending the set of characters that can be represented. One is to simply add more character codes. The second uses sequences of codes from a small set to make a larger number of codes. These two approaches may also be used in combination.

INTERESTING FACT:

The first electrical communications systems, such as the telegraph, were designed to carry capital letters in the English language, and to do so as quickly as possible. In order to transmit in other languages, additional letters, lowercase letters, and more punctuation marks were added. The size of the character sets used in communications systems had to grow to meet the demand.

The implications of adding new characters to the character set are different for the methods already discussed. For all methods except the serial binary method, adding new codes requires fundamental changes in the transmitter and receiver capabilities. For example, with the one-tone-per-character method, the addition of new character codes requires the addition of a new tone for each new character. If we assume that we want to keep all of the tones within the same frequency range, the addition of new tones means that the frequency separation between tones must be reduced and both the number of filters and the precision of the filters must be increased. Both of these factors increase the implementation cost of the system.

The same considerations apply to the parallel binary method, but fewer new tones would be needed. The number of additional tones would increase with the base-two logarithm of the number of characters for the parallel binary method. Since the number of unique tone combinations we can make with P tones is $2^P - 1$, we roughly double the number of characters we can represent every time we add a tone. If the character set contained 127 characters, the single-tone system would require 127 tones, while the parallel binary system would require 7 tones, since $2^7 - 1 = 128 - 1 = 127$.

Only the serial binary (FSK) method continues to use the same two tones as the number of characters increases. If the size of the character set is increased from 31 to 128, then the serial binary method simply increases the number of tone bursts for each character from five to seven. This change causes the transmission time for each character to

INTERESTING APPLICATION

From Morse Code to the Phonograph (Record Player)

The importance of the invention of the telegraph and the ways in which it changed how people worked and lived are discussed at the end of Section 7.1. In addition to these direct consequences, there were many significant, less direct, consequences, such as the invention of the telephone and the development of power systems. American inventor Thomas Alva Edison (1847–1931) held 1093 patents and is probably best known for inventing the light bulb and the first power system. He recorded Morse code messages as patterns of indents on paper tape or rotating wax drums for later retransmission. Edison noticed that when these tapes or cylinders were read back to resend the stored message, they made interesting sounds. From that observation he had the idea of making tinfoil-covered cylinders with patterns of indents for the explicit purpose of storing and re-creating the sound of the human voice. This idea led directly to his invention of the record player, which became very popular for re-creating music for entertainment purposes rather than for business office use. The record player was his favorite invention. The CD and DVD have direct roots in Edison's phonograph.

International Morse Code: The first important method of electrical communications was the telegraph. It uses a code devised by Samuel Morse. It employed neither tones nor ones and zeros. It used "dashes" and "dots," or long and short electrical pulses, respectively, in a code devised to carry the English language as fast as possible.

Baudot Code: A five-bit code used in the transmission of messages in various languages. A keyboard, representing 32 different symbols, each containing five bits, was used to transmit international communications until the mid-1970s.

ASCII Code: American Standard Code for Information Interchange, which is currently used on most computers, can represent 128 different symbols, using seven bits of information.

increase in proportion to the number of bits representing the character, but no new filters are needed by the receiver.

Because the sender and receiver must be using the same method and the same character codes in order to communicate successfully, standard coding methods are extremely important for communications systems. The **International Morse code** standard evolved from the code first used by Morse in the United States in the early 1840s. It consists of 50 codes for the uppercase Roman alphabet characters, numbers, punctuation, and control codes. The five-bit **Baudot code** used in Tables 7.5, 7.6, and 7.7 was developed in 1875 for international communications and was widely used for more than 100 years.

In 1963, a new seven-bit code was defined for computer communications systems. It increased the number of symbols from 31 to 127 and explicitly represented lowercase characters. The **American Standard Code for Information Interchange (ASCII)**, which is currently used on most computer systems, is shown in Table 7.9. From this table, the seven-bit code for the letter *A* is 1000001, and the code for *a* is 1100001. Codes below 0100000 are used to control communication connections and printing. Only selected codes are shown in this table. For example, BS represents backspace, FF represents form feed or new page, and EOT represents end of transmission.

Although the Roman character set was historically important in the development of computers, the 127 character codes are insufficient to represent the characters used in the various languages throughout the world. For effective global communication, a new standard was needed. The **Unicode** is a standard 16-bit code adopted by many communications and computer equipment manufacturers. In version 3.0, 57,709 of the 65,535 possible codes have been assigned, which includes 49,194 codes representing characters and symbols from languages worldwide.

Table 7.9 Seven-Bit ASCII Character Codes

	000	001	010	011	100	101	110	111
0000–	NUL				EOT			BEL
0001–	BS	HT	LF	VT	FF	CR		
0010–								
0011–				ESC				
0100–	SP	!	"	#	$	%	&	'
0101–	(	)	*	+	,	–	.	/
0110–	0	1	2	3	4	5	6	7
0111–	8	9	:	;	<	=	>	?
1000–	@	A	B	C	D	E	F	G
1001–	H	I	J	K	L	M	N	O
1010–	P	Q	R	S	T	U	V	W
1011–	X	Y	Z	[	\	]	^	_
1100–	'	a	b	c	d	e	f	g
1101–	h	i	j	k	l	m	n	o
1110–	p	q	r	s	t	u	v	w
1111–	x	y	z	{	|	}	~	DEL

Unicode: A 16-bit code that can represent 65, 535 different symbols from international and historical alphabets.

Binary Data Streams

How would we modify a communications system if the objective were not the transmission of a single text message from one human to another, but rather the continuous transmission of a very long stream of binary data representing images, sound, or other nontext data?

A simple, and perhaps obvious, answer is to use one of the communication methods already described, but with the minor modification that the incoming bits are grouped together into sets, and then each set of bits is sent as a "character." The character sequence would have no meaning to us as text, but it would represent the complete bitstream for the audio or video. At the receiver, each transmitted character is determined, and then the bits defining that character are added to the data stream that is passed on to the consumer of the data at the destination.

For example, Table 7.10 is a typical table for the single-tone-per-character method. At the receiver, the presence of each pulse is detected,

INTERESTING FACT:

Even though the ASCII codes require only seven bits for each character, eight bits are often used. This eighth bit, called the *parity bit,* is selected to guarantee that the number of ones received is always either even or odd. This condition provides the receiver with the ability to detect that an error has been made during the transmission.

Table 7.10 A Table for Translating Sets of Binary Bits into a Tone Burst's Frequency

Binary Pattern	Frequency in Hz	Binary Pattern	Frequency in Hz	Binary Pattern	Frequency in Hz
00000	300	01011	1400	10110	2500
00001	400	01100	1500	10111	2600
00010	500	01101	1600	11000	2700
00011	600	01110	1700	11001	2800
00100	700	01111	1800	11010	2900
00101	800	10000	1900	11011	3000
00110	900	10001	2000	11100	3100
00111	1000	10010	2100	11101	3200
01000	1100	10011	2200	11110	3300
01001	1200	10100	2300	11111	3400
01010	1300	10101	2400		

KEY CONCEPT

If more than two different symbols can be sent by the transmitter and distinguished by the receiver, then the data rate can be increased by having each symbol "carry" more than one bit.

the tone frequency is determined, and the corresponding set of bits is sent on to the user. For example, if the binary sequence 0101000100 were transmitted, it would first be divided into the two five-bit sequences of 01010 and 00100. From Table 7.10, the frequency for the tone burst for the first sequence is determined to be 1300 Hz. After the signal at 1300 Hz is transmitted, the frequency for the second five-bit sequence is found to be 700 Hz and then is sent. At the receiver, the 1300-Hz signal is interpreted as 01010. When the second tone burst at 700 Hz arrives, it is interpreted as 00100, and this set of five bits is appended to the first set of five bits to re-create the original 10-bit binary sequence of data.

This scheme can be extended to handle longer and longer streams of bits, carried five bits at a time on each single-tone symbol.

Transmission Rate: The binary transmission rate of a communications system is given by the product of the number of symbols sent per second and the number of bits carried on each symbol.

EXAMPLE 7.4 Determining Binary Transmission Rate

A common radio telegraph system uses FSK transmission and sends 100 symbols per second. What is the binary **transmission rate** R?

Solution

The transmission rate, the rate at which a communications system can send information, is measured in bits per second. It is computed as

$$R \text{ (bits per second)} = r \text{ (symbols per second)} \times d \text{ (bits per symbol)} \quad (7.3)$$

Here, $r = 100$, from the statement of the problem and $d = 1$, because FSK uses only two tones to specify the value of only one bit. Thus we have

$$R \text{ (bits per second)} = 100 \text{ (symbols per second)} \times 1 \text{ (bit per symbol)} = 100 \text{ bits per second}$$

If all methods use the same time duration for tone bursts, then the single-tone-per-character method will transmit the binary data five times faster than the serial binary method, since five bits are carried with each tone burst. However, the serial binary method would seem to be a more natural match for a stream of binary input data, because, as discussed in the previous section, that method can send bitstreams longer than five bits without requiring significant modifications.

Now, faced with the objective of transmitting large streams of data via already existing character-oriented communications systems, we need to divide our bitstream into character-sized patterns and transmit the data as if they were a text message, regardless of the original source of the bitstream. In this case, we are converting binary data into equivalent text messages in order to use available transmitters and receivers. This concept is the basis of modern high-speed data communications. It works because the communications systems are designed to accurately transmit one of a fixed number of symbols during a specified time interval, but it does not matter to the communications system whether we interpret that bit pattern as a textual character, a number, or part of an image.

Figure 7.16 illustrates this process graphically. On the left-hand side are two different types of inputs to a data communications system; on the right-hand side are two different ways that the data are commonly conveyed over the transmission channel. On the left-hand side, the data to be transmitted are often provided as a file of characters—the letters in the English

Infinity Project Experiment: How a Facsimile Works

The function of a facsimile machine is simple: Take a black-and-white image, scan it pixel by pixel, and transmit a one or a zero for each pixel, depending on its color. A demonstration of the fax machine's operation shows how the transmission works. Verify that the transmitter works by watching the scanning window and observing that a tone burst is created whenever a dark pixel is exposed. Use the sliders to increase the scanning rate, and observe the increase in transmission rate. Substitute your own binary image as a file, and listen to it as it is transmitted. Can you deduce any characteristics of the image by how the output sounds?

language, for example. In other cases, they are presented as pure binary data—a long list of ones and zeros. On the right-hand side, some communications channels permit the transmission of symbols that can have many values, such as tone bursts with many different frequencies, while others can reliably convey only ones and zeros. An example of the latter is the FSK transmission scheme discussed in Section 7.4. Another important example is an optical fiber that carries only the presence or absence of light.

Figure 7.16 illustrates that either type of source on the left can be used with either type of transmission system on the right, if the designer makes the proper adjustments. Character-oriented data, which are destined to be carried over a binary transmission system like the Internet, must be decomposed into bits before transmission. The serial binary method could do this task. Conversely, a transmission system capable of carrying multivalued symbols, such as tone bursts with multiple frequencies, can be used to carry native binary data by grouping bits together before transmission. The incentive is that all the bits in each group are carried on a single symbol, making this scheme much faster than ones in which only one bit is carried on each symbol. For example, if a symbol can have 64 possible values, then it can "carry" six bits and thereby transport those bits six times faster than it could have if each symbol had required its own two-valued binary pulse.

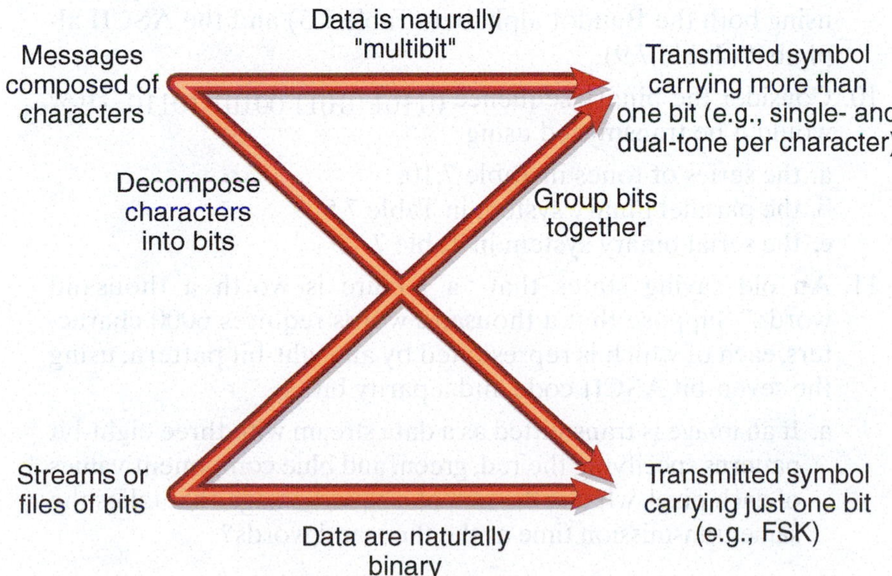

Figure 7.16 Interrelationships between the nature of the data to be transmitted and the type of transmissions used.

EXERCISES 7.5

Mastering the Concepts

1. How many bits are used by the American Standard Code for Information Interchange (ASCII) to represent each character? Why are that many needed?
2. What is the ASCII code for the letter *B*?
3. Use Table 7.9 to determine the letter corresponding to the binary pattern 1011010.
4. How many more bits are needed to specify an ASCII character than a Baudot character? How many more characters could be specified?

Try This

5. Assuming that the burst lengths are the same, how much longer does it take to send a serial binary representation of an ASCII character than a parallel binary version?
6. Suppose that we want to transmit all possible ASCII characters by using the single-tone-per-character method. How many distinct tone frequencies will we need?
7. Suppose we want to transmit a very long stream of binary data by using the single-tone-per-character technique of Table 7.10 to map any pattern of binary bits into characters.
   a. What is the tone sequence generated for the input data 00100 01110 10110 10111?
   b. What is the bit sequence corresponding to the received tone sequence 700 Hz, 300 Hz, 1200 Hz, 1700 Hz, 2900 Hz, 3400 Hz?
   c. If the tone bursts are sent out at a rate of 100 bursts, or pulses, per second, what is the transmission rate of the system in units of bits per second?
8. A dial-up modem, popular with PC users in the late 1980s, sent 600 symbols per second and could carry four bits per symbol. What was the binary transmission rate for this modem?
9. Translate the phrase QUICK BROWN FOX into binary form, using both the Baudot alphabet (Table 7.5) and the ASCII alphabet (Table 7.9).
10. Consider the binary sequence 01101 01011 00101 10110. How would it be transmitted using
    a. the series of tones in Table 7.10;
    b. the parallel binary system in Table 7.5;
    c. the serial binary system in Table 7.7?
11. An old saying states that "a picture is worth a thousand words." Suppose that a thousand words requires 6000 characters, each of which is represented by an eight-bit pattern, using the seven-bit ASCII code and a parity bit.
    a. If an image is transmitted as a data stream with three eight-bit patterns specifying the red, green, and blue component values of each pixel, what is the size of a square image that takes the same transmission time as the thousand words?

b. Suppose that digitized speech is transmitted as a data stream with eight-bit values representing the amplitude of each data sample. If the speech is sampled at 8000 samples per second, what is the duration of a speech segment that takes the same transmission time as the thousand words?

In the Laboratory

12. How is the size of a digital image, in pixel width and pixel height, related to the time it takes to transmit it? What are the trade-offs involved in sending higher quality faxes (i.e., more pixels per linear dimension)?

Back of the Envelope

13. In what situation might the Baudot representation have an advantage over the ASCII or Unicode representations?
14. Access information about Unicode through its Web site (www.unicode.com), and find the codes for two languages that do not use the same characters as English. How many codes are used for each character set?
15. In the ASCII character set, what is the difference between the code for an *A* and for an *a*? Between that for a *B* and for a *b*? Find a rule that determines whether a code represents an uppercase character or a lowercase character.
16. Morse code uses a variable-length code, with short-duration codes corresponding to common letters. Suppose that the following proportions specify the time intervals for Morse code:
    - The dash time is equal to three dot times.
    - The time between dots and dashes within a character is equal to one dot time.
    - The time between characters is equal to three dot times.
    - The time between words is equal to five dot times.

    Pick a two- or three-sentence passage from a book, and compute the number of dot times needed to send it via Morse code. If the same message is sent in the same amount of time using eight-bit ASCII codes, how does the amount of time required for the transmission of one bit compare with the dot time? (Remember that ASCII has an eight-bit code for the space between words, so that every character and space takes eight-bit times.)

INTERESTING FACT: The first electrical and optical communications systems were designed to carry text messages. The most reliable ones actually broke the text into binary ones and zeros. From this basis, it was easy to extend these communications systems to carry black-and-white image pixels as well. Shortly after the telegraph was introduced in the early 1840s, the first *facsimile machine* was patented. It mechanically scanned an image line by line, transmitting current for each "dark" pixel and opening the circuit for each "white" pixel. Today, fax machines are used to communicate everything from important business contracts to an overseas letters from family and friends.

Figure 7.17 A microwave radio relay tower and its reflection in a lake.

7.6 Other Transmission Channels

The communications system used as an example in this chapter sends the characters of a message as discrete tone bursts through the air (the "acoustic channel") to the intended receiver. There are many variations on this idea, all of which conform to Figure 7.1. Our first example used a single frequency to represent an individual letter; we then branched out to try other approaches. In practice, many combinations of frequencies, amplitudes, and phases are used in the quest to transmit multimedia data faster and faster.

Similarly, practical communications systems use other channels for carrying their signals. So far, we have focused on sending our signals as sound waves through the air. This approach resembles the way that humans communicate, but in reality it is not the fastest way to send information. Most practical communications systems send their signals as electrical waves through wire, as electromagnetic (radio) waves through the air or space (Figure 7.17), or as higher frequency electromagnetic waves that we call *light*. These light waves can be conveyed through air or space (like the infrared signal from your TV remote control) or guided through optical fibers from the transmitting laser to the receiving detector located miles away (as in telephone transmissions).

If we examine these different kinds of transmission channels in more detail, we will find that, in all cases, the fundamental concepts are the same. The input data are applied to the transmitter, which is designed to best convey those data over the transmission channel that has been chosen by the design engineer. The receiver collects the signal and recovers the transmitted data as best it can in the presence of noise and interference.

Master Design Problem

Let's consider a different design for a communications system. The communications systems developed in this chapter are based on sending signals representing characters from an alphabet of 26 to 256 possible characters, using five to eight bits per character. Section 7.5 extends the design by considering the Unicode, with 16-bit characters, and considers methods of sending binary data streams of audio or video, for example, using character-based systems.

Consider an alternative communications system design approach based on words rather than characters. This approach is analogous to a written language like Chinese, which uses many complex characters, each of which usually represents a word. Although there are potentially an infinite number of words in a language, since new words are created every day, we can place some limits on how many words our system will represent. According to popular wisdom, a vocabulary of 500 words is the minimum needed to communicate at a basic level when learning a foreign language. A typical college graduate has a working vocabulary of about 3000 words. We will start with a specification that our system will represent 1024 words. We will use a fixed-length code that has the same number of bits for each word, and the bit patterns will be sent like the data-stream methods of Section 7.5.

- Determine how many bits will be needed to represent 1000 words and 20 special codes. Compare the average time to send a word by using this system with the average time to send a word by using seven bits per character, with an average of five characters per word. (Remember that we need to represent the space between words, so we actually need to send six characters per word.)
- Consider how the 1024 words might be chosen and how the bit patterns might be assigned.
- Consider how a user might enter a message via this system. Would a keyboard with 1024 word keys be easy to use? Modify the block diagram of Figure 7.1 to show more explicitly the data encoding and decoding at the transmitter and receiver. Could the system accept normal keyboard entry and display normal characters, but still do transmission based on words? How? What blocks would need to be added? What purpose would a spell checker serve? What prearrangement would be needed between the sender and the receiver?
- Now consider what has been omitted.
  1. Would we normally need to represent the space between words explicitly if each code represents a word? How would we send numbers, punctuation, and other special symbols? How many of our 1024 codes would be needed for these cases?
  2. What would we do for words that are not on the list of 1024 words? Should we dedicate 27 of the word codes to individual letters in order to allow the spelling of words not on the list? (This is how American Sign Language handles proper names and words for which there is no standard sign.)
- Review your completed design, and then make a detailed block diagram for it.
- Consider some additional benefits of this design. For example, consider implementing a system that transmits data in English, but has to convert the data to German at the receiver. What would the translating block have to do for the system sending English words one letter at a time? What would it have to do for the new system, which sends data one word at a time?
- Now reconsider the three initial basic specifications:
  1. We specified that the system represents 1024 words. What is the impact on the design and the performance of the system if we change this requirement to 4096 words?
  2. We specified that all words would have the same number of bits. How would it impact the system's design and performance if we were to assign a smaller number of bits to the most common words and a longer bit pattern to the less commonly used words? Carefully consider how the receiver would be able to tell whether a pattern is a short-bit pattern or a long-bit pattern. How would you decide which words should have short code? Would this decision depend on the type of messages being sent?
  3. We specified that the system would be based on words, but then found it necessary to reserve some of the words to be letters, numbers, punctuation, and special symbols. Would it make sense to replace some of our word codes with codes for common phrases? Explore how you would decide what phrases to use. Would the most common phrases used be different for you and your friends than they would be for your grandparents?

Big Ideas

Math and Science Concepts Learned

In this chapter, we defined the three basic components of any communications system: the transmitter, the channel, and the receiver. As we developed a simple acoustic system for text communication to illustrate the operation of each component, we learned the following:

- It is possible to reliably move text information, that is, communicate the information between two different locations.
- The transmitter is designed to put the data into a form so that the electronic receiver can distinguish it from all other patterns of data.
- The receiver must know the characteristics of the transmitted signal so that it can do its job properly.
- Noise, interference, weak signals, and other factors can degrade the received signal to the point that it can no longer be recovered accurately by the receiver.

With these basic points established, we extended them in two ways:

- We found alternatives to the single-tone-per-character scheme and determined that some carry data faster than others, some are more tolerant of noise and interference than others, and some are cheaper than others. Engineers evaluate alternative schemes and their capabilities to find the best one for each particular application.
- Many communications systems need to send very long streams of binary data, such as digitized voice and images, rather than just short messages using the Roman alphabet. We found, however, that this binary data can be sent as a sequence of "characters" or symbols by grouping the bits together into bundles, and then mapping those bundles into characters that the transmitter is capable of sending and the receiver is capable of accurately detecting.

In the next chapter, we turn our attention to using the communications systems we've examined here to build networks that efficiently can provide multimedia telecommunications services to many users.

Important Equations

The general mathematical representation of a tone is given by Equation (7.1):

$$s(t) = a \times \cos(2\pi f t + \phi)$$

Transmission speed, the rate at which a communication system can send information, is measured in bits per second. It is computed by:

$$R(\text{bits per second}) = r(\text{symbols per second}) \times d(\text{bits per symbol})$$

Building Your Knowledge Library

Matlin, Margaret W. and Foley, Hugh J., *Sensation and Perception*. Allyn and Bacon, Needham Heights, Massachusetts, 1997.

This is one of many text books which describe human vision and hearing.

Lebow, *Information Highways and Byways*. IEEE Press, 1995.

This is a very readable description of the development of modern communications systems and many of the individuals who contributed to that development. It describes the economic implications and constraints and the social impact of the changes resulting from innovations in communications.

Standage, Tom, *The Victorian Internet.* Walker and Company, New York, 1998.

The Victorian Internet tells the story of the telegraph's creation and remarkable impact and of the visionaries, oddballs, and eccentrics who pioneered it. The book shows that the excitement about the telegraph in the late 1900s was very similar to that associated with the Internet.

PONY EXPRE

chapter 8

Networks from the Telegraph to the Internet

In Chapter 7, we learned how to build reliable signal transmission systems that can carry multimedia data from one location to another. We now want to figure out the best way to use such transmission systems to interconnect many users at the same time and to do so as inexpensively as possible. We will discover in this chapter that in order to accomplish this goal, we need to abandon our simple broadcast and point-to-point concepts. In their place, we will design an interconnected set of transmission links called a network.

OUTLINE

Network: A group of interconnected individuals. In the world of telecommunications, a network is a set of communications equipment and relays required in order to permit communications between individuals or devices.

INTERESTING FACT:

In 2002, it cost approximately $1 billion to build a fiber optic network large enough to serve the continental United States.

INTERESTING FACT:

The Internet is based on the same principles as those of the telegraph networks of 1850. However, the Internet uses electronics instead of humans to send messages and operates about a billion times faster.

Building Networks

Before we begin our investigation of **networks**, let's once again define our objectives from the perspective of engineering design:

- **What problem are we trying to solve?** Just as with our simple communications systems (Chapter 7), we want to move information from one location to another. As before, we would like to move it as quickly as possible, receive it as accurately as possible, and accomplish this task as cheaply as possible. The introduction of our new network is intended to reduce costs as much as possible while offering communications services to many users instead of just a few.
- **How do we state the underlying engineering design problem?** The potential user of a communications network will have the same objectives as he or she would for a simple communications link. These objectives typically include speed, accuracy, and cost, but may also include other factors, such as the delay between transmission and reception, the security of the communications system, and its ease of use.
- **What are the rewards if the designed communications system satisfies all of the user's needs?** Engineering design is often driven by economic objectives. This condition holds true especially in the development of communications networks, since the purpose of a network is to minimize the cost to each of the connected users.
- **How will you test your design?** From the user's perspective, the services offered by a network are the same as those that would be provided by separate transmission links, except that the former are cheaper and have a broader reach to other potential users. As a result, the list of specifications for a network closely resembles that for a simple communications system. If those specifications are met, then the happy customer should pay for use of the network. In the case of multimedia communications systems, whether simple or a network, the parameters needing to be tested in order to verify that the specifications have been met will typically include some of the following:

    1. *Peak data rate*—the maximum rate at which a short message can be sent.
    2. *Average data rate*—the sustained rate at which a long message can be sent.
    3. *Maximum error rate*—the maximum number of errors that can be tolerated as a fraction of the total number of bits transmitted.
    4. *Error-free seconds*—the number of seconds for which no data are received with an error.
    5. *Delay*—the maximum and minimum delay between transmission and successful reception.

    For a network, it is also common to add requirements that specify how many other users can be reached, with what difficulty, and at what cost.
- **What do we need to learn in order to understand and design communications networks?** In this chapter, we first examine the economic rationale for developing a network. Then we develop several approaches for sharing the transmission resources on which a network depends. Based on these ideas, we'll explore the first important data communications system, the telegraph, and then we'll move up to the Internet.

8.1 What is a Network?

A network, by its dictionary definition, is a group of interconnected individual elements. When used in the communications business, the word "network" means an arrangement that permits any of many users to communicate with any of the other users, typically through a set of interconnected transmission links.

How is a network constructed? How would the design differ if the objective were to carry electronic messages rather than voice messages? Further, how could we minimize the cost of service to each of the network users? To answer these questions, we need to reexamine the types of communications systems we've discussed in previous chapters and then determine how they need to be adapted for operation as a network.

INTERESTING FACT:

A single television transmitter in New York City can broadcast to more than three million television receivers.

Network Basics

The communications systems we looked at in Chapter 7 fall into one of two categories: broadcast or point to point. **Broadcast systems** use one transmitter to send information to many receivers, as shown in Figure 8.1. These systems traditionally use radio rather than wire transmission, and, in most cases, the users who receive the signal do not send any information back to the transmitter. Obvious examples are television and FM radio broadcast systems. A less obvious example is the Global Positioning System (GPS), whose satellites transmit digital information constantly to any and all GPS receivers, so that they can determine their locations. Cable television, as originally implemented, also falls into this category, although it uses wired connections instead of radio transmission links.

Broadcast Systems: Communications systems that use one transmitter to send information to many receivers.

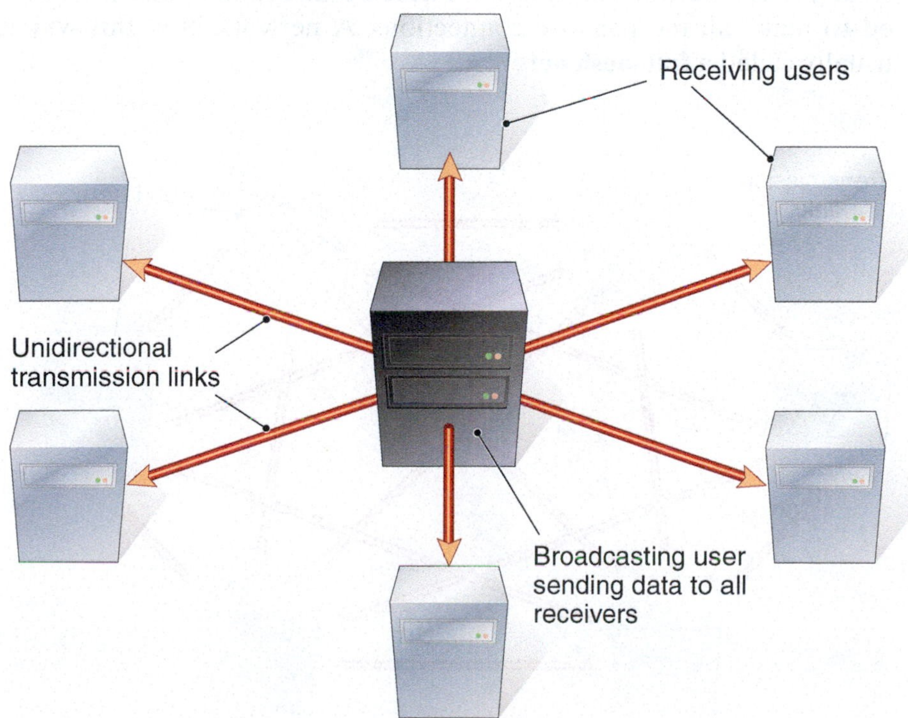

Figure 8.1 A communications network in which one user "broadcasts" to six others, using transmission links that send in only one direction.

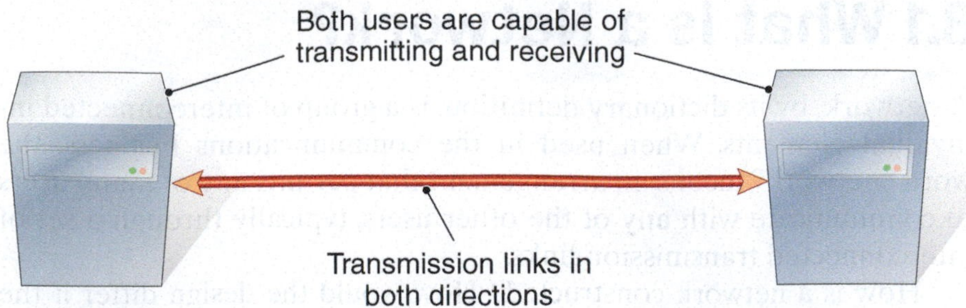

Figure 8.2 A point-to-point communications system in which both users are able to send information to each other.

Point-to-Point Systems: Systems in which two users are connected directly to each other by a wire or radio transmission system.

Full Mesh (fully meshed) Network: A network in which all of the users are directly connected to one another with transmission links.

The second category is **point-to-point systems,** in which two users are connected directly to one another, using a radio or wire transmission system. This type of system is shown in a simple form in Figure 8.2. The presence of arrows on both ends of the link connecting the two users indicates the ability to transmit in both directions. While there are exceptions, point-to-point systems usually permit communication in both directions, allowing the two users to respond to each other's messages and to start communications whenever they are ready.

How would these two approaches be used to build a communications network? The broadcast approach is very difficult to adapt, as it assumes implicitly that one user does the talking and all others do the listening. We will assume, for our design purposes, that we would like all of the network users to be able to both talk and listen. With this assumption, the point-to-point approach can be used to meet our objective. We simply construct point-to-point transmission links between every pair of users. Such an arrangement is shown in Figure 8.3. This structure is a network of six users. Fifteen bidirectional links are needed to make all the pairwise connections. A network built this way is usually called a **full mesh network.**

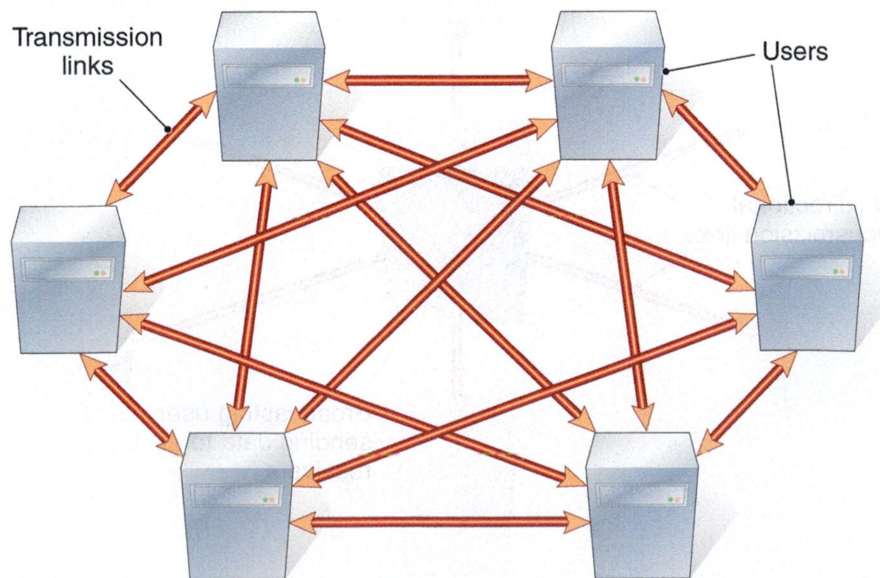

Figure 8.3 A first attempt in the development of a communications network involves direct interconnection between all pairs of users into a "full mesh".

Infinity Project Experiment: Multiple-User Network with Meshed Connections

In the laboratory, we can set up a network that simulates the characteristics of a multiple-user network in which each user has a direct link to every other user. How easy is it to establish this network? Once you have the network up and running, how simple is it to transmit information to other users?

With this network design, we have achieved our first system specification goal. Any user can send data to any other user by simply picking the transmission link that connects the two and then sending the bits comprising the data. However, as we'll see shortly, this full mesh network is an expensive way to meet our objective; we might want to refine our specifications and consider other alternatives. For this approach, we must build two transmission links between each pair of users, because we need one for each direction of transmission. Since each link is used only for traffic between the connected users, most of the links are likely to be idle for much of the time. Our new objective is to figure out how we can still connect any pair of users, but do it less expensively by making more efficient use of transmission resources.

Reducing the Cost of a Network

As an alternative to the network design shown in Figure 8.3, consider the approach shown in Figure 8.4. In this case, there is only one two-way link to each user. Instead of connecting to some other user, however, the link from each user connects to a central **relay point**. This type of network is often called a **star configuration**. The relay's role is to accept messages or bits from any user and "relay" them on to the intended destination. If the relay is capable of performing this function, and if it can perform at the rate required by all of the users, then this network should still be able to meet its primary objective—the delivery of information from any user to any other user. For this design, six bidirectional links are needed for the six users.

Relay Point: A person or device capable of receiving, directing, and transmitting a signal.

Star Configuration: A network design in which all of the users are connected directly to a single relay point, using individual transmission links.

What would make this approach more attractive when compared with the network of direct user-to-user links shown in Figure 8.3? The simple answer is that, in most practical situations, even though the relay might be expensive, the whole network costs less. To understand this concept,

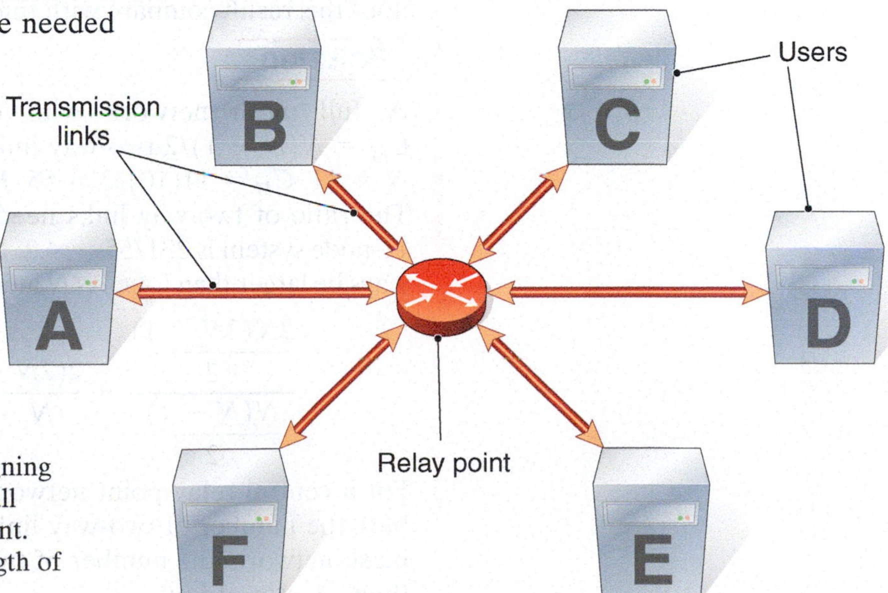

Figure 8.4 Our second attempt in designing a network involves two-way links from all users joining at a single central relay point. This design reduces the number and length of transmission links.

INTERESTING FACT:

While a star network might cost less than a fully meshed network, it may perform poorly if the relay is not fast enough to handle all of the messages sent by the users of the network.

INTERESTING FACT:

With this full mesh design of a multiple-user network, for N users, it would take $N(N-1)/2$ links. So, to connect 1000 users, it would take roughly 500,000 links—probably far too expensive to build!

assume, for the moment, that the relay point costs nothing and that the cost of each transmission link is proportional to the link's length. By simply counting the number of links and the total length of the links in both Figures 8.3 and 8.4, you can see that the network using the single relay is a lot less expensive to build than the one that relies completely on point-to-point links. If the actual cost of the relay is less than this difference, then the star configuration is the cheaper of the two.

The relative advantage of a network designed with a relay point increases as the number of users grows. To find the number of links needed for a fully connected network (as in Figure 8.3) with N users, we need to count the number of possible combinations of N users taken two at a time to form a link. A little bit of mathematics shows that the number of two-way links for the full mesh network, C_M, is given as

$$C_M = \frac{N(N-1)}{2} \tag{8.1}$$

For example, a fully connected network of 1000 users would require 499,500 links. In contrast, if the central relay-based scheme is used for N users, then the number relay-based links, C_R, is given as

$$C_R = N \tag{8.2}$$

because each of the N users is directly connected to the central relay. As the number N grows very large, the relay scheme's advantage is $C_M/C_R = (N-1)/2$. For large N, this expression is approximately $N/2$. The ratio of costs, which depends on the total length of the links, goes up similarly. Consider, for example, the relative advantage for a telephone network if there are roughly 1 billion telephones in the world! Even for modest-sized networks, we can reasonably conclude that the relay scheme is a cheaper way to build a network than the fully connected, or full mesh, method.

EXAMPLE 8.1 Comparison of a Full Mesh Network and a Central Relay-Point Network

A full mesh network with 11 nodes must double its number of nodes to 22. By what factor is the number of two-way links increased? How does this result compare with that for a central relay-point network?

Solution

A full mesh network with C_M links and N users requires $C_M = N(N-1)/2$ two-way links, as defined in Equation (8.1). For $N = 11$, $C_M = 11(10)/2 = 55$. For $N = 22$, $C_M = 22(21)/2 = 231$. The ratio of two-way links needed for a 22-node system versus an 11-node system is $231/55 = 4.2$. Note that, in general, this ratio will always be larger than four. As N increases, the ratio will get closer to four:

$$\frac{\dfrac{2N(2N-1)}{2}}{\dfrac{N(N-1)}{2}} = \frac{2(2N-1)}{N-1} > \frac{2(2N-2)}{N-1} = 4$$

For a central relay-point network, when the number of nodes doubles, the number of two-way links, C_R, also doubles, but for the full mesh network, the number of two-way links, C_M, increases by more than a factor of four.

Infinity Project Experiment: Multiple-User Network Using a Single Router

A router is often added to the design of a data network in order to lower the cost of connecting all of the network's users. These types of networks don't perform as well, but they are cheaper to implement. In the laboratory, we can set up a simulation of a multiple-user network in which a single node acts as a relay. How easy is it to establish this network? Once you have the network up and running, how simple is it to transmit information to other users? How much work does the router have to do compared with that of the individual transmitters? How does this network's speed compare with that of the full mesh network?

The relay method will be less expensive, but we should still explore ways to improve it so that the central relay does not become a bottleneck and so that costs can be further reduced. Figure 8.5 shows a network with more than one relay. Specifically, each user is connected to a relay point, and those relays are connected to others. Is this configuration cheaper? If the total length of the transmission links is still the indicator of total network cost, as it is in many cases, and if the relays can be trained to deliver the users' messages properly to their intended recipients, then this network is clearly cheaper than either of the previous two we examined. In addition, not all of the messages have to go through R4 as they would through the central relay of Figure 8.4.

Although this multirelay network is cheaper in terms of transmission links, we achieved these savings only by requiring that the relay points be a little "clever." An example will help illustrate this point.

Our objective is to be able to send messages from any user to any other user. Suppose we would like to send a message from user A to user B in Figure 8.5. The two users, A and B, use the leftmost relay, marked

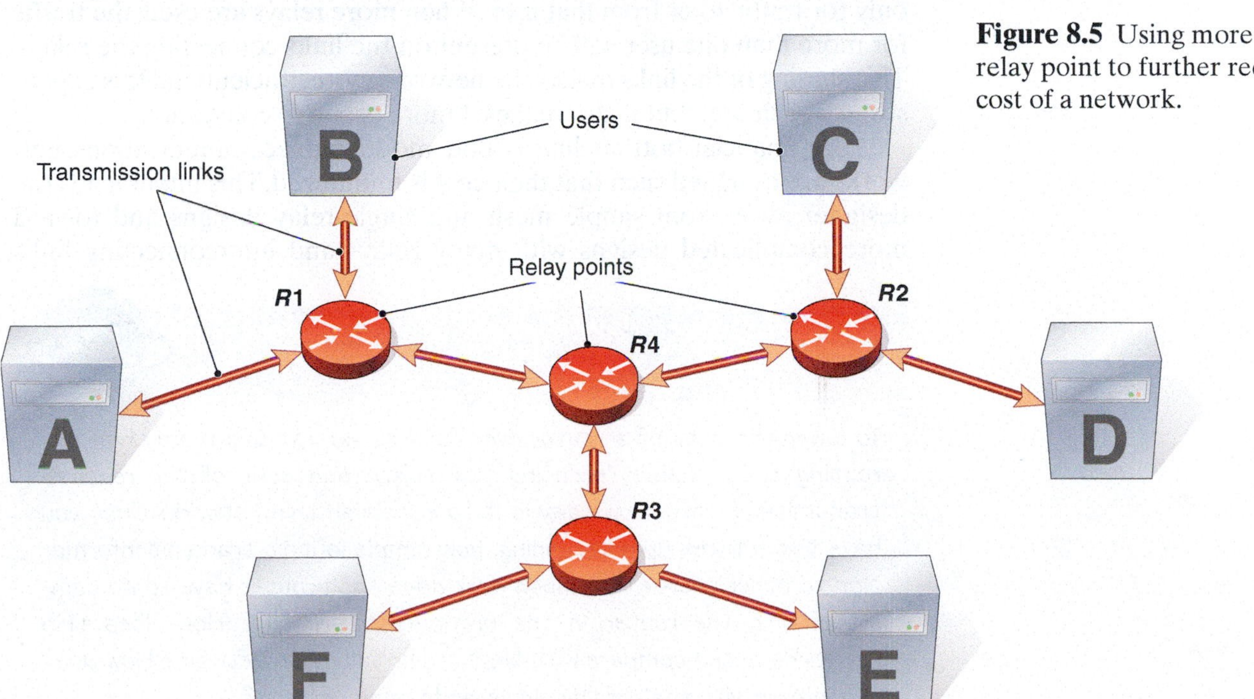

Figure 8.5 Using more than one relay point to further reduce the cost of a network.

INTERESTING FACT:

The cable for a modern fiber optic network costs about $25,000 per mile. It also costs a considerable amount of money to install, operate, and maintain this cable.

R1. A message from A intended for B is sent to relay R1. The relay then forwards the message to user B. In this case, the relay does not need to do anymore than it did in the one-relay design of Figure 8.4.

Suppose now that we want to send a message from user A to user D in Figure 8.5. User D is not connected directly to relay point R1, so the operation of relay R1 must become more complicated. In this case, relay R1 must decide which link it should use to forward the message to another relay point rather than directly to the intended recipient. User D is served by relay point R2. If user A's relay point, R1, sends the message over its link toward relay R2, then relay R2 can deliver the message to user D once it receives it. If this "forwarding" mechanism is available at all of the relays, then it should be clear that any user can send messages to any other user. In the case of this example, R1 must relay the message to R4, which then sends the message on to R2 for final delivery to user D.

Since the multirelay method reduces the total cost of the expensive transmission links, we should then consider other costs of the system and determine whether it will always be the least expensive network solution. If transmission costs are the dominant expense, then it probably will be. In general, however, there is no simple answer, because the best design depends on many factors. Specific applications may differ greatly from each other in terms of the number and location of the users, the amount of message traffic to be sent between each pair, the price of the transmission systems, the cost of the relays, and more.

Local Area Network (LAN): A computer network (or data communications network) that is confined to a limited geographical area.

Despite these many variable factors, several general observations can be made. In geographically diverse networks with many users, it is common to use many relay points to minimize the cost of the system, since that cost tends to be dominated by the costs of transmission. In contrast, in geographically confined networks within a single building, such as **local area networks (LANs)**, the tendency is to use only one relay, as shown in Figure 8.4, since, in this case, the cost of the relay might be more than the cost of the transmission system. In both the full mesh network of Figure 8.3 and the central-relay network of Figure 8.4, the links to each user are used only for traffic to or from that user. When more relays are used, the traffic for more than one user will be present on the links connecting the relays. This sharing of the links makes the network more efficient and less expensive for each user, but it also makes it more prone to congestion.

The financial bottom line is that modern telecommunications networks are designed such that their cost is minimized. This intent leads the designer away from simple mesh and single-relay designs and toward more complicated designs with many relays and interconnecting links

Infinity Project Experiment: Multiple-User Network Using Several Routers

To lower the cost of a router even further, we can use more than one routing node, thereby reducing the length and cost of the required transmission links. How easy is it to establish this network? Once you have the network up and running, how simple is it to transmit information to other users? How much work does each router have to do compared with the router in the previous experiment? How does this network's speed compare with that of the full mesh network? How does it compare with that of the single-node relay network?

that carry the traffic of many users. These principles apply to communications networks like a telephone network and the Internet, as well as to highways, railroads, and airline systems.

EXERCISES 8.1

Mastering the Concepts

1. Classify the following communications systems as one way or two way:
   a. FM radio
   b. cell phones
   c. cable TV
   d. GPS
   e. dispatch system for taxis, limos, or the fire department
   f. walkie-talkies
2. Classify the following communications systems as point to point or broadcast:
   a. FM radio
   b. cell phones
   c. cable TV
   d. GPS
   e. dispatch system for taxis, limos, or the fire department
   f. walkie-talkies
3. List at least five of the costs that should be examined when considering alternative network designs.
4. The usual goal of a network is to connect the users with minimal cost. What are three other goals that might determine which type of network you would choose?

Try This

5. For a full mesh network with eight nodes, how many links are needed? Are the links one way or two way?
6. Compute the number of two-way transmission links needed to build a full mesh communications network connecting 4, 10, 100, and 1000 users, respectively. How many links are needed to connect the same number of users when using the star configuration shown in Figure 8.4?
7. The number of two-way links, L, needed to connect N users in a full mesh network is given by $L = N(N - 1)/2$. Graph this equation for values of N from 1 to 20. What is this type of function called by mathematicians?
8. Consider the construction of a network to serve six users who are distributed evenly about a circle that has a radius of 10 miles. Assume that cable containing a pair of wires is used to make the connections.
   a. How much cable, measured in miles, is needed to build a full mesh network like the one shown in Figure 8.3?
   b. How much cable is required to build the network if we use the star configuration seen in Figure 8.4 and assume that the relay point is located at the exact center of the circle?

c. Now assume that we use the multirelay design shown in Figure 8.5. Also assume that the center relay is located at the center of the circle and that the three other relays are located on a concentric circle with a radius of five miles. How many miles of cable are needed for this network design?
d. Assume that the wire and its installation cost $10,000 per mile and that a relay costs $100,000 to build and install. What is the total cost for each system? Which is the cheapest of the three to build?
e. Assume that the networks and the price of a relay stay the same as in part (d), but that the price of the cable falls to $100 per mile. How does this scenario change the costs of the three approaches? Which scheme is now the cheapest?

9. Suppose that a network is built in the form of the multirelay approach shown in Figure 8.5. Suppose further that every user wants to send a total of 100,000 bits per second to the other five destinations and that this "load" is split evenly, that is, each user sends at the same rate—20,000 bits per second—to each of the other five users. (*Hint*: Do the problem first with only user A transmitting. Then add user B's transmissions. Continue to add the other four users.)
   a. Compute the number of bits that must be carried on each of the transmission systems used in the network.
   b. Compute the number of bits per second that each of the relays must be able to handle in order to allow complete flow of the user's data.
   c. Suppose the central relay is capable of handling only 300,000 bits per second. How will this condition affect the users?
   d. Suppose that the transmission links connecting users with their associated relay can carry only 150,000 bits per second in each direction. How will this condition affect the network's operation?
   e. Suppose instead that the links connecting the central relay with the outer relays can carry only 150,000 bits per second in each direction. What effect does this condition have?
   f. So far, we've assumed that the "load," that is, the amount of data transmitted, is the same from and to all users. How do the answers to parts (a)–(e) change (qualitatively) if the load is not the same for all users?

Back of the Envelope

10. Draw a full mesh network that connects you to four of your close friends. Figure out the rough distances between your locations (homes, seats in the classroom, or similar placements), and calculate the total amount of cable required to build a wired network.
11. Another type of network is called a *ring network*. This network connects each node to its nearest neighbor, and any node has only two links to two other nodes.
    a. Suppose the arrangement of users is similar to that in Figure 8.3. Draw a picture of the ring network for this configuration.

b. How much cable is required in order to link all of the users in a ring network if each pair of linked users is separated by one kilometer (km)? How does this amount compare with the amount of cable required for the network in Figure 8.3? How does it compare with the amount required for the network in Figure 8.4?
c. What is the maximum number of nodes that a message would have to pass through in order to get to its destination? Compare the speed of this network with that of the single-relay network in Figure 8.4.
d. Describe how messages would be routed in this network. What decision does a transmitter of a message have to make? What decisions do each of the nodes have to make when receiving a message?

8.2 Relays

In the previous section, we found that the ability to relay a message from one communications link to another was key to minimizing the cost of building a network. We learned that the relay must be able to accept a message and then send it on toward its intended destination. In this section, we examine how this process might be done and how we may need to modify our communications procedures in order to make it work.

There are many types of relays used in telecommunications systems. We will explore the two most important types in this chapter. We start with a simple, but very important, case—the design of relays to move a message from one user, the originator, to another user at the destination. The relay must be able to accept a message and then send it on toward its intended destination. In order to meet that objective, several requirements must be met.

When a relay receives a message, it must first determine to whom the message is directed. Then the relay must find a way to send the message on a path that will lead to the intended recipient. If the relay is connected directly to the destination, it will send the message directly there. However, if the relay is not connected directly to the destination, the relay must determine the best path available toward that destination and then send the message along that path.

These basic requirements assume that there is no other traffic carried in the network. But there almost always is other traffic, and so additional requirements are needed. When a relay has received a message destined for a user, but the link toward that user is already occupied with another message, the relay should hold onto the message by storing it until the link to the destination is available. When the link does become available, the relay should send or "forward" the stored message.

To make the network reliable in the face of equipment failures and times of heavy traffic load, we will add one more requirement: When the relay has received a message and determined the best path for it, but finds that the route is unavailable, the relay should consider the use of another route, even though the alternative route normally would not be considered the best route to take.

KEY CONCEPT

Switching and Routing: Networks consist of communications links and the relays. When a customer's data signal arrives at a node, a junction of two or more links, it must be relayed from the incoming link to one of the outgoing links. The direction in which it is sent might be arranged before the data are sent, using a technique called *switching*. Alternatively, the direction may be decided when the data packet actually arrives at the node, a technique called *routing*.

INTERESTING FACT:

A good telegrapher could relay Morse code at a rate of 30 words per minute or more. At that rate, the typical message consisting of 1000 letters would take about five minutes to send.

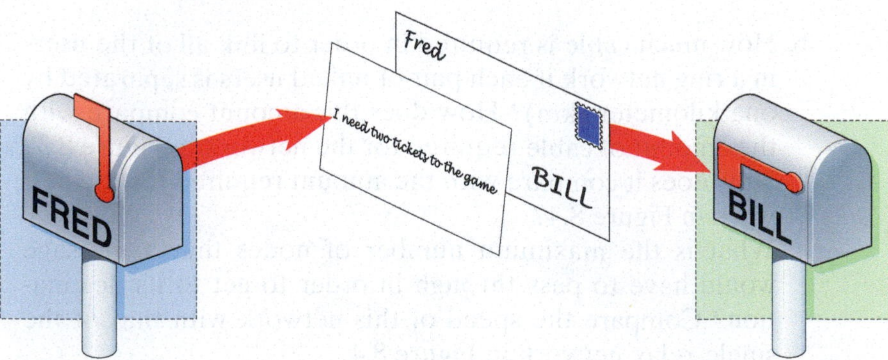

Figure 8.6 A common method of adding address information to a message: "prepending" the destination and source addresses to the beginning of the message.

The first step in this process does not occur in the relay device, but rather back at the origination point. In addition to sending the message, the originator must also send information that identifies the destination of the message. All relays involved in the transmission of the message will use this address information to route the message to its ultimate destination.

There are several ways to send the address, but we will describe the one that is used most frequently. When using this method, the message itself is extended in length by adding two pieces of information: an address, or identification, for the destination, and an address, or identification, for the originator. These pieces of information are usually *prepended* to, or placed in front of, the message text. An example is shown in Figure 8.6.

INTERESTING FACT:

Many believe that data networking is a new business. In fact, the Western Union Company, which grew to be the biggest telegraph company in the United States, was one of the country's largest companies in 1876—the year that the telephone was invented.

When each relay receives this new extended message, it extracts the destination identifier from the front of the message ("BILL") and uses it to determine which way the message should be sent. When it sends out the message, it sends the full, extended message so that any succeeding relays will have all of the address information they need in order to deliver the message.

It is obvious that the destination address must be included with the message, but why send the address of the source? The answer is that once the message finds its way through a network that uses even one relay, the destination user has no way of knowing who the originator was unless he or she is told. It is not possible for the user to respond without this information. Therefore, the source address attached to the message provides this information.

Store-and-Forward Network: A strategy for relaying a message in which a network point, by receiving messages, selects an outgoing transmission link based on the address information carried in the message; stores the message temporarily in a queue, if necessary; and then sends it on to the prescribed link as soon as its turn comes in the queue.

Store-and-Forward Networks

The network structure described in the preceding section is commonly termed a **store-and-forward network**, since each relay point stores each message until it is ready to forward it to either its ultimate destination

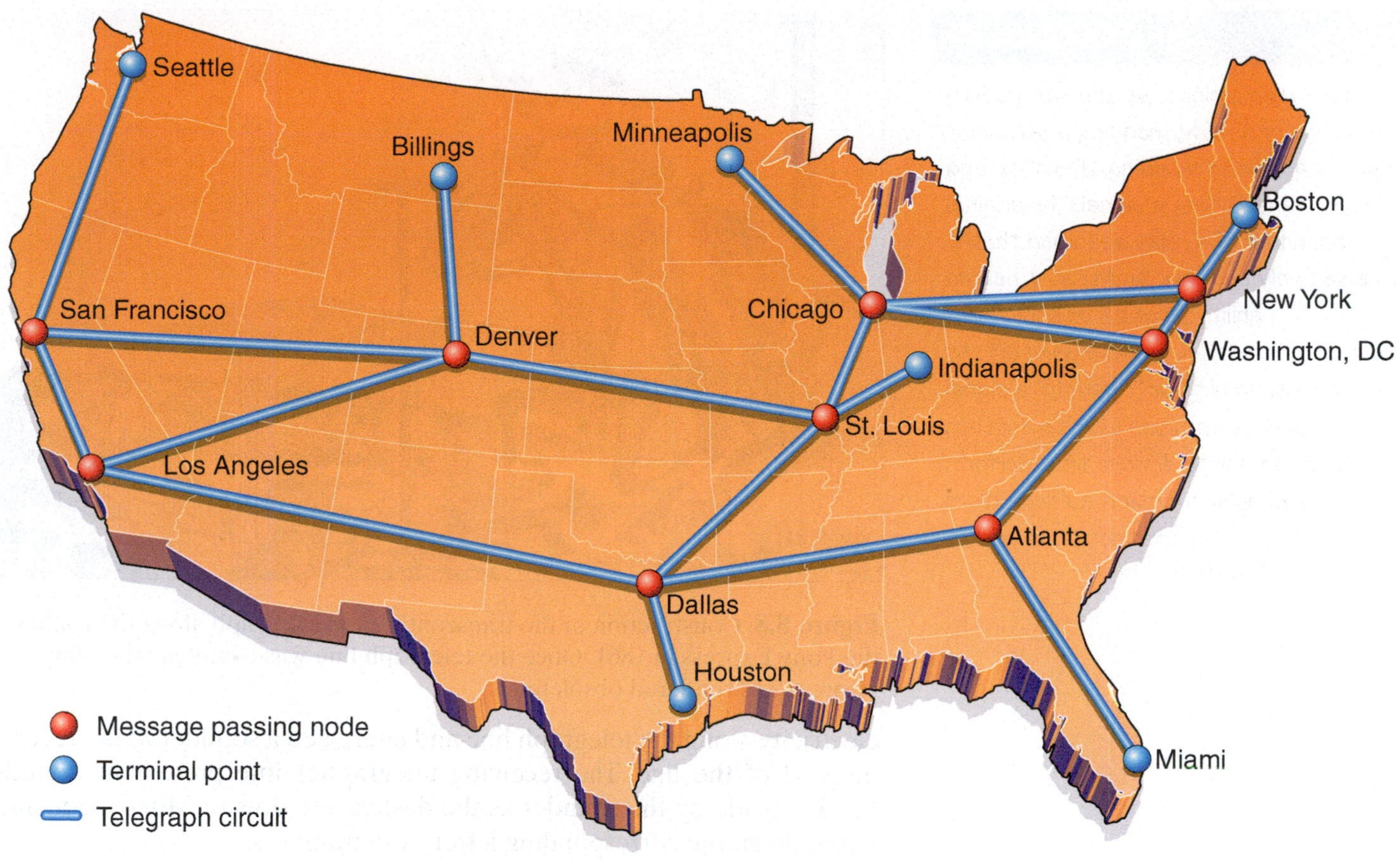

Figure 8.7 The configuration of a simple "store-and-forward" network.

or to another relay point. This method of constructing a communications network is a famous one. The original telegraph networks of the 1800s were built in this way. What might be more surprising is that this method also provides an accurate description of the Internet.

Figure 8.7 shows a simple example of what such a network might look like on a broad geographical scale. The blue dots show terminal points where traffic can be originated or terminated. The red dots are relay points where traffic may be forwarded, originated, or terminated. The lines connecting the relay and terminal points are the transmission links. In the 1800s, these links were telegraph wires (Figure 8.8) carrying messages at about five bits per second. With today's Internet, they are more likely to be fiber optic lines operating at billions of bits per second. The speed and scale have changed, but the concepts have not.

> **INTERESTING FACT:**
>
> Thomas Edison is famous the world over for his inventions of the electric light bulb and the phonograph. Some of his most lucrative inventions, however, were those that multiplexed more than one telegraph signal through a single line, making it possible for the telegraph companies to serve more customers and make more money without adding new lines. Modern telecommunications systems use the same ideas to multiplex many thousands of customers onto the same optical fiber.

How is the Relay Implemented?

In the days of the telegraph, relay points were operated by people. Let's look at what these operators did in some detail so that we can better understand what modern networks do.

At a telegraph relay point, all but one of the workers operated the telegraph lines, sending and receiving messages, using telegraph keys of the type shown in Figure 8.9. The telegraph key controlled the flow of

KEY CONCEPT

In our discussions, we use the modern definition of a *relay*, meaning a person or device capable of receiving, directing, and then retransmitting a signal. The original definition of a relay assumed that it served only two links and that it had no ability to route or schedule a retransmission. Relays were used only to boost, amplify, or "clean up" a signal traveling from one user to another. We now refer to devices that perform this simpler function as *repeaters* rather than relays.

Figure 8.8 Construction of the transcontinental telegraph along the route of the Pony Express in 1861. Once the telegraph line was complete, the Pony Express was rendered obsolete.

electricity along the telegraph line and energized a sounder at the receiving end of the line. The receiving telegrapher interpreted the sounds (clicks) made by the sounder as the dashes and dots of Morse code and wrote down the corresponding letters and numbers.

The remaining person at the telegraph relay point was the dispatcher. This person constantly was making decisions regarding how each received message was to be handled. For each incoming message, the dispatcher first determined whether the message terminated at that office or whether it should be relayed to another office. If the destination of the message was not the local office, the message needed to be routed to another relay point. The dispatcher had to decide where it should be sent next. Usually there was a preferred path to reach a particular

Figure 8.9 A telegraph key used to transmit messages in the 1800s.

destination, but that line might have been down or the queue of messages waiting for that line might have been long. Either of these conditions might have led the dispatcher to select an alternative path to the destination. In more modern systems, the information that the dispatcher uses to make his or her decisions is called the **routing table** and is stored in an electronic form. In the days of the telegraph, however, this information was stored in the dispatcher's personal memory!

Routing Table: A table held by a router that instructs the router as to the best transmission path for sending a message to its ultimate destination.

Finally, the dispatcher had to make a decision about the relative priority of each incoming message. Of all the messages leaving the relay point on a particular transmission link, which one should go first? Messages for which transmission is running late should go before messages that are not late, and messages for which priority treatment had been paid should go earlier than messages without priority. The human dispatcher used his or her knowledge of how the network was configured in order to make these decisions. Since the length of time needed to send out each message was a few minutes, decisions had to be made at a rate of one decision every minute or so.

Router: A network relay point that operates by receiving messages, selecting an outgoing transmission link; storing the message temporarily in a queue, if necessary; and then sending it out via the prescribed link as soon as its turn comes in the queue.

The Internet uses all of the same concepts, but operates at much faster rates. Fiber optics support higher data transmission rates. The human dispatcher is replaced by a special-purpose computer called a **router**, so named because it chooses the route for each of the outgoing messages. A photograph of modern routers from 2002 is shown in Figure 8.10. They were built by Cisco Systems, Inc., and accept messages at data rates of up to 40 billion bits per second. The dispatcher function, choosing routes for the messages, must be done at the rate of almost 1 million decisions per second!

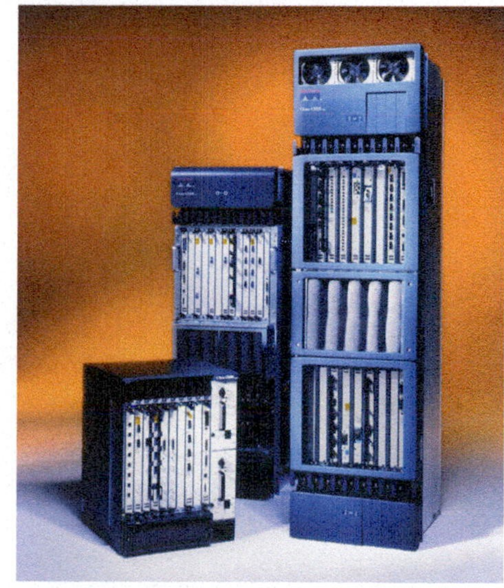

Figure 8.10 Cisco 12016 gigabit Switch Routers, each capable of accepting messages from up to 16 streams, each operating at 2.5 Gigabits per second.

Characteristics of Store-and-Forward Networks

Let's summarize the characteristics of the store-and-forward network structures. First, each message must have its destination address attached so that each dispatcher or router can choose its next transmission link. Although the message has the originator's address attached, the originator has no idea whether a message reaches the intended destination unless a separate message is sent back to acknowledge the receipt. The general strategy applied by the dispatcher or router is to find a path for each message that is likely to result in the least amount of delay in reaching the destination. The time it takes to send a message between the same origination and destination points can be different when attempted at different times of the day. The time it takes will depend on the specific path the message takes and on the amount of time it spends in queues awaiting transmission on each link chosen.

INTERESTING FACT:

Cisco Systems, Inc., started in 1986, is the world's leader in Internet routers.

From an economic perspective, the store-and-forward method of operating a network is very efficient. By allowing messages to be stored in a **queue** at each relay point, the network's operator can keep the transmission links, or "data pipes," full much of the time. Since these transmission links historically have been the most expensive part of the system, using them efficiently means that the whole system operates at its most efficient. This interest in cost efficiency is why both the Internet and the U.S. Postal Service are designed in the same way.

Queue: In the networking field, a queue is a stream of message packets lined up for attention at a server.

Infinity Project Experiment: Multiple-User Network with Choice of Transmission Path

The reliability and throughput of a router-based network can often be improved by adding extra transmission links between the nodes. Set up a multiple-user network for which everyone is served by two different routers. What choices does each transmitter have when sending a message? Can you think of schemes that would increase the performance of this network?

Switching—Another Way to Relay Data

The store-and-forward method is not the only practical way to build a communications network. In some important cases, it is not even the best method.

Instead of sending written messages from one user to another, suppose the objective is to send streams of digitized voice and to use the network to carry telephone calls. When we send a written message through a network, our expectation of service is that we will receive the message with reasonable reliability and within a reasonable amount of time. As an Internet user, we are unsatisfied if we do not receive the message at all, but we accept the fact that the amount of time to get the message to us varies widely.

With a telephone, however, our expectations are very different. When we speak, we expect to be heard immediately and completely. When the person we are talking to speaks, we expect to hear that person immediately, and we expect to hear everything that is said. Delays destroy the rhythm of an interactive conversation and make effective communication extremely difficult for human speakers. In short, our expectation for **quality of service** is different for voice than it is for written messages, files, broadcast television, or other types of data.

Quality of Service: A term used in data networks to describe the performance attributes of a network.

How then should a network be designed to handle real-time interactive media like voice and video teleconferencing? The answer often will not be the store-and-forward method, because of the variable delay it introduces, as well as some other problems pertaining to quality. The most common approach in these cases is a method called **switching**. Switching makes use of a switch at the relay points, instead of a dispatcher or a router. The switch reserves a certain amount of data-transferring capacity for each "connection" between a pair of users and moves their data through the switch quickly when it arrives. In this way, a switch can guarantee that each pair of users obtains the quality of service that it wants. To do so, the switch must be able to handle data at the peak rate applied, since delaying transmission is not acceptable.

Switching: A strategy for relaying messages in which a network relay point receives a message and immediately sends it on to its destination because the route has already been arranged.

INTERESTING FACT:

Switch networks provide better quality of service than routed networks for real-time interactive multimedia.

Although networks based on switches seem best for real-time interactive media, this is not the best structure for all networks. Switched networks are not as efficient as routed networks when measured in terms of how full the transmission "pipes" are kept. Reserving capacity for each pair of talkers so that they can always get through means that far too many links will be idle for some portion of the time, typically 50%. This situation is like waiting in line to get money from an ATM: Banks like it when people line up for the ATMs, keeping the ATMs efficiently occupied. However, the ATM users like it when there are extra ATMs, so that one is always idle when the user arrives. The bank sees the latter condition as less efficient, although the customers prefer it.

INTERESTING FACT:

Switched networks are not as efficient as routed networks when measured in terms of how full the transmission links are kept.

Routing versus Switching

Now that we've seen both routing and switching networks, which is best? The answer, as in most engineering problems, is that the way we measure "best" depends on the problem to be solved. The Internet may not be the best way to carry voice and other interactive multimedia, where "best" is measured in low delay. The telephone is not the "best" way to carry delayable message-based services like e-mail, where "best" is measured in terms of low cost. However, some general observations apply to both types of networks:

INTERESTING FACT:

The switches used for the backbone of large long-distance telephone networks can carry more than 60,000 calls at once and make more than four million connections per hour. Such a switch typically costs more than 10 million dollars to purchase.

- The goal of a communications network is to permit any of its users the ability to communicate with any other user in a timely fashion.
- If cost is no object, the "best" network would be fully meshed, consisting of a direct link between every pair of users. This approach is usually cost prohibitive, however, driving us to create different designs. To make the network cheaper, we can introduce the concept of relays, which accept data from one link and place them on another link. To make this process possible, the data must be augmented with information that indicates their destination.
- Relays can be implemented in two important ways:
  1. Switches are often used when low delay and reliable transmission are important. The switches guarantee the allocation of bandwidth from one link to the next and introduce very little delay. They are commonly used for telephone networks and interactive multimedia.
  2. The other important way to implement a relay point is with a router—a special-purpose computer that stores incoming data and uses the attached address to determine which output path to send the message on, once that path is available. Networks using routers make very efficient use of the transmission bandwidth and are often the most cost-effective solution when carrying traffic, such as e-mail and data files, that can tolerate variable delay in its delivery. The original telegraph, the Internet, and the U. S. Postal Service are all examples of routed networks.

EXERCISES 8.2

Mastering the Concepts

1. The information added to a message for use by the relay dispatcher is called the *header*. What pieces of information constitute the header in the message shown in Figure 8.6?
2. Does the relay dispatcher need the source address during normal operation? Who does need the source information? Can you think of an abnormal situation where the relay dispatcher could use the source address?
3. If you were acting as the dispatcher at a telegraphic relay station and had two paths open to send a message on toward its destination, how would you choose? What information would you need to make that choice?
4. Consulting the network diagram in Figure 8.7, determine a reasonable path from Seattle to Miami. Repeat this procedure for a path from Houston to Billings. In each case, is there more than one reasonable path?
5. Using the network diagram in Figure 8.7, find the longest (and presumably worst) path from Minneapolis to Indianapolis that does not reuse a link or pass through a node more than once. Now find the shortest path.
6. Suppose we send a message to a friend in Boston. How do we know that the friend got it?
7. What was the function of the dispatcher at a telegraphic relay point? What do we call this same function when referring to the Internet?
8. Messages received by a store-and-forward node, such as an Internet router, are placed into a *queue* until they can be sent out. Under what circumstances will the queue be empty? Under what circumstances will the queue be full? What do you imagine happens when the queue is full and yet another message arrives? (Extra credit: If you were designing the system, how would you handle a "full queue" situation?)
9. Why build a communications network with relays at all? How would it be designed if cost were no object and the very best service were desired?
10. Is your postal delivery person a router or a switch?
11. What are the two types of relays? Give an example of each.
12. Zip codes are used for routing mail. What do the first three digits of your zip code tell your post office about the destination of a letter sent to your house? What area is covered by your five-digit zip code? (*Hint*: The telephone directory may have zip-code maps.) What area is covered by you nine-digit zip code?
13. Look at the simple network in Figure 8.7, and list seven different ways a message could travel from Boston to Houston. Are there more? Which of these paths would you actually consider using?

Try This

14. The modern fiber optic transmission lines used by the Internet can carry data streams at 10 gigabits per second, while the old telegraph systems operated at about 5 bits per second.

a. How many telegraph lines would be needed to match the transmission capabilities of a single fiber optic link?
b. If the cross-section of a fiber is 0.001 inch (in) and the cross-section of a telegraph wire is 0.10 in, what is the total cross-section of all the telegraph wires needed to carry the same data at the same rate as that of a single fiber?

15. Most routers use a routing table to determine how to handle each incoming message. Such a table consists of a list of all possible destinations for the message and the most appropriate transmission path for each. The accompanying table would be appropriate for router R1 in Figure 8.5.

Ultimate Destination	Next Hop
User A	User A
User B	User B
User C	Router R4
User D	Router R4
User E	Router R4
User F	Router R4

The information in this table can be summarized in the following way: A message arriving at router R1 and addressed to User A or B will be sent directly there, since R1 is connected directly to both of them. If the destination is not user A or B, the router R1 will send the message on to router R4, which will then pass them on in the right direction. Build a routing table for routers R2 and R4 from Figure 8.5.

16. Redraw the network shown in Figure 8.5 to include a two-way transmission link directly connecting relays R1 and R2.
a. Assuming the availability of this new link, what is the best way to send messages from user A to user D? What is the best way to send traffic from user A to user F?
b. There are now two ways to send traffic from R1 to R2: directly and via relay point R4. Revise the routing table for R1 shown in Exercise 8.2.15 to include alternative paths where they exist. The table should now have columns labeled as follows: |Ultimate Destination| Preferred Next Hop |Alternative Next Hop|
c. When might the alternative path be used instead of the preferred direct path to send traffic between R1 and R2? Who would make the decision of which path is used?

17. Suppose that the traffic arriving at a relay point is "bursty" (i.e., many messages arrive during a certain time interval, while fewer arrive at other times). Let's assume that a data transmission link entering a relay brings 50 messages per minute for two minutes, none for a minute, 200 message per minute during the next minute, and then none for two minutes. Assume further that this pattern repeats continuously.
a. How many messages must a store-and-forward processor (a router) be able to process per minute in order to keep up with the traffic load brought by this transmission link?

b. When the system is operating at a rate at which it is barely able to keep up with traffic, what is the maximum delay encountered by any message, assuming that messages are handled on a first-come, first-served basis by the router? How big must the router's message buffer be for it to avoid losing any messages?
c. How fast must the router be able to process messages in order to ensure the smallest amount of delay between the arrival and successful dispatch of a message?
d. When the router operates fast enough to reduce the delay to its minimum, how much time does the router spend idle, waiting for the next burst of traffic?
e. Suppose that we use a switch at the relay point instead of a router. How many messages per second must the switch be able to handle in order to avoid losing messages? What percentage of the time over a six-minute cycle is it idle?
f. Compare the results from parts (a) and (e), and comment on the probable difference in implementation cost between the barely adequate router and the barely adequate switch.
g. How do all of these comparisons change if the traffic flow on the incoming transmission link is constant at 50 messages per second?

In the Laboratory

18. A *ring network* connects each node to its nearest neighbor, and any node has only two links, each to one other node. Set up a ring network in the laboratory, and have each node user send messages to other users in the network. How is a node in a ring network like a relay node?

Back of the Envelope

19. How would you modify the message format in Figure 8.6 to include information about the priority of a message? You might consider two types of priority information. One would indicate a class of service paid for by the originator, like indicating the difference between third-class and first-class mail. The second type of priority information might be added by the system, based on how the network is doing in meeting its delivery time commitment.
20. A network that uses switches to make connections between users will take a certain amount of time at the beginning of each data exchange to "set up the call." This amount of *call setup time* depends on the network type, the number and type of switches, and several other considerations. As an experiment, measure the amount of time required for the network switches to set up the following types of connections:
    a. a local telephone call, measured by the time between pressing the last digit in the phone number and hearing the first ringing signal;
    b. a long-distance telephone call, measured in the same way as the local telephone call;
    c. a cellular phone call, measured in the same way as the local telephone call;
    d. a dial-up modem connection, measured from the time that the last digit is dialed to the time that the modem is connected and ready to carry data.

8.3 The Internet

In this section, we will explore the **Internet** in more detail. It is not the most widely accessible data network in the world (that's the telephone network!), but it is growing rapidly, and more and more people are gaining access to it daily.

Internet: The name for a networking technology that permits the interconnection of many smaller data networks.

The Origin of the Internet

In the late 1960s, the Advanced Research Projects Agency (ARPA) of the U.S. government was confronted with a vexing problem. It was running a very successful program that had, as its goal, the development of supercomputers—computers capable of operating hundreds or thousands times faster than those commercially available the time. A few of these supercomputers were built, but they were very expensive. More researchers wanted to be involved in the project, but either they had to move to where the supercomputers were located or ARPA needed to buy more supercomputers. The researchers didn't want to leave their home laboratories and universities, and ARPA didn't have the money to buy everyone a supercomputer.

INTERESTING FACT:

While the Internet is the most popularly known data network, many types of data networks have been built and are still in use.

INTERESTING FACT:

In the early days of the Internet, expensive computers performed the routing functions. By placing these functions in smaller, special-purpose computers called routers, the cost of the system was reduced, allowing many more users to join the network.

A potential solution was suggested. Engineers and computer scientists were already developing something called "data networks" for both military and commercial applications. ARPA decided that the same concept could be used to build a communications network that would let many physically dispersed researchers "reach out and touch" the agency's small number of supercomputers. Figure 8.11(a) illustrates the situation, and Figure 8.11(b) shows ARPA's approach to a solution. ARPA started a research program to design and build such a network. It was termed the "ARPAnet" and ended up having much more of an effect on how modern data networks are designed than anyone ever imagined.

The distant users were attached to the ARPAnet by using the smaller computers located in their own labs and universities. Special routers were built to relay the messages sent to and from the supercomputers. The transmission links were rented from the Bell System, America's only telephone company at the time. **Protocols**, which are the procedures by which a network operates, were developed, tested, and improved by a large number of engineers and computer scientists who were encouraged by ARPA and, later, by the National Science Foundation (NSF) to do so.

Protocols: Sets of rules that all parties in a network use to format and communicate data.

Another view of the ARPAnet's architecture is shown in Figure 8.12. On the left are the users, in the middle is the communications network itself, and on the right are the supercomputers. In modern terms, users are given **access** to the network, which then connects them to the **servers**, those devices that provide the services desired by the users. In the case of the ARPAnet, the servers were the supercomputers themselves, and the service that the users desired was the ability to send commands to the supercomputers (and get results back) as if they were sitting right next to them. The modern Internet takes this same idea and extends it to a much wider range of access and services.

Access: A user's entry point into a data network.

Servers: Devices that provide the services desired by the users. A server typically is a network-accessible computer augmented with software and additional equipment that help it perform its job.

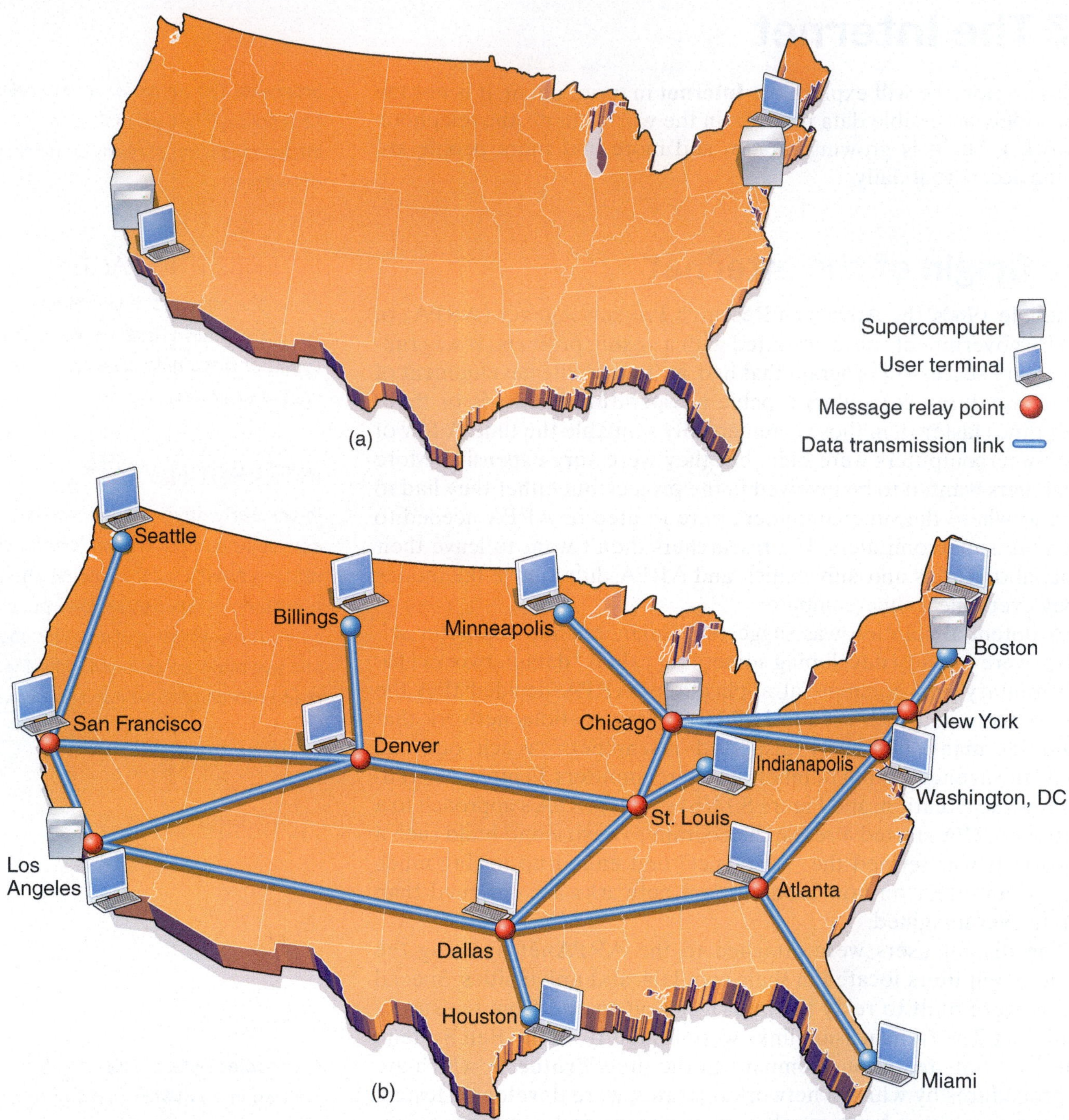

Figure 8.11 The motivation for developing the ARPAnet. (a) The few expensive supercomputers could be used only by those researchers physically close to them. (b) The development of a data network permitted researchers at other universities and labs to "log on" from a distance.

How the Internet Was Built

As computers became cheaper and data transmission systems got faster, the concepts used to build the ARPAnet were commercialized and spread rapidly. We call this network the "Internet" today because it was specially designed to permit networks owned and operated by many universities and laboratories to interconnect smoothly with each other. This concept of "internetworking" was shortened to become simply the "Internet."

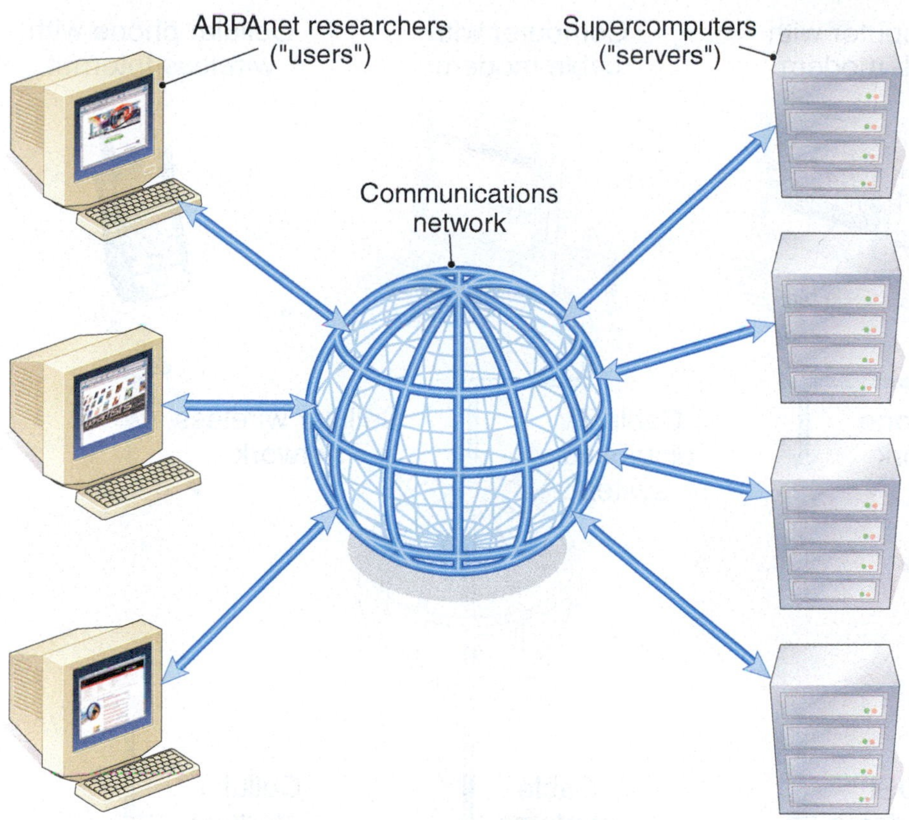

Figure 8.12 A simple version of the ARPAnet, highlighting the separate roles of the users, the network itself, and the supercomputers.

Design of a Local Internet Service Provider Figure 8.13 shows a block diagram of a local **Internet Service Provider (ISP)**. (Caution: We'll find out as we proceed that the term "internet service provider" can mean many things!) As we dissect this diagram, we find, just as with the ARPANet, that this ISP can be broken into three basic segments:

1. access;
2. connectivity; and
3. service.

Internet Service Provider: An organization or business that provides users with access to the Internet, connection with the various services available on the Internet, or the services themselves.

The ISP's users gain access to the ISP through telephone modems, digital subscriber loop (DSL) modems, cable modems, cellular phones, and personal digital assistants (PDAs). Once they've gained access, a router sends their messages to whichever service they are seeking. Thus, the router provides the connectivity between the users and the services they desire. Examples of these services are e-mail, file storage, video games, and more that we'll discuss shortly.

Following the diagram in Figure 8.13 from the top, we see that computers equipped with telephone dial-up modems (such as the V.90 modems found in most modern personal computers) connect to modems at the ISP through a telephone network. The bank of modems located at the ISP is called an **access server**, because it is accessible by, or "serves," many users. Similarly, the next computer to the right, equipped with a DSL modem, connects to the ISP, again via the telephone network, into a DSL access server. Clearly then, the entry or access into the ISP from any user occurs via an access server of the appropriate type.

Access Server: A set of equipment located at an Internet service provider through which users gain access to the network and its services.

The **content servers** are shown along the bottom row of Figure 8.13. These servers are computers that are augmented with special software and, sometimes, hardware to perform their specified tasks.

Content Servers: Network-connected computers that have the ability to provide informational services to network users upon request.

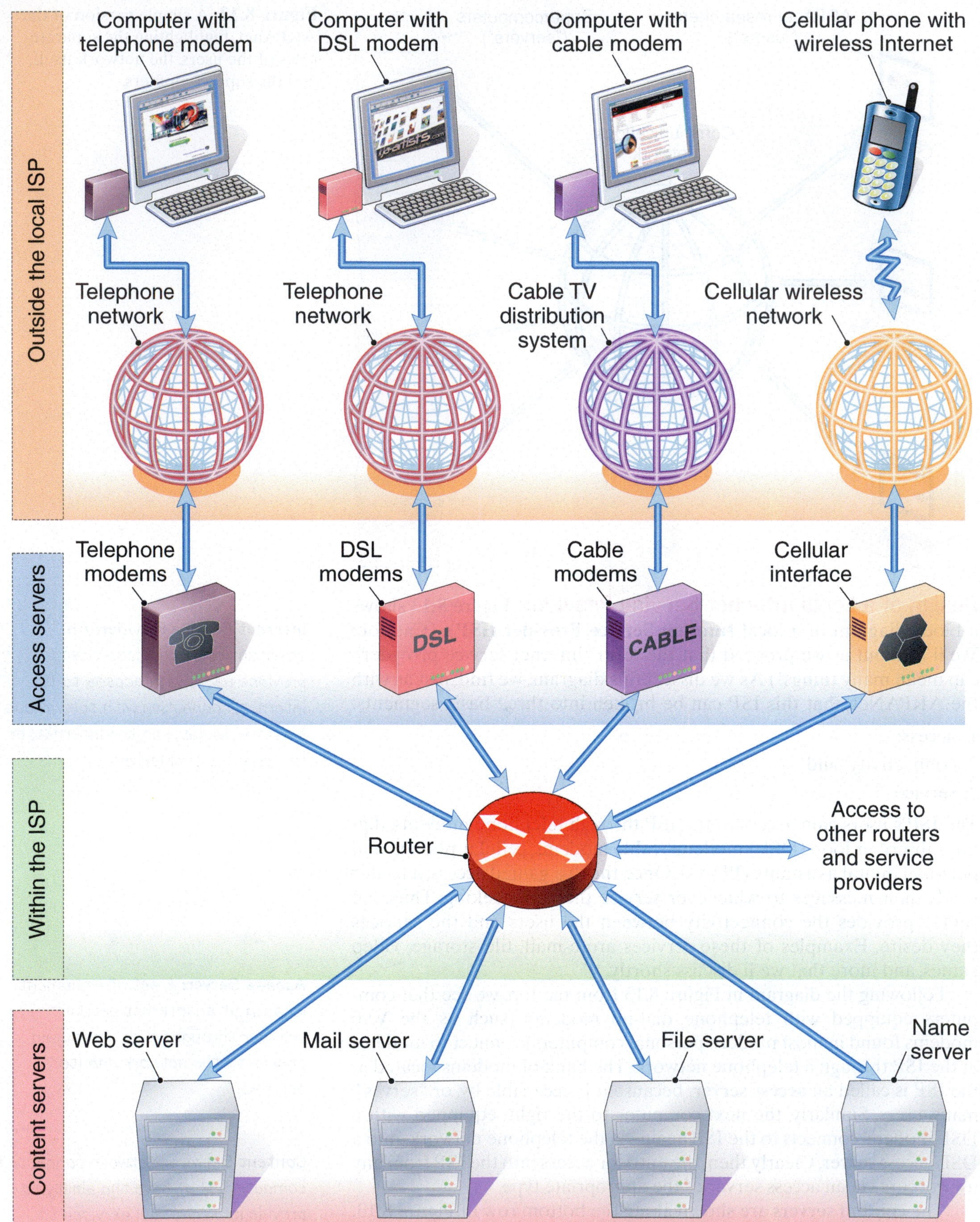

Figure 8.13 An Internet Service Provider (ISP) with a broad range of access technologies and content services.

Some of the content servers that one might commonly find at an ISP include the following:

- a *mail server* to store and retrieve e-mail for the ISP's clients;
- a *Web server* to send Web pages to anyone who asks for them;
- a *multimedia server* to provide audio or movies to customers;
- a *game server* to provide access to interactive games;
- a *name server* to "find names" (this function will be described later in this chapter).

As new services are added, new servers will be added to provide them. In principle, the access servers and the router do not need to change in order to handle new services. Sometimes, however, the increase in data flow caused by the new services requires that they be upgraded to handle it. For example, simple dial-up modems and a low-capacity router can handle all of the data needed to send e-mail between users. As soon as the ISP's customers want to obtain streaming video (for example, movies), however, faster access and faster routing will be required.

INTERESTING FACT:

In 2002, an average of three trillion bits per second flowed in the Internet between American cities during the busiest times of the day. This number is expected to grow indefinitely into the future.

Interconnecting Two Local ISPs So far, we've assumed that all of the services that a user needs can be provided by his or her own ISP. What if this is not true? A simple example would be when a user wants to send e-mail to a user of another ISP. Another example would be when a user wants to surf World Wide Web pages held on a remotely located server.

The first step in solving this problem is shown in Figure 8.14. Here, we consider two ISPs, where we want to make any of the services (such as e-mail) available on either of them available to the users of both. This goal can be accomplished by setting up a two-way (full duplex) data communications link between the routers used in the two ISPs. If the routing tables in these two routers are set up properly, then a user who gains access to the ISP on the left-hand side of Figure 8.14 can send e-mail to the mail server located in the ISP on the right-hand side, and vice versa. This procedure happens as follows:

- The user of ISP #1 gains access through an access server. He or she then writes the e-mail message and asks that it be sent to the mail server that holds incoming mail for the intended recipient. We assume here that the recipient is served by the mail server in ISP #2.
- When the user's computer sends the mail message, it adds to it the address of the recipient and the user's own address, as shown in Figure 8.6 and discussed in Section 8.2. Once the addresses are added, it sends the message on to its router.

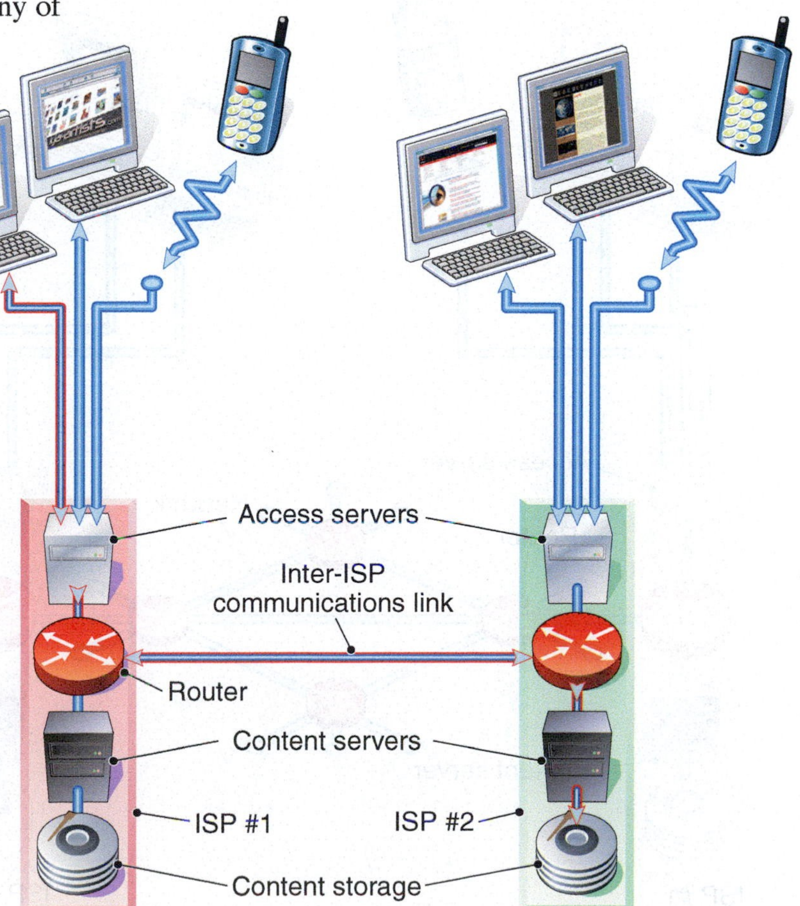

Figure 8.14 Reaching out to obtain content service from another local ISP.

Intranet: A data network that uses the technology developed for the Internet, but is not connected to it. The networks used in companies, schools, and the military are often intranets.

Internetwork: A set of networks that are interconnected and can act as a whole in communicating data from a user to any other user or service.

Backbone Network: An Internet service provider whose sole function is to connect other Internet service providers to each other. A backbone network tends to use high-capacity transmission systems and very fast routers in order to handle the traffic load imposed by the numerous client ISPs.

- The router serving the originating user examines the address and consults its routing table. It finds that the message is to be sent not to its own mail server, but rather to the one in ISP #2. The routing table instructs the router to send the message on to its counterpart in ISP #2. That router again examines the address in the message, determines that ISP #2 is the proper destination, and then sends the message on to its own mail server.
- At some later time, the intended recipient of the e-mail gains access to ISP #2 and sends a message to ISP #2's mail server, asking if there is "new mail." A message would be returned informing the recipient of the waiting message.

We've focused here on the transmission of electronic mail, but it should be clear that this same scheme will work with any of the users of either of the two ISPs and with any of the services offered by either. In particular, a Web page available on the Web server on ISP #1 can be browsed by any user of either of the two ISPs so long as a good communications link exists between the two routers and the routers are properly instructed as to where to find the desired services.

Using a Network of Links and Routers to Connect Two Local ISPs In Figure 8.14, we used a dedicated transmission link between the two local ISPs in order to let the users of the two ISPs have access to any of the services offered on either of them. From Sections 8.1 and 8.2, it should be clear that it is often more economical to use a network to connect the two, rather than a dedicated link. This concept is shown in Figure 8.15. On the left and right are the two local ISPs, as before. Instead of their two routers being connected directly, however, the ISPs are connected through a network of digital communications links and routers. If we assume the network can provide the same transmission rate, accuracy, and response time as the dedicated link, then the users of the two ISPs can retain all of the services and quality they had, while reducing the cost of the whole operation.

Before we go on, we should observe that we have built an **internetwork**. Both of the ISPs are networks in their own right. We have now connected them to each other, using yet another network (often called a **backbone network**). This kind of interconnection requires that all of the networks use the same rules for addressing data and

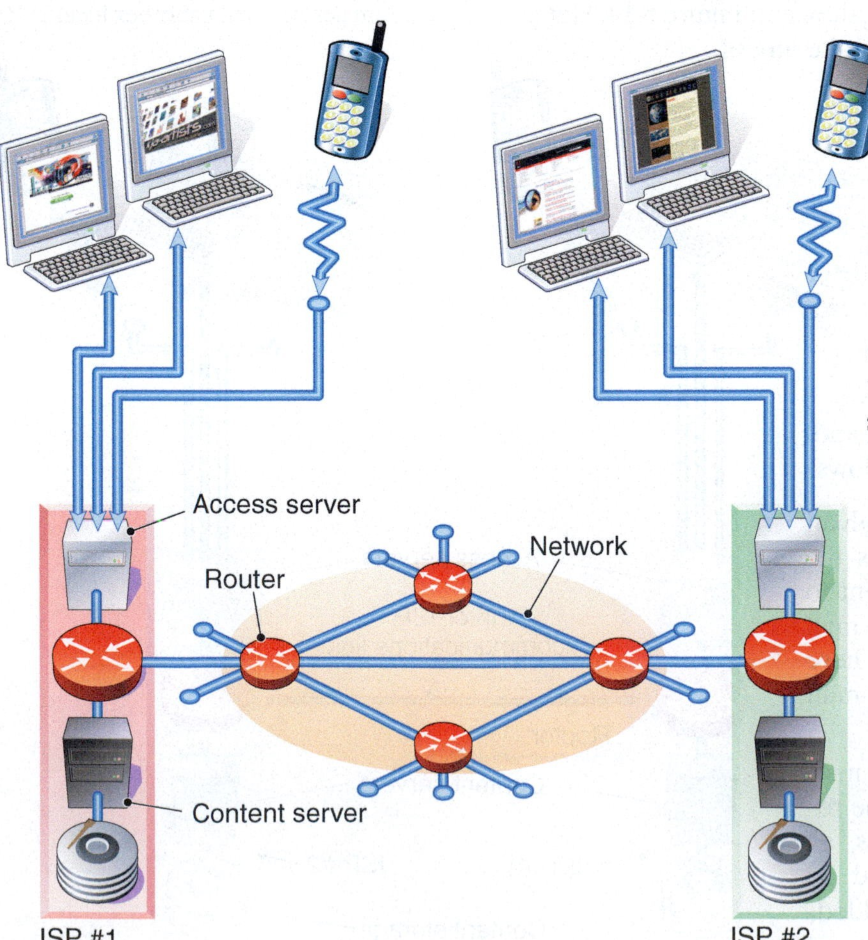

Figure 8.15 Using a network rather than a dedicated communications link to connect two local ISPs.

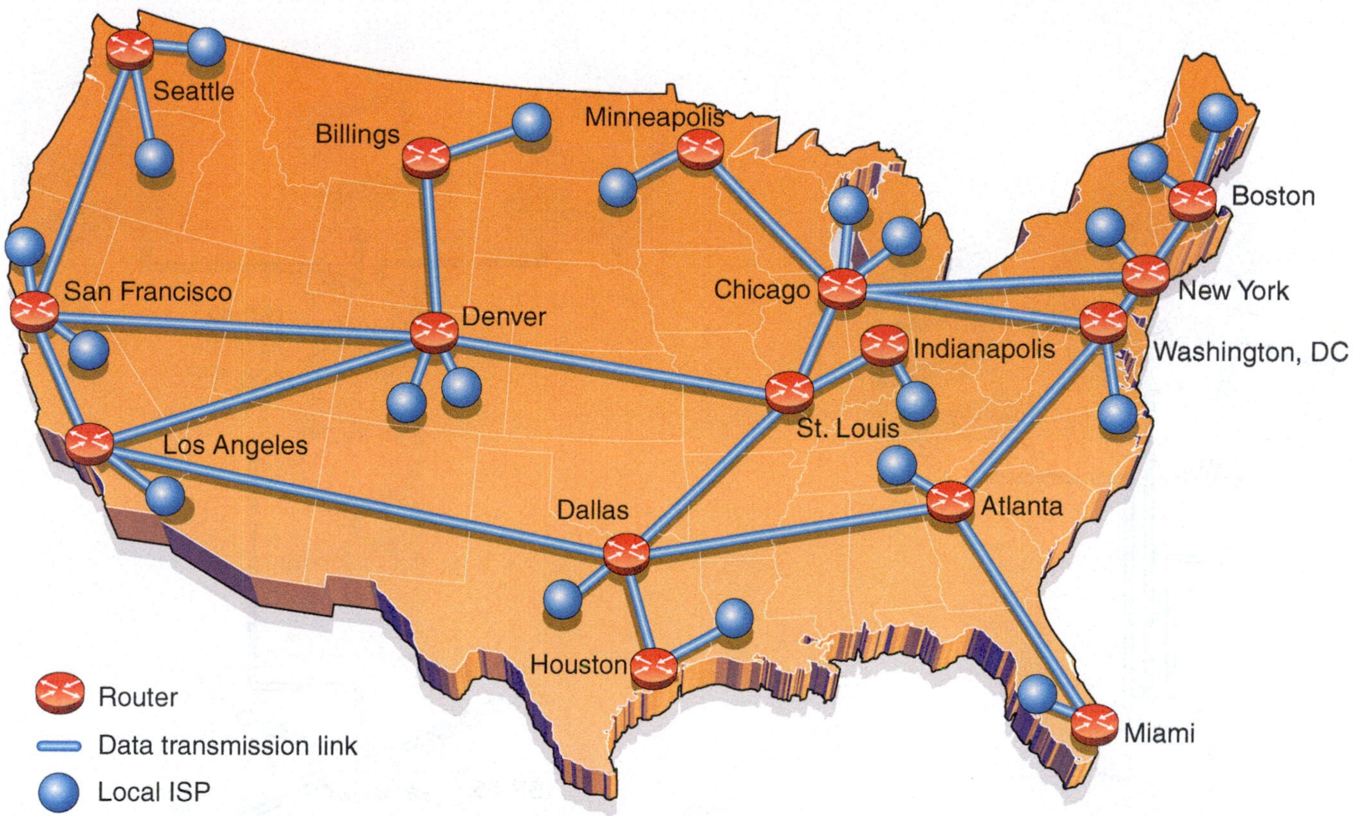

Figure 8.16 A geographical view of local and regional ISPs interconnected by a backbone transport network.

sending the data through routers, even though all three networks typically are owned and operated by different people or organizations. Again, it is from this ability to internetwork that the Internet gets its name.

Interconnecting Many Local ISPs—The Internet It is a simple, but important, step to go from connecting two local ISPs to connecting many of them by the routher symbols. This structure is shown by the router symbols in Figure 8.16. Each blue circle represents all or part of an ISP of the type shown in Figure 8.13. Each serves a local area or perhaps a region of the United States. All of them are connected to each other with a network of high-speed routers and digital communications links—a backbone, or transport, network. It is called this to emphasize that its principal function is moving data from ISP to ISP, rather than providing access to individual users or providing content services.

INTERESTING FACT:

A network of high-speed routers and fast digital communications links used to move large volumes of data is often called a backbone, or transport, network.

Revisiting the Definition of an ISP One of the things that often confuses people who are new to the Internet is the fact that almost anyone and anything can be called an Internet Service Provider. The reason for this confusion is that some companies and organizations have specialized, and rather than offering all aspects of what we consider to be the role of an ISP, they have focused on just one or two. The backbone ISP is an example of such a specialized service. This type of ISP simply provides data transport services between other ISPs. This concept of specialization can be understood using the diagram in Figure 8.17.

INTERESTING FACT:

The average e-mail message traveling from one person to another across the country is likely to pass through more than 15 Internet routers.

The ISPs marked ISP #1 and #4 in Figure 8.17 are the "full-service" ISPs of the type shown in Figure 8.13. They provide access to their

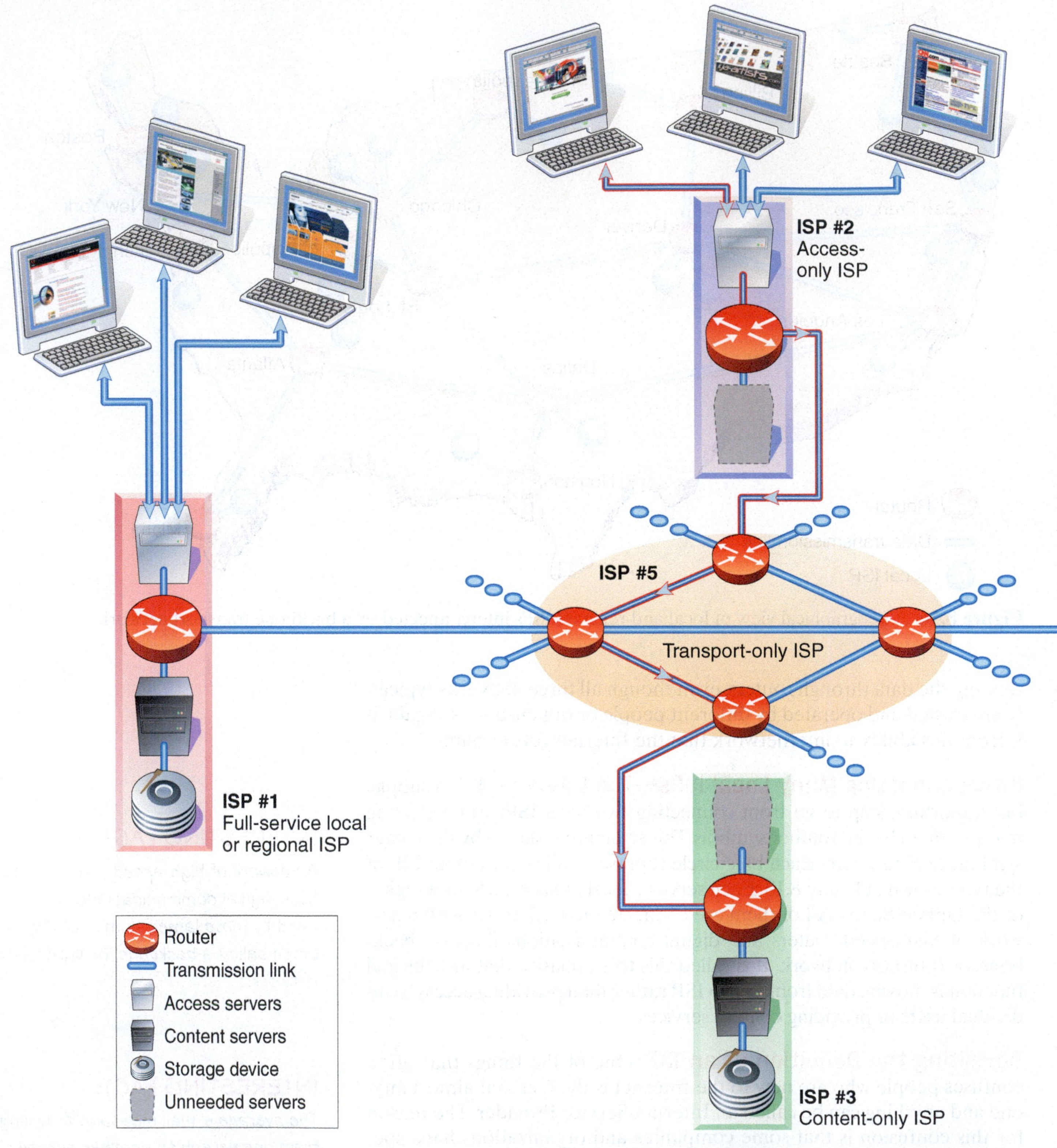

customers, some of the services needed by their customers, and the connectivity needed to provide their services to any other "Internet citizen." ISP #5 is a transport, or backbone, ISP, connecting all of the other ISPs. ISP #2 provides access, but no content. Its customers must obtain any content services they desire by "reaching out" to other ISPs through the backbone network. ISP #3 is just the opposite. It has no customers who directly access it, but rather just serves content over the network instead.

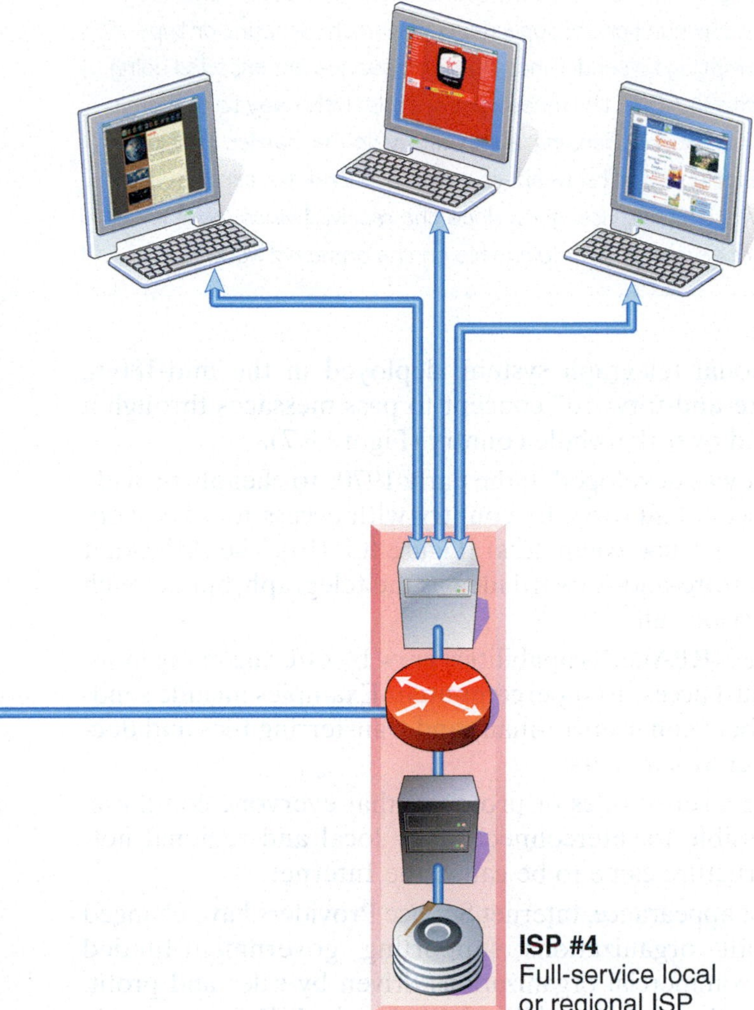

Figure 8.17 Illustrating the role of specialized ISPs: ISPs #1 and #4 provide full service, ISP #2 provides only access to users, ISP #3 only serves content and, ISP #5 only transports data packets between the other four ISPs.

Most industrial-grade Web servers take the form of ISP #3. To build one of these Web servers, many large computers equipped with large amounts of semiconductor and disk memory (that is, the content servers) are connected through very fast routers to one or more transport ISPs. These "server farms" can provide Web access to thousands of users at the same time. The same concept is used for video and game servers as well.

A Quick Review—Building the Internet in Big Steps It is useful to step back and reexamine how we reached this design for the modern Internet:

- In Section 8.1, we stated the objective of sending messages between any pair of many possible users who are distributed over a wide geographical area. We examined several approaches and decided that a network of nodes and short communications links (Figures 8.4 and 8.5) is often less expensive than a full mesh (Figure 8.3) scheme that provides direct links between all possible users.

Infinity Project Experiment: Internet Performance

The rate at which information can be sent over a network depends on the design of the network, the number of people using it, and the amount or type of data they are attempting to send. Since typed messages are encoded using small numbers of bits in ASCII, the messages are relatively easy to transmit quickly on modern networks. Sending audio and video is harder. Set up a one-way communications link between you and a friend to send a video stream, using a Web camera. How good does the received video look? What happens if there are several people doing this on the same network?

- The first national telegraph system, deployed in the mid-1800s, used the "store-and-forward" concept to pass messages through a network spread over the whole country (Figure 8.7).
- The ARPAnet was developed in the early 1970s to cheaply provide researchers located all over the country with access to a few very expensive servers (supercomputers) [Figure 8.11(b)]. The ARPAnet used the same store-and-forward idea as the telegraph, but at much faster transmission rates.
- Once built, the ARPAnet's capabilities grew by offering many more services than just access to supercomputers. Examples include sending messages between users (e-mail) and transferring files and documents between researchers.
- By developing a set of rules or protocols that everyone could use, it became possible to interconnect many local and regional networks. This structure came to be called the Internet.
- Since their first appearance, Internet Service Providers have changed from nonprofit organizations supporting government-funded research into commercial organizations driven by sales and profit. With this change has come specialization, whereby ISPs focus on only the portion of the network where they "add value" and can make money. For some, this focus means only providing user access, transporting messages, or serving content (such as Web pages and movies).

Protocols and Their Use in the Internet

A key to the successful development of the Internet was a general agreement to use a standard set of rules regarding how messages are built and how services are provided. These rules are called protocols. (We defined protocols earlier in this section.) In this discussion, we examine what a protocol is, why we call it that, and why we want them. We also describe a few of the most important ones used in the Internet.

INTERESTING FACT:

Protocols are the rules and procedures that define all aspects of data communications. Scientists and engineers designed all the protocols used in the Internet.

The term "protocol" comes originally from the conduct of diplomacy between two nations. Rules and procedures were developed so that heads of state, like kings, could deal with each other without misunderstanding. When engineers began developing communications systems, it quickly became clear that the lack of clear agreement on the rules for sending messages led to chaos and confusion. Thus rules, or protocols, were born. They define how messages are built, when they are sent, how they are acknowledged, and all other aspects of data communications. Many engineers and computer scientists worked together to develop the protocols used on the Internet.

There are literally hundreds of protocols that define how the Internet operates. We will describe a few of the most important ones here.

TCP/IP—Making Communication Reliable and Fair When a computer linked to the Internet has a message that is ready to be sent to another computer linked to the Internet, the two computers frequently use a pair of protocols called the **Transport Control Protocol (TCP)** and the **Internet Protocol (IP)** to send the message and ensure its reliable delivery. The operation of the protocols can be summarized in three steps:

1. At the transmitter, software using the TCP breaks the message to be sent into **packets** of smaller length and places an address header of the type seen in Figure 8.6 on each of them according to the IP.
2. The IP packets are delivered from the transmitting computer to the receiving computer via one or more routers, which use the address information in the header to seek out the proper destination.
3. At the receiver, software using the TCP reassembles the received packets into a complete and accurate message and then delivers it to the user.

There are many motivations for and benefits to breaking up each message or file into packets, but one of the most important is to make the network "fair" in the sense that many users are not delayed by one user who is sending a very long message. Breaking all messages into packets allows the routers to intermix traffic from many users in a relatively fair way.

TCP/IP: Transport Control Protocol/Internet Protocol are a pair of protocols used in the Internet to fairly and reliably carry information between any two computers. TCP breaks a user's message into packets and then uses IP to sends each of them to the destination computer.

Packets: Segments of a complete message or file broken down to permit smooth and fair sharing of a data communications network by many users.

SMTP—Sending and Receiving Electronic Mail **Simple Mail Transfer Protocol (SMTP)** is the procedure by which most computers send and receive electronic mail (e-mail). It works in a very straightforward fashion. We will illustrate the method via an example in which a student at SMU is sending an e-mail to a friend at State University. Here are the steps that happen after the sender clicks "Send":

1. A computer at SMU with an e-mail to send informs the recipient's mail server at State University that it has such mail to send. The mail server responds with a message (delivered with TCP and IP) of the form

   stateuniv.edu SMTP service ready

   where statuniv.edu is the general domain name of all computers at State University.
2. To that, the originator at SMU sends

   HELO smu.edu

3. The receiving mail server then responds with

   stateuniv.edu says hello to smu.edu

4. With this "handshake" complete, the computer at SMU states its purpose and sends the following line of ASCII characters:

   MAIL FROM: somebody@smu.edu

Simple Mail Transfer Protocol (SMTP): SMTP is used to carry e-mail from the mail server in one local computer network to the mail server in another. A different protocol, the Post Office Protocol, Version 3 (POP3), provides the rules and procedures for moving the e-mail from the local mail server to your computer so that you can read it.

In this same back-and-forth fashion, the computer at SMU identifies the user to whom the e-mail is directed and then sends the mail message. The receiving computer at State University accepts the message, and then the two computers agree that the transmission is complete.

MIME (Multipurpose Internet Mail Extension): A set of procedures for organizing the binary data produced by multimedia sources such as voices, music, pictures, and movies and attaching them to electronic mail (e-mail) messages for transmission. Since e-mail uses only ASCII or Unicode for transmission, it is necessary to modify the multimedia data appropriately so that they can be attached and carried with an e-mail message.

An e-mail composed purely of text is sent easily using this line-by-line scheme, but what if the e-mail users want to send multimedia data like music or pictures? A scheme called **Multipurpose Internet Mail Extensions (MIME)** was developed for just this purpose. It works by taking strings of bits and mapping them six at a time into ASCII characters. These characters are sent using SMTP as if they were a textual message. The recipient's mail viewer recovers the bits and then displays or plays them appropriately.

HTTP—Finding and Retrieving Web Pages The World Wide Web (WWW) was developed in the early 1990s as a way of using the Internet to reach out and obtain information held anywhere in the "web" of computers present on the network. The information is prepared for display on other people's computers by converting it into hypertext, a format that uses ASCII characters to describe not only the textual information, but also the placement of text and figures on the display screen. Once one has found the Web server containing the desired information, it is necessary to transport the information to the user who wants it. For this purpose, a protocol called **Hypertext Transfer Protocol (HTTP)** was developed.

Hypertext Transfer Protocol (HTTP): A protocol used to transfer hypertext from a Web server to a client computer.

The software that performs the functions defined by HTTP works very much like the software that handles the Simple Mail Transfer Protocol (SMTP). The "calling" computer uses TCP and IP to establish a connection with the computer containing the information of interest to the caller. In a four-step exchange similar to that of the SMTP protocol, the "called" computer responds to the request by finding the data file containing the desired hypertext and then sending it, line by line, to the calling computer. The Web browser at the calling computer then interprets the hypertext and displays it appropriately.

INTERESTING FACT:

The World Wide Web (WWW) was begun in 1989 at CERN, the European Center for Nuclear Research, as a method for easily and cheaply sharing the results of scientific research. Its inventor, Tim Berners-Lee, is a physicist.

FTP—Sending and Receiving Data Files Many activities among computers in a network rely on the ability to reliably transfer data files from one computer to another. A simple example is the case where a computer user wishes to print a document. In the early days of personal computers, each computer had its own printer. Now it is common for the computers in a laboratory or office to be "networked" together and to share the use of a single "print server"—a printer that has been augmented with a built-in computer that accepts files from the client computers, stores them, and then sends them to the attached printer when it becomes available. A necessary part of this scheme is the ability for the client computers to reliably send the document to be printed to the print server. To make this and many similar cases very simple, the **File Transfer Protocol (FTP)** was developed.

File Transfer Protocol (FTP): A protocol used to reliably transfer disk files from one computer to another over the Internet.

The computer originating the transfer begins by establishing a connection with its counterpart, using TCP. Over this "control connection," the two computers identify the file or files to be transferred and then start and stop the transfer. The software that performs the FTP functions actually opens a second connection between the two computers in order to carry the data. Once the data transfer is complete and TCP confirms that it is accurate, then the two computers agree to break down the connections and continue with their other tasks.

As mentioned earlier, the protocols discussed in this section are but a few of the hundreds of protocols used by the Internet. There are other types of data networks that also have their own sets of protocols. The details of these protocols are not important here, but the fact that they exist and the reasons for their existence are important: Without an orderly and well-understood set of rules, it would be impossible to reliably interconnect two computers, much less millions of them.

Domain Names and Finding Computers in the Internet

Before we leave our description of the Internet, there is one more protocol that needs to be described. It performs a vital function: It converts the name of a computer or service to an actual binary address that the Internet Protocol (IP) uses to route and deliver the packets of any data transfer. To understand why this protocol is necessary, we must first understand how computers or services are recognized and named.

INTERESTING FACT:

Every computer with access to the Internet has a 32-bit address. A new version of the Internet Protocol, called "Version 6," will use 128-bit addresses instead. This new version supplies enough bits to provide a separate address for every grain of sand on every beach in the world.

Internet Protocol Addresses When the Internet Protocol was first defined, the engineers involved decided to use the scheme shown in Figure 8.6. Each data packet, composed of a message from one user to another, would be "prepended" with two addresses—the first being that of the destination computer and the second being that of the source computer. The engineers also decided that each address would be 32 bits long. Since there are about 4 billion different combinations of 32 bits, it was felt at the time (around 1982) that this length provided more than enough addresses to handle all of the computers that may ever exist. Even though there are efforts underway now to increase the address length to 128 bits, virtually all computers on the modern Internet still use these 32-bit addresses.

Thus, each computer in the network has its own unique 32-bit address. It is also true that all devices composing the Internet, including routers and servers, also must have unique IP addresses.

While computers easily accommodate 32-bit binary numbers, the humans who administer them are not so flexible. For them, the addresses are usually written in "three-dot" form. The 32 bits are broken into four blocks of eight bits, and then each of the blocks is converted into a decimal number. Those numbers are separated by dots. For example, the 32-bit address given in binary form by

11000000 00110101 11011110 00000111

would be written by humans as [192.53.222.7].

Uniform Resource Locators (URLs) Even with the IP address simplified to a string of decimal numbers, it is inconvenient for human computer users to remember how to address a particular computer or resource. To solve this problem, an even more "human friendly" method was developed. Computers and the services that they provide were thought of as operating in "domains," and a scheme of "domain names," or, more formally, **Uniform Resource Locators (URLs)**, was

Uniform Resource Locator (URL): The method of naming computers and servers on the Internet or within an intranet so that each has a unique name and can be reached by all others. Each computer is a member of a "domain" and, within that, possibly, a sub- and a subsubdomain.

developed. All Internet users were placed into sets. In the original system, a computer might be part of an educational domain, a government domain, a military domain, a nonprofit organization, or a part of the network itself. At that time, these domains were designated by the letters EDU, GOV, MIL, ORG, and NET, respectively. (Recently, many more of these "root domains" were added to accommodate the rapid growth in the number of users of the Internet.) Each of these large domains contains many members. Southern Methodist University (SMU), for example, is a member of the educational domain EDU. The member's name is placed in front of the domain name. Accordingly, the URL for SMU is SMU.EDU.

This "prepending" scheme can go on indefinitely. There are different organizations at SMU that have separate computers and need separate addresses. This configuration is indicated by adding these organizational or functional designators in front of the URL for SMU. For example, the school of engineering at SMU can be reached at

www.engr.smu.edu,

while the university's Web server is addressed with the URL

www.smu.edu.

This type of addressing scheme is termed "hierarchical," since there is a clear hierarchy in the names for each computer or service. The relationships can be seen easily using Figure 8.18. The Internet is split into "root domains" that each include many members. Each of those members can be composed of many submembers, and, progressively, each of the submembers can have many parts. The URL describes the progression from the smallest member up to the root domain.

INTERESTING FACT:

The way that a Uniform Resource Locator (URL) is built can be thought of in terms of sets and subsets. Every URL is in a set such as .COM or .EDU. It may also be in a subset, such as SMU.EDU, or even in a subsubset, such as ENGR.SMU.EDU.

Domain Name Service (DNS): DNS was developed to provide a simple and automatic way to convert Uniform Resource Locators (URLs), which are easily read and remembered by human beings, into the 32-bit binary addresses used by computers and routers, which aren't so easily remembered.

Relating the Two Addressing Schemes

Computers easily handle 32-bit addresses, but work less well with variable-length ASCII URLs. Humans are just the opposite. How can the two addressing schemes be used together? The answer comes in the creation of a new service and the definition of a new protocol. The new service translates a URL into a 32-bit IP address. The computer that offers this service is called a *name server*. The new protocol is called **Domain Name Service (DNS)**. It is used by any computer that needs to have a URL translated. Figure 8.19 shows how this process works.

Infinity Project Experiment: Exploring the Internet

Network professionals often use a utility (a type of program) named TRACEROUTE to determine the path that data packets take from their computer to a network service such as a Web server or an e-mail server. Almost all personal computers have a mechanism for running this program from any user or account. Run this program with the IP address 129.119.4.101, which is the IP address for SMU, and see what happens. Repeat this task with the URL SMU.EDU. Is the result different? Would the result of TRACEROUTE be different if you were in a different physical location? Why?

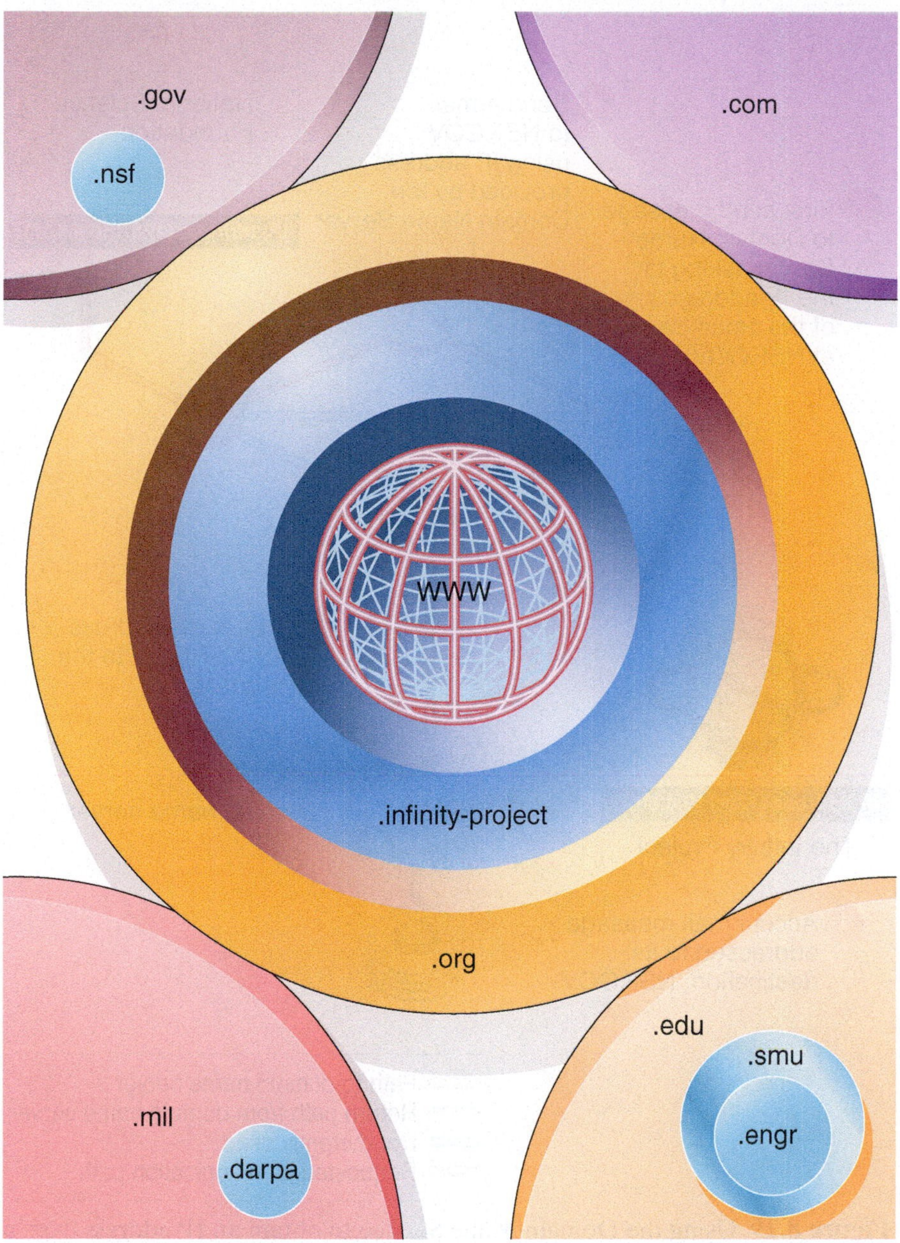

Figure 8.18 Using domain names to specify destinations in the Internet.

Suppose a user of a computer on the left-hand side of Figure 8.19 writes an e-mail message and directs it to a mail server located in the domain on the right-hand side. This e-mail cannot be sent until the sending computer knows the IP address of the mail server. It might have the address stored in its **cache**, a temporary memory for IP addresses, but, if not, it sends a message defined by the Domain Name Service (DNS) to its name server. This name server might be in the user's ISP, or it might be a more complete one to which it has been referred. In either case, the name server looks up the IP address of the computer corresponding to the URL in a lookup table and returns it to the requesting computer. Now equipped with the proper IP address, the mail transfer can proceed, using SMTP and TCP/IP.

Cache: A temporary memory that can be used for IP addresses.

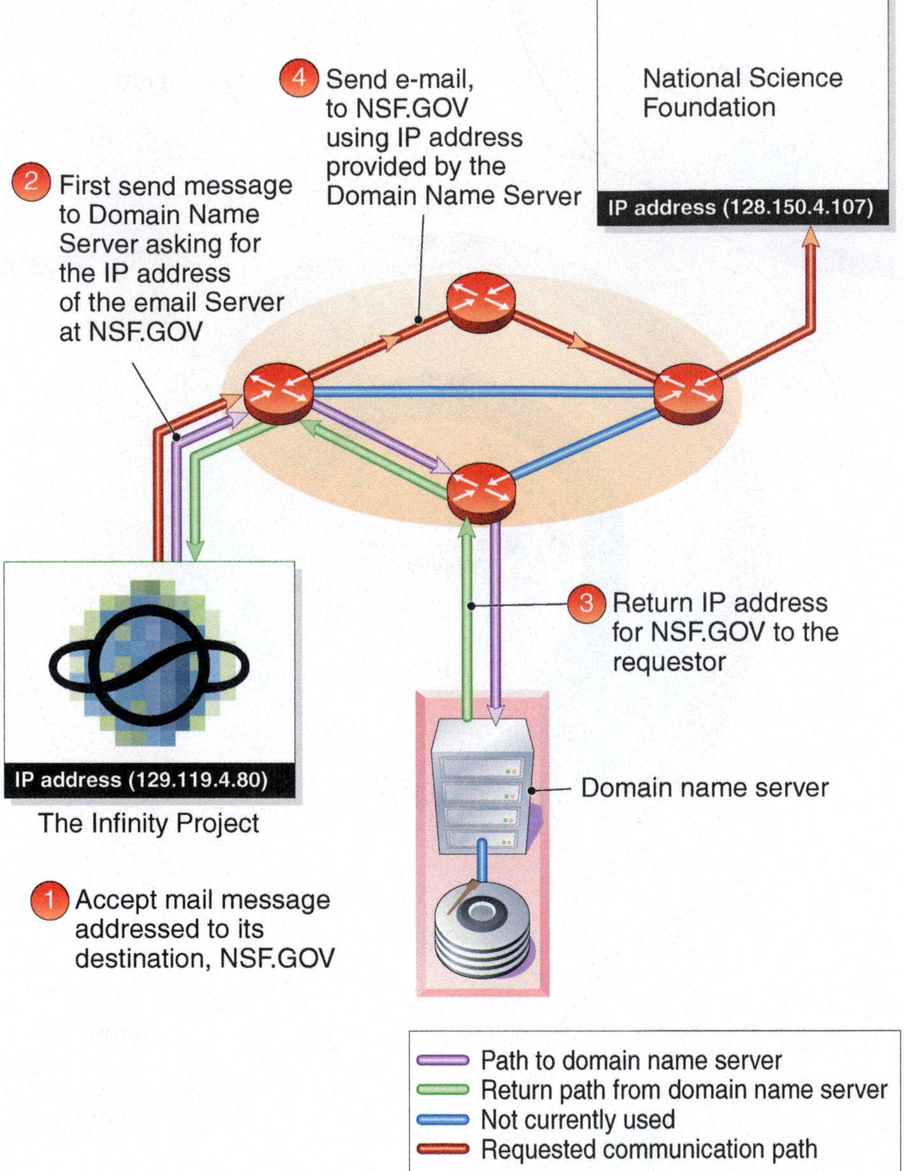

Figure 8.19 Using the Domain Name Service to obtain an IP address from a name server.

Just as the domain-naming convention is hierarchical, so are the name server conventions. If a computer asks for a name unknown to its local name server, it can "go up the hierarchy" to get the answer. For example, the name server at SMU can be expected to "know" the IP addresses for all of the departments and services at SMU, but might not "know" the IP address for the mail server at another university. To find it, the name server at SMU can ask the master name server for the entire EDU domain for the name of the name server at some other university, using DNS. Armed with that answer, it will send a message to the other university's name server asking for the IP address for the mail server. Using this scheme, any computer in the Internet can, with successive queries using DNS, find the IP address of any computer or service on the Internet. This functionality is the glue that holds the Internet together.

EXERCISES 8.3

Mastering the Concepts

1. What is a server?
2. What type of service does an access server provide? Provide three examples of access servers.
3. What is a backbone network, and what type of service does it provide?
4. The original ARPAnet was built to permit access to a few expensive supercomputers. Do supercomputers fit into one of the standard types of servers, or are they something different?
5. Is every data network designed like the Internet? If yes, state why. If no, give an example of how the designs might differ.
6. Can the technology used for the Internet be used to build a network for private use? If so, what would you call such an "internal Internet"? Name two examples of organizations that might build such a network. Why might they build this network instead of simply being a part of the public Internet?
7. Define an ISP. What types of services do ISPs offer? Must every ISP offer every type of service?
8. What type of server would store digitized movies waiting for customers to view them?
9. How does a router know where to send a message?
10. The Internet consists of many interconnected networks. What determines the fastest rate that data can flow from one user to another?
11. Where does the name "Internet" come from, and why is it appropriate?
12. Does the Internet rely on switches or routers at its relay nodes? Why was that choice made?
13. A friend tells you to check out a really interesting IP address of 128.376.2.1, but you know immediately without trying it out that the address is not correct. How do you know?

Try This

14. Convert the following bit strings into "three-dot" decimal IP addresses:
    a. 11101010 00100100 11111000 00110011
    b. 00010100 11100001 01010101 0000001
15. Convert the following "three-dot" sequences into 32-bit IP addresses:
    a. [192.34.208.56]
    b. [36.1.23.255]
    c. [255.255.255.255]
16. Suppose that 10,000 gamers were playing at the same time in a city and that a game server needed to send screen updates to each of them 60 times a second in order to make the game

look sufficiently realistic. If each screen required 5000 bytes of information to be sent, what is the total rate, in bits per second, that the game server must be capable of providing?

In the Laboratory

17. Using a personal computer with an Internet connection, run the TRACEROUTE program on any URL that you know. (Use SMU.EDU if you do not know one.) The first response from TRACEROUTE is the IP address corresponding to the URL.
    a. Write down the IP address in both three-dot and binary form.
    b. What protocol was used by TRACEROUTE to convert the URL?
    c. Find the path to any three universities. Determine as best you can from the cryptic responses to each call of TRACEROUTE how the Internet would send messages to the respective destinations.
18. Use the TRACEROUTE command to find the path to each of the following addresses:
    a. SMU.EDU
    b. ENGR.SMU.EDU
    c. WWW.ENGR.SMU.EDU
    d. WWW.SMU.EDU

    Does the name server identify the same computer for each of these URLs? Would you expect it to? Are the routing paths different? Would you expect them to be?

Back of the Envelope

19. What is the Internet Engineering Task Force (IETF), and what does it do?
20. Find a line of hypertext, and explain what it causes a Web browser to do.
21. Examine a map of the interstate highway system, which handles vehicle traffic rather than data traffic. Find three reasonable routes from San Francisco to New York.
22. People who use walkie-talkies employ a "protocol" even though they are often not aware of it. From your own experience, what are the meanings of the following words when used over a radio link:
    a. Over
    b. Roger
    c. Out
    d. Extra credit: Wilco
23. The Multipurpose Internet Mail Extensions (MIME) provides a set of schemes for converting binary voice and image data into ASCII characters so that they can be attached to a text-based e-mail. To see how this procedure can be done, translate the binary string

    0000 1111 0001 1110 0010 1101 0011 1100 0100
    1011 0101 1010 0110 1001 0111 1000

    into an ASCII string, using the hexadecimal conversion of $A = 0000$, $B = 0001$, $C = 0010$, and so forth.

Master Design Problems

After studying all of the concepts in this chapter, it is reasonable to ask how a new network should be designed. The answer to that question depends on what the network will be expected to do. Specifically, we ask questions like the following:

- How much traffic must be transmitted?
- How many users are there?
- How much delay can be tolerated?
- How much are the users willing to pay?
- How secure must the system be?
- How much data loss can be tolerated?
- What will government regulators permit?
- Must equipment from a specific vendor be used?
- Must the new system work with older "legacy" equipment and software?

What we find, of course, is that there is no single "best" design for a network. The definition of "best" depends on the application and the needs of the customer. The engineer's design problem is to match most effectively the available technology with the desires and needs of that customer.

- Design a *Star Trek* communicator system for a small environment such as a school or a shopping mall. Determine what the range of the communicators should be. Then determine the spacing for the relay points, and decide how they will be fixed into existing structures such as walls or benches. Consider how the users will be identified and how transmissions will be routed to them.
- Design a communications network to support the International Space Station (ISS).

  It must
  1. be in contact with the ISS at all times;
  2. be in real-time contact with mission management centers in Houston, Moscow, Frankfort, and Tokyo;
  3. provide contact to other scientific centers, as needed;
  4. be capable of carrying voice, images, video, messages, and telemetry, as well as the control signals needed to run the ISS;
  5. provide teleconferencing for both ground and space;
  6. be 0.9999% reliable;
  7. not cost a fortune;
  8. be resistant to malevolence;
  9. be easy for people of multiple cultures and who speak various languages to use.

  Assume the following:
  1. The ISS has some type of local network already.
  2. Each of the various ground facilities has its own telephone and data networks.
  3. This system will be used to interconnect all of the other local networks.

  Questions:
  1. What type of transmission link would you use to reach the ISS?
  2. How many ground-to-space communications points do you need in order to keep in constant contact?

3. What type of transmission links would you use to interconnect all of the ground facilities?
4. Would you build new links or try to take advantage of the transmission systems that already exist?
5. How much data have to be moved? With what quality of service must the data be treated?
6. Should this network be a "network of networks" with no one in charge, or should it be a "managed network" with a network administrator who has control over all of the network's components?

Big Ideas

Math and Science Concepts Learned

There are already many different types of communications networks in the world, and their numbers continue to grow as computers and microprocessors become a part of every home, business, automobile, and appliance. The designers of microprocessors are now using the concepts of data networking to design how the various parts of a single silicon chip communicate with each other. While the applications of networks will become more widespread and diverse, the fundamental principles will remain the same as those examined in this chapter. They can be summarized in the following points:

- The objective of a communications network is to permit any users of the network to communicate information with any or all other users.
- There are many ways to design a network, and the choices made depend on the users' requirements. These requirements might include speed of delivery, accuracy of delivery, reliability of delivery, instant availability, the ability to obtain immediate responses, or any combination of the foregoing.
- Cost is the requirement that most frequently drives the design of a communications network. Minimizing costs is key to all good designs.
- We analyzed a case in which minimizing cost was the key issue. The network in this case had to operate over long distances, so the costs of transmission quickly dominated the total cost of the system. In this case, we found that the cheapest way to build a network is not to connect all users directly to each other with their own dedicated links, but rather to share transmission links. This was done by connecting links to each other with "relay points." When one of these relays receives data destined for a user, it selects, by some strategy, an outgoing transmission link, which carries the data to, or closer to, the intended recipient.
- Two key strategies have emerged over the past 200 years for relaying messages.
  1. The first is termed "routing" and uses information, called the "address," added to the message by the outgoing user to determine the next best step in the network toward the intended recipient. Since routers are permitted to temporarily store messages until the best output link is available, this method makes very efficient use of expensive transmission systems. The U.S. Postal Service, the telegraph, and the Internet use this strategy in their relays.
  2. The second key strategy is termed "switching." In this scheme, all of the switches involved in a planned data exchange first send

messages to each other to determine a path through the network for the data and to ensure that there is enough network capacity for the transmission. Once this prearrangement, termed "call setup," is done, the data can then flow with assurance and reliability—two aspects of what is commonly called "quality of service." The telephone network is an important example of a switched network.

- The technology used to implement electrical communications networks has improved markedly from its first appearance in the 1830s. Back then, it operated over a few miles and at rates of about 5 bits per second, using human transmitters and receivers. Modern versions send data at up to 10 billion bits per second and use high-speed electronic routers to send, receive, and dispatch the data. The concepts are identical to those used nearly two centuries ago, but the implementation has improved enormously.
- Modern networks still use both approaches—routing and switching. The modern version of the router-based scheme is called the Internet, so called because the rules, or "protocols," developed for its use permit the interconnection of networks owned and operated by many different organizations and companies.
- The Internet's ability to "internetwork" has allowed specialization in "Internet Service Providers (ISPs)" and the services that they offer. Some ISPs specialize in providing users with access to the Internet, some provide content like Web pages and video, and some provide the communications paths between the users and the servers that they wish to use.

The role of network engineers is the same as that of other types of engineers. They must determine what the users really want and what constraints, such as cost and performance, apply. They must then develop a number of alternative designs and compare them in order to determine the best one. They supervise the construction of the system, test it once it is complete, and verify that it will operate as designed. Engineers typically also contribute to the maintenance of the network and to the long-term support of its users.

Important Equations

The number of connections, or links, for a fully meshed network with N users is given by

$$C_M = \frac{N(N-1)}{2}$$

The number of connections, or links, for a central relay network with N users is given by

$$C_R = N$$

Building Your Knowledge Library

Standage, Tom, *The Victorian Internet*, Walker and Company, New York, 1998.

The Victorian Internet tells the story of the telegraph's creation and remarkable impact and of the visionaries, oddballs, and eccentrics who pioneered it. The book shows that the excitement about the telegraph in the late 1900s was very similar to that associated with the Internet in the 1990s and 2000s.

Tanenbaum, Andrew J., *Computer Networks,* 3rd Ed., Prentice Hall, Inc., Upper Saddle River, New Jersey, 1996.

This textbook is often used in universities to teach the principles of data networks. It is very well written and provides lucid explanations of most of the protocols used in the Internet.

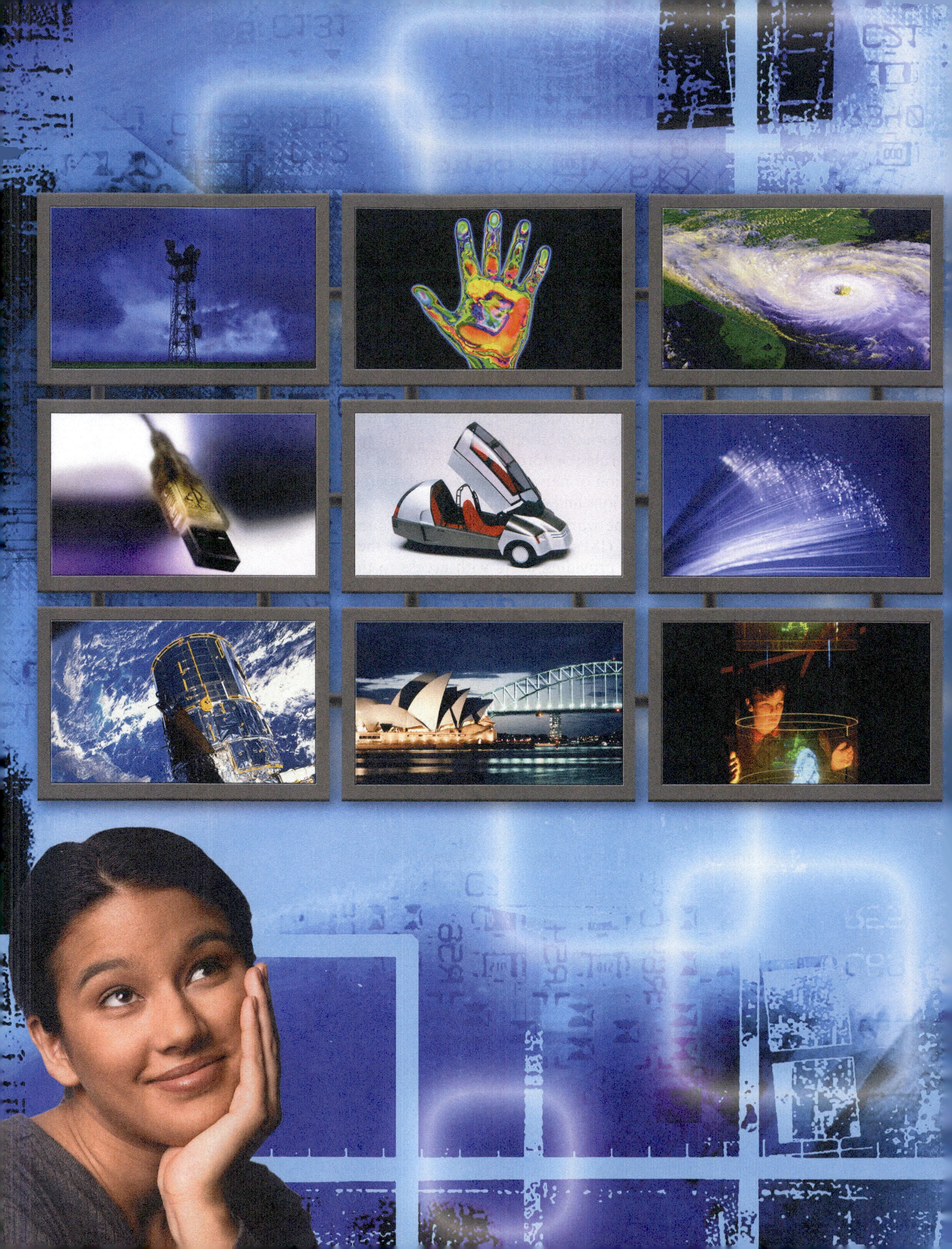

The Big Picture of Engineering

You have probably heard the stereotype: Engineers don't really care about changing people's lives—they just love designing and building new gadgets.

Well, history has shown that this couldn't be further from the truth. Yes, some engineers love the challenge of creating something new for the shear sake of being the first, but most engineers who have been successful throughout time have understood their special role as society's great problem solvers.

When the world needed to find a new means of transportation for global travel, they looked to engineers for the answer. When the world needed new sources of energy, new means of communications, new structures to live in, and even new forms of entertainment, society looked to engineers for the answers.

Throughout history, engineers have always been deeply connected to the challenges and opportunities of their time. Today is no different. Right now, engineers all over the globe are rushing to find new ways to use the vast power of technology to make the world a better place—in thousands of innovative ways only now being fully realized.

OUTLINE

What will the future bring? Robots the size of a few molecules, medical treatments that extend our lifespan by decades, vast supplies of essentially free energy, virtual reality that is not so virtual, and even space travel to nearby planets. All are real possibilities—especially with engineers on the job.

9.1 Engineers—Society's Problem Solvers

Throughout history, the progress and achievements of human society have been influenced by the creativity and efforts of engineers. From the water we drink to the food we eat, from the transportation we use to the buildings in which we live, from the ways in which we do business to the ways in which we play, engineering has played a major role.

Throughout this book, we have focused heavily on applying math, science, and the engineering design approach to problems that can be solved using modern digital technologies. However, engineers work on a much wider array of social and technical problems than just those that involve digital technologies. In fact, the world of engineering includes such a diverse set of topics that it would be impossible to cover them in one lifetime, much less in one book. This is what makes engineering so special and so interesting—we can always find an engineering topic or problem that excites us.

The problem-solving tools we've learned in this book can be applied to every area of engineering. Look around you. Everything made by humans is the product of an engineer's or group of engineers' vision. The chair you are sitting on, the room you are in, the lights that light the room, even the book you are holding in your hands are the result of the engineering design and manufacturing process. Although we focused on emerging digital technologies to show you how engineering is done, there is much more to engineering than computers, sounds, pictures, and the Internet.

To give you a sense of how engineering has helped improve human lives throughout history, we shall take a brief look at 10 different and important engineering achievements. In each case, we shall see how the developers of these engineering feats used the engineering design process to fulfill the most important human needs at that time. For each achievement, we will examine:

- What were the human and social needs of the time?
- In response to these needs, what problems were engineers trying to solve?
- What were the immediate and long-term impacts of these engineering achievements?

9.2 Ten Engineering Feats that Changed the World

Picking just ten great engineering feats isn't an easy task. Engineers have been working for centuries solving thousands of problems. To give you a sense of how important engineers have been to society, we have selected just ten engineering feats that have helped change the world.

For each feat, we will describe what the social or technical problem was at the time of the innovation, how the problem was solved, and how this solution changed the world.

Watch the sidebars. Here we describe various fields of engineering. Think about things that interest you, then look for the field(s) of engineering that relate to your interest.

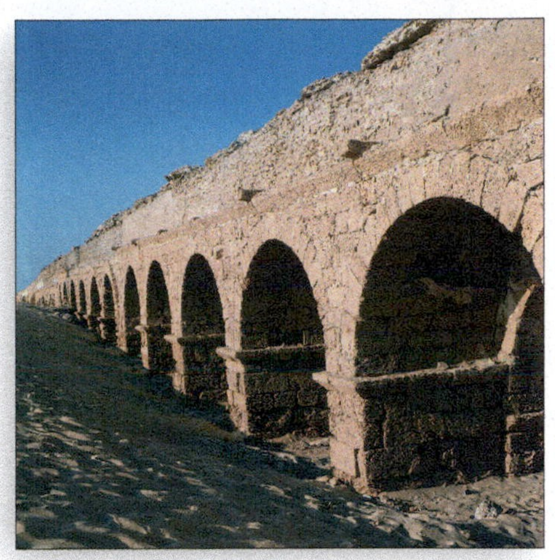

Water, Water Everywhere and Not a Drop to Drink—The Roman Aqueducts

What Was Happening at the Time? One of the world's first great engineering achievements occurred in Rome during the year 312 B.C. At that time, Rome was becoming a major power in the world. In order to continue its expansion, it was important that Rome's commerce and military have easy access to the sea. You might recall that, at this time in history, ships were the primary means of moving things and people over large distances. So, it was absolutely critical that powerful and important cities were located by the sea. Unfortunately, salt water can't be consumed by humans. Therefore, Rome required a source of fresh water to support its expanding population.

The Engineering Solution There was fresh water within 50 or 60 miles of Rome. With a population in the millions, the city required lots of water for daily use. Enter the Roman engineers. They set out to design and build aqueducts to bring water to the city—lots of it. Just to give you a sense of the scale of the problem, a typical human requires over a cubic meter of water per day for drinking, cleaning, cooking, and bathing!

CIVIL ENGINEERING:

Civil Engineering is the oldest branch of engineering. Civil Engineers design buildings, bridges, dams, roads, pipelines, and the like. If you enjoy working on BIG things, consider Civil Engineering.

Designing and building the Roman aqueducts was no easy task. The Roman engineers needed to find a way to bring the water from a long distance, and it had to flow on its own because there were no automatic pumps back then. It had to travel through hilly terrain and it had to work day and night.

The solution was a 3-foot-wide by 6-foot-deep channel, big enough for workers to enter to do regular maintenance. The channel followed the contours of the land and it dropped roughly 2 feet for every 200 feet of length, which ensured that the water didn't flow too quickly. In some places, it traveled under hills and through holes bored in rocks. If the rock was porous, it was lined with concrete so it wouldn't leak.

The water was delivered to the public via fountains no more than 100 meters apart. The wealthy received water via lead pipes directly to their homes. The first aqueduct was installed around 312 B.C., the last one (number 11) was built around 226 A.D. In this 500+ year span, engineers continuously studied the system and made many improvements.

The Long-Term Impact The most visible impact of this huge engineering project was that Rome was able to continue to grow and flourish. Because of this, it remained a major power for hundreds of years. It wasn't until the 6th century A.D. that the Goths cut off the aqueducts and besieged Rome. The longer term impact has been the development of the field of Civil Engineering and the building of infrastructure for the greater good of society. The most tangible outcome of the Roman aqueduct's design is indoor plumbing. It was invented back then, but has become so common place that we take it for granted.

ARCHITECTURAL ENGINEERING:

Are you interested in buildings and architecture? Then take a look at architectural engineering. Amongst other things, you can design the lighting used in stadiums, develop efficient indoor cooling and heating systems, investigate better structural systems for new building designs, and even manage building construction.

INDUSTRIAL ENGINEERING:

If you like working with people and machines, then you should consider industrial engineering. Industrial engineers figure out how to do things better—from designing the admissions procedure at a hospital to designing a bar coding system for identifying and transporting an air passengers' luggage to ensure that it does not get lost. Industrial Engineers are always at work making things better.

A Guiding Light—The Great Lighthouse of Alexandria

You have probably heard of the Seven Wonders of the Ancient World. These include such engineering feats at the Great Pyramid of Giza and the Hanging Gardens of Babylon. We want to focus here on the Great Lighthouse of Alexandria, which turned out to have a major impact on history.

What Was Happening at the Time? Alexander the Great had just taken over Egypt in 332 B.C. and, like all leaders, he needed a capital. He selected Alexandria's location because it was 20 miles or so west of the Nile. Because of its distance from the river, Alexandria's harbors wouldn't get filled with mud and silt. Just north of Alexandria was a small island called Pharos. This island was attached to the mainland by a small dike that divided the great harbor. The location avoided the silt of the Nile, but was situated along a flat coastline that made for very dangerous sailing conditions.

The Engineering Solution Sailors needed a way to find the entrance to Alexandria's harbors—by either the East Harbor, coming from the Mediterranean Sea, or the West Harbor, coming from the Nile. The entrance needed to be visible far away, both day and night.

The Alexandrian engineers' solution was to build a lighthouse (the first ever) on the island of Pharos that was some 117 meters (384 feet) tall so it could be seen from a great distance. The tower was lit by a fire at night and during the day it was lit by the sun via a large mirror. The exterior of the tower was built of white marble to withstand the salt water.

The Long-Term Impact The Lighthouse of Alexandria, and other smaller ones like it, reliably guided sailors for some 1500 years, making seaborne commerce practical around the Mediterranean Sea. Historic records say the lighthouse succumbed to an earthquake some time in the 1300s. Except for the great pyramid, the Lighthouse was the longest standing of the Seven Wonders of the Ancient World. It opened a new area of Architectural Engineering and spawned the building of many similarly engineered structures around the Mediterranean.

The Low-Cost Transfer of Information—The Gutenberg Printing Press

Reading and writing helps all of us understand the world and communicate with each other. However, it wasn't until the mid-1400s that an engineer named Johannes Gutenberg changed reading and writing from something reserved for the elite of society to something widely available to the common person.

What Was Happening at the Time? Printing text on paper was slow and tedious when Gutenberg started his work. Each page had to be custom built, often carved out of wood. Although the Chinese had used movable type hundreds of years earlier, their techniques apparently had been forgotten by Gutenberg's time. Gutenberg had to figure out whether to make the type out of wood or metal, what to use for ink, and how to make impressions of type on paper. He also had to figure out where to get the money to do all this. He became one of history's first entreprenurial engineers.

The Engineering Solution Through experimentation, Gutenberg found a metal alloy that would melt at a sufficiently low temperature so that he could then pour it into letter molds. He also found inks that would transfer the impressions from the metal type to the paper cleanly.

He adapted a wine press to control the pressure of the type on the paper. In 1448, Gutenberg convinced Johann Fust, a lawyer and goldsmith, to invest in the press. Seven years later, his printed Latin Bible (now called the Gutenberg Bible) appeared on the market at a trade fair. (A perfect copy of the Gutenberg bible sold in 1978 for $2.4 million.)

The Long-Term Impact Some say the printing press is the earliest example of mass production. It is estimated that Gutenberg's printing press cut the cost of transferring and duplicating information by a factor of a thousand. Now people could learn from one another without being in the same place or time. Its first impact was that it put Bibles in the hands of common people, which, in turn, spurred the Christian Reformation. Later, the cheaper and easier dissemination of Greek and Roman writings caused a revival in classical learning that, in turn, led to the Renaissance. Before the printing press, only the wealthy could afford to read. Afterward, books were more affordable. Literacy made it possible to advance socially. Gutenberg's printing press has had such an impact that some 500 years later we can't imagine a world without print.

A Cool Idea—Refrigeration

You may think of a refrigerator as simply a nice way to keep your drinks cool, or air conditioning as a way of making the hot summer more comfortable. Both are true—but their impact on society has been so much greater. Refrigeration and its cousin air conditioning have changed the world forever.

What Was Happening at the Time? Prior to the 1870s, it was well understood that refrigeration was a good way to preserve food. Refrigeration permitted foods to be transported over greater distances without spoiling, thus allowing the places where food was produced to be farther from those who consumed it. However, natural ice was the major method of cooling. (Even Thoreau's Walden Pond was producing 1000 tons of ice per day in 1847.) Most was scraped from ponds, but clean ponds were getting harder to find.

The Engineering Solution Refrigeration is achieved by circulating a special chemical in a closed system; this chemical absorbs heat when it evaporates to a gas. It then releases the heat in another place when it condenses back to a liquid. This allows us to move heat from one place to

MATERIALS ENGINEERING:

Materials Engineers are on the cutting edge of technology in nearly every field. They work with scientists to develop the materials with the right kinds of properties that make other advances possible—from innovative shoe soles to special semiconductors—materials engineers are helping with nearly every engineering project today.

AGRICULTURAL ENGINEERING:

Where does all that food come from? Ask an Agricultural Engineer. They bring food from the farm to the table by designing sophisticated agricultural systems like hydroponics and new food products like chips, breakfast cereals, and healthy food supplements. Next time you have a great meal, thank an agricultural engineer.

CHEMICAL ENGINEERING: Chemical engineers put chemistry into action. They take raw materials and turn them into valuable products including medicines, plastics, and even special fibers and fabrics used in clothing.

another—thus allowing us to create cool locations. Two of the problems the engineers of the time had to solve were how to make the special chemicals, called refrigerants, circulate efficiently and how to find a refrigerant that was safe to use. The solution, unlike Gutenberg's press, is the result of innovations of thousands of engineers working on many different parts of this problem. By 1920, the electrical motor was reliable—so it was used to pump the refrigerant around. In 1928, Freon 12® was developed and found to be a safe refrigerant.

The Long-Term Impact Refrigeration technology led to the ability to keep foods fresh longer, which generally improved health and the quality of life as well as increased the availability of food to an ever-growing world population. Air conditioning reversed a long trend of people leaving southern U.S. cities for cooler climates. Many of the largest cities in the U.S. are today found in the south and have been made more liveable by air conditioning. Temperature-controlled rooms have made us less dependant on the weather for both our work and play can occur since we can now control much of our indoor environment.

However, together with the good there can occur a bad consequence of engineering innovations. An unfortunate long-term impact was the discovery in the late 1980s that the fluorocarbons, which make up Freon®, were destroying the Earth's ozone layer. Once again engineers and scientists have been called to action to find a replacement that won't do such damage.

Going Farther—The Automobile

AUTOMOTIVE ENGINEERING: So you like cars? Then think about being an automotive engineer. They dream up, design, and help build nearly everything—from the brakes to the engine—found on a car, truck or motorcycle.

People have always wanted to have faster, more reliable travel. Back in 1769, Frenchman Nicolas-Joseph Cugnot built the first car, a steam powered tricycle. While steam power dominated automobile development for years, it had many problems. It was not until the emergence of the internal combustion engine together with other engineering innovations that automobiles became a part of our world in the early 1900s.

What Was Happening at the Time? By the early 1900s, various ways of powering cars had been tried, including horses, steam, springs, electricity, and even the wind. But each had its own shortcomings. Boiler explosions made steam dangerous. Electricity was safer, but large banks of storage bat-

teries were needed. Range and speed were limited to 10 to 20 mile per hour and 50 miles between charges. Also, at that time, to drive a car you really needed to be a mechanic. Standard repair kits carried in all cars had over 50 items ranging from tape to repair fuel lines to extra hoses, spark plugs, and grease. Tire repairs were sometimes needed every 10 or 20 miles.

In the early 1900s, each car was handmade and cost around $1550. Only the wealthy could afford a car at this price, since the average worker's wage was under $13 per week.

The Engineering Solution Automotive engineers found internal combustion engines were much safer than steam engines and had better speed and range than electric cars. It was Ransom Olds and then Henry Ford who found faster and cheaper ways to make cars out of this technology. This drove the price down to where the common person could afford a car. In 1906, you could by a Model T Ford for $400; in 1916, it cost only $290.

Engineers were also able to significantly increase the reliability of cars. Today, a car can run 50,000 to 100,000 miles with little more maintenance than adding fuel and changing the oil.

The Long-Term Impact In 1900, there were some 8000 cars registered in the United States. Today, there are a half a billion cars in the world, one third of those are in the United States. Cars are everywhere. The automobile industry makes up more than a quarter of all retail trade in the United States. Japan and Western Europe are quickly approaching this level.

Cars have changed our social patterns. It's been said that the Henry Ford freed the common people from the limitations of geography, creating social mobility on a scale previously unknown. The same technology, in the form of tractors and other machinery, has revolutionized farming. Heavy goods are carried from city to city by trucks—yet another spin off of the automobile. All this is made possible by the automotive engineer and/or the transportation engineer.

TRANSPORTATION ENGINEERING:

Transportation engineers design streets, highways, and other transit systems that allow people and goods to move safely and efficiently. For example, before constructing a new sports stadium, city officials rely on transportation engineers to plan traffic patterns that will prevent major tie-ups before and after a game.

A Shocking Idea—Electricity

Can you name all the things you have touched today that use electricity? How many did you get? 25? 50? Maybe it would be easier to name the few things you used that didn't need electricity. Electricity has

ELECTRICAL ENGINEERING:

Electrical Engineers are the largest group of engineers working today, and for good reason. If it has electricity flowing through it—cell phones, computers, videogames, CAT scans, etc.—then electrical engineers are part of the design team.

clearly made a major impact on the world today. At the end of the 20th century, the National Academy of Engineering asked people what the 20 most important technological achievements were for the 20th century. Electrification was picked as the *most* important. Here's why.

What Was Happening at the Time? Back in 1882, Thomas Edison developed the first commercial power plant. Unfortunately, he was using direct current, which couldn't be distributed efficiently over long distances. Even 40 years later, folks in the countryside were using hand-wound clocks for time, candles or lanterns for light, and drinking hand-pumped water. So why were so many things still being done by hand when power plants had been around so long? The engineers still had many problems to solve and money had to be raised to make it possible to build generating plants and transmission lines all over the country.

The Engineering Solution The problems to be solved were both electrical and mechanical. Most electricity was generated by water wheels or burning coal. And, once a power plant was built, engineers had to get the power from the plant to the users. Engineers such as Nikola Tesla and Charles Steinmetz discovered that using alternating current like that found in your wall sockets today allowed higher-voltage power to be sent over much longer distances. Once this was known, factories could be built farther from the sources of power and power plants could be placed many miles from the cities they served. Of course, thousands of other engineers were involved to solve the other problems of making electricity safer and more reliable.

The Long-Term Impact What hasn't been improved by the availability of electricity? It's been said that the ready availability of electricity has liberated humans from the drudgery of manual chores. Clean industrialization became a reality since the source of power did not have to be where the power was being used. Electric-powered, labor-saving machinery also has led to a more highly educated populace, because electrical power has significantly reduced the time required to do the thousands of tedious tasks at home and work. This has left us more time for family, leisure activities, and exploration.

NUCLEAR ENGINEERING:
Nuclear Engineers design and develop processes to extract a wide variety of social benefits from nuclear energy. These include developing new nuclear power sources for spacecraft or various industrial and medical uses for radioactive materials, including equipment to diagnose and treat medical problems.

BIOMEDICAL ENGINEERING:
Bioengineers utilize a wide variety of science, mathematics, and engineering principles to improve the heath and well-being of humans and animals. They work on an extremely broad array of problems from the molecular level all the way to complete organ systems. Biomedical engineers work hand in hand with medical doctors to push the frontiers of medicine.

A Big Shortcut—The Panama Canal

What Was Happening at the Time? It's the early 1900s and you need to move goods from the East Coast of the United States to the West Coast. How would you do it? You could put the cargo on a railroad or send it by ship. If it was really big, it had to go by ship, and that ship had to journey all the way around the tip of South America. This took a long time, to say the least.

In the mid 1800s, engineers began talking about creating a shortcut using a canal through Nicaragua or Panama. Sometime around 1882, a French company began to cut a canal through Panama; however, after six years they decided to give up. Some 10,000 to 20,000 workers died from mosquito-carried malaria and other diseases. Even if the disease problem could be solved, there were still many other engineering challenges, such as where and how to build the locks so they could survive enemy attacks, runaway ships, and even inept lock operators.

The Engineering Solution At the start of the project, doctors did not know that malaria was carried by the common mosquito. Once it was understood that the mosquitoes were carrying the diseases, it was easy for the engineers to have still bodies of water drained or sprayed. The locks were located inland a few miles from each ocean so that it would be hard for enemy naval ships to launch an attack. The locks were built with two gates at each end. That way if one gate was damaged or needed maintenance, the other gate would keep the lock functional. Also, the ships do not traverse the locks under their own power; railroad-type engines hold and move the ships through the 50-mile canal. Americans took over the work on the canal around 1904, and 10 years later, in 1914, the first ship passed through!

The Long-Term Impact During its 90 years of continuous service, the Panama Canal has made goods cheaper by reducing the cost of transporting them. Initially, traffic was around 2000 ships per year and now it has reached approximately 13,000 ships per year. The canal is used by most interoceanic travel from the Atlantic to the Pacific. What was thought to be impossible 30 years before the American construction of the canal began, is now commonplace thanks to the engineers.

OCEANOGRAPHIC ENGINEERING:

Do you care about the health of beaches, coastal waterways, and the ocean? Then think about becoming a Coastal and Oceanographic Engineer. They work closely with civil engineers to design systems that protect our beaches and estuaries, as well as design structures such as offshore drilling platforms.

Visit the Stars—Space Exploration

"If you talk about going to the moon, they call you a poet or a dreamer. If you actually go there, they call you an engineer."[1] For as long as humans have looked at the stars, they have dreamed of visiting them. For many centuries, it was only a dream, but just a few decades ago, it became a reality thanks to the engineers.

What Was Happening at the Time? Just a year after the Panama Canal was finished in 1915, Robert Goddard proved rockets could work in a vacuum. However, it wasn't until the early 1960s that humans traveled into space—first, in 1961, Yuri Gagarin, a Russian, and then later that year Alan Shepard, an American. Engineers had to make many advances in energy, communications, materials, and computers before we could put a person on the moon.

AEROSPACE/AERONAUTICAL ENGINEERING:

Are you interested in aviation or space travel? Then look into Aerospace/Aeronautical Engineering. You can learn how to design airplanes, or maybe spaceships. Who knows, you might even get to work on the International Space Station.

The Engineering Solution Many problems had to be resolved in many fields in order to safely land and return a person from the surface of the Moon. For example, small, lightweight computers were needed to guide the spacecraft. Unfortunately, at the time, computers usually filled a entire room. In the biomedical field, we needed to understand the effects of space travel and weightlessness on people. New sources of energy were needed to escape the Earth's gravity. New, lighter, stronger materials had to be developed, so less energy would be needed to get into orbit. These, and many other problems, were solved and, in 1969, Neil Armstrong and Buzz Aldrin became the first humans to walk on the Moon.

The Long-Term Impact Of all the great achievements, space travel probably has generated more new innovations and products than any other. It's estimated that some 60,000 new products have come

[1]Advertisement slogan.

ENVIRONMENTAL ENGINEERING:

If you are concerned about the environment, consider Environmental Engineering. Environmental Engineers design water, land, and air purification systems, as well as help create manufacturing plants and processes that are environmentally friendly.

from the space program and more are being developed every day. For example:

- The need for smaller, fast computers fueled today's PC industry, which led to many of the devices we use today (i.e., computers, MP3 players, digital cameras, etc.).
- The development of sophisticated control systems that could guide rockets and spacecraft led to modern control systems for commercial aircraft and a wide variety of robotic applications.
- New sensors were developed to monitor the astronauts while in flight. Those sensors are now used routinely by doctors to monitor Earthbound patients.
- Plastics and polymers developed for space use are now used in many everyday products.
- The first demonstration of a commercial communications satellite, Telstar®, was designed and built by AT&T and launched in 1962. Today, satellites handle millions of telephone conversations, Internet traffic, and television channels.

Of course, the broadest impact is that we better understand the universe and our place in it from studying and being in space.

Television—What a Sight to See

Can you imagine having no television? Or just three channels? Or worse, television in black and white? The television we take for granted today is the result of many engineers' visions.

What Was Happening at the Time? Back in 1920, KDKA in Pittsburgh, PA started broadcasting the first commercial radio signal. It wasn't long before radio was the main source of entertainment in most of the households in America. The limitation of radio was that you could only hear broadcasts. In order to send visual images,

engineers had to discover how to capture an image, transmit it, and then display it. To do this properly, engineers studied, among other things, the human vision system to better understand how it works.

COMPUTER ENGINEERING:

Computer Engineers work with all aspects of computer or digital systems. This may be the computer sitting on your desk, or the computer in your cell phone, MP3 player, TV clicker, DVD player, or any number of devices that can contain computer technologies.

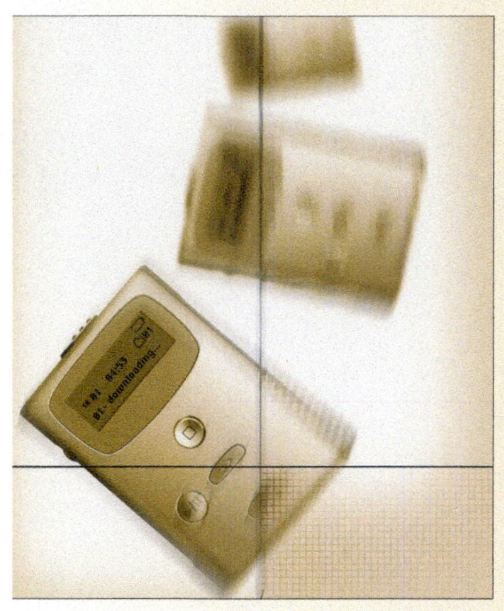

The Engineering Solution In 1927, Philo T. Farnsworth transmitted the first television image. First, he had to invent the scanning cathode ray tube, which could convert a two-dimensional image into a signal. What's remarkable is that Farnsworth was only 21-years old at the time. Back then, the pictures were on a round display, they were black and white, the cameras weren't very good, and the cameras needed a lot of light.

In 1930, Vladimir Zworykin of RCA created the first good television camera. The rectangular television tube was perfected in the late 1950s. In the 1950s, some 1 million black-and-white televisions were sold, and the next decade saw the transition to color television. The first coast to coast live colorcast showed the Rose Parade on New Year's day in 1954. Today one million televisions are sold every few hours.

The Long-Term Impact It wasn't long after the rectangular picture tube was perfected that the cost of a TV dropped to $200. Within 10 years, another 45 million televisions were sold. It's now estimated that 106.7 million households in the United States have at least one television.

Radio and television have been major facilitators of social change in the last 50 years. Television has improved communication around the world, making the world more connected so that people can see what is happening in other countries in real time. In addition, people living in remote areas can use television for education and entertainment. For many people, the only sporting events or live concerts they see are through television.

Television has also led to routine use of live displays from remote sites for applications such as underwater exploration, medical imaging, or even industrial inspection. It is not clear that all the uses of television are socially beneficial, but people are always looking for new ways to use it and to take advantage of its tremendous capabilities. It is clearly an understatement to say that television has changed us all forever.

Integrated Circuit—Small Makes a Big Impact

Most of the great engineering feats we have chosen have been big, very visible projects. The integrated circuit is successful because it is so small. In fact, this invention has made possible many other great scientific and engineering feats.

SOFTWARE ENGINEERING:

Do you love to program? Then think about Software Engineering. Software engineers work with all kinds of engineers to design, write, and test the billions of lines of software that is running the computer systems around the globe.

What Was Happening at the Time? Back in 1955, a high-speed computer would weigh three tons, consume 50 kilowatts of power, and cost $200,000. It would fill a whole room, but could do only hundreds or thousands of multiplications per second. Such a computer was powered by vacuum tubes, which used a stream of current inside a glass tube to switch circuits on and off. A typical tube was about the size of your thumb. The 1950s computer unfortunately was too big, used too much

power, cost too much money, and ran too slowly—but was the best that could be created at the time.

The Engineering Solution The transistor was invented in 1947 by three Bell Labs® engineers and scientists. The transistor allowed a small electrical current to control a larger current, like the vacuum tube; however, the transistor was much smaller than a tube and it used much less power. It took many engineers several years to really learn how to use the transistor.

In 1958, just 11 years later, Jack Kilby at Texas Instruments invented the integrated circuit (IC). The IC allowed several transistors (and other components) to be manufactured together on the same device. This allowed even more transistors to be put in a smaller area—leading to today's digital revolution.

The Long-Term Impact The long-term impact of the integrated circuit has been amazing. In the 1950s, a transistor cost between \$5 and \$45 to make. Today, the cost of a transistor on an IC is one hundred-thousandth (10^{-5}) of a cent. They are so cheap, you can think of them as almost free.

Because of the integrated circuit, computers today weigh 10 pounds or less, consume 100 watts of power, and cost under \$500. They easily fit on your desk and can do one billion multiplications (or more) per second.

But that's only a big computer. Digital technologies, made possible by the integrated circuit, have become so small that they are in many things that you don't even think about. For example, you can find integrated circuits in your watch, your calculator, CD/MP3 player, microwave, cell phone, hearing aids, and even in your car.

EXERCISES 9.2

1. List some innovations that interest you, then list the fields of engineering that relate to those things.
2. There are many more engineering feats we could list that have impacted the world over the centuries. List a few major engineering feats that we haven't discussed. Which on your list do you think are as important to society as the ones listed above? Why?
3. List a few *minor* engineering feats. Are these important too? Why?
4. In this chapter, we have outlined the short- and long-term impact of each of the 10 feats. List other impacts these feats have had. Are these impacts intended by the engineers who designed and built the feats? Are they always beneficial to society?
5. Look at all the feats again and try to find out where the money came from to make them happen and keep them going.
6. Pick a feat and list all the fields of engineering needed to develop, manufacture, and support it.

9.3 What Most People Don't Know About Engineering

The engineering process presented in this book is a method of using science, technology, and mathematics in an orderly way to create solutions to a problem. The people who do this are called "engineers" and they have the opportunity to do some of the most creative work in the world. This said, there are a lot of people who don't understand engineering and the men and women that do it. We're here to "set the record straight" and demonstrate how and why engineers do their jobs the way they do.

Engineering and Science Are Different Scientists work to understand the natural world, while engineers use scientific and mathematical knowledge to build things that people need. Even though engineers need to know science in order to do their work, and scientists often need to know some engineering in order to build their experimental equipment, the methods by which the two approach their jobs are quite different. Engineering is focused on creating things for people and society, rather than analyzing and observing basic physical laws.

You Don't Need to Have Perfect Grades in Math and Science to be an Engineer A knowledge of math and science is certainly important for engineering, but the best engineers are the most creative ones, not necessarily the ones who've aced the most classes. Once engineers are in a business environment, they often use their judgment and experience to guide their efforts. Math and science principles provide the basis upon which much engineering is performed, but it is the creative aspect of engineering that leads to new products, services, and benefits for society.

Engineers Do More than Design Things We've focused on the engineering design process in this book. However, many engineers do not design products. Some make sure that they are manufactured properly, some repair and maintain products and systems, some train the product's users, some help raise money to start an engineering business, and some manage people and projects. Some engineers work with potential customers to figure out what is needed in the first place! In all of these activities, the basic engineering problem-solving approach helps lead to successful efforts.

Students Who Earn an Engineering Degree Have a Broad Choice of Career Paths An engineering degree is an excellent preparation for a wide variety of careers. In addition to the traditional engineering career options, many other career paths can benefit from engineering knowledge and skills. Lawyers interested in patent law or intellectual property law or business leaders interested in managing high-tech companies will find an undergraduate engineering degree useful in their graduate studies. In addition, many new fields, such as biotechnology, rely on nontraditional mixtures of science, engineering, and instrumentation.

Engineering Is a Collaborative Process Very few important products or projects are completed by one person working alone. Thomas Edison, one of the greatest inventors who ever lived, worked with a large group of engineers, technicians, and business people. When designing complex systems, teams of people are required to get the job done. In fact, engineering is one of the most collaborative fields that one can imagine, and working with others is an important asset and skill for the task.

Most Engineering Projects Are Not Perfectly Planned Virtually no project or product turns out exactly as planned at the beginning. Inevitably unexpected problems arise that must be solved, or new ideas occur that can be used to make the result even better than originally thought. The ability to accept these changes and constantly revise a design is the hallmark of an experienced engineer. The practice of engineering, by its very nature, is an iterative process.

Engineers Approach Their Work with Passion No group of professionals is limited to one stereotypical set of personality traits. While engineers are not unemotional by nature, we expect their professional judgments to be objective. Their decisions should be based on the best facts and analysis that are available. But engineers have choices as to what they work on, and these choices are driven by their concerns and caring about the problem at hand. That is one reason why engineering is such a fulfilling profession. Engineers find joy in their work and satisfaction in knowing that, directly or indirectly, the results of their work lead to improved products or services that people want or need. Nothing makes an engineer happier than when all of the pieces fall into place to make that fantastic product, achievement, or benefit happen that changes peoples lives.

Engineers Work in a Very Wide Variety of Areas In this book, we have focused on the engineering process in the area of emerging digital technologies. Engineers, however, work in many other fields, ranging from the depths of the ocean to heights of the sky (and beyond). Pick anything you are interested in and you can find an engineer that works in that area.

EXERCISES 9.3

1. Identify three well-known engineers and describe their contributions to society.
2. Identify three famous people who are engineers by training and are doing things other than traditional engineering.
3. Identify a new technology that interests you. Describe the engineering team that came up with the design.

9.4 Getting Ready to Change the World

In Section 9.2, we explored ten major engineering breakthroughs that changed the course of humankind. Although not every engineering effort has such a major impact, it's easy to see that even without pursuing engineering as a career, the engineer's approach to problem solving can prove useful through life.

Do you want to be an engineer or, at least, know when and how to approach problems like an engineer? Well, the key elements to being a great engineer are to focus on society's problems, work creatively with your colleagues, be a good communicator, and be a great problem solver.

If you plan on pursuing a career in engineering, you should take all of the math and science courses that you can. You should also learn how to communicate your ideas effectively. English, public speaking, creative writing, and business writing courses will give you some basics in effective communications.

What types of mathematics should you learn? Different types of engineering tend to rely on different types of mathematics. Virtually all fields of engineering take advantage of algebra, trigonometry, and beginning calculus. From there, the required background depends on the engineering field. For example, the design of computer circuitry depends heavily on a knowledge of logic and discrete mathematics, while calculus and linear algebra are needed for electrical and mechanical engineering. Engineering students are encouraged to learn as much math as possible, in both high school and college, so that they can pursue any type of engineering.

What science do you need to know? Most fields of engineering are tightly associated with a field of science. Chemistry, for example, provides the basic knowledge of nature that gives chemical engineers the starting point for the development of new products. In a famous trademark of the 1960s, the chemical manufacturer E.I. Dupont called this process the creation of "better things for better living." Similarly, physics is the scientific starting point for telecommunications systems, for nuclear engineering, and even for mechanical engineering. Mining engineers are expected to know about chemistry, physics, and geology. Some of the most promising fields of the future, such as biotechnology, will require multi-disciplinary teams, in which individuals have both a solid education in a particular area, such as chemistry or electrical instrumentation, and a good understanding of the specialization areas of the other team members. Just as with mathematics, potential engineers are encouraged to learn as much of science as possible, both in high school and in college.

What else is important? Truly innovative solutions to problems often rely on knowledge beyond that taught in math and science classes. Knowledge of the history, language, and culture of an area can make the difference between success and failure of a project in the eyes of the people for whom the engineering project was conceived. Ultimately, most engineering endeavors count on the availability of money and on the economic viability of the project at hand. Some knowledge of finance and economics is valuable for any engineer. But in all cases, communicating your ideas depends on your ability to write clearly and speak to groups of technical and nontechnical people.

EXERCISES 9.4

1. What future engineering feat would you like to be involved with?
   a. What problem will you be solving?
   b. Where will you get the financial resources?
   c. Where will you find the engineering and management talent?
   d. What engineers will you need to help you?
   e. What new fields of engineering will you have to invent?
   f. What will be the short and long-term impact of the feat? Can you imagine any unintended consequences?
   g. If you solve this problem, what other new applications of this technology might your solution be able to address?

9.5 Looking to the Future

The examples of Section 9.2 have shown that engineering has had a tremendous impact on the world over the past centuries. The work of engineers can be found everywhere in our daily lives. This will continue into the future, using types of engineering that we can't even imagine now.

Engineering is a creative process that uses math, science, and all of the other knowledge that you can gather for solving problems that matter to people. The engineering process can be applied to a range of problems far broader than just the technical ones explored in this book.

The world of today is due in large part to the efforts of engineers. And the marvels of tomorrow will be made possible by the efforts of our future engineers—you.

Appendix

Constant Growth Rates, Logarithms, and Exponentials

The Exponential Function

If something grows at a constant rate, its value can be computed using the following *exponential function*:

$$y = f(x) = a^x \qquad (A.1)$$

The base of the exponential function is a. The base, which is greater than 0, determines the growth factor for y when x increases by 1, since $f(x + 1) = a \times f(x)$, as shown in Equation (A.2):

$$f(x + 1) = a^{x+1} = a^x \times a^1 = a^1 \times f(x) \qquad (A.2)$$

When $a > 1$, we say that y is *growing exponentially*, because whenever x increases by 1, y gets larger by a factor of a. When $0 < a < 1$, we say that y is *decaying exponentially*, because whenever x increases by 1, y gets smaller by a factor of a. If $a = 1$, then $y = 1$ for all values of x. Table A.1 shows some values of y computed for five different bases. Although the values of x in this table are integers, x in Equation (A.1) can be any real number. For example, if $a = 2$ and $x = 0.5$, then $y = 2^{0.5} = 1.414\ldots$. If x is a real number and a is a real number greater than 0, then y is a real number that is greater than 0.

Table A.1 Values for Exponential Functions with Five Different Bases for Selected Values of x (Although the values of x shown here are integers, x can be any real number.)

Base	$a = 2$	$a = 10$	$a = 16$	$a = 0.5$	$a = 0.1$
x	$y = f(x) = 2^x$	$y = f(x) = 10^x$	$y = f(x) = 16^x$	$y = f(x) = 0.5^x$	$y = f(x) = 0.1^x$
−1	0.5	0.1	0.0625	2.0	10.0
0	1.0	1.0	1.0000	1.0	1.0
1	2.0	10.0	16.0000	0.5	0.1
2	4.0	100.0	256.0000	0.25	0.01
3	8.0	1000.0	4096.0000	0.125	0.001
4	16.0	10,000.0	65,536.0000	0.0625	0.0001
5	32.0	100,000.0	1,048,576.0000	0.03125	0.00001

The Logarithm

The logarithm function is the inverse of the exponential function. When y is defined by the exponential function in Equation (A.1), then we can use the following equation, the inverse of the exponential function, to find x when we know the value of y:

$$x = f^{-1}(y) = \log_a(y) \qquad (A.3)$$

The base of the logarithm, a, is the same as the base of the exponential function. Values for the logarithm of any positive value of y can be found using series expansion, precomputed tables of values, a calculator, or a computer function.

To gain some intuition about the logarithm, consider the special case where x is a positive integer. If $y = a^x$ and the integer value x is known, then it is easy to compute the value of y. Start by setting y equal to 1 and then perform x multiplications of y by a to get $y = a^x$. If the value of y is known but the value of x is not, x can be computed by dividing y by a until the value of y gets down to 1. The number of divisions needed to get the value of y down to 1 will be x. This simple example shows the inverse relationship between the logarithm and the exponential, and it helps demonstrate that the base a is the constant growth factor.

Logarithms have many applications, because they make comparing different growth rates very easy. Estimating the rate of growth from the logarithms of a sequence of values is much easier than estimating the growth rate from the values themselves. For this reason, logarithmic plotting scales routinely are used to analyze such things as financial growth, technology growth, or population growth.

Logarithms can have any base, although the most commonly used base is 10. Functions to raise 10 to a power or compute a logarithm with a base of 10 are found on mathematical calculators and computers. Base-2 logarithms often are used to compute the number of bits needed for a binary number that will have a specified range of values. Many calculators and computers can also compute logarithms using the base $e = 2.718\ldots$, which has very interesting mathematical behavior. When this base is used, the logarithms produced are often called *natural logarithms*. A few examples of base-2 and base-10 logarithms are shown in Table A.2.

Graphs of the exponential function and the logarithm for base $a = 2$ are shown in Figure A.1 The values from the base-2 column of Table A.1 are marked on the graphs with circles. Note that in Figure A.1(a), when x is 0, y has a value of 1, not 0. As x becomes more negative, y will decrease in value, but it will always be greater than 0. We have not allowed negative values of y for the same reason that the square root of a negative number is not defined when we can use only real numbers. This restriction can be removed for logarithms and square roots if imaginary numbers with $i = \sqrt{-1}$ are used.

Table A.2 Values for Some Base-2 and Base-10 Logarithms

Base $a = 2$		Base $a = 10$	
$y = 2^x$	$x = \log_2(y)$	$y = 10^x$	$x = \log_{10}(y)$
0.25	−2	0.001	−3
1	0	1	0
2	1	2	0.30103
8	3	8	0.90309
10	3.322	10	1
256	8	100	2
1000	9.966	1000	3
1024	10	1024	3.0103
2048	11	2000	3.30103

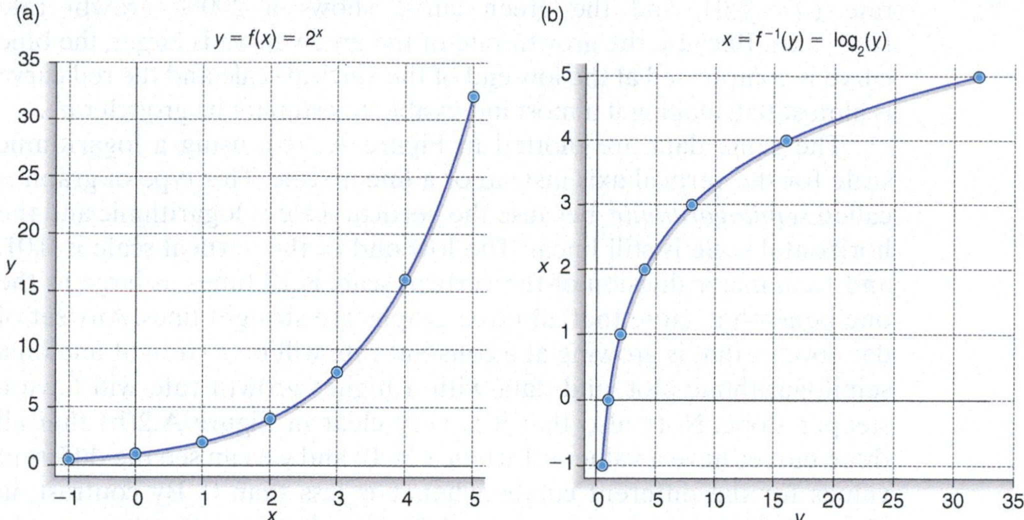

Figure A.1 Plots of $y = f(x) = 2^x$ and $x = f^{-1}(y) = \log_2(y)$ for $0 \leq x \leq 5$. The circles show integer values of x.

Plotting Exponential Functions on Linear and Semilogarithmic Plots

It is difficult to determine by visual inspection whether the graph in Figure A.1(a) has a constant growth rate. When x is small, the values of y are difficult to read accurately, because they are compressed at the low end of the vertical scale compared with the values of y for larger x. The problem is even worse when data from processes with different growth rates are plotted on the same axes for comparison.

In Figure A.2(a), the red curve shows a constant growth rate of 50% for each increment of x ($a = 1.5$), the blue curve shows a 100% growth

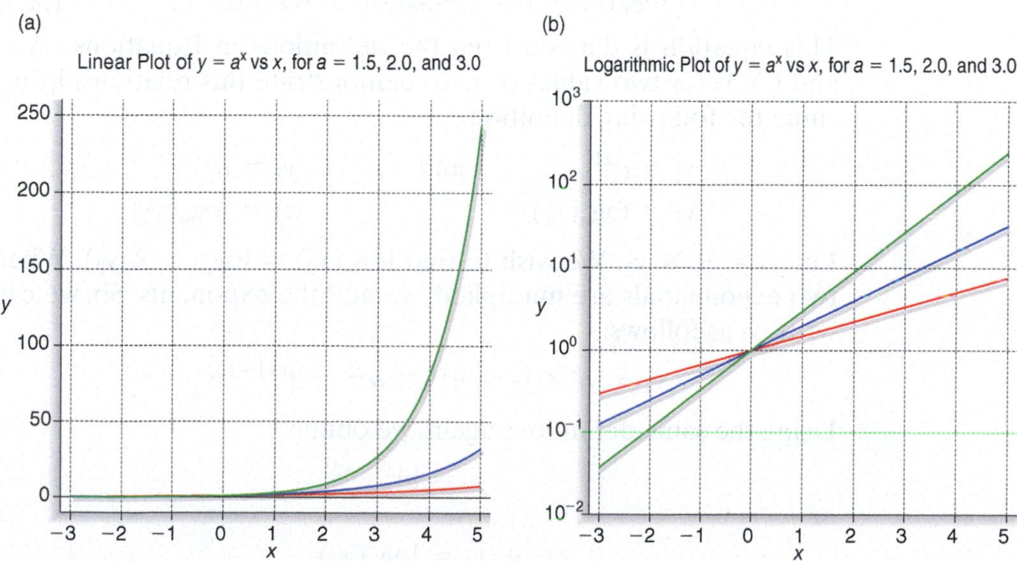

Figure A.2 The graphs show three quantities growing at constant rates of 50% (red), 100% (blue), and 200% (green). In (a), the vertical scale is linear, and in (b) the vertical scale is logarithmic.

rate ($a = 2.0$), and the green curve shows a 200% growth rate ($a = 3.0$). Because the growth rate of the green curve is larger, the blue curve is compressed at the low end of the vertical scale and the red curve is almost flat, making it almost impossible to estimate its growth rate.

The same data are plotted in Figure A.2(b), using a logarithmic scale for the vertical axis instead of a linear scale. This type of graph is called *semilogarithmic*, because the vertical scale is logarithmic and the horizontal scale is still linear. The low end of the vertical scale is 0.01, and each major division of the vertical scale is 10 times as large as the one beneath it. Note that all three graphs are straight lines. Any set of data points that is growing at a constant rate will be a straight line on a semilogarithmic plot, and data with a higher growth rate will have a steeper slope. Note also that it is very clear in Figure A.2(b) that all three curves have a value of 1 when $x = 0$; and we can see the different values for the different curves when x is less than 0. By contrast, in Figure A.2(a) we can not see any difference between the three curves when x is negative. Because the semilogarithmic plot can show such a wide range of values, it is ideal for plotting data that are growing at a constant rate or for comparing different growth rates.

Useful Relationships

- **Simple relationships for 0 and 1:**

$$a^0 = 1 \qquad 0 = \log_a(1) \qquad \text{(A.4)}$$
$$a^1 = a \qquad 1 = \log_a(a) \qquad \text{(A.5)}$$
$$a^{-1} = 1/a \qquad -1 = \log_a(1/a) \qquad \text{(A.6)}$$

- **The logarithm of the product of two values is the sum of the logarithms of the values:**

$$\log_a(y_1 \times y_2) = \log_a(y_1) + \log_a(y_2) \qquad \text{(A.7)}$$

This equation is derived from the definitions in Equations (A.1) and (A.3) for two values of x. To demonstrate this relationship, assume the following definition

$$y_1 = a^{x1} \qquad \text{and} \qquad y_2 = a^{x2}$$
$$x_1 = \log_a(y_1) \qquad\qquad x_2 = \log_a(y_2)$$

Let $y_3 = y_1 \times y_2$. We wish to find $\log_a(y_3) = \log_a(y_1 \times y_2)$. When two exponentials are multiplied, we add the exponents. So, we can write y_3 as follows:

$$y_3 = y_1 \times y_2 = a^{x1} \times a^{x2} = a^{(x1+x2)}$$

Using the same definitions again, we obtain

$$y_3 = a^{(x1+x2)}$$

or

$$x_1 + x_2 = \log_a(y_3)$$

and substituting for x_1 and x_2 gives us

$$\log_a(y_1) + \log_a(y_2) = \log_a(y_3)$$

Several examples can be taken from Table A.2. If $y_1 = 2$ and $y_2 = 1000$, then

$$y_3 = 2 \times 1000 = 2000$$

and

$$\begin{aligned}\log_{10}(2000) &= \log_{10}(y_1) + \log_{10}(y_2) \\ &= 0.30103 + 3 = 3.30103\end{aligned}$$

- **The logarithm of the quotient of two values is the difference of the logarithms of the values:**

$$\log_a(y_4/y_5) = \log_a(y_4) - \log_a(y_5) \tag{A.8}$$

This equation follows directly from Equation (A.7) if we let $y_4 = y_1$ and $y_5 = (y_2)^{-1} = a^{-x2}$. When $y_4 = 1$, we have

$$\log_a(1/y_5) = -\log_a(y_5) \tag{A.9}$$

because we know from Equation (A.4) that $\log_a(y_4) = \log_a(1) = 0$.

- **The logarithm of a value raised to the power *m* is *m* multiplied by the logarithm of the value:**

$$\log_a(y_6{}^m) = m \times \log_a(y_6) \tag{A.10}$$

This equation is derived from the definitions in Equations (A.1) and (A.3). Assume that

$$y_6 = a^{x6}$$

and

$$x_6 = \log_a(y_6)$$

Raising y_6 to the power m, we can write

$$y_6{}^m = (a^{x6})^m = a^{x6\times m}$$

This means that, using the same definitions again, we obtain

$$\begin{aligned}y_6{}^m &= a^{x6\times m} \\ x_6 \times m &= \log_a(y_6{}^m)\end{aligned}$$

Substituting for x_6, we get

$$\log_a(y_6) \times m = \log_a(y_6{}^m)$$

Several examples can be taken from Table A.2. If $y_6 = 2$ and $m = 10$, then

$$y_6{}^m = 2^{10} = 1024$$

and

$$\begin{aligned}\log_{10}(1024) &= 10 \times \log_{10}(2) \\ &= 10 \times 0.30103 = 3.0103\end{aligned}$$

Similarly, if $m = 3$, then $\log_{10}(8) = 3 \times \log_{10}(2) = 0.90309$.

- **Logarithms from one base can be used to make computations in another base:** We can compute base-10 logarithms by using calculators or computer functions. However, there may be certain situations where we may need to use logarithms that employ another base that is not so readily available. Using the following relationship, we can find logarithms for any base a, using the base-10 logarithms:

$$\log_a(y) = \frac{\log_{10}(y)}{\log_{10}(a)} \tag{A.11}$$

This equation is also derived from the definitions in Equations (A.1) and (A.3), using the basic definition of x and the base-10 logarithm of the base a that we want to use. Assume that

$$y = a^x \qquad \text{and} \qquad a = 10^q$$
$$x = \log_a(y) \qquad\qquad q = \log_{10}(a)$$

We can now substitute 10^q for a:

$$y = a^x = (10^q)^x = 10^{(q \times x)}$$

This form makes it easy to write an expression for $\log_{10}(y)$, because y is already expressed as a power of 10:

$$\log_{10}(y) = \log_{10}\left(10^{(q \times x)}\right) = q \times x$$

Solving for x and then substituting for q, we have

$$x = \log_{10}(y)/q = \log_{10}(y)/\log_{10}(a)$$

Since we wanted $x = \log_a(y)$, we now have the expression found in Equation (A.11), and we can compute x using only base 10 logarithms.

We can use Equation (A.11) to compute $\log_2(1000)$ in Table A.2. Since $y = 1000$,

$$\log_{10}(1000) = 3$$

and

$$\log_{10}(2) = 0.30103$$

so

$$\log_2(1000) = 3/0.30103 = 9.966$$

The derivation of Equation (A.11) also shows us an often useful approximation:

$$2^{10} \approx 10^3$$

From Table A.2, we can see that $2^{10} = 1024 \approx 10^3$. We can find the precise equality by using

$$2 = 10^q$$

or

$$q = \log_{10}(2) = 0.30103$$

Raising 2 to the power x, we have

$$2^x = (10^q)^x = 10^{qx} = 10^{0.30103x} \tag{A.12}$$
$$2^{10} = 10^{3.0103}$$

This relationship gives us a way to quickly estimate the value of large powers of 2 in terms of a power of 10.

Glossary

Access: A user's entry point into a data network. (8.3)

Access Server: A set of equipment located at an Internet Service Provider through which users gain access to the network and its services. An access server is commonly composed of a large number of voice-band or DSL modems, a controlling computer, and a router that aggregates the data from the modems and sends them into the network. (8.3)

Additive Synthesis: A synthesis technique that creates a periodic signal by adding sinusoids together. (2.4)

Algorithm: A step-by-step process to achieve a goal. In the case of this book, our goal is to describe and illustrate how to design new products or devices. (1.1)

Aliasing: An undesired effect, caused by under-sampling, where one signal can masquerade as another. (5.2)

American Standard Code for Information Interchange (ASCII): American Standard Code for Information Interchange, which is currently used on most computer systems. This binary representation of the letters and other characters on a keyboard can represent 128 different symbols, using 7 bits of information. (5.3, 7.5)

Amplitude: The size of a signal. (2.2)

Analog: Used to describe an electrical circuit or system in which the signals are direct analogs of physical processes such as distance, sound pressure, and velocity. It implies a continuously evolving or continuously changing signal, since that is the way that the physical processes are found in nature. (1.2)

Analog-to-Digital Conversion: Sampling of an analog signal followed by quantization of the samples to binary numbers. Also called digitization. (5.4)

Animation: Creation of moving images by creating individual frames that produce the illusion of smooth motion when shown in quick succession. (3.1)

Antialiasing Filter: An analog low-pass filter preceding the sampler to assure that the analog signal is band limited prior to sampling. (5.2)

Approximation: A calculation or procedure that produces a value close to the actual or perfectly computed one. (2.2)

Audio Sampling: The process of recording values (samples) of a signal at distinct points in time or space. (5.2)

Average: The average of a set of values is the sum of all the values, divided by the number of values. It gives us a general idea about the values in the set, but it does not tell us any specific value. (4.3)

Average Codeword Length: The average number of coded symbols it takes to store an original symbol sequence, using a particular code. (6.3)

Backbone Network: An Internet Service Provider whose sole function is to connect other Internet Service Providers to each other. A backbone network tends to use high-capacity transmission systems and very fast routers in order to handle the traffic load imposed by the numerous client ISPs. (8.3)

Band-Limited Signal: A signal whose highest frequency falls below some finite value. (5.2)

Bandpass Filter: A filter that has a strong response to frequencies in a specific range or band and has minimal response to all other frequencies; an "ideal" bandpass filter would have zero response to all frequencies outside the particular band. (7.2)

Bandwidth: The difference between the highest and lowest frequencies contained in a signal. Often, the lowest frequency is zero. In this case, the bandwidth is just the highest frequency in the signal. (5.2)

Bass Clef: Also called F clef. Indicates that the note on the fourth line is F below middle C. (2.2)

Baudot Code: A five-bit code (Tables 7.5, 7.6, and 7.7) used with a keyboard for international communications until the mid-1970s; represents 32 different symbols by using five bits of information per symbol or character. (7.5)

Binary: A term referring to a quantity that takes on one of two different values. Often used when referring to base-two, or binary, numbers. (1.2)

Binary Number System: An arithmetic system for representing numbers as a series of bits. (5.3)

Binary Numbers: Mathematical representation of numbers by using the base-2 system of numbers rather than the familiar base-10 system. (1.2)

Binary Point: Notation used to separate the integer and fractional parts of a binary number. (5.3)

Bit: Short for "*bi*nary digi*t*." A bit takes on only the value 0 or 1. (1.2)

Block Diagram: A graphical description of the overall operation of some system or design. Typically, the inputs to the system are on the left-hand side and the outputs are on the right-hand side. (1.4)

Broadcast System: Communications systems that use one transmitter to send information to many receivers. (8.1)

Byte: Eight consecutive bits. (1.2)

Change Detection: A process that looks for changes in pixel intensities found at the same location in two different images of the same scene taken at different times. After change detection, the resulting image shows the objects that have moved, but not the objects that have stayed in the same place. (4.1)

Chord: A collection of simultaneously played tones. (2.2)

Chromakey: A method of extracting a foreground subject from one image and placing it on a background that comes from another image. The foreground subject is usually photographed against a solid blue background so that the subject and background are easily separated. For this reason, Chromakey is also called a blue-screen effect. (4.1)

Clef: The sign written at the beginning of a staff to indicate pitch. (2.2)

Clipping: A form of signal distortion that results when the signal amplitude entering a digitizer exceeds the range of available quantization levels. Clipping results in the tops and bottoms of the signal being "clipped off." (5.4)

Cochlea: A fluid-filled tube in the ear that responds to different frequencies at different points along its length. (7.2)

Code: The rules that define the meanings of a set of bits—for example, the ASCII code that assigns 7-bit binary numbers to letters, spaces, punctuation, and other computer characters. (6.1)

Codebook Dictionary: The table that describes the translation of important symbols to their binary representation. (6.1)

Codebook Size: The number of entries in a particular codebook. (6.1)

Codeword: An encoded version of an important symbol. Represented by a bit string. (6.1)

Codeword Length: The number of bits in a particular codeword. (6.1)

Color Plane: Every color image can be created using three single-color images: one red, one green, and one blue. These three images are placed on top of each other and added together. These single-color components are the color planes. (3.2)

Colormap: A list of specific colors used in an image. Each color is defined by its red, blue, and green content, typically with 8 bits of resolution for each color. When using colormaps, a pixel value simply identifies the position of its color in the list. (3.4)

Communication Channel: The physical medium that carries a communications signal from the transmitter to the receiver—for example, telephone wires, optical fiber, or air. (7.1)

Compression: An encoding method that tries to reduce the number of bits needed to store or transmit information. (6.2)

Compression Ratio: The ratio of the number of bits of the compressed version of a signal to the number of bits in the original representation of the signal. (6.2)

Computer Graphics: The means by which digital images are created from descriptions or models of the image content. (4.1)

Constraints: Limits that are placed on a design problem. (1.1)

Content Server: A network-connected computer that has the ability to provide informational services to network users upon request. (8.3)

Cryptography: A form of coding whose goal is to make information difficult to read or be understood by others. (6.6)

Decoding: The act of recovering meaningful symbols through the translation of sets of bits. (6.1)

Decryption: In cryptography, a generic name for the decoding process. (6.6)

Digital: Describing technology or phenomena that are characterized by numbers, most typically expressed in base-two mathematics. (1.2)

Digital Age: The era born with the creation of the transistor. The digital age is generally thought to have reached maturity at the time that computers gained widespread use in the 1980s. It also implies that most new devices being produced today and in the future will rely heavily on digital technology. (1.2)

Digital Imaging: The technique of representing images (or movies) as a matrix of numbers. (3.1)

Digital Signal: Set of sampled values represented in binary form as bits. (5.1)

Digital Signal Processing (DSP): The manipulation or processing of digital signals. (5.2)

Digital-to-Analog Conversion (D/A): The process of creating an analog signal from a set of samples. (5.2)

Digitization: Combined operations of sampling and quantization. Also called analog-to-digital conversion. (5.2)

Domain Name Service (DNS): Protocol used on the Internet to obtain from a name server the translation from a uniform resource locator (that is, a domain name) of a computer on the network to its associated 32-bit IP address. (8.3)

Dynamic Range: The difference between the largest and smallest amplitudes or intensities that a signal takes on. (5.4)

Edge Detection: The process of looking for intensity differences between pixels that are near each other in an image. After edge detection, the resulting image shows the edges of objects and patterns, but not the areas where the color or intensity does not change. (4.1)

Element: An element of a matrix is a single value of the matrix. Each element is identified by its row number and column number. (4.2)

Encoding: The act of translating meaningful symbols into sets of bits. (6.1)

Encryption: An encoding method that tries to hide information from others. (6.1)

Engineering Design Algorithm: A nine-step process that is followed by engineers: (1) Identify the problem or design objective; (2) define the goals and identify the constraints; (3) research and gather information; (4) create potential design solutions; (5) analyze the viability of solutions; (6) choose the most appropriate solution; (7) build or implement the design; (8) test and evaluate the design; (9) repeat *all* steps as necessary. (1.1)

Entropy: In compression, the fundamental limit to the amount that a particular signal can be compressed without any loss. (6.3)

Envelope: A description of a signal's general size or amplitude over time. (2.4)

Error Correction: A joint encoding–decoding procedure that tries to fix unknown errors in the encoded information. (6.1)

Exclusive-OR (X-OR) Operation: A means of combining two sequences of bits into one, used for both encryption and decryption of information. (6.8)

Exponential Function: A waveform that grows or decays in a particular increasing or decreasing pattern. (2.4)

Field of View: When applied to an imaging system, the area of the scene that will be captured by the camera. (3.3)

File Transfer Protocol (FTP): Protocol used to reliably transfer disk files from one computer to another over the Internet. (8.3)

Filtering: Filtering an image produces pixel values for an output image based on the values of a group of pixels in the input image. Filtering is a neighborhood operation and is used, for example, to sharpen or smooth an image. (4.2)

Frame: An individual image in a sequence that, when displayed in time, constitutes a movie. If we push the pause button on a DVD player, the stationary image we see on the screen is a frame. (3.1)

Frequency Shift Keying (FSK): A method of representing a character by a series of bits whose value can be 0 or 1 and transmitting the character by sending the bits one at a time, using one frequency to represent a 0 and a different frequency to represent a 1. (7.4)

Fundamental Frequency: A mathematical quantity describing the repetition rate of a periodic signal. (2.2)

Gray Scale: Levels of gray tones covering ranges from all white to all black. (3.1)

Guaranteed Service: An attribute of a network capable of ensuring all users that they will receive the quality of service for which they have asked. This quality might be measured in terms of bit rate, maximum error rate, delay, variation in delay, reliability, or a combination of these factors. (8.3)

Halftone Image: Hertz: Units of frequency or repetition rate, measured in periods per second. (3.4)

Hertz: Units of frequency or repetition rate, measured in periods per second. (2.3)

Hue: Hue is the familiar perception-based definition of color. Each hue, such as orange, is associated with particular relative combinations of red, green, and blue values. (4.3)

Hypertext Transfer Protocol HTTP: A protocol used to reliably transfer hypertext from a Web server to a client computer. (8.3)

Image Enhancement: Processing an image to improve its appearance. (3.1)

Image Processing: Mathematical computation using pixel values to create new images or to compute information about objects in an image. (4.1)

Image Sampling: Measuring a continuously varying image at uniformly separated points in space and assigning a value that corresponds to the average light intensity value within a box surrounding the points. (3.2)

Image Segmentation: A process that divides an image into distinct areas that each correspond to an object in the image or a specific region in the image. For example, segmentation might find some pixels that belong to a book and other that belong to the carpet-covered region of the floor. (4.1)

Impulsive Noise: A disruptive process by which randomly chosen pixels in an image are set to the brightest or darkest values. Impulsive noise in a signal affects randomly chosen samples by setting them to the most positive or negative values. (4.1)

Input: Instructions or data used by a system to carry out a task. For example, the disk and the controls are both inputs to a CD player. (1.4)

Integrated Circuit (IC): A single computer chip that is built from many different components. Typically, nearly all of the individual components on an IC are transistors. (1.2)

Interference: Signals or disturbances unrelated to the signal you want to receive that can make it more difficult for the receiver to operate correctly. (7.3)

Internet: The name for a networking technology that permits the *inter*connection of many smaller data *net*works. (8.3)

Internet Service Provider (ISP): An organization or business that provides users with access to the Internet, connection with the various services available on the Internet, or the services themselves. Some ISPs can provide all three of these services, while others specialize in just one or two. (8.3)

Internetwork: A set of networks that are interconnected and can act as a whole in communicating data from a user to any other user or service. (8.3)

Joint Pictures Experts Group (JPEG): An international standards committee for still-image compression. Also, a particular lossy image-compression method. (6.3)

Key: In cryptography, a unique numeric or symbolic sequence used to decrypt or encrypt important information. (6.6)

Least Significant Bit: In a binary number, the rightmost 0 or 1. (5.3)

Lossless Compression: A class of compression techniques in which nothing about the original set of numbers or symbols is lost in the compression process. (6.2)

Lossy Compression: A class of compression techniques that throw away information about the sets of numbers or symbols being compressed. (6.2)

Loudspeaker: A device that turns electrical energy into sound energy. (2.2)

Mapping: A rule changing one value into another, using a table, a graph, or table. Mapping is used in image and signal processing to methodically modify the values of pixels or sample values and to turn bits or symbols in to signals for transmission in digital communications systems. (4.2), (7.2)

Mask: An image matrix that is used to select part of an image and remove the rest. A mask contains only two values—a zero value and a nonzero value, which is usually one. When an image is multiplied by a mask image, the parts of the image corresponding to zero-valued pixels in the mask are set to zero, so they are removed or "masked out." (4.2)

Masking: A psychoacoustic effect of the human hearing system whereby soft sounds are made inaudible by louder sounds that are close in frequency. (6.4)

Matrix: A two-dimensional array of numbers, called elements. The numbers are arranged in rows and columns. (4.2)

Median: The median value of a set of numbers is found by putting the numbers in a list and then sorting the numbers in ascending order. The median value is the value in the center of the list. (4.3)

Melody: A sequence of notes that make up a piece of music. (2.2)

Memory: Physical devices, typically digital, used to store information such as data, music, pictures, video, or computer programs. (1.2)

Microphone: A device that turns sound energy into electrical energy. (2.2)

Modulo-N Operation: The process of calculating the remainder after dividing an integer through by another integer. For example, the modulo-6 value of 26 is 2, because 26 divided by 6 is 4 with a remainder of 2. (6.8)

Moore's Law: An insightful observation by Gordon Moore in the 1960s stating that the number of transistors on a computer chip doubles every two years. (1.3)

Morphing: A slow change over many image frames that converts an object in an image to another object. This term is a shortened form of "metamorphosing." Each intermediate frame shows an object with a reasonable structure. In contrast, blended images, which look like double exposures, simply add two images without considering the structure of objects in the image. (4.1)

Morse Code: A code devised by Samuel F. B. Morse (1791–1872) to send characters by telegraph, the first commercially successful electrical communication system; uses combinations of short intervals (dots) and long intervals (dashes) to represent characters and can be used with electrical signals, lights, or sounds. (7.5)

Most Significant Bit: In a binary number, the leftmost 0 or 1. (5.3)

MP3: An acronym for Audio Layer 3 of the MPEG1 Multimedia Coding Standard, a popular Internet audio-coding standard. (6.4)

Motion Picture Experts Group (MPEG): An international standards committee for multimedia. (6.2)

Musical Instrument Digital Interface (MIDI): A specification for storing and transmitting music information between digital services. (2.3)

Negative: The negative of an image is created by subtracting each pixel value from the highest possible value so that black becomes white and white becomes black. The name comes from film negatives used to make photographic prints, but negatives are also frequently used in the filmless digital darkroom. (4.2)

Neighborhood Operations: Operations that compute output pixel values based on a group of input pixels with row and column numbers close to the output pixel's location. These neighboring pixels used in the computation define a neighborhood. (4.2)

Network: The dictionary definition: A group of interconnected individuals. In the world of telecommunications, a network is a set of communications equipment and the associated relays needed to permit communications between individuals or devices. (8.1)

Noise: Disturbances in signals that can make it more difficult for the receiver to operate correctly. (7.3)

Note: In music, a single sound with a definite pitch. (2.2)

Nyquist Rate: Minimum sampling rate needed for a signal or image. It equals twice the bandwidth of the analog signal. (5.2)

Nyquist Sampling Theorem: Specifies precisely how fast we must sample a signal, such as music or a movie, to ensure that we can re-create the original signal from the samples. (5.2)

Output: The final product of a system or device. For example, the electrical version of audio sent to a stereo amplifier is the output of a CD player. (1.3)

Packet: A segment of a complete message or file. Messages are commonly broken into packets in order to permit smooth and fair sharing of a data communications network by many users. (8.3)

Palette: The selection of colors available for a digital color image; this concept is similar to the selection of colors on an artist's palette. (3.4)

Parallel Transmission: Sending communication units such as bits or tones all at the same time. (7.4)

Period: The repeating interval of a periodic signal. (2.2)

Periodic Signal: A signal that exactly repeats at regular intervals. (2.2)

Permutation Encoding: An encryption method for encoding letters in English text by using a shuffling operation. (6.7)

Pitch: The perceived frequency of a sound. (2.2)

Pixel: (3.1) Contraction of "picture element." A pixel is the smallest detail of a digital image that can be changed, so it must correspond to the smallest detail in the image that one wants to preserve.

Point-to-Point Systems: A system in which two users are directly connected to each other by a wire or radio transmission system. (8.1)

Protocol: A set of rules that all parties in a network use to format and communicate data. (8.3)

Prototype: An original model of a design. Engineers use prototypes of systems to prove that the systems actually work. The final stage of production would be to convert the prototype into a version that could be sold to the general public. (1.4)

Pseudo-Random-Number Generator: A mathematical device for generating long strings of random-looking bits. (6.8)

Public-Key Cryptography: An encryption—decryption method that does not require secret information to be shared between the sender and the receiver. (6.8)

Quality of Service (QOS): A term used in data networks to describe the performance attributes of a network. (8.2)

Quantization: The process of taking a set of continuous values and mapping them into a finite number of discrete steps represented by integers. (3.2, 5.4)

Quantization Noise: The error or noise introduced into signal samples through the process of quantization. (5.4)

Queue: A line in which messages wait to be transmitted. (8.2)

Radians: A particular unit of angular measure, for which 2π corresponds to one revolution. (2.3)

Random Noise: Variation in a signal that is not related to the signal or an image. Each pixel can have a different random variation added to its value. (4.1)

Rate of Decay: A number that describes the way an exponential function decreases over time. (2.4)

Receiver: A device that recovers transmitted information from a signal and converts it into a form that the recipient can use. (7.1)

Relative Frequency: The number of observations of a particular event, divided by the total number of observations of all events. (6.3)

Relay Point: A point in a network that can accept messages from one user over a transmission link and then forward them to other users on other transmission links. (8.1)

Rotational Encoding: An encryption method for encoding letters in English text by using a shift operation. (6.7)

Router: A network relay point that operates by receiving messages; selecting an outgoing transmission link, based on address information carried in the message; storing the messages temporarily in a queue, if necessary; and then sending them out the prescribed link as soon as their turn comes in the queue. A router executes a strategy termed "store and forward." (8.2)

Routing Table: A table held by a router that instructs the router as to the best (and, possibly, an alternative) transmission path to be taken out of the router to send a message toward its ultimate destination. (8.3)

Run-Length Coding: A lossless compression method that is efficient for binary signals that contain long strings of contiguous 1's or 0's. (6.3)

Sampling Artifact: The distortion arising in a sampled image when the sampling rate is too low to adequately capture the finest details in the original continuous image. (3.2)

Sampling Period (T_s): Spacing in time between two adjacent samples. $T_s = 1/f_s$. (5.2)

Sampling Rate or Sampling Frequency (f_s): Number of samples per second. $f_s = 1/T_s$. (5.2)

Scalar: A scalar is a matrix with only one row and one column, so it has only one element. A scalar is just a simple number. (4.2)

Scientific Method: The five-step process by which scientists explain the universe: (1) Observe some aspect of the universe; (2) invent a tentative description (hypothesis) consistent with what you have observed; (3) use the hypothesis to make predictions; (4) test those predictions by experiments or further observations, and modify the hypothesis in light of your results; (5) repeat steps 2, 3, and 4 until there are no discrepancies between theory and experiment or observation. (1.1)

Score: Notation showing all parts or instruments. (2.2)

Seed: The starting value in a pseudo-random-number generator. (6.8)

Sequence: A set of numbers. A sequence of samples is the set of numerical sample values of a signal. (5.2)

Server: A network component that performs its service for many users, possibly for many at the same time. A server is typically a network-accessible commuter that has been augmented with software and additional equipment in order to perform its function. Servers typically fall into two classes, those that support a user's access to the network (that is, an access server) and those that provide information at a user's request (that is, a content server). (8.3)

Sign–Magnitude Form: A binary representation for numbers where the leftmost bit indicates the sign of the number. This bit is 0 for positive numbers and 1 for negative numbers. (5.3)

Signal: A pattern or variation that contains information, usually denoted as $s(t)$. (2.1)

Signal-to-Noise Ratio (SNR): Ratio of maximum signal level to maximum noise level, usually expressed in a logarithmic form in decibels (dB). For quantized digital signals where the quantization noise is to be measured in decibels dB, $\text{SNR} = 6B$, where B is the number of bits per sample. (5.4)

Simple Mail Transfer Protocol SMTP: A protocol used to reliably transfer electronic mail from a user's computer to a mail server located somewhere else on the Internet. (8.3)

Sinusoid or Sinusoidal Function: A simple oscillating waveform created from the sine or cosine function. (2.3)

Sound Signal: A pattern or variation in air molecules that a sound makes. (2.2)

Sound Synthesis: The creation of useful and complicated sounds from more basic sounds. (2.4)

Spatial Aliasing: Literally means "also appearing under a different name." For imaging, it means that the motion in a sampled movie takes on a completely different appearance than the original scene. (3.2)

Spatial Sampling: The process of assigning a single value to a pixel, which represents a rectangular area of an image. (3.2)

Spatial Sampling Rate: The number of samples of an image taken per unit of physical length of that image. (3.2)

Spectrogram: A two-dimensional image describing the spectrum of a sound over time. (2.3)

Spectrum: A plot of a periodic signal's sinusoidal components. (2.3)

Spectrum Analyzer: A device used to determine the amplitudes and frequencies of the sinusoids of any signal. (2.3)

Square Wave: A periodic function that is 1 for half its period and −1 for the other half. (2.4)

Standard: A description for a method or process that a group of people has agreed to use. By establishing a standard, people can use, enjoy, and even build on other people's work. (2.3)

Star: A network design in which all of the users are directly connected to a single relay point, using individual transmission links. (8.1)

Store and Forward: A strategy for relaying a message whereby a network relay point, upon receiving a message, selects an outgoing transmission link, based on address information carried in the message; stores the message temporarily in a queue, if necessary; and then sends it out on the prescribed link as soon as its turn comes in the queue. (8.2)

Subband Coding: A compression method whereby the sound signal is broken up into different frequency bands and each frequency-band signal is assigned its own bit sequence. (6.4)

Switching: A strategy for relaying messages in which a network relay point receives a message and immediately sends it on toward its destination. To make this procedure possible, capacity on both the switch and the selected transmission link must be reserved before the data transmission can begin. (8.2)

Synthesis: The creation of useful and complicated items from more basic ones. (2.4)

Tempo: The speed of a piece of music. (2.2)

Texture: Texture in an image refers to variations in the pixel values caused when light is reflected from surfaces that are not smooth such as woven fabrics, mowed grass, or sand on a beach. (4.3)

Threshold: A threshold is a value used to separate pixels in an image into two groups. All pixels with values above the threshold value are set to one value, and all pixels with values at or below the threshold value are set to a second value. These two values are often 0 and 1, or 0 and the maximum value. (4.2)

Threshold of Feeling: The upper limit of the human hearing system. Describes the loudest sounds that humans can possibly hear. (6.4)

Threshold of Quiet: The fundamental lower limit of the human hearing system. Describes the softest sounds that humans can hear. (6.4)

Tone: A signal that has a constant frequency and amplitude—a sinusoidal signal. (7.2)

Transistor: A switch made from semiconducting material that regulates the flow of voltage or current through electrical circuits. As such, the transistor is the basic building block of all digital and electronic technology. When used in digital applications, transistors are devices that operate in only two states corresponding to the familiar 0 and 1 bit representation. Today, transistors found in computer chips typically are much smaller than 1 millionth of a meter on a side. The transistor is thought by many to be the most important invention of the 20th century. (1.2)

Transmission Speed: The rate at which information can be communicated, measured in communication units per time interval, such as words per minute, characters per second, or bits per second. (7.4)

Transmitter: A device or circuit that converts a communication signal into a form that can be conveyed to a distant physical location. For example, a signal might be converted into sound vibrations in the air or an electrical signal on a wire. (7.1)

Transport Control Protocol/Internet Protocol TCP/IP: A pair of protocols used on the Internet to fairly and reliably carry information between any two computers. TCP breaks a user's message into packets and then sends each of them to the destination computer, using IP. The destination computer reassembles the original message after it has received all of the packets. (8.3)

Treble Clef: Also called G clef. Indicates that the note on the second line of the staff is G' (G above middle C). (2.2)

Unicode: A 16-bit code that can represent 65,535 different symbols from international and historical alphabets. (7.5)

Unsharp Masking: Unsharp masking adds a blurred negative to an image in order to make the edges more crisp and noticeable. (4.3)

Upper Frequency Threshold: The frequency above which humans cannot hear sounds. (6.4)

Vacuum Tube: An early technology that was used in nearly every piece of electronics, primarily to amplify signals. It is rare to find vacuum tubes used today in any application other than very high-end audio systems, guitar amplifiers, and radar devices. (1.2)

Waveform Synthesis: A synthesis technique that stores a characteristic signal of one specific note and then shrinks or stretches it in time in order to play different notes with different fundamental frequencies. (2.4)

Photo Credits

Chapter 1
Figure 01-01: Photo Researchers, Inc.; Figure 01-02: NASA/Science Photo Library/Photo Researchers, Inc.; Figure 01-03: MIT Museum; Figure 01-04: Corbis; Figure 01-04.01: Bettmann/Corbis; Figure 01-05.01: Getty Images, Inc.–Hulton Archive Photos; Figure 01-05a: Getty Images, Inc.; Figure 01-05b: Property of AT&T Archives, reprinted with permission of AT&T; Figure 01-06a: Courtesy of Texas Instruments; Figure 01-06b: Andrew Syred/Science Photo Library/Photo Researchers, Inc.; Figure CO-01: National Archives and Records Administration.

Chapter 2
Figure 02-02a: prudkov/Fotolia; Figure 02.02b: Getty Images, Inc.–EyeWire, Inc.; Figure 02-03: Getty Images, Inc.; Figure 02.09: Index Stock Imagery, Inc.

Chapter 3
Figure 03.01a: Corbis; Figure 03.01b: Corbis; Figure 03.01c: Corbis; Figure 03.01d: Corbis; Figure 03-01e: The Metropolitan Museum of Art, Gift of John Stewart Kennedy, 1987. Photograph© 1992 The Metropolitan Museum of Art; Figure 03.01f: Getty Images, Inc.; Figure 03-02a: Getty Images, Inc.–Image Bank; Figure 03-02c: Getty Images, Inc.; Figure 03-02d: Index Stock Imagery, Inc.; Figure 03-03a: Getty Images, Inc.–Taxi; Figure 03-03b: Photo Researchers, Inc.; Figure 03-03c: © Yorgos Nikas/GettyImages, Inc.–Stone Allstock; Figure 03-03d: Getty Images, Inc.–Stone Allstock; Figure 03-03e: Getty Images, Inc.–Stone Allstock; Figure 03-03f: NASA Headquarters; Figure 03-04: Getty Images, Inc.–Taxi; Figure 03-12: NASA/Jet Propulsion Laboratory; Figure 03-14a: NASA, ESA and Erich Karkoschka (University of Arizona); Figure 03-14b: NASA/Johnson Space Center; Figure 03-15: Corbis; Figure 03-23a: Getty Images, Inc.–Stone Allstock; Figure 03-26: The Granger Collection, New York; Figure 03-38: Courtesy of OSA (National Organization for Automotive Safety and Victim's Aid).© 2003. All rights reserved. Reprinted with permission. Figure 03-41: The Granger Collection; Figure 03-46: Getty Images, Inc.–Taxi; Figure CO-03: Omni-Photo Communications, Inc.

Chapter 4
Figure 04-02: AP/Wide World Photos; Figure 04-07: Bonnie Kamin/PhotoEdit; Figure 04-09: Mary Kate Denny/PhotoEdit; Figure 04-10: Motion Picture & Television Photo Archives/MPTV; Figure 04-11: Getty Images, Inc.–Stone Allstock.

Chapter 5
Figure 05-01: Getty Images, Inc.; Figure 05-09: Getty Images, Inc.; Figure 05.1.01: Photo Researchers, Inc.; Figure 05.1.02: Getty Images, Inc.; Figure 05.2.01: Getty Images, Inc.–Stone Allstock; Figure CO-05: Getty Images, Inc.–Photodisc; Figure CO-05-1: Getty Images, Inc.

Chapter 6
Figure 06-06: Getty Images, Inc.; Figure 06-09: Property of AT&T Archives, reprinted with permission of AT&T; Figure 06-14: Dorling Kindersley Media Library.

Chapter 7
Figure 07-02: Richard Green/Alamy; Figure 07-15a: Corbis Digital Stock; Figure 07-17: Getty Images, Inc.–Image Bank.

Chapter 8
Figure 08-08: Corbis; Figure 08-09: The Granger Collection; Figure 08-10: Cisco Systems.

Chapter 9
Figure 09-01: Dorkling Kindersley Media Library; Figure 09-02: Corbis; Figure 09-03: Courtesy of the Library of Congress; Figure 09-04: Property of AT&T Archives, reprinted with permission of AT&T; Figure 09-05: Library of Congress; Figure 09-06: Corbis; Figure 09-07: Getty Images, Inc.–Hulton Archive Photos; Figure 09-08: NASA/Johnson Space Center; Figure 09-09: Digital Vision Ltd.; Figure 09-10: Wonderfile Royalty Free; Figure 09-11: Property of AT&T Archives, reprinted with permission of AT&T; Figure CO-9a: Courtesy of Ford Motor Company; Figure CO-9b: Getty Images, Inc.–Image Bank.

Index

J

K

L

M

T

U